The Role of Reactive Oxygen Species in Human Health and Disease

Pranav Kumar Prabhakar
Research and Development Cell, Parul University, India

IGI Global
Scientific Publishing
Publishing Tomorrow's Research Today

Vice President of Editorial	Melissa Wagner
Director of Acquisitions	Mikaela Felty
Director of Book Development	Jocelynn Hessler
Production Manager	Mike Brehm
Cover Design	Phillip Shickler

Published in the United States of America by
IGI Global Scientific Publishing
701 East Chocolate Avenue
Hershey, PA, 17033, USA
Tel: 717-533-8845
Fax: 717-533-7115
Website: https://www.igi-global.com E-mail: cust@igi-global.com

Library of Congress Cataloging-in-Publication Data

Names: Kumar Prabhakar, Pranav editor
Title: The role of reactive oxygen species in human health and disease / edited by Pranav Kumar Prabhakar.
Description: Hershey, PA : IGI Global Scientific Publishing, [2025] | Includes bibliographical references and index. | Summary: "The primary objective of this book is to provide a comprehensive and up-to-date resource on the roles of ROS in human biology"-- Provided by publisher.
Identifiers: LCCN 2024055589 (print) | LCCN 2024055590 (ebook) | ISBN 9798369379196 hardcover | ISBN 9798369379202 paperback | ISBN 9798369379219 ebook
Subjects: LCSH: Active oxygen--Pathophysiology | Oxidative stress--Pathophysiology
Classification: LCC RB170 .R663 (print) | LCC RB170 (ebook) | DDC 616.07--dc23/eng/20250430
LC record available at https://lccn.loc.gov/2024055589
LC ebook record available at https://lccn.loc.gov/2024055590

British Cataloguing in Publication Data
A Cataloguing in Publication record for this book is available from the British Library.

Table of Contents

Detailed Table of Contents

Reactive oxygen species (ROS) are highly reactive molecules that are constantly generated as natural byproducts of cellular metabolism. While traditionally considered detrimental due to their potential to cause oxidative damage, recent research has unveiled their intricate involvement in cellular signaling pathways. This book chapter provides a comprehensive overview of ROS generation mechanisms, encompassing enzymatic and non-enzymatic processes and their pivotal roles in redox signaling pathways. Additionally, it delves into the various functions of reactive oxygen species (ROS) in cellular signaling, highlighting their influence on redox balance, gene expression, and stress reactions. This chapter tries to establish the foundation for understanding ROS biology and their intricate functions in both health and illness by clarifying some basic concepts.

Reactive oxygen species (ROS) are highly reactive molecules generated as a byproduct of normal metabolism and through exposure to external factors, such as radiation, pollutants, and certain medications. Overproduction of ROS leads to oxidative stress. Stress can damage vital biomolecules like DNA, proteins, and lipids, leading to chronic diseases. This chapter discusses the response of cells to oxidative stress. Roles and regulatory mechanisms of superoxide dismutase, catalase, the glutathione system, peroxiredoxin, and thioredoxin, in ROS defense have been discussed. The benefits of non-enzymatic antioxidants such as vitamins C and E, and minerals like selenium, and zinc have been highlighted. The synergistic interactions between enzymatic and non-enzymatic components have been detailed. This chapter highlights the potential of antioxidants to prevent chronic diseases, such as cardiovascular disorders, cancer, and neurodegeneration. The chapter concludes by discussing the intricacies of antioxidant supplementation, including challenges and future directions.

Reyed M. Reyed, Genetic Engineering and Biotechnology Research Institute (GEBRI) & City of Scientific Research and Technological Applications, Egypt
P. K. Prabhakar, Lovely Professional University, India & Nagaland University, India

This chapter discusses the roles of reactive oxygen species (ROS) in cells, including their role in oxidative stress and redox balance. It explores how dietary antioxidants affect redox homeostasis, their sources, mechanisms of action, and antioxidant capacity. The chapter also discusses how antioxidants affect ROS-mediated signaling cascades and endogenous antioxidant defense mechanisms. It discusses the use of antioxidants in pharmacological approaches to treat chronic illnesses like cancer, neurological disorders, and cardiometabolic diseases. It also addresses the antioxidant paradox, where excessive antioxidant supplementation can disrupt redox equilibrium, emphasizing the importance of balancing pro- and antioxidative forces.

Ansari Rashid, Department of Pharmacology, Sanjay Education Society's Institute of Pharmacy, Dhule, India
Pawan Gupta, Department of Pharmaceutical Chemistry, Shri Vile Parle Kelavani Mandal's Institute of Pharmacy, Dhule, India
Yogesh Chaudhari, Department of Pharmaceutical Chemistry, Shri Vile Parle Kelavani Mandal's Institute of Pharmacy, Dhule, India
Thangaraj Devadoss, Department of Pharmaceutical Chemistry, Karpagam Academy of Higher Education, Coimbatore, India
Rahul Kumar Garg, Asian College of Pharmacy, Udaipur, India
Rewati Wanjari, Department of Pharmaceutical Chemistry, Shri Vile Parle Kelavani Mandal's Institute of Pharmacy, Dhule, India

Exposure to environmental pollutants/toxins has progressively been identified as a major cause of different health concerns. Air pollution, occupational and radiation exposures, and agricultural chemicals are significant environmental factors contributing to oxidative stress (OS) and adverse health outcomes. Air pollution (particulate matter/toxic gases) is worsening respiratory and cardiovascular diseases. The occupational exposures to harmful substances can create chronic oxidative conditions which affects overall health. Radiation exposures (medical procedures /environmental sources) and agricultural chemicals cause the cellular damage, increasing the risk of cancer, affecting endocrine and neurological functions. Overall impact of these exposures is required rigorous monitoring and regulation to mitigate health risks. Addressing these environmental challenges is essential in protecting health and promoting public awareness to reduce harmful exposures and improve overall well-being.

ROS are molecules of the cell signalling, produced during oxidative metabolism and causes various human disease as its accumulation leads to deregulation of antioxidant defence mechanism (Wang et al. 2021). ROS regulates several cell signalling pathways which are responsible for apoptosis, survival of the cell, growth and maintenance of metabolic activities of the cell. This chapter focuses on the role of reactive oxygen species in cell signalling. Cell signalling is the process through which cell communicate each other and responds to environment and stimuli through a cascade of events, ultimately leading to production of desired protein as a result. Therefore, a comprehensive understanding of mechanism of cell signalling enhances our knowledge of intrinsic cellular process and also provide insight for development therapeutic strategies for curing various diseases.

This chapter explores the intricate relationship between oxidative stress and metabolic diseases, specifically diabetes and obesity. It begins with an overview of these conditions, emphasizing their global impact and associated metabolic dysfunctions. The discussion shifts to the biochemical mechanisms of oxidative

has evolved mechanisms to survive oxidative stress by enhancing its antioxidant defences like superoxide dismutase and trypanothione reductase, which neutralize these reactive species. This ability to counteract ROS and RNS not only aids in the parasite's survival within host cells but also contributes to chronic inflammation and tissue damage, ultimately influencing disease progression and severity. This chapter explores the intricate role of ROS and RNS, highlighting their dual function in regulating the parasite and contributing to disease pathology, therapeutic potential of targeting ROS/RNS pathways.

Karan Goel, M.M College of Pharmacy, Maharishi Markandeshwar University (Deemed), India
Isha Chawla, M.M College of Pharmacy, Maharishi Markandeshwar University (Deemed), India
Garima Malik, Chitkara College of Pharmacy, Chitkara University, India
Meenakshi Dhanawat, Amity Institute of Pharmacy, Amity University, India

Cardiovascular disorders (CVDs) include atherosclerosis, cardiac failure, hypertension, myocardial infarction, and oxidative stress, which are increasingly recognized as essential factors for the onset and advancement of these conditions. he imbalance between the formation of reactive oxygen species (ROS) and the antioxidant defense system is the molecular basis of oxidative stress, which leads to oxidative damage to lipids, proteins, and DNA. This potential damage promotes inflammatory responses, disrupts endothelial function, and speeds up atherogenesis. The nitric oxide synthase uncoupling, mitochondrial dysfunction, and NADPH-oxidizing enzymes are important pathways that enhance ROS production. Antioxidants including polyphenols, vitamins E and C, and others have shown conflicting outcomes in clinical studies. These novel methods aim to improve natural antioxidant defenses, alter redox-sensitive signaling pathways, and target specific ROS sources.

Odangowei Inetiminebi Ogidi, Bayelsa Medical University, Nigeria
Simon Akpomemera Erho, Bayelsa Medical University, Nigeria
Okiemute Omu, Federal Polytechnic, Ekowei, Nigeria

Reactive oxygen species (ROS), which are extremely bioactive chemicals, have been extensively researched in many forms of cancer. ROS are regarded as typical consequences of several biological processes. Cancer cells generally demonstrate

Reactive oxygen species (ROS), are a class of unstable chemicals produced by almost every subcellular organelle in cell. The ROS are highly reactive molecules that play essential roles in cell signaling and homeostasis but can also cause damage to cellular components when present in excess. The elevated levels of ROS, are causing damage to DNA, proteins, and lipids, promoting genetic instability, number of neurological illnesses, including, Alzheimer's, and Parkinson's disease. Precision medicine is an innovative approach to adapting medical treatment to the individual characteristics of each patient. This involves treating every patient as an individual and focusing on alternative healthcare treatment to every patient's unique characteristic. It relies on understanding how a person's genetic makeup, environment, and lifestyle affect their health and disease. The precision medicine enables early detection, prognostication, and monitoring of disease progression, allowing for timely and tailored interventions to mitigate ROS-mediated damage and improve patient outcomes.

Reactive oxygen species (ROS) are crucial for immune defense and cellular signaling, but their overproduction can damage cells and contribute to diseases. Understanding this duality is essential for developing ROS-based therapies. Future treatments might target specific ROS types or their production mechanisms, offering promise

Preface

Reactive oxygen species (ROS) have emerged as central players in the intricate balance between health and disease, serving as both essential signaling molecules and agents of oxidative stress. This duality places ROS at the forefront of research aimed at decoding the complexities of oxidative biology. Maintaining redox homeostasis through a delicate balance between ROS generation and antioxidant defenses is critical for cellular function and overall health. When this equilibrium is disrupted, oxidative stress contributes to a broad spectrum of diseases, including cancer, neurodegenerative disorders, cardiovascular diseases, metabolic dysfunctions, and aging-related conditions. In an era marked by the rising prevalence of lifestyle diseases, environmental challenges, and the increasing focus on personalized medicine, understanding ROS has never been more relevant. Their multifaceted roles as regulators of cellular signaling and mediators of pathology underscore their significance in developing targeted therapeutic and diagnostic strategies. The study of ROS not only provides insights into fundamental biological processes but also offers translational opportunities to mitigate oxidative stress and its consequences.

This book, *The Role of Reactive Oxygen Species in Human Health and Disease*, offers a comprehensive exploration of ROS, delving into their generation, signaling pathways, and dual roles in physiological and pathological processes. The chapters are carefully curated to present the latest advancements in ROS research, with a focus on translational applications that bridge basic science and clinical practice. By addressing the complexities of ROS, this book aims to equip researchers, clinicians, and students with the knowledge to tackle current health challenges and pave the way for innovative approaches in therapeutic and diagnostic development.

This book caters to a diverse audience, including academic researchers, clinical practitioners, biotechnology professionals, and postgraduate students in life sciences, medicine, and allied fields. By addressing fundamental mechanisms alongside translational opportunities, the content bridges the gap between foundational science and clinical applications.

CHAPTER HIGHLIGHTS

Each chapter delves into a specific aspect of ROS research, providing valuable insights into their roles across diverse biological contexts:

1. Reactive Oxygen Species: Generation and Signaling Pathways

This chapter explores the biochemical mechanisms underlying ROS generation and the signaling pathways they regulate. It sets the stage for understanding how ROS influence cellular communication and function.

2. Antioxidant Defense Mechanisms: Enzymatic and Non-Enzymatic

A detailed discussion on the body's defense systems that neutralize ROS, highlighting the interplay between enzymatic (e.g., superoxide dismutase) and non-enzymatic antioxidants (e.g., glutathione).

3. Impact of Dietary Antioxidants on Redox Homeostasis and Its Effects on Health and Disease

This chapter examines how dietary components like polyphenols and vitamins impact ROS levels, emphasizing their role in maintaining redox balance and preventing diseases.

4. Environmental Exposure and Oxidative Stress Effects on Human Health

Addressing how environmental pollutants and lifestyle factors contribute to oxidative stress, this chapter connects external stressors to human health outcomes.

5. Role of Reactive Oxygen Species in Cellular Signaling

Delving into the mechanisms by which ROS modulate cellular signaling pathways, this chapter connects molecular processes to therapeutic opportunities for various diseases.

6. Mechanisms, Impact, and Therapeutic Strategies for Oxidative Stress in Metabolic Diseases

A focused look at oxidative stress in metabolic conditions like diabetes and obesity, discussing innovative strategies to mitigate its effects.

7. Aging and Reactive Oxygen Species: Implications for Cellular Senescence and Longevity

This chapter highlights how ROS contribute to cellular aging and explores their role in longevity and senescence.

8. Intricate Role of Reactive Oxygen Species and Reactive Nitrogen Species in Leishmaniasis

A unique perspective on how ROS and RNS interact in infectious diseases like leishmaniasis, providing insights into mechanisms and potential therapies.

9. Oxidative Stress in Cardiovascular Disorders: Mechanisms and Therapeutic Implications

Exploring the relationship between oxidative stress and cardiovascular conditions, this chapter focuses on how ROS contribute to heart diseases and therapeutic interventions.

10. Reactive Oxygen Species in Cancer

A detailed analysis of the role of ROS in tumor progression and the therapeutic potential of targeting ROS in oncology.

11. Skin Diseases: Mechanisms and Treatments

Discussing the impact of oxidative stress on dermatological conditions, this chapter bridges molecular mechanisms to clinical applications.

12. Reactive Oxygen Species in Neurodegenerative Disorders

Focusing on diseases like Alzheimer's and Parkinson's, this chapter explores how ROS influence neurological health and emerging therapeutic options.

13. Reactive Oxygen Species in Precision Medicine and Lifestyle Diseases

Connecting ROS research to personalized medicine, this chapter highlights tailored approaches for managing lifestyle-related conditions.

14. Personalized Approaches for Neurological Disease Management

A deep dive into the potential of personalized therapeutic strategies for neurological disorders, integrating ROS-focused insights.

15. Future Directions in Reactive Oxygen Species Research

The concluding chapter envisions the future of ROS research, emphasizing translational opportunities and the challenges that lie ahead.

The insights presented in this book underscore the pivotal role of ROS in shaping human health and disease. By addressing both fundamental mechanisms and translational applications, this volume contributes significantly to the field, paving the way for innovative approaches in diagnostics and therapeutics. The collaborative efforts of leading experts have made this book a valuable resource for advancing the understanding of oxidative stress and its implications, fostering a foundation for further research and clinical breakthroughs.

Pranav Kumar Prabhakar
Nagaland University. India

Chapter 1
Reactive Oxygen Species: Generation and Signaling Pathways

Anam Khan
Department of Biochemistry, All India Institute of Medical Sciences Bhopal, India & Department of Biotechnology, Barkatullah University, Bhopal, India

Ragini Gothalwal
Department of Biotechnology, Barkatullah University, Bhopal, India

Ashwin Kotnis
https://orcid.org/0000-0002-0995-6266
Department of Biochemistry, All India Institute of Medical Sciences, Bhopal, India

ABSTRACT

Reactive oxygen species (ROS) are highly reactive molecules that are constantly generated as natural byproducts of cellular metabolism. While traditionally considered detrimental due to their potential to cause oxidative damage, recent research has unveiled their intricate involvement in cellular signaling pathways. This book chapter provides a comprehensive overview of ROS generation mechanisms, encompassing enzymatic and non-enzymatic processes and their pivotal roles in redox signaling pathways. Additionally, it delves into the various functions of reactive oxygen species (ROS) in cellular signaling, highlighting their influence on redox balance, gene expression, and stress reactions. This chapter tries to establish the foundation for understanding ROS biology and their intricate functions in both health and illness by clarifying some basic concepts.

DOI: 10.4018/979-8-3693-7919-6.ch001

A. INTRODUCTION:

Highly reactive molecules known as Reactive Oxygen Species (ROS) are produced during the cellular metabolism in the form of byproducts. This oxygen containing natural byproduct species include non-radical species like hydrogen peroxide (H_2O_2) and free radicals such as superoxide (O_2^-) and hydroxyl (OH·) radicals (Jakubczyk et al., 2020). Although reactive oxygen species (ROS) are commonly perceived as harmful byproducts of cellular metabolism that can cause oxidative damage to proteins, lipids, and DNA, their function in cellular physiology is significantly more complex (Saikolappan et al., 2019). The dual nature of ROS has been elucidated by recent research, which highlights their vital functions in defense mechanisms, homeostasis, and cell signalling (Sies et al., 2022).

Under typical physiological settings, a variety of cellular functions produce ROS under control. For example, during the ATP-producing electron transport chain, ROS are produced by mitochondria (Rabinovich-Nikitin et al., 2021; Yin & O'Neill, 2021; Auger et al., 2021). In a similar vein, ROS are produced on purpose by NADPH oxidases (NOX) and other enzymes as signaling molecules. These compounds participate in redox signaling, whereby they oxidize particular protein targets in a reversible manner to modify cellular functions. Maintaining cellular homeostasis and adapting to environmental changes depend on this redox control (Bedard & Krause, 2007; Vermot et al., 2021; Jaquet et al., 2009; Pecchillo Cimmino et al., 2023). The concept of ROS functioning as signaling entities marks a paradigm shift from the traditional view of ROS solely as damaging agents. ROS function as secondary messengers in intracellular signaling cascades that affect a variety of physiological processes, such as immunological responses, cell proliferation, and differentiation, at low to moderate concentrations. They modulate the function of multiple transcription factors, including Nrf2 and NF-κB, which govern the production of genes linked to cellular stress responses, antioxidant defense, and inflammation (Xue et al., 2020; Huang et al., 2022).

Nonetheless, it's crucial to strike a careful balance between the generation and removal of ROS (Figure 1).

Figure 1. ROS effects and cellular dysfunction

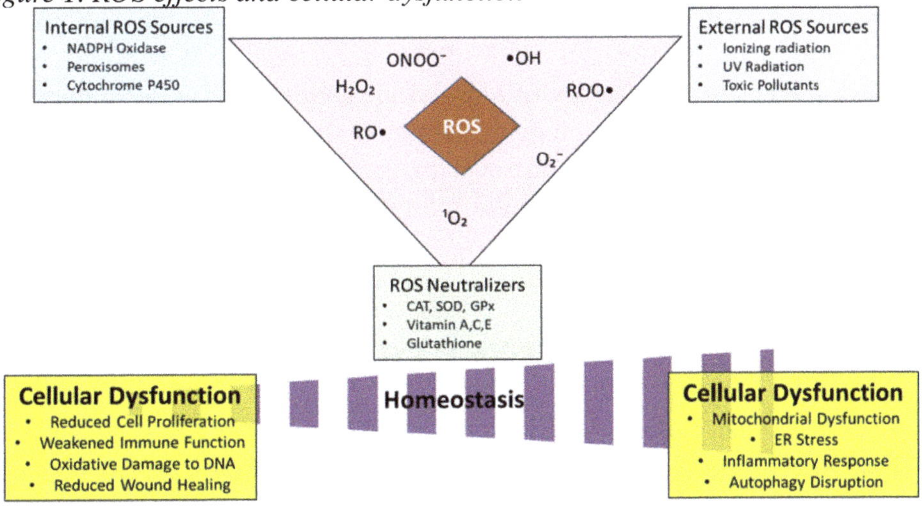

Oxidative stress is caused by too many ROS, which can arise from increased production or weakened antioxidant defenses. Numerous diseases, including cancer, cardiovascular disease, neurological disorders, and chronic inflammatory ailments are linked to this imbalance in their etiology. By permanently altering lipids, proteins, and nucleic acids, oxidative stress can impair cellular processes and ultimately result in cell death (Jomova et al., 2023; Hussain et al., 2021). Cells have developed complex antioxidant defense mechanisms to prevent the potentially damaging effects of ROS. ROS are enzymatically changed into less reactive molecules by enzyme-catalase, glutathione peroxidase, and superoxide dismutase (SOD). Glutathione, flavonoids, and the vitamins C and E are examples of non-enzymatic antioxidants that are essential for neutralizing reactive oxygen species (ROS) (Blokhina et al., 2003; Association of glutathione S-transferase M1/T1 polymorphisms, 2024). Together, these antioxidant systems preserve redox equilibrium and shield cells from oxidative damage. It is crucial to comprehend the dual functions of reactive oxygen species (ROS) in cellular physiology and disease while developing therapeutic approaches that aim to regulate ROS levels. Treating disorders linked to oxidative stress may be possible with targeted regulation of ROS and antioxidant pathways. For instance, ischemia-reperfusion injury and neurodegenerative illnesses can be treated with reduced oxidative damage by utilizing exogenous antioxidants or enhancing endogenous antioxidant defences (Spoelstra-de Man et al., 2018; Mitochondrial dysfunction and oxidative stress in Parkinson's disease, 2024). On the other hand, a prospective method for cancer therapy involves deliberately raising ROS levels in cancer cells in order to take advantage of their increased oxidative stress, leading to the capability to cause cell death.

Our goal in this chapter is to give a thorough review of the processes that produce reactive oxygen species (ROS), how they function in redox signaling, and the cellular antioxidant systems that control ROS concentrations. We will talk about treatment strategies that make use of ROS's dual nature as well as the implications of ROS dysregulation in the etiology of disease. We intend to establish the groundwork for the next investigations and novel therapeutic developments in the field of redox biology by clarifying the complex roles that ROS plays in cellular signaling and homeostasis.

B. GENERATION OF REACTIVE OXYGEN SPECIES:

Reactive oxygen species (ROS) are chemically reactive molecules containing oxygen, which are produced within cells through both enzymatic and non-enzymatic processes (Figure 2). ROS have significant roles in cellular signaling, immune response, and homeostasis, but an excess of ROS can lead to oxidative stress, causing cellular damage and contributing to pathological conditions such as cancer, neurodegenerative diseases, and aging (Zhang et al., 2016; Sinenko et al., 2021). Understanding the mechanisms behind ROS generation is essential for deciphering how these molecules function under physiological and pathological conditions.

Figure 2. Effects of increased ROS

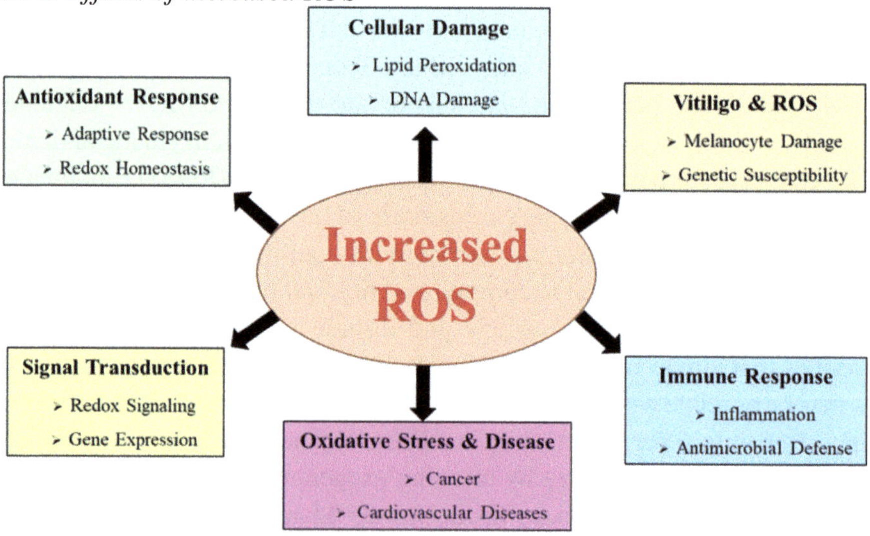

a) Enzymatic Processes:

Enzymatic sources of ROS involve the activity of specific enzymes that produce ROS as part of cellular processes. These enzymes include NADPH oxidases (NOX), mitochondrial electron transport chains, peroxisomal enzymes, and cytochrome P450 enzymes.

i. NADPH Oxidases (NOX):

NADPH oxidases are a family of membrane-bound enzymes that generate superoxide (O_2•−) by transferring electrons from NADPH to molecular oxygen (8) (Vermot et al., 2021). The NOX family consists of several isoforms, each with distinct tissue distributions and functions, including NOX1, NOX2, NOX3, NOX4, NOX5, DUOX1, and DUOX2. NOX2 (also known as gp91phox) is predominantly expressed in phagocytic cells such as neutrophils and macrophages (Table 1). It plays a crucial role in the immune response by generating superoxide, which is converted into more reactive species to neutralize invading pathogens during the "respiratory burst." NOX2 activation involves the assembly of cytosolic subunits with the membrane-bound complex, which leads to the rapid production of ROS as a defense mechanism. Other NOX isoforms are expressed in non-phagocytic cells and have roles beyond the immune response. For instance, **NOX4** produces H_2O_2 continuously in response to stimuli such as growth factors and is involved in the regulation of gene expression, cell proliferation, and differentiation (Bedard & Krause, 2007; Cipriano et al., 2023; Frontiers, 2016).

Table 1. Characteristic features of NOX isomers

NOX Isoform	Location	ROS generated	Primary Function	Components	Activation Mechanism	Role in Disease
NOX1	Colon, vascular smooth muscle, uterus, prostate, pancreas	Superoxide (O2•−)	Host defense, regulation of cellular signaling, and cell proliferation	Catalytic core (NOX1), regulatory subunits (p22phox, NOXO1, NOXA1)	Activated by RAC1 and NOXO1, and NOXA1	Implicated in hypertension, inflammatory bowel disease, and cancer
NOX 2	Neutrophils, macrophages, microglia, B lymphocytes	Superoxide (O2•−)	Phagocytic host defense, oxidative burst in immune response	Catalytic core (NOX2/gp91phox), regulatory subunits (p22phox, p47phox, p67phox, p40phox)	Activated by RAC2, p47phox, p67phox, and p40phox	Defects cause Chronic Granulomatous Disease (CGD), linked to neuroinflammation
NOX 3	Inner ear, fetal kidney, vestibular system	Superoxide (O2•−)	Generation of ROS for balance and hearing function	Catalytic core (NOX3), regulatory subunits (p22phox, NOXO1, NOXA1)	Constitutively active with modulation by NOXO1 and NOXA1	Mutations associated with sensorineural hearing loss, balance disorders

continued on following page

Table 1. Continued

NOX Isoform	Location	ROS generated	Primary Function	Components	Activation Mechanism	Role in Disease
NOX 4	Kidney, vascular endothelial cells, cardiac fibroblasts, lung, osteoclasts	Hydrogen peroxide (H2O2)	Cellular signaling, oxygen sensing, regulation of apoptosis, and differentiation	Catalytic core (NOX4), often forms a complex with p22phox	Activated by TGF-β, shear stress, angiotensin II, hypoxia	Linked to pulmonary fibrosis, cardiovascular diseases, renal dysfunction, and cancer
NOX 5	Testis, spleen, lymph nodes, vascular cells, endothelial cells	Superoxide (O2•−), H2O2	Calcium-dependent ROS production, involved in cellular proliferation and apoptosis	Catalytic core (NOX5), lacks p22phox, contains EF-hand calcium-binding domains	Directly activated by intracellular calcium	Associated with atherosclerosis, autoimmune diseases, and certain cancers
DUOX 1	Thyroid gland, airway epithelium, gastrointestinal tract, salivary glands	Hydrogen peroxide (H2O2)	Thyroid hormone biosynthesis, mucosal host defense, epithelial cell differentiation	Catalytic core (DUOX1), regulatory subunit (DUOXA1), peroxidase homology domain	Activated by DUOXA1, calcium, and TSH in the thyroid	Implicated in thyroid dysfunction, chronic respiratory diseases (e.g., asthma, COPD)
DUOX 2	Thyroid gland, airway epithelium, gastrointestinal tract, pancreas	Hydrogen peroxide (H2O2)	Thyroid hormone synthesis, defense against gut pathogens, regulation of intestinal microbiota	Catalytic core (DUOX2), regulatory subunit (DUOXA2), peroxidase homology domain	Activated by DUOXA2, calcium, and TSH	Mutations cause congenital hypothyroidism, linked to inflammatory bowel disease (IBD), chronic pancreatitis

ii. Mitochondria:

Mitochondria are a major source of ROS, especially during oxidative phosphorylation, the process by which mitochondria generate ATP. The electron transport chain (ETC), located in the inner mitochondrial membrane, transfers electrons from electron donors (NADH and $FADH_2$) to molecular oxygen through a series of complexes (I-IV). During this electron transfer, some electrons leak from the chain, primarily at complexes I and III, and reduce oxygen to form superoxide (O_2•−). The superoxide produced is rapidly dismutated to hydrogen peroxide (H_2O_2) by mitochondrial superoxide dismutase (SOD2)(6). H_2O_2 is relatively more stable and can diffuse across membranes, acting as a signaling molecule. However, if not adequately detoxified, it can be further converted into more reactive species, such as hydroxyl radicals (OH•), leading to oxidative stress. The detoxification of H_2O_2 is managed by antioxidant enzymes like glutathione peroxidase and catalase. These enzymes convert H_2O_2 into water and oxygen, thus protecting the cell from ROS-induced damage (Scialò et al., 2017; Mitochondrial electron transport chain, ROS generation and uncoupling, 2024)

While mitochondrial ROS are crucial for cellular signaling, particularly in processes such as autophagy and apoptosis, excessive mitochondrial ROS production can lead to mitochondrial dysfunction, a hallmark of various diseases (Mitochondrial quality control in human health and disease, 2024; Mukherjee et al., 2024).

iii. Peroxisomes:

Peroxisomes are small organelles involved in fatty acid β-oxidation and the detoxification of hydrogen peroxide. In peroxisomes, enzymes like acyl-CoA oxidase produce H_2O_2 as a byproduct of lipid metabolism, while xanthine oxidase generates ROS during purine catabolism. To prevent oxidative damage, peroxisomes contain high levels of catalase, an enzyme that converts H_2O_2 into water and oxygen. This balance is crucial because an accumulation of H_2O_2 can damage cellular components, leading to oxidative stress. By maintaining ROS levels, peroxisomes contribute to cellular redox homeostasis and lipid metabolism. (Del Río & López-Huertas, 2016; Fransen et al., 2012).

iv. Cytochrome P450 Enzymes:

Cytochrome P450 enzymes are a family of heme-containing enzymes that participate in the metabolism of both endogenous compounds (such as steroids) and xenobiotics (such as drugs and toxins). During the catalytic cycle of P450 enzymes, molecular oxygen is reduced, and one oxygen atom is incorporated into the substrate, while the other forms superoxide or hydrogen peroxide. ROS produced by cytochrome P450 enzymes can participate in redox signaling; however, excessive ROS production during drug metabolism or toxin processing can lead to oxidative stress, liver damage, and other pathological conditions (Mao et al., 2019; Valente et al., 2012; Kiyohara et al., 2011).

b) Non-Enzymatic Processes:

Non-enzymatic generation of ROS occurs through various chemical reactions that involve environmental factors, transition metals, and endogenous metabolites.

i. Transition Metals:

Transition metals, such as iron (Fe) and copper (Cu), can catalyze ROS formation through redox cycling. The Fenton reaction is one of the primary non-enzymatic sources of ROS, in which ferrous iron (Fe^{2+}) reacts with hydrogen peroxide (H_2O_2) to produce hydroxyl radicals (OH•) and ferric iron (Fe^{3+}). The Fenton reaction can be summarized as:

Fe2+ + H2O2 → Fe3+ + OH• + OH−

The Haber-Weiss reaction is another ROS-generating process in which superoxide reduces ferric iron back to ferrous iron, perpetuating the cycle of ROS production. These metal-catalyzed reactions are tightly regulated within cells, as excessive ROS production can cause severe oxidative damage to proteins, lipids, and DNA. The Haber-Weiss resaction can be summarized as:

O2- + H2O2 → OH + OH- + O2

To prevent uncontrolled ROS generation, cells use iron-binding proteins such as ferritin and transferrin, which sequester free iron and prevent its participation in the Fenton reaction (Hajam et al., 2022; Manta & Gladyshev, 2017).

ii. Environmental Factors:

Various environmental factors, such as ultraviolet (UV) radiation, ionizing radiation, and pollutants like cigarette smoke and heavy metals, can induce ROS production in cells. UV radiation promotes ROS generation through the photolysis of cellular water, resulting in the formation of hydroxyl radicals. These radicals can damage cellular DNA, leading to mutations and potentially contributing to carcinogenesis. Ionizing radiation interacts directly with biological molecules, causing ionization and the formation of free radicals, which initiate ROS cascades. Pollutants, including cigarette smoke and heavy metals, can disrupt cellular redox balance and promote oxidative stress by increasing intracellular ROS level (Jena et al., 2023; Shu et al., 2023).

iii. Endogenous Metabolites:

Certain endogenous metabolites can also contribute to non-enzymatic ROS generation. For instance, catecholamines, flavones, and thiols can undergo auto-oxidation, leading to the production of superoxide and hydrogen peroxide. In conditions such as diabetes, ROS generation is exacerbated by the auto-oxidation of glucose and the formation of advanced glycation end products (AGEs). These AGEs can trigger oxidative stress, promoting inflammation and vascular complications characteristic of diabetes (Jomova et al., 2023).

C. SIGNALING PATHWAYS INVOLVING REACTIVE OXYGEN SPECIES

Highly reactive compounds generated from oxygen are known as reactive oxygen species (ROS). They consist of both non-radical molecules like hydrogen peroxide (H_2O_2) and free radicals such superoxide anions (O_2^-) and hydroxyl radicals (OH•) (Figure 3) (Parascandolo & Laukkanen, 2019). ROS are byproducts of aerobic metabolism that, depending on their amounts and the context of the cell, can be both harmful and necessary signaling molecules. Although the electron transport chain in the mitochondria is the main source of ROS, peroxisomes and NADPH oxidases are also cellular sources (Gray et al., 2004).

Under controlled conditions, ROS are involved in redox biology, which is the complex equilibrium between oxidation and reduction processes that supports homeostasis and cell signalling (38). This equilibrium gives rise to the idea of redox signaling, which describes how cells use ROS to control metabolic pathways and synchronize reactions to cytokines, growth hormones, stress, and other stimuli (39). Nevertheless, oxidative stress, in which an excess of ROS destroys cellular constituents like proteins, lipids, and DNA, can result from dysregulation of ROS levels (40,41). Numerous diseases, such as cancer, neurological diseases, and cardiovascular disorders, have pathophysiology that is rooted in this damage (Simpson & Oliver, 2020; Perillo et al., 2020; Aldosari et al., 2018).

Figure 3. Antioxidants decreases as ROS increases in the cell

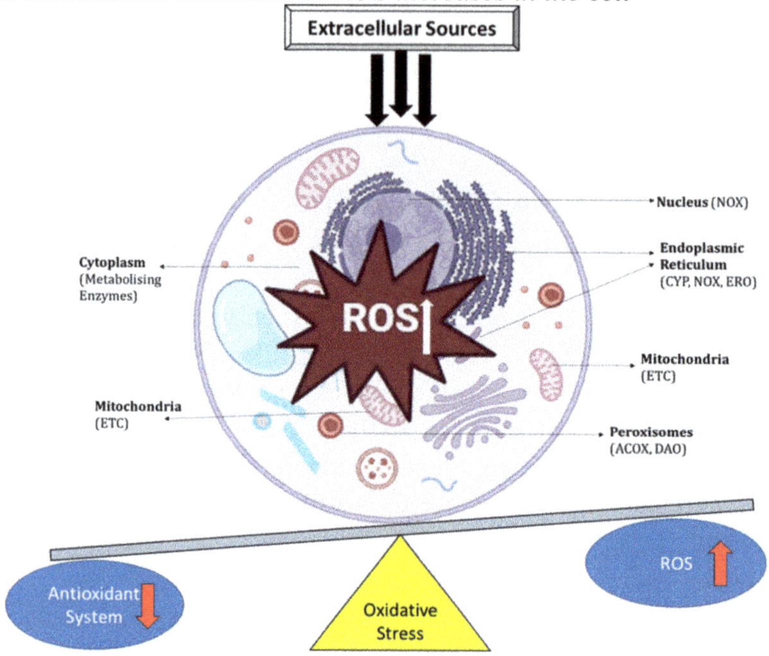

a) Redox Signaling Mechanisms

When ROS are produced, mostly by mitochondrial action, redox signaling starts. Especially at complexes I and III, electrons can prematurely reduce oxygen to produce superoxide in the mitochondrial electron transport chain (ETC), a primary source of reactive oxygen species (ROS) (Mitochondrial electron transport chain, ROS generation and uncoupling, 2024; Chenna et al., 2022; Xia et al., 2022). The enzyme superoxide dismutase (SOD), which is essential for reducing ROS levels and oxidative stress, dismutates superoxide into H_2O_2. In reaction to growth hormones, cytokines, and stress signals, non-mitochondrial sources such NADPH oxidases (NOX) also produce ROS at the plasma membrane (Vermot et al., 2021; Eid et al., 2022; Lee et al., 2020).

Antioxidants like as catalase, glutathione peroxidase (GPx), and peroxiredoxins, which convert $H2O_2$ to water, are involved in the enzymatic regulation of ROS. Since unregulated ROS accumulation can result in irreversible oxidative damage, these antioxidant mechanisms are crucial for preserving cellular redox balance (49,50). Redox signaling depends on low, strictly controlled ROS levels, which enable reversible signaling molecule alterations and enable cells to quickly adjust to environmental changes (51,52).

The selective oxidation of cysteine residues on target proteins makes redox signaling extremely specific. Sulfenylation, disulfide bond formation, and S-glutathionylation are among the changes that might result from the reversible oxidation of these cysteine residues by ROS due to their thiol groups (53,54). A dynamic method for signal transmission is provided by these alterations, which change the activity of proteins by either activating or inhibiting them (55). They are reversible when antioxidants are present.

i. MAPK Pathway:

ROS are important for the mitogen-activated protein kinase (MAPK) pathway, which consists of three main cascades: p38 MAPK, c-Jun N-terminal kinases (JNKs), and extracellular signal-regulated kinases (ERKs). Stress, growth factors, and cytokines all cause these cascades to become active. H_2O_2, for instance, increases ERK signaling by blocking phosphatases that deactivate ERK1/2, hence modulating the MAPK pathway. Cellular reactions like survival, differentiation, and proliferation follow from this (55–57).

ii. PI3K/Akt Pathway:

ROS, particularly $H2O_2$, can affect the phosphatidylinositol 3-kinase (PI3K)/Akt pathway. By preventing apoptotic processes, Akt phosphorylation, which results from PI3K activation, encourages cell growth and survival. Through their impact on the oxidation state of PTEN, a phosphatase that negatively controls PI3K, ROS modify this pathway. ROS-induced PTEN inactivation increases PI3K/Akt signaling, which strengthens cells' resistance to damage and stress (58–60).

iii. The NF-κB Pathway:

A redox-sensitive transcription factor, nuclear factor kappa B (NF-κB) controls genes related to inflammation, immunity, and cell survival. By encouraging the phosphorylation and destruction of its inhibitor, IκB, ROS under oxidative stress activate NF-κB (61). This makes it possible for NF-κB to get into the nucleus and promote the production of genes related to the inflammatory response, which, depending on the situation, can be both beneficial and harmful (62).

iv. Keap1-Nrf2 Pathway:

Nrf2 is crucial to preserving redox homeostasis and is a master regulator of antioxidant responses. Under normal circumstances, Nrf2 is attached to Keap1, its inhibitor, which directs Nrf2's destruction. ROS cause Keap1's cysteine residues to oxidize, which breaks down its connection to Nrf2 and permits Nrf2 to build up in the nucleus. Nrf2 initiates transcription to combat oxidative stress by binding to antioxidant response elements (ARE) in the promoters of genes encoding antioxidant enzymes, such as SOD, catalase, and glutathione-related enzymes (63).

b) ROS-Mediated Modulation of Gene Expression and Cellular Responses

ROS are important regulators of gene expression and cellular responses, not just metabolic byproducts. ROS affect several physiological functions, including as inflammation, autophagy, apoptosis, and proliferation, by oxidatively altering transcription factors and other regulatory proteins. This modification affects gene expression profiles and cellular consequences and is accomplished via ROS-mediated transcription factor activation or inhibition (64).

i. ROS and Gene Expression

a. Nrf2 and Antioxidant Response:

As previously stated, Nrf2 controls genes involved in antioxidant defense, but it also affects larger gene expression networks connected to metabolism, cell survival, and proliferation. Through the upregulation of genes related to glutathione production, heme oxygenase-1, and NAD(P)Hoxidoreductase-1, the Nrf2 pathway guarantees that cells can mount a strong defense against oxidative stress. This transcriptional response provides cytoprotective benefits, especially to organs like the brain, liver, and lungs that are subjected to long-term oxidative stress (65).

b. NF-κB and Inflammation:

ROS levels are tightly associated with NF-κB activation, especially in immune cells. When ROS activates NF-κB, it moves to the nucleus and stimulates the production of pro-inflammatory cytokines like TNF-α and IL-6, as well as enzymes like inducible nitric oxide synthase (iNOS). While these elements are necessary for launching an immune response, long-term NF-κB activation is linked to conditions like atherosclerosis, rheumatoid arthritis, and even some types of cancer (66).

c. p53 and Apoptosis:

Through direct alteration of the p53 protein or DNA damage, ROS can trigger the tumor suppressor protein p53. Programed cell death results from activated p53, which stimulates the transcription of pro-apoptotic genes. This function is essential for stopping damaged cells from proliferating, which could otherwise result in cancer. On the other hand, chronic p53 activation brought on by excessive or protracted ROS generation can cause tissue damage and play a role in aging and degenerative illnesses (67).

ii. Cellular Responses Mediated by ROS

a. Apoptosis and Survival:

Due to their varying amounts and duration, ROS have two roles in apoptosis. Cell survival is frequently enhanced by low to moderate ROS levels, which activate survival pathways such as PI3K/Akt. By activating apoptotic pathways through p53 and JNK, high ROS can cause damage to cellular components and ultimately result in cell death. ROS has two important functions in cancer biology: they can either encourage tumor cell survival or cause cancer cells to undergo apoptosis (68).

b. Autophagy:

ROS have a key role in controlling autophagy, a cellular mechanism that breaks down and recycles damaged proteins and organelles. Autophagy-related proteins, such as AMP-activated protein kinase (AMPK) and the mammalian target of rapamycin (mTOR) complex, are activated by ROS during oxidative stress. By eliminating damaged mitochondria, autophagy acts as a defense against oxidative stress and stops the generation of more ROS. Autophagy dysregulation has been connected to neurodegenerative illnesses, where cellular failure is made worse by improper removal of damaged proteins and organelles (69).

c. Cellular Senescence:

Cellular senescence—a irreversible cell cycle arrest linked to aging and age-related disorders—is facilitated by ROS. Because it triggers DNA damage response pathways like ATM/ATR and p53/p21, persistent oxidative stress can cause cells to enter senescence. The senescence-associated secretory phenotype (SASP), which is caused by pro-inflammatory substances secreted by senescent cells, can worsen tissue inflammation and malfunction, aggravating aging and chronic illnesses (70).

d. Immune Response:

The immune response's generation of ROS is essential, particularly when the body is defending against a disease. The NADPH oxidase complex is used by phagocytic cells, such as neutrophils and macrophages, to generate ROS in order to eradicate infections. In addition to aiding in the eradication of pathogens, this "respiratory burst" can harm tissue if left unchecked. In autoimmune disorders, where ROS-mediated tissue damage and immune cell activation perpetuate inflammatory cycles, chronic ROS-induced inflammation is linked (71).

ROS's significance in both health and illness is highlighted by its involvement in cellular signaling. Pathological ROS levels cause oxidative stress, which contributes to diseases like cancer, cardiovascular disease, and dementia, whereas physiological ROS levels are necessary for preserving cellular function (72). Malignant transformation and the advancement of cancer are fueled by ROS-induced DNA damage and oxidative alterations of oncogenes or tumor suppressor genes. On the other hand, ROS-targeted treatments face difficulties since cancer cells frequently use antioxidant systems like Nrf2 to thwart ROS-mediated apoptosis (73).

Elevated ROS in cardiovascular disorders cause inflammation, endothelial dysfunction, and vascular remodeling, which results in diseases like atherosclerosis and hypertension (74). ROS-induced harm to neurons and glial cells is also linked to neurodegenerative illnesses like Parkinson's and Alzheimer's, aggravating protein misfolding and mitochondrial malfunction (75). Therefore, in an effort to restore redox equilibrium and slow the progression of disease, therapeutic strategies that target ROS signaling—such as antioxidants or medications that modulate ROS—are being investigated (76–78).

Redox signaling, which is mediated by ROS, is essential to many physiological functions and allows cells to react to both internal and external stimuli. Gene expression, apoptosis, autophagy, and immunological responses are just a few of the cellular processes that ROS control by reversibly oxidizing signaling proteins and transcription factors. However, oxidative stress, which is the basis for the pathophysiology of many diseases, is caused by persistent or high ROS levels. Comprehending the dual function of ROS in both health and illness offers valuable perspectives on possible treatment approaches for ailments associated with oxidative stress, presenting encouraging opportunities for focused therapies that restore redox equilibrium (79).

D. REGULATION OF REACTIVE OXYGEN SPECIES LEVELS

Although necessary for regular cellular functions, reactive oxygen species (ROS) can be extremely harmful when they are present in excess. These compounds are continuously generated as byproducts of aerobic metabolism and include hydroxyl radicals (OH•), hydrogen peroxide (H_2O_2), and superoxide anions (O_2^-). They may also be produced when cells react to outside stimuli, like immunological or inflammatory responses (80). As part of signaling networks, ROS influence protein function, gene expression, and cellular functions like differentiation, proliferation, and death (81). But excessive production of ROS can harm cellular structures, resulting in oxidative stress, which is a factor in the etiology of a number of illnesses, such as cancer, heart disease, neurological conditions, and aging (82,83).

Cells have developed extremely complex antioxidant defense systems that control ROS levels and preserve redox equilibrium in order to mitigate the negative consequences of ROS accumulation. The survival and healthy operation of cells depend on this delicate balance between ROS production and neutralization. We will go over the different antioxidant defense systems in this part that support the maintenance of cellular redox equilibrium and the control of ROS levels (84).

a) Antioxidant Defense Mechanisms:

By scavenging ROS or by strengthening the body's ability to neutralize these reactive species, antioxidants are chemicals that prevent oxidative damage. Both enzymatic and non-enzymatic processes are used by cells' antioxidant defense mechanisms. Together, these processes prevent oxidative stress and inadequate signaling by ensuring that ROS are maintained within a healthy range (85).

i. Enzymatic Antioxidants:

a. Superoxide Dismutases (SODs):

The mitochondria and cell membranes are the main sites for the production of the superoxide anion (O_2^-), one of the most reactive ROS. Since they catalyze the conversion of O_2^- into hydrogen peroxide (H_2O_2) and molecular oxygen, superoxide dismutases (SODs) are the first line of defense against ROS (86). Three kinds of SOD exist: extracellular SOD (SOD3), which is found in the extracellular matrix; MnSOD (SOD2), which is found in the mitochondria; and Cu/Zn SOD (SOD1), which is found in the cytoplasm. SOD enzymes aid in the detoxification of ROS and inhibit cellular damage by neutralizing superoxide radicals (87).

b. Catalase:

After SOD produces H_2O_2, catalase can further decompose it into oxygen and water. The enzyme catalase is mostly found in peroxisomes, where it works to neutralize $H2O_2$, one of the most hazardous ROS, which can otherwise take part in damaging processes including lipid peroxidation or the production of hydroxyl radicals. Catalase's capacity to change $H2O_2$ into innocuous molecules is essential for shielding cells from oxidative damage, especially in areas with high oxidative metabolism (88).

c. Glutathione Peroxidases (GPx):

A family of enzymes known as glutathione peroxidases uses reduced glutathione (GSH) as a cofactor to decrease lipid hydroperoxides and hydrogen peroxide ($H2O_2$). Water and oxidized glutathione (GSSG) are the products of this process. Among the many isoforms of the GPx family, GPx1 is the most abundantly expressed and cytosolically active. GPx is essential for lowering oxidative stress and averting cellular damage because it turns $H2O_2$ into water. Glutathione reductase is responsible for restoring GSH from its oxidized state, preserving the cell's antioxidant potential (89,90).

d. Peroxiredoxins (Prx):

Another class of antioxidant enzymes called peroxiredoxins aids in the reduction of organic peroxides and $H2O_2$, which helps control the amount of ROS in the environment. Thioredoxin converts the conserved cysteine residue found in these enzymes back to its active state when peroxides oxidize it to a sulfenic acid group (91). Widely expressed peroxiredoxins are essential for preserving the redox equilibrium within cells. They are especially crucial in protecting red blood cells and other cells that are subjected to varying ROS concentrations (92).

e. Thioredoxins (Trx):

Small proteins known as thioredoxins have a thiol group that can decrease other proteins, controlling the redox state of different cellular targets. Peroxiredoxins and other oxidized proteins are reduced by them (93). By either directly lowering ROS or by influencing other antioxidant mechanisms, thioredoxins play a vital role in shielding cells from oxidative damage. When it comes to preserving cellular homeostasis and shielding cells from oxidative stress, the thioredoxin system is essential (94).

ii. Non-Enzymatic Antioxidants:

In addition to enzymatic antioxidants, cells also rely on a wide range of non-enzymatic antioxidants to scavenge ROS and maintain redox balance.

a. Glutathione (GSH):

The most prevalent intracellular antioxidant is glutathione, which can be found in both reduced (GSH) and oxidized (GSSG) forms. The reduced form of glutathione shields cellular macromolecules from oxidative damage by acting as a strong scavenger of reactive nitrogen species (RNS) and ROS (95). One important measure of cellular redox state is the GSH to GSSG ratio. GSH is converted to GSSG when ROS levels are high, and the enzyme glutathione reductase, which requires NADPH as a cofactor, restores the GSH pool. Through S-glutathionylation, glutathione also regulates protein function and aids in the detoxification of toxic electrophilic substances (96).

b. Vitamin C (Ascorbic Acid):

A water-soluble antioxidant, vitamin C directly scavenges reactive oxygen species (ROS), especially singlet oxygen and hydroxyl radicals. By donating electrons to neutralize these reactive species, ascorbate shields cellular constituents from oxidative damage. Along with helping to maintain the activity of enzymes involved in collagen formation, immunological response, and neurotransmitter processing, vitamin C also regenerates other antioxidants, including vitamin E. Its antioxidant properties are crucial for defending against oxidative stress in tissues like the skin, eyes, and central nervous system (17,97).

c. Vitamin E (Tocopherol):

A fat-soluble antioxidant, vitamin E mainly inhibits lipid peroxidation by targeting lipid peroxyl radicals. Because unsaturated fatty acids are vulnerable to oxidative damage in cell membranes, this is especially crucial. Lipid peroxyl radicals are neutralized and the integrity of cell membranes is preserved when vitamin E, in its reduced form, provides electrons to them. Vitamin E has antioxidant qualities, but it also affects inflammation and immune cell activity (97,98).

d. Coenzyme Q10 (Ubiquinone):

An antioxidant that is lipid-soluble and found in the mitochondria, coenzyme Q10 is involved in the electron transport chain. Coenzyme Q10 not only contributes to the creation of cellular energy but also scavenges free radicals, especially in the mitochondria, which are the site of the highest concentration of ROS generation. Its antioxidant effect is essential for mitochondrial lifetime and function because it shields proteins, lipids, and DNA from oxidative damage (99,100).

e. Flavonoids and Polyphenols:

Strong antioxidants that scavenge a broad range of ROS include flavonoids, polyphenols, and other substances produced from plants. It has been demonstrated that these substances, which are present in tea, fruits, and vegetables, lessen oxidative damage to tissues and cells. Quercetin and catechins are examples of flavonoids that neutralize free radicals and guard against inflammatory reactions, potentially providing therapeutic benefits for oxidative stress-related disorders like cancer and cardiovascular disease (101,102).

b) Maintenance of Redox Homeostasis:

The equilibrium between the generation and neutralization of reactive oxygen species (ROS) in a cell is known as redox homeostasis. Because oxidative stress can result from both excessive ROS and antioxidant depletion, maintaining this equilibrium is essential for cellular function. Oxidative stress can harm cells and contribute to a number of disorders (103). To maintain redox equilibrium and guard against oxidative damage, cells use a variety of strategies, such as feedback loops, redox buffer systems, and the control of antioxidant gene expression.

i. Redox Buffer Systems:

Redox buffer systems regulate the ratio of oxidation to reduction reactions, which is essential for preserving cellular redox equilibrium. The thioredoxin and glutathione systems are two important systems in this process.

a. Glutathione System:

One of the most significant cellular antioxidant systems is the glutathione system. The redox pair GSH/GSSG (reduced glutathione to oxidized glutathione) is involved. Small molecule glutathione (GSH) directly scavenges reactive oxygen species (ROS), whereas its oxidized version, GSSG, indicates the oxidative state of the cell. The cellular redox equilibrium is indicated by the GSH to GSSG ratio (104). By employing NADPH as a cofactor to transform GSSG back into its reduced form, GSH, enzymes like glutathione reductase (GR) are essential for preserving this equilibrium. GSH regeneration helps preserve cellular integrity and guarantees ongoing defense against oxidative stress (105).

b. Thioredoxin System:

Redox regulation also heavily relies on the thioredoxin (Trx) system. A protein called thioredoxin can directly break down disulfide connections in proteins and other molecules when it is lowered, preserving the components' ability to function (106). To keep thioredoxin in its active, reduced state, the enzyme thioredoxin reductase (TrxR) requires NADPH to decrease it. By preserving a decreased cellular environment—which is essential for cellular function and survival under oxidative stress—the Trx/TrxR system works in concert to regulate protein function and stability(107).

ii. Regulation of Antioxidant Gene Expression:

Redox-sensitive transcription factors enable cells to upregulate the expression of antioxidant genes in response to oxidative stress. Numerous proteins involved in detoxification, antioxidant defense, and repair processes have their expression activated by these transcription factors (108).

iii. Nuclear Factor Erythroid 2–Related Factor 2 (Nrf2):

An important modulator of the antioxidant response is Nrf2. Normally, Nrf2 is cytoplasmically attached to its inhibitor, Keap1. Nrf2 is released from Keap1, moves to the nucleus, and attaches itself to antioxidant response elements (ARE) in target gene promoters when ROS levels rise (109). Genes related to detoxification, antioxidant defense (including SOD, catalase, and glutathione peroxidase), and redox homeostasis maintenance are activated by this binding. Nrf2 helps cells adjust to oxidative stress by upregulating these defense genes (110).

iv. Activating Transcription Factor 4 (ATF4):

The oxidative stress response is also influenced by the transcription factor ATF4. Stressors such as oxidative stress cause it to become active, and it helps control genes related to amino acid metabolism and antioxidant defense. To assist cells in overcoming oxidative damage and preserving homeostasis, ATF4 coordinates with other stress-response mechanisms (111,112).

c) Feedback Mechanisms:

To maintain redox homeostasis, cells employ feedback mechanisms that tightly regulate ROS production and antioxidant responses. These mechanisms include:

i. Modulation of ROS-Producing Enzymes:

The cellular redox status can be used by cells to modify the activity of ROS-producing enzymes, such as NADPH oxidases (NOX). It is possible to reduce the generation of ROS in response to oxidative stress by either downregulating these enzymes or by triggering antioxidant defences (113).

ii. Activation of Antioxidant Responses:

In response to ROS levels surpassing a threshold, feedback mechanisms trigger the production of antioxidant enzymes, including those controlled by Nrf2 and ATF4. By neutralizing ROS and lowering their levels, these antioxidants aid in avoiding cellular harm (114).

iii. Repair or Removal of Oxidatively Damaged Molecules:

Cellular repair systems fix or eliminate damage caused by ROS to proteins, lipids, and DNA. Damaged proteins are broken down by the proteasome or autophagy pathways, whereas base excision repair pathways fix oxidative DNA damages. Cellular function is restored and the buildup of oxidized molecules is prevented by this ongoing cycle of removal and repair (115).

Cells can respond to changes in their redox environment by means of these feedback loops and interdependent systems, which keep ROS levels within a range that promotes regular cellular function while averting the negative effects of oxidative stress.

E. IMPLICATIONS OF REACTIVE OXYGEN SPECIES DYSREGULATION

Highly reactive molecules with oxygen, reactive oxygen species (ROS) are important for metabolism and signaling within cells. As byproducts of metabolic processes, particularly during cellular respiration in the mitochondria, ROS are created under normal physiological conditions. Their dysregulation, on the other hand, can lead to oxidative stress when production surpasses the cellular antioxidant defenses' capacity (116). This can play a role in the pathophysiology of a wide range of disorders. In both acute and chronic conditions, such as autoimmune diseases, diabetes, cancer, cardiovascular disease, neurodegenerative illnesses, and skin conditions including vitiligo, dysregulation of ROS production has been linked. In addition to discussing the therapeutic implications of addressing oxidative stress and ROS, this article investigates the role that ROS imbalance plays in disease development (117,118).

a. Role of ROS Imbalance in Disease Pathogenesis:

ROS are produced by both enzymatic and non-enzymatic activities and are involved in a number of cellular processes. Under typical circumstances, ROS act as signaling molecules that control immunological responses, apoptosis, and cell proliferation. However, oxidative stress can harm cellular macromolecules (such as lipids, proteins, and DNA) if ROS production is excessive or if the antioxidant mechanisms are unable to neutralize these molecules. Pathological reactions brought on by this damage aid in the emergence of disease. The function of ROS imbalance in different diseases is covered here (119,120).

i. Cancer:

The role that ROS dysregulation plays in carcinogenesis is among its most researched effects. ROS have the ability to harm DNA, resulting in alterations that encourage the development and spread of cancer. Tumors frequently have elevated ROS levels because of increased inflammation and metabolic activity (121). These increased ROS levels have the ability to trigger oncogenic signaling pathways, including those mediated by hypoxia-inducible factors (HIF) and nuclear factor-kappa B (NF-κB), which support tumor growth, metastasis, and apoptosis resistance. Furthermore, ROS can change the tumor microenvironment by encouraging angiogenesis and inflammation, which aids in the advancement of cancer (122). However, ROS also contribute to the effectiveness of some cancer treatments. ROS are produced as part of the mechanism of action of chemotherapeutic drugs like doxorubicin and cisplatin. But controlling ROS levels is a therapeutic problem; too much can cause tissue toxicity and oxidative damage, while too little can lessen treatment effectiveness (122,123).

ii. Cardiovascular Diseases:

One important factor in the emergence of cardiovascular diseases (CVD) is oxidative stress. One of the main early events in atherosclerosis is endothelial dysfunction, which can be caused by ROS damaging endothelial cells. One important stage in the development of plaque is the oxidation of low-density lipoprotein (LDL), which is facilitated by ROS. Furthermore, ROS can exacerbate atherosclerosis by triggering inflammatory pathways that encourage immune cell penetration into arterial walls. ROS further contributes to cardiovascular events like heart attacks and strokes by influencing vascular smooth muscle cells and encouraging vascular remodelling (124,125). Furthermore, ROS have a role in the etiology of other cardiovascular diseases such heart failure and hypertension. Increased vascular resistance and decreased nitric oxide bioavailability have been linked to elevated ROS levels, which can cause vasoconstriction and hypertension. In heart failure, oxidative stress can impair cardiac contractility and increase myocardial injury (126).

iii. Neurodegenerative Diseases:

Oxidative stress has a significant role in neurodegenerative illnesses such as Huntington's disease, Parkinson's disease, and Alzheimer's disease. ROS can harm glial and neuronal cells, causing neuroinflammation and compromising biological processes. For instance, ROS play a role in the buildup of tau tangles and amyloid plaques in Alzheimer's disease, which impair neuronal communication and cause cognitive impairment (127). Excess ROS generation from mitochondrial malfunction in Parkinson's disease damages dopaminergic neurons in the substantia nigra.

The activation of apoptotic pathways and neuroinflammation, which worsen neuronal death, are linked to oxidative stress in these disorders. As a result, treatment approaches for neurodegenerative illnesses have focused on reducing ROS generation or strengthening antioxidant defense systems (128).

iv. Diabetes and Metabolic Disorders:

Type 2 diabetes and other metabolic diseases are significantly influenced by oxidative stress. Insulin resistance is a defining feature of type 2 diabetes that is exacerbated by excessive ROS generation in organs such the liver, skeletal muscle, and pancreas. ROS hinder cells' ability to react to insulin by interfering with insulin signaling pathways. Moreover, inflammation and the activation of pathways that hinder glucose metabolism are encouraged by oxidative stress. Diabetic neuropathy, retinopathy, and nephropathy are among the consequences of diabetes that ROS also contribute to. Advanced glycation end products (AGEs) are created when oxidative stress and elevated blood sugar levels combine to further harm tissues and worsen diabetic consequences (129,130).

v. Autoimmune Diseases:

One of the main causes of immune system malfunction in autoimmune illnesses such as vitiligo, lupus, and rheumatoid arthritis is oxidative stress. The autoimmune response is triggered by immune cells including T cells and macrophages, which are activated in part by ROS. ROS, for instance, can alter self-antigens in lupus, which results in the generation of autoantibodies. Oxidative stress stimulates the release of pro-inflammatory cytokines in rheumatoid arthritis, resulting in long-term inflammation and joint degeneration (131,132). By harming melanocytes, the skin's pigment-producing cells, ROS play a critical part in the pathophysiology of vitiligo. Melanocyte apoptosis is caused by dysregulated ROS levels, which also contribute to the depigmentation observed in vitiligo. Furthermore, oxidative damage to skin cells caused by ROS can trigger immunological reactions, which exacerbates pigmentation loss (133).

b. Therapeutic Implications

The recognition of ROS dysregulation in various diseases has led to the development of therapeutic strategies targeting oxidative stress. Below, we explore the potential therapeutic approaches for managing ROS-related diseases.

i. Antioxidant Therapy:

The application of antioxidants is one of the main strategies for regulating ROS levels in illness. Antioxidants are compounds that stop oxidative damage by neutralizing ROS. Endogenous substances, like glutathione, catalase, and superoxide dismutase (SOD), are created by the body, whereas exogenous substances, such flavonoids, polyphenols, and vitamins C and E, are obtained from outside sources (134,135). In diseases such as cancer, neurological disorders, and cardiovascular diseases, antioxidant supplements have demonstrated potential in lowering oxidative stress and mitigating symptoms (136). Because it inhibits LDL oxidation, vitamin E, for example, is frequently utilized in cardiovascular disorders. Likewise, has shown anti-inflammatory and antioxidant qualities in the treatment of neurological illnesses including Parkinson's and Alzheimer's (137).

However, there is some debate around the use of antioxidants as medicinal agents. Although they have shown potential in lab conditions, the results of clinical trials have been inconsistent. Overconsumption of antioxidants may disrupt the body's normal ROS signaling pathways, which could harm cellular processes vital to tissue repair and immunological defense. As a result, creating antioxidant-based medicines requires careful thought (138).

ii. Targeting ROS-Producing Enzymes:

Inhibiting the enzymes that produce ROS is another therapeutic strategy. ROS in a variety of disorders are primarily produced by enzymes such as cyclooxygenase, xanthine oxidase, and NADPH oxidase. Oxidative damage can be lessened, and ROS production can be decreased by specifically blocking these enzymes. For example, the possibility of using NADPH oxidase inhibitors to treat neurodegenerative and cardiovascular diseases has been investigated (139,140). Similarly, there may be therapeutic potential in addressing mitochondrial dysfunction, a significant source of ROS in numerous illnesses. Researchers are looking into medications that enhance mitochondrial activity and antioxidants that target the mitochondria in conditions like Parkinson's and Alzheimer's (141).

iii. Gene Therapy and ROS Modulation:

ROS levels may be genetically modulated by gene therapy. With this strategy, genes encoding antioxidant enzymes are sent to the afflicted tissues (142). Delivering genes for catalase or superoxide dismutase (SOD), for instance, may strengthen the antioxidant defense system in illnesses where the body produces too many reactive oxygen species. The potential to rectify genetic abnormalities that contribute to ROS dysregulation is another benefit of genome editing tools like CRISPR/Cas9 (143).

iv. ROS Scavengers and Nanomedicine:

One possible strategy for addressing ROS in disease is nanomedicine. At the location of damage or illness, nanoparticles can be designed to scavenge ROS directly (144). The ROS-neutralizing capabilities of these nanoparticles can be improved by functionalizing them with antioxidants or enzymes like SOD and catalase. Additionally, medicinal substances can be delivered via nanoparticles in a focused and regulated manner, increasing treatment efficacy and lowering side effects (145).

Numerous diseases, from cancer and heart disease to neurological conditions and autoimmune diseases, are largely caused by ROS dysregulation. Oxidative stress, which harms cellular components and starts disease processes, is caused by an imbalance between the generation of ROS and antioxidant defences (26). Although enzyme inhibition and antioxidant therapy are viable therapeutic options, more investigation is required to improve these methods and handle the difficulties associated with practical translation. The creation of innovative treatment approaches, such as gene therapy and nanomedicine, has enormous potential for more successfully addressing disorders linked to ROS (144). Enhancing therapeutic strategies and patient outcomes will require an understanding of the intricate role that ROS play in disease.

CONCLUSION

Reactive oxygen species (ROS) are highly reactive molecules that play dual roles in cellular physiology. While they can cause oxidative damage and contribute to disease pathogenesis, they also serve as crucial signaling molecules involved in maintaining cellular homeostasis

ROS are produced through enzymatic processes, such as those involving NADPH oxidases, mitochondria, and peroxisomes, as well as non-enzymatic processes influenced by transition metals and environmental factors (125). They participate in redox signaling by modulating protein function through reversible oxidative modifications. They activate key signaling pathways, including MAPK, PI3K/Akt, and Nrf2, and influence calcium signalling, along with transcription factors such as NF-κB, AP-1, and Nrf2, leading to changes in gene expression that regulate inflammation, stress responses, and antioxidant defenses. Cells employ a range of antioxidant defense mechanisms to neutralize excess ROS. These include enzymatic antioxidants (e.g., SOD, catalase, GPx) and non-enzymatic antioxidants (e.g., glutathione, vitamins C and E, flavonoids). Cellular redox homeostasis is maintained through redox buffer systems like the glutathione and thioredoxin systems, regulation of antioxidant gene expression, and feedback mechanisms that modulate ROS levels. Even though our understanding of ROS's roles in cellular signaling and disease has advanced significantly, there are still a number of areas that need more research. For example, determining the specificity of ROS signaling requires an understanding of the exact temporal and spatial dynamics of ROS production and signaling within cells, which is essential for defining the roles that ROS play in physiological and pathological contexts. To fully understand ROS's context-dependent effects, including how various cell types and tissue contexts affect ROS signaling and results, more research is required.

REFERENCES

Agita, A., & Alsagaff, M. T. (2017, April). Inflammation, Immunity, and Hypertension. *Acta Medica Indonesiana, 49*(2), 158–165. PMID: 28790231

Al-Qahtani, A. A., Pantazi, I., Alhamlan, F. S., Alothaid, H., Matou-Nasri, S., Sourvinos, G., Vergadi, E., & Tsatsanis, C. (2022, October 31). SARS-CoV-2 modulates inflammatory responses of alveolar epithelial type II cells via PI3K/AKT pathway. *Frontiers in Immunology, 13*, 1020624. DOI: 10.3389/fimmu.2022.1020624 PMID: 36389723

Aldosari, S., Awad, M., Harrington, E., Sellke, F., & Abid, M. (2018, January 16). Subcellular Reactive Oxygen Species (ROS) in Cardiovascular Pathophysiology. *Antioxidants, 7*(1), 14. DOI: 10.3390/antiox7010014 PMID: 29337890

Ansari, M. Y., Ahmad, N., & Haqqi, T. M. (2020, September). Oxidative stress and inflammation in osteoarthritis pathogenesis: Role of polyphenols. *Biomedicine and Pharmacotherapy, 129*, 110452. DOI: 10.1016/j.biopha.2020.110452 PMID: 32768946

Ansari, S. A., Pendurthi, U. R., & Rao, L. V. M. (2017, November 28). The lipid peroxidation product 4-hydroxy-2-nonenal induces tissue factor decryption via ROS generation and the thioredoxin system. *Blood Advances, 1*(25), 2399–2413. DOI: 10.1182/bloodadvances.2017010132 PMID: 29296890

Antioxidant Therapy and Antioxidant-Related Bionanomaterials in Diabetic Wound Healing | Semantic Scholar [Internet]. [cited 2024 Nov 12]. Available from: https://www.semanticscholar.org/paper/Antioxidant-Therapy-and-Antioxidant-Related-in-Zhang-Chen/ffeaea57cc35b39beacc180c4fb4dd571cd356ab

Asantewaa, G., & Harris, I. S. (2021, April). Glutathione and its precursors in cancer. *Current Opinion in Biotechnology, 68*, 292–299. DOI: 10.1016/j.copbio.2021.03.001 PMID: 33819793

Ashafaq, M., Hussain, S., Alshahrani, S., Siddiqui, R., Alam, M. I., Elhassan Taha, M. M., Almoshari, Y., Alqahtani, S. S., Jali, A. M., & Aljohani, H. M. (2023, March 4). Neuroprotective Effects of Nano-Curcumin against Cypermethrin Associated Oxidative Stress and Up-Regulation of Apoptotic and Inflammatory Gene Expression in Rat Brains. *Antioxidants, 12*(3), 644. DOI: 10.3390/antiox12030644 PMID: 36978892

Auger, C., Vinaik, R., Appanna, V. D., & Jeschke, M. G. (2021, May). Beyond mitochondria: Alternative energy-producing pathways from all strata of life. *Metabolism: Clinical and Experimental, 118*, 154733. DOI: 10.1016/j.metabol.2021.154733 PMID: 33631145

Baev, A. Y., Vinokurov, A. Y., Novikova, I. N., Dremin, V. V., Potapova, E. V., & Abramov, A. Y. (2022, February 17). Interaction of Mitochondrial Calcium and ROS in Neurodegeneration. *Cells, 11*(4), 706. DOI: 10.3390/cells11040706 PMID: 35203354

Bauer, G. (2021, October 9). Inhibition of Membrane-Associated Catalase, Extracellular ROS/RNS Signaling and Aquaporin/H2O2-Mediated Intracellular Glutathione Depletion Cooperate during Apoptosis Induction in the Human Gastric Carcinoma Cell Line MKN-45. *Antioxidants, 10*(10), 1585. DOI: 10.3390/antiox10101585 PMID: 34679719

Bedard, K., & Krause, K. H. (2007, January). The NOX family of ROS-generating NADPH oxidases: Physiology and pathophysiology. *Physiological Reviews, 87*(1), 245–313. DOI: 10.1152/physrev.00044.2005 PMID: 17237347

Benipal, S. S., Liu, T., & Knowlton, A. (2017, April). Repetitive ROS Inhibits Nrf2 Antioxidant Defense in Ischemic Heart Failure. [Internet]. *The FASEB Journal, 31*(S1). Advance online publication. Retrieved November 12, 2024, from https://faseb.onlinelibrary.wiley.com/doi/10.1096/fasebj.31.1_supplement.1080.19. DOI: 10.1096/fasebj.31.1_supplement.1080.19

Bermejo-Bescós, P., Jiménez-Aliaga, K. L., Benedí, J., & Martín-Aragón, S. (2023, March 2). A Diet Containing Rutin Ameliorates Brain Intracellular Redox Homeostasis in a Mouse Model of Alzheimer's Disease. *International Journal of Molecular Sciences, 24*(5), 4863. DOI: 10.3390/ijms24054863 PMID: 36902309

Bhattacharjcc, S., & Bhattacharjcc, S. (2019). ROS and oxidative modification of cellular components. Reactive Oxygen Species in Plant Biology, 81-105.

Blanchetot, C., & Boonstra, J. (2008). The ROS-NOX connection in cancer and angiogenesis. *Critical Reviews in Eukaryotic Gene Expression, 18*(1), 35–45. DOI: 10.1615/CritRevEukarGeneExpr.v18.i1.30 PMID: 18197784

. Blokhina O, Virolainen E, Fagerstedt KV. Antioxidants, oxidative damage and oxygen deprivation stress: a review. Ann Bot. 2003 Jan;91 Spec No(2):179–94.

Bolduc, J. A., Collins, J. A., & Loeser, R. F. (2019, February 20). Reactive oxygen species, aging and articular cartilage homeostasis. *Free Radical Biology & Medicine, 132*, 73–82. DOI: 10.1016/j.freeradbiomed.2018.08.038 PMID: 30176344

Brägelmann, J., Lorenz, C., Borchmann, S., Nishii, K., Wegner, J., Meder, L., Ostendorp, J., Ast, D. F., Heimsoeth, A., Nakasuka, T., Hirabae, A., Okawa, S., Dammert, M. A., Plenker, D., Klein, S., Lohneis, P., Gu, J., Godfrey, L. K., Forster, J., & Sos, M. L. (2021, September 17). MAPK-pathway inhibition mediates inflammatory reprogramming and sensitizes tumors to targeted activation of innate immunity sensor RIG-I. *Nature Communications*, *12*(1), 5505. DOI: 10.1038/s41467-021-25728-8 PMID: 34535668

Chasara, R. S., Ajayi, T. O., Leshilo, D. M., Poka, M. S., & Witika, B. A. (2023). Exploring novel strategies to improve anti-tumour efficiency: The potential for targeting reactive oxygen species. *Heliyon*, *9*(9).

Chenna, S., Koopman, W. J. H., Prehn, J. H. M., & Connolly, N. M. C. (2022, July 1). Mechanisms and mathematical modeling of ROS production by the mitochondrial electron transport chain. *American Journal of Physiology. Cell Physiology*, *323*(1), C69–C83. DOI: 10.1152/ajpcell.00455.2021 PMID: 35613354

Cipriano, A., Viviano, M., Feoli, A., Milite, C., Sarno, G., Castellano, S., & Sbardella, G. (2023, September 14). NADPH Oxidases: From Molecular Mechanisms to Current Inhibitors. *Journal of Medicinal Chemistry*, *66*(17), 11632–11655. DOI: 10.1021/acs.jmedchem.3c00770 PMID: 37650225

Cui, Q., Wang, J. Q., Assaraf, Y. G., Ren, L., Gupta, P., Wei, L., Ashby, C. R.Jr, Yang, D.-H., & Chen, Z.-S. (2018, November). Modulating ROS to overcome multidrug resistance in cancer. *Drug Resistance Updates*, *41*, 1–25. DOI: 10.1016/j.drup.2018.11.001 PMID: 30471641

Del Río, L. A., & López-Huertas, E. (2016, July). ROS Generation in Peroxisomes and its Role in Cell Signaling. *Plant & Cell Physiology*, *57*(7), 1364–1376. DOI: 10.1093/pcp/pcw076 PMID: 27081099

Dias, V., Junn, E., & Mouradian, M. M. (2013). The role of oxidative stress in Parkinson's disease. *Journal of Parkinson's Disease*, *3*(4), 461–491. DOI: 10.3233/JPD-130230 PMID: 24252804

Diaz-Vivancos, P., de Simone, A., Kiddle, G., & Foyer, C. H. (2015, December). Glutathione—Linking cell proliferation to oxidative stress. *Free Radical Biology & Medicine*, *89*, 1154–1164. DOI: 10.1016/j.freeradbiomed.2015.09.023 PMID: 26546102

Dong, M., Shan, B., Han, X., Zhao, X., Wang, F., Zhu, L., Ou, Q., Ma, X., & Pan, Y. (2022, April 4). Baseline Mutations and Up-Regulation of PI3K-AKT Pathway Serve as Potential Indicators of Lack of Response to Neoadjuvant Chemotherapy in Stage II/III Breast Cancer. *Frontiers in Oncology*, *11*, 784985. DOI: 10.3389/fonc.2021.784985 PMID: 35480699

Dontaine, J., Bouali, A., Daussin, F., Bultot, L., Vertommen, D., Martin, M., Rathagirishnan, R., Cuillerier, A., Horman, S., Beauloye, C., Gatto, L., Lauzier, B., Bertrand, L., & Burelle, Y. (2022, April 12). The intra-mitochondrial O-GlcNAcylation system rapidly modulates OXPHOS function and ROS release in the heart. *Communications Biology*, *5*(1), 349. DOI: 10.1038/s42003-022-03282-3 PMID: 35414690

Eid, S. A., Savelieff, M. G., Eid, A. A., & Feldman, E. L. (2022, September 1). Nox, Nox, Are You There? The Role of NADPH Oxidases in the Peripheral Nervous System. *Antioxidants & Redox Signaling*, *37*(7–9), 613–630. DOI: 10.1089/ars.2021.0135 PMID: 34861780

Elzagallaai, A. A., Sultan, E. A., Bend, J. R., Abuzgaia, A. M., Loubani, E., & Rieder, M. J. (2020, March). Role of Oxidative Stress in Hypersensitivity Reactions to Sulfonamides. *Journal of Clinical Pharmacology*, *60*(3), 409–421. DOI: 10.1002/jcph.1535 PMID: 31709574

Eman Ali Abdelrazik, E. A. A., Hend Mohammed Mohammed Hassan, H. M. M. H., Zienab Abdallah Mahmoud, Z. A. M., Alshimaa Magdy Yousef, A. M. Y., & Eman Abdo Elsayed, E. A. E. (2022, December 16). Renoprotective effect of N-acetylcysteine and vitamin E in bisphenol A-induced rat nephrotoxicity; Modulators of Nrf2/NF-κB and ROS signaling pathway. *Acta Biomed Atenei Parm.*, *93*(6), e2022301.

Engelhardt, J. F., Sen, C. K., & Oberley, L. (2001). Redox-modulating gene therapies for human diseases. *Antioxidants & Redox Signaling*, *3*(3), 341–346.

Fatehi, D., Moayeri, A., Rostamzadeh, O., Rostamzadeh, A., & Kebria, M. M. (2018). Reactive Oxygenated Species (ROS) in Male Fertility; Source, Interaction Mechanism and Antioxidant Therapy. *Research Journal of Pharmacy and Technology*, *11*(2), 791. DOI: 10.5958/0974-360X.2018.00150.6

Fernandes, I. A., Mattos, J. D., Campos, M. O., Rocha, M. P., Mansur, D. E., Rocha, H. M., & Nóbrega, A. C. (2021). Reactive oxygen species play a modulatory role in the hyperventilatory response to poikilocapnic hyperoxia in humans. *The Journal of Physiology*, *599*(16), 3993–4007.

Fernández-Sánchez, A., Madrigal-Santillán, E., Bautista, M., Esquivel-Soto, J., Morales-González, A., Esquivel-Chirino, C., Durante-Montiel, I., Sánchez-Rivera, G., Valadez-Vega, C., & Morales-González, J. A. (2011). Inflammation, oxidative stress, and obesity. *International Journal of Molecular Sciences*, *12*(5), 3117–3132. DOI: 10.3390/ijms12053117 PMID: 21686173

Fransen, M., Nordgren, M., Wang, B., & Apanasets, O. (2012, September). Role of peroxisomes in ROS/RNS-metabolism: Implications for human disease. *Biochimica et Biophysica Acta*, *1822*(9), 1363–1373. DOI: 10.1016/j.bbadis.2011.12.001 PMID: 22178243

Frontiers | NOX Activation by Subunit Interaction and Underlying Mechanisms in Disease [Internet]. [cited 2024 Oct 1]. Available from: https://www.frontiersin.org/journals/cellular-neuroscience/articles/10.3389/fncel.2016.00301/full

Fu, Y., Yang, Y., Chen, S., Ning, N., & Hu, H. (2019, April 25). Arabidopsis IAR4 Modulates Primary Root Growth Under Salt Stress Through ROS-Mediated Modulation of Auxin Distribution. *Frontiers in Plant Science*, *10*, 522. DOI: 10.3389/fpls.2019.00522 PMID: 31105724

Gao, P., Du, X., Liu, L., Xu, H., Liu, M., Guan, X., & Zhang, C. (2021, January 18). Astragaloside IV Alleviates Tacrolimus-Induced Chronic Nephrotoxicity via p62-Keap1-Nrf2 Pathway. *Frontiers in Pharmacology*, *11*, 610102. DOI: 10.3389/fphar.2020.610102 PMID: 33536919

Gęgotek, A., & Skrzydlewska, E. (2023). Ascorbic acid as antioxidant. *Vitamins and Hormones*, *121*, 247–270. DOI: 10.1016/bs.vh.2022.10.008 PMID: 36707136

Ghahremani, F., Sabbaghzade, R., Ebrahimi, S., Javid, H., Ghahremani, J., & Hashemy, S. I. (2021, April). Pathogenic role of the SP/ NK1R system in GBM cells through inhibiting the thioredoxin system. [Internet]. *Iranian Journal of Basic Medical Sciences.*, *24*(4). Advance online publication. Retrieved November 12, 2024, from. DOI: 10.22038/ijbms.2021.52902.11945 PMID: 34094032

Gray, G. R., Maxwell, D. P., Villarimo, A. R., & McIntosh, L. (2004, December). Mitochondria/nuclear signaling of alternative oxidase gene expression occurs through distinct pathways involving organic acids and reactive oxygen species. *Plant Cell Reports*, *23*(7), 497–503. DOI: 10.1007/s00299-004-0848-1 PMID: 15322810

Grzeszczak, K., Łanocha-Arendarczyk, N., Malinowski, W., Ziętek, P., & Kosik-Bogacka, D. (2023, December 9). Oxidative Stress in Pregnancy. *Biomolecules*, *13*(12), 1768. DOI: 10.3390/biom13121768 PMID: 38136639

Guo, C. L. (2022). Self-Sustained Regulation or Self-Perpetuating Dysregulation: ROS-dependent HIF-YAP-Notch Signaling as a Double-Edged Sword on Stem Cell Physiology and Tumorigenesis. *Frontiers in Cell and Developmental Biology*, *10*, 862791. DOI: 10.3389/fcell.2022.862791 PMID: 35774228

Hajam, Y. A., Rani, R., Ganie, S. Y., Sheikh, T. A., Javaid, D., Qadri, S. S., Pramodh, S., Alsulimani, A., Alkhanani, M. F., Harakeh, S., Hussain, A., Haque, S., & Reshi, M. S. (2022, February 5). Oxidative Stress in Human Pathology and Aging: Molecular Mechanisms and Perspectives. *Cells*, *11*(3), 552. DOI: 10.3390/cells11030552 PMID: 35159361

Hao, Z., Zhang, M., Chen, X., Zhu, M., Han, B., He, Y., Yi, H., & Tang, S. (2023, April). Genetic Variants of the Nuclear Factor Erythroid 2-Related Factor 2/Antioxidant Reaction Element Pathway on the Risk of Antituberculosis Drug-Induced Liver Injury: A Systematic Review. *Pharmacogenomics*, *24*(6), 345–357. DOI: 10.2217/pgs-2023-0040 PMID: 37166414

Hassan, S. S. U., Samanta, S., Dash, R., Karpiński, T. M., Habibi, E., Sadiq, A., Ahmadi, A., & Bungau, S. (2022, October 10). The neuroprotective effects of fisetin, a natural flavonoid in neurodegenerative diseases: Focus on the role of oxidative stress. *Frontiers in Pharmacology*, *13*, 1015835. DOI: 10.3389/fphar.2022.1015835 PMID: 36299900

Hatzivassiliou, G., Song, K., Yen, I., Brandhuber, B. J., Anderson, D. J., Alvarado, R., Ludlam, M. J. C., Stokoe, D., Gloor, S. L., Vigers, G., Morales, T., Aliagas, I., Liu, B., Sideris, S., Hoeflich, K. P., Jaiswal, B. S., Seshagiri, S., Koeppen, H., Belvin, M., & Malek, S. (2010, March). RAF inhibitors prime wild-type RAF to activate the MAPK pathway and enhance growth. *Nature*, *464*(7287), 431–435. DOI: 10.1038/nature08833 PMID: 20130576

Henkel, R., Sandhu, I. S., & Agarwal, A. (2019, February). The excessive use of antioxidant therapy: A possible cause of male infertility? *Andrologia*, *51*(1), e13162. DOI: 10.1111/and.13162 PMID: 30259539

Huang, J. C., Yue, Z. P., Yu, H. F., Yang, Z. Q., Wang, Y. S., & Guo, B. (2022, June 14). TAZ ameliorates the microglia-mediated inflammatory response via the Nrf2-ROS-NF-κB pathway. *Molecular Therapy. Nucleic Acids*, *28*, 435–449. DOI: 10.1016/j.omtn.2022.03.025 PMID: 35505966

Hussain, T., Murtaza, G., Metwally, E., Kalhoro, D. H., Kalhoro, M. S., Rahu, B. A., Sahito, R. G. A., Yin, Y., Yang, H., Chughtai, M. I., & Tan, B. (2021). The Role of Oxidative Stress and Antioxidant Balance in Pregnancy. *Mediators of Inflammation*, *2021*, 9962860. DOI: 10.1155/2021/9962860 PMID: 34616234

Ida, C. M., Johnson, D. R., Nair, A. A., Davila, J., Kollmeyer, T. M., Minn, K., Fadra, N. M., Balcom, J. R., Fung, K.-M. A., Kim, D. K., Kaufmann, T. J., Kipp, B. R., Halling, K. C., Jenkins, R. B., & Giannini, C. (2021, September 27). Polymorphous Low-Grade Neuroepithelial Tumor of the Young (PLNTY): Molecular Profiling Confirms Frequent MAPK Pathway Activation. *Journal of Neuropathology and Experimental Neurology*, *80*(9), 821–829. DOI: 10.1093/jnen/nlab075 PMID: 34363682

Jakubczyk, K., Dec, K., Kałduńska, J., Kawczuga, D., Kochman, J., & Janda, K. (2020, April 22). Reactive oxygen species - sources, functions, oxidative damage. *Pol Merkur Lek Organ Pol Tow Lek.*, *48*(284), 124–127. PMID: 32352946

Jambrovics, K., Póliska, S., Scholtz, B., Uray, I. P., & Balajthy, Z. (2023, June 30). ATO Increases ROS Production and Apoptosis of Cells by Enhancing Calpain-Mediated Degradation of the Cancer Survival Protein TG2. *International Journal of Molecular Sciences*, *24*(13), 10938. DOI: 10.3390/ijms241310938 PMID: 37446117

Jaquet, V., Scapozza, L., Clark, R. A., Krause, K. H., & Lambeth, J. D. (2009, October). Small-molecule NOX inhibitors: ROS-generating NADPH oxidases as therapeutic targets. *Antioxidants & Redox Signaling*, *11*(10), 2535–2552. DOI: 10.1089/ars.2009.2585 PMID: 19309261

Jena, A. B., Samal, R. R., Bhol, N. K., & Duttaroy, A. K. (2023, June). Cellular Red-Ox system in health and disease: The latest update. *Biomedicine and Pharmacotherapy*, *162*, 114606. DOI: 10.1016/j.biopha.2023.114606 PMID: 36989716

Jiang, H., Wang, H., & De Ridder, M. (2018, December 1). Targeting antioxidant enzymes as a radiosensitizing strategy. *Cancer Letters*, *438*, 154–164. DOI: 10.1016/j.canlet.2018.09.004 PMID: 30223069

Jomova, K., Raptova, R., Alomar, S. Y., Alwasel, S. H., Nepovimova, E., Kuca, K., & Valko, M. (2023, October). Reactive oxygen species, toxicity, oxidative stress, and antioxidants: Chronic diseases and aging. *Archives of Toxicology*, *97*(10), 2499–2574. DOI: 10.1007/s00204-023-03562-9 PMID: 37597078

Kang, H., & Kim, B. (2023, February 22). Bioactive Compounds as Inhibitors of Inflammation, Oxidative Stress and Metabolic Dysfunctions via Regulation of Cellular Redox Balance and Histone Acetylation State. *Foods*, *12*(5), 925. DOI: 10.3390/foods12050925 PMID: 36900446

Kang, Q., & Yang, C. (2020, October). Oxidative stress and diabetic retinopathy: Molecular mechanisms, pathogenetic role and therapeutic implications. *Redox Biology*, *37*, 101799. DOI: 10.1016/j.redox.2020.101799 PMID: 33248932

Karvandi, M. S., Sheikhzadeh Hesari, F., Aref, A. R., & Mahdavi, M. (2023, March 6). The neuroprotective effects of targeting key factors of neuronal cell death in neurodegenerative diseases: The role of ER stress, oxidative stress, and neuroinflammation. *Frontiers in Cellular Neuroscience*, *17*, 1105247. DOI: 10.3389/fncel.2023.1105247 PMID: 36950516

Kishi, S., Nagasu, H., Kidokoro, K., & Kashihara, N. (2024, February). Oxidative stress and the role of redox signalling in chronic kidney disease. *Nature Reviews. Nephrology*, *20*(2), 101–119. DOI: 10.1038/s41581-023-00775-0 PMID: 37857763

Kiyohara, C., Miyake, Y., Koyanagi, M., Fujimoto, T., Shirasawa, S., Tanaka, K., & Nagai, M. (2011). APOE and CYP2E1 polymorphisms, alcohol consumption, and Parkinson's disease in a Japanese population. *Journal of Neural Transmission (Vienna, Austria)*, *118*, 1335–1344.

Lee, D., Moawad, A. R., Morielli, T., Fernandez, M. C., & O'Flaherty, C. (2017, February 10). Peroxiredoxins prevent oxidative stress during human sperm capacitation. *Molecular Human Reproduction*, *23*(2), 106–115. PMID: 28025393

Lee, S. R., An, E. J., Kim, J., & Bae, Y. S. (2020, January 1). Function of NADPH Oxidases in Diabetic Nephropathy and Development of Nox Inhibitors. *Biomolecules & Therapeutics*, *28*(1), 25–33. DOI: 10.4062/biomolther.2019.188 PMID: 31875663

Lennicke, C., & Cochemé, H. M. (2021, September 16). Redox metabolism: ROS as specific molecular regulators of cell signaling and function. *Molecular Cell*, *81*(18), 3691–3707. DOI: 10.1016/j.molcel.2021.08.018 PMID: 34547234

Li, W., Khor, T. O., Xu, C., Shen, G., Jeong, W. S., Yu, S., & Kong, A.-N. (2008, December 1). Activation of Nrf2-antioxidant signaling attenuates NFkappaB-inflammatory response and elicits apoptosis. *Biochemical Pharmacology*, *76*(11), 1485–1489. DOI: 10.1016/j.bcp.2008.07.017 PMID: 18694732

Li, Z., Chen, C., Ying, P., Ji-jun, G., Lian-jie, D., Dan, H., Xuan-min, Z., Tian-yue, G., Chao, Z., & Jia-hu, H. (2023, December). Bisphenol A and its analogue bisphenol S exposure reduce estradiol synthesis via the ROS-mediated PERK/ATF4 signaling pathway. *Food and Chemical Toxicology*, *182*, 114179. DOI: 10.1016/j.fct.2023.114179 PMID: 37944787

Liu, B. H., Xu, C. Z., Liu, Y., Lu, Z. L., Fu, T. L., Li, G. R., & Geng, Q. (2024). Mitochondrial quality control in human health and disease. *Military Medical Research*, *11*(1), 32.

Liu, L., Wang, M., Gong, N., Tian, P., & Deng, H. (2021, September 3). Se improves GPX4 expression and SOD activity to alleviate heat-stress-induced ferroptosis-like death in goat mammary epithelial cells. *Animal Cells and Systems*, *25*(5), 283–295. DOI: 10.1080/19768354.2021.1988704 PMID: 34745435

Liu, T., Sun, L., Zhang, Y., Wang, Y., & Zheng, J. (2022, January). Imbalanced GSH/ROS and sequential cell death. *Journal of Biochemical and Molecular Toxicology*, *36*(1), e22942. DOI: 10.1002/jbt.22942 PMID: 34725879

Loffredo, L., Spalice, A., Salvatori, F., De Castro, G., Guido, C. A., Zicari, A. M., Ciacci, P., Battaglia, S., Brindisi, G., Ettorre, E., Nocella, C., Salvatori, G., Duse, M., Violi, F., & Carnevale, R. (2020, December). Oxidative stress and gut-derived lipopolysaccharides in children affected by paediatric autoimmune neuropsychiatric disorders associated with streptococcal infections. *BMC Pediatrics*, *20*(1), 127. DOI: 10.1186/s12887-020-02026-8 PMID: 32188439

Lu, J., & Holmgren, A. (2014, January). The thioredoxin antioxidant system. *Free Radical Biology & Medicine*, *66*, 75–87. DOI: 10.1016/j.freeradbiomed.2013.07.036 PMID: 23899494

Lu, L., Wu, W., Tu, Y., Yang, Z., He, L., & Guo, M. (2014). Association of glutathione S-transferase M1/T1 polymorphisms with susceptibility to vitiligo. *Gene*, *535*(1), 12–16.

Luengo, E., Trigo-Alonso, P., Fernández-Mendívil, C., Nuñez, Á., Campo, M. D., Porrero, C., García-Magro, N., Negredo, P., Senar, S., Sánchez-Ramos, C., Bernal, J. A., Rábano, A., Hoozemans, J., Casas, A. I., Schmidt, H. H. H. W., & López, M. G. (2022, February). Implication of type 4 NADPH oxidase (NOX4) in tauopathy. *Redox Biology*, *49*, 102210. DOI: 10.1016/j.redox.2021.102210 PMID: 34922273

Mannucci, C., Casciaro, M., Sorbara, E. E., Calapai, F., Di Salvo, E., Pioggia, G., Navarra, M., Calapai, G., & Gangemi, S. (2021, February 8). Nutraceuticals against Oxidative Stress in Autoimmune Disorders. *Antioxidants*, *10*(2), 261. DOI: 10.3390/antiox10020261 PMID: 33567628

Manta, B., & Gladyshev, V. N. (2017, August). Regulated methionine oxidation by monooxygenases. *Free Radical Biology & Medicine*, *109*, 141–155. DOI: 10.1016/j.freeradbiomed.2017.02.010 PMID: 28229915

Mao, J., Dai, W., Zhang, S., Sun, L., Wang, H., Gao, Y., Wang, J., & Zhang, F. (2019, May). Quinone-thioether metabolites of hydroquinone play a dual role in promoting a vicious cycle of ROS generation: In vitro and in silico insights. *Archives of Toxicology*, *93*(5), 1297–1309. DOI: 10.1007/s00204-019-02443-4 PMID: 30976846

Moloney, J. N., & Cotter, T. G. (2018, August). ROS signalling in the biology of cancer. *Seminars in Cell & Developmental Biology, 80*, 50–64. DOI: 10.1016/j.semcdb.2017.05.023 PMID: 28587975

Mukherjee, A., Ghosh, K. K., Chakrabortty, S., Gulyás, B., Padmanabhan, P., & Ball, W. B. (2024, June 8). Mitochondrial Reactive Oxygen Species in Infection and Immunity. *Biomolecules, 14*(6), 670. DOI: 10.3390/biom14060670 PMID: 38927073

Nakahira, K., Haspel, J. A., Rathinam, V. A., Lee, S. J., Dolinay, T., Lam, H. C., & Choi, A. M. (2011). Autophagy proteins regulate innate immune responses by inhibiting the release of mitochondrial DNA mediated by the NALP3 inflammasome. *Nature Immunology, 12*(3), 222–230.

PDF Naturally occurring small molecules with dual effect upon inflammatory signaling pathways and endoplasmic reticulum stress response | Semantic Scholar [Internet]. [cited 2024 Nov 12]. Available from: https://www.semanticscholar.org/paper/Naturally-occurring-small-molecules-with-dual-upon-Silva-Valent%C3%A3o/9614767df03b762c0eab8d8cd3406d9851ecbf15

Newsholme, P., Cruzat, V. F., Keane, K. N., Carlessi, R., & De Bittencourt, P. I. H.Jr. (2016, December 15). Molecular mechanisms of ROS production and oxidative stress in diabetes. *The Biochemical Journal, 473*(24), 4527–4550. DOI: 10.1042/BCJ20160503C PMID: 27941030

NF-κB signaling in inflammation | Signal Transduction and Targeted Therapy [Internet]. [cited 2024 Nov 12]. Available from: https://www.nature.com/articles/sigtrans201723

Ochoa, C. D., Wu, R. F., & Terada, L. S. (2018, October). ROS signaling and ER stress in cardiovascular disease. *Molecular Aspects of Medicine, 63*, 18–29. DOI: 10.1016/j.mam.2018.03.002 PMID: 29559224

Onyango, I. G. (2008). Mitochondrial dysfunction and oxidative stress in Parkinson's disease. *Neurochemical Research, 33*, 589–597.

Panth, N., Paudel, K. R., & Parajuli, K. (2016, September 28). Reactive Oxygen Species: A Key Hallmark of Cardiovascular Disease. *Advances in Medicine, 2016*, 9152732. DOI: 10.1155/2016/9152732 PMID: 27774507

Paramasivan, P., Kankia, I. H., Langdon, S. P., & Deeni, Y. Y. Emerging role of nuclear factor erythroid 2-related factor 2 in the mechanism of action and resistance to anticancer therapies. Cancer Drug Resist [Internet]. 2019 [cited 2024 Nov 12]; Available from: https://www.oaepublish.com/articles/cdr.2019.57

Parascandolo, A., & Laukkanen, M. O. (2019, January 20). Carcinogenesis and Reactive Oxygen Species Signaling: Interaction of the NADPH Oxidase NOX1–5 and Superoxide Dismutase 1–3 Signal Transduction Pathways. *Antioxidants & Redox Signaling*, *30*(3), 443–486. DOI: 10.1089/ars.2017.7268 PMID: 29478325

Pecchillo Cimmino, T., Ammendola, R., Cattaneo, F., & Esposito, G. (2023, January 20). NOX Dependent ROS Generation and Cell Metabolism. *International Journal of Molecular Sciences*, *24*(3), 2086. DOI: 10.3390/ijms24032086 PMID: 36768405

Perillo, B., Di Donato, M., Pezone, A., Di Zazzo, E., Giovannelli, P., Galasso, G., Castoria, G., & Migliaccio, A. (2020, February). ROS in cancer therapy: The bright side of the moon. *Experimental & Molecular Medicine*, *52*(2), 192–203. DOI: 10.1038/s12276-020-0384-2 PMID: 32060354

Phoswa, W. N., & Khaliq, O. P. (2021). The role of oxidative stress in hypertensive disorders of pregnancy (preeclampsia, gestational hypertension) and metabolic disorder of pregnancy (gestational diabetes mellitus). *Oxidative Medicine and Cellular Longevity*, *2021*(1), 5581570.

Prajapati, V. D., Prajapati, R. V., Purohit, V. B., Avalani, J. R., Kamani, R. D., Sapariya, N. H., & Raval, D. K. (2022). CuI-catalyzed sulfenylation of 1-aryl trifluoromethyl pyrazolones: Direct formation of CSC bond using aryl iodides and carbon disulfide. *Journal of Fluorine Chemistry*, *253*, 109920.

Qasim, N., Arif, A., & Mahmood, R. (2023, February 1). Hyperglycemia enhances the generation of ROS and RNS that impair antioxidant power and cause oxidative damage in human erythrocytes. *Biochemistry and Cell Biology*, *101*(1), 64–76. DOI: 10.1139/bcb-2022-0008 PMID: 36379031

Quintero-Espinosa, D. A., Sanchez-Hernandez, S., Velez-Pardo, C., Martin, F., & Jimenez-Del-Rio, M. (2023, June 22). LRRK2 Knockout Confers Resistance in HEK-293 Cells to Rotenone-Induced Oxidative Stress, Mitochondrial Damage, and Apoptosis. *International Journal of Molecular Sciences*, *24*(13), 10474. DOI: 10.3390/ijms241310474 PMID: 37445652

Rabinovich-Nikitin, I., Rasouli, M., Reitz, C. J., Posen, I., Margulets, V., Dhingra, R., Khatua, T. N., Thliveris, J. A., Martino, T. A., & Kirshenbaum, L. A. (2021, November). Mitochondrial autophagy and cell survival is regulated by the circadian Clock gene in cardiac myocytes during ischemic stress. *Autophagy*, *17*(11), 3794–3812. DOI: 10.1080/15548627.2021.1938913 PMID: 34085589

Ray, P. D., Huang, B. W., & Tsuji, Y. (2012, May). Reactive oxygen species (ROS) homeostasis and redox regulation in cellular signaling. *Cellular Signalling*, *24*(5), 981–990. DOI: 10.1016/j.cellsig.2012.01.008 PMID: 22286106

Ross, D., & Siegel, D. (2021, May). The diverse functionality of NQO1 and its roles in redox control. *Redox Biology, 41*, 101950. DOI: 10.1016/j.redox.2021.101950 PMID: 33774477

Saikolappan, S., Kumar, B., Shishodia, G., Koul, S., & Koul, H. K. (2019, June 28). Reactive oxygen species and cancer: A complex interaction. *Cancer Letters, 452*, 132–143. DOI: 10.1016/j.canlet.2019.03.020 PMID: 30905813

Saima, S., Equbal, D., Lavekar, A. G., & Sinha, A. K. (2016). Cooperative catalysis by bovine serum albumin–iodine towards cascade oxidative coupling-C(sp^2)–H sulfenylation of indoles/hydroxyaryls with thiophenols on water. *Organic & Biomolecular Chemistry, 14*(25), 6111–6118. DOI: 10.1039/C6OB00930A PMID: 27251465

Scialò, F., Fernández-Ayala, D. J., & Sanz, A. (2017, June 27). Role of Mitochondrial Reverse Electron Transport in ROS Signaling: Potential Roles in Health and Disease. [Internet]. *Frontiers in Physiology, 8*, 428. Retrieved October 1, 2024, from https://www.frontiersin.org/journals/physiology/articles/10.3389/fphys.2017.00428/full. DOI: 10.3389/fphys.2017.00428 PMID: 28701960

Shahina, Z., Al Homsi, R., Price, J. D. W., Whiteway, M., Sultana, T., & Dahms, T. E. S. Rosemary essential oil and its components 1,8-cineole and α-pinene induce ROS-dependent lethality and ROS-independent virulence inhibition in Candida albicans. Khalaf RA, editor. PLOS ONE. 2022 Nov 16;17(11):e0277097.

Sharifi-Rad, J., Rayess, Y. E., Rizk, A. A., Sadaka, C., Zgheib, R., Zam, W., Sestito, S., Rapposelli, S., Neffe-Skocińska, K., Zielińska, D., Salehi, B., Setzer, W. N., Dosoky, N. S., Taheri, Y., El Beyrouthy, M., Martorell, M., Ostrander, E. A., Suleria, H. A. R., Cho, W. C., & Martins, N. (2020). Turmeric and Its Major Compound Curcumin on Health: Bioactive Effects and Safety Profiles for Food, Pharmaceutical, Biotechnological and Medicinal Applications. *Frontiers in Pharmacology, 11*, 01021. DOI: 10.3389/fphar.2020.01021 PMID: 33041781

Shu, P., Liang, H., Zhang, J., Lin, Y., Chen, W., & Zhang, D. (2023). Reactive oxygen species formation and its effect on CD4+ T cell-mediated inflammation. *Frontiers in Immunology, 14*, 1199233. DOI: 10.3389/fimmu.2023.1199233 PMID: 37304262

Sies, H. (1997, March). Oxidative stress: Oxidants and antioxidants. *Experimental Physiology, 82*(2), 291–295. DOI: 10.1113/expphysiol.1997.sp004024 PMID: 9129943

Sies, H., Belousov, V. V., Chandel, N. S., Davies, M. J., Jones, D. P., Mann, G. E., Murphy, M. P., Yamamoto, M., & Winterbourn, C. (2022, July). Defining roles of specific reactive oxygen species (ROS) in cell biology and physiology. *Nature Reviews. Molecular Cell Biology*, *23*(7), 499–515. DOI: 10.1038/s41580-022-00456-z PMID: 35190722

Simpson, D. S. A., & Oliver, P. L. (2020, August 13). ROS Generation in Microglia: Understanding Oxidative Stress and Inflammation in Neurodegenerative Disease. *Antioxidants*, *9*(8), 743. DOI: 10.3390/antiox9080743 PMID: 32823544

Sinenko, S. A., Starkova, T. Y., Kuzmin, A. A., & Tomilin, A. N. (2021, August 6). Physiological Signaling Functions of Reactive Oxygen Species in Stem Cells: From Flies to Man. [Internet]. *Frontiers in Cell and Developmental Biology*, *9*, 714370. Retrieved October 1, 2024, from https://www.frontiersin.org/journals/cell-and-developmental-biology/articles/10.3389/fcell.2021.714370/full. DOI: 10.3389/fcell.2021.714370 PMID: 34422833

Slika, H., Mansour, H., Wehbe, N., Nasser, S. A., Iratni, R., Nasrallah, G., Shaito, A., Ghaddar, T., Kobeissy, F., & Eid, A. H. (2022, February). Therapeutic potential of flavonoids in cancer: ROS-mediated mechanisms. *Biomedicine and Pharmacotherapy*, *146*, 112442. DOI: 10.1016/j.biopha.2021.112442 PMID: 35062053

Spoelstra-de Man, A. M. E., Elbers, P. W. G., & Oudemans-Van Straaten, H. M. (2018, August). Vitamin C: Should we supplement? *Current Opinion in Critical Care*, *24*(4), 248–255. DOI: 10.1097/MCC.0000000000000510 PMID: 29864039

Stieg, D. C., Wang, Y., Liu, L. Z., & Jiang, B. H. (2022, June 16). ROS and miRNA Dysregulation in Ovarian Cancer Development, Angiogenesis and Therapeutic Resistance. *International Journal of Molecular Sciences*, *23*(12), 6702. DOI: 10.3390/ijms23126702 PMID: 35743145

Stockwell, B. R., Friedmann Angeli, J. P., Bayir, H., Bush, A. I., Conrad, M., Dixon, S. J., Fulda, S., Gascón, S., Hatzios, S. K., Kagan, V. E., Noel, K., Jiang, X., Linkermann, A., Murphy, M. E., Overholtzer, M., Oyagi, A., Pagnussat, G. C., Park, J., Ran, Q., & Zhang, D. D. (2017, October). Ferroptosis: A Regulated Cell Death Nexus Linking Metabolism, Redox Biology, and Disease. *Cell*, *171*(2), 273–285. DOI: 10.1016/j.cell.2017.09.021 PMID: 28985560

Sun, J., Yang, F., Wang, L., Yu, H., Yang, Z., Wei, J., Vasilev, K., Zhang, X., Liu, X., & Zhao, Y. (2023, May). Delivery of coenzyme Q10 loaded micelle targets mitochondrial ROS and enhances efficiency of mesenchymal stem cell therapy in intervertebral disc degeneration. *Bioactive Materials*, *23*, 247–260. DOI: 10.1016/j.bioactmat.2022.10.019 PMID: 36439087

Sun, L., Wang, H., Yu, S., Zhang, L., Jiang, J., & Zhou, Q. (2022, February). Herceptin induces ferroptosis and mitochondrial dysfunction in H9c2 cells. *International Journal of Molecular Medicine*, *49*(2), 17. DOI: 10.3892/ijmm.2021.5072 PMID: 34935058

Sun, S., Zhang, C., Zhang, Q., Li, C., Huang, D., Ding, R., Cao, J., & Hao, J. (2023, September). Role of ROS-mediated PERK/ATF4 signaling activation in extracorporeal tube formation injury of human umbilical vein endothelial cells induced by cooking oil fume PM2.5 exposure. *Ecotoxicology and Environmental Safety*, *263*, 115332. DOI: 10.1016/j.ecoenv.2023.115332 PMID: 37611476

Tahir, A., Ashfaq, A., Aqeel, A. K., Nuzhat, Y., Sobia, A. R., & Iram, M. (2015, January 1). Imbalance of serum trace elements in acute leukemia contributes to pro and anti-oxidative stress via ROS and SOD regulation: A population based study. *Trace Elements and Electrolytes*, *32*(01), 22–27. DOI: 10.5414/TEX01358

Valente, M. J., Carvalho, F., Bastos, M. L., de Pinho, P. G., & Carvalho, M. (2012). Bastos M d L, de Pinho PG, Carvalho M. Contribution of oxidative metabolism to cocaine-induced liver and kidney damage. *Current Medicinal Chemistry*, *19*(33), 5601–5606. DOI: 10.2174/092986712803988938 PMID: 22856659

van der Pol, A., van Gilst, W. H., Voors, A. A., & van der Meer, P. (2019, April). Treating oxidative stress in heart failure: Past, present and future. *European Journal of Heart Failure*, *21*(4), 425–435. DOI: 10.1002/ejhf.1320 PMID: 30338885

Varela-Eirín, M., & Demaria, M. (2022, May 23). Cellular senescence. *Current Biology*, *32*(10), R448–R452. DOI: 10.1016/j.cub.2022.04.003 PMID: 35609537

Vermot, A., Petit-Härtlein, I., Smith, S. M. E., & Fieschi, F. (2021, June 1). NADPH Oxidases (NOX): An Overview from Discovery, Molecular Mechanisms to Physiology and Pathology. *Antioxidants*, *10*(6), 890. DOI: 10.3390/antiox10060890 PMID: 34205998

Wille, V., Vorhauser, J., & Sandbichler, A. M. (2022, February). Hypoxia in mouse and zebrafish fibroblasts: A differential role of peroxisomes in cellular redox balance. *Free Radical Biology & Medicine*, *180*, s99. DOI: 10.1016/j.freeradbiomed.2021.12.230

Wyckelsma, V. L., Venckunas, T., Brazaitis, M., Gastaldello, S., Snieckus, A., Eimantas, N., Baranauskiene, N., Subocius, A., Skurvydas, A., Pääsuke, M., Gapeyeva, H., Kaasik, P., Pääsuke, R., Jürimäe, J., Graf, B. A., Kayser, B., Place, N., Andersson, D. C., Kamandulis, S., & Westerblad, H. (2020, September 17). Vitamin C and E Treatment Blunts Sprint Interval Training–Induced Changes in Inflammatory Mediator-, Calcium-, and Mitochondria-Related Signaling in Recreationally Active Elderly Humans. *Antioxidants*, *9*(9), 879. DOI: 10.3390/antiox9090879 PMID: 32957522

Xia, T., Guo, J., Zhang, B., Song, C., Zhao, Q., Cui, B., & Liu, Y. (2022, July 4). Bisphenol A Promotes the Progression of Colon Cancer Through Dual-Targeting of NADPH Oxidase and Mitochondrial Electron-Transport Chain to Produce ROS and Activating HIF-1α/VEGF/PI3K/AKT Axis. *Frontiers in Endocrinology*, *13*, 933051. DOI: 10.3389/fendo.2022.933051 PMID: 35860704

Xu, S., Chen, T., Tian, M., Rahantaniaina, M. S., Zhang, L., Wang, R., Xuan, W., & Han, Y. (2022). Genetic Manipulation of Reactive Oxygen Species (ROS) Homeostasis Utilizing CRISPR/Cas9-Based Gene Editing in Rice. *Methods in Molecular Biology (Clifton, N.J.)*, *2526*, 25–41. DOI: 10.1007/978-1-0716-2469-2_3 PMID: 35657510

Xue, D., Zhou, X., & Qiu, J. (2020, November). Emerging role of NRF2 in ROS-mediated tumor chemoresistance. *Biomedicine and Pharmacotherapy*, *131*, 110676. DOI: 10.1016/j.biopha.2020.110676 PMID: 32858502

Yang, Y., Bazhin, A. V., Werner, J., & Karakhanova, S. (2013). Reactive oxygen species in the immune system. *International Reviews of Immunology*, *32*(3), 249–270.

Yin, M., & O'Neill, L. A. J. (2021, December). The role of the electron transport chain in immunity. *The FASEB Journal*, *35*(12), e21974. DOI: 10.1096/fj.202101161R PMID: 34793601

Zakaria, S., Elshazly, A. M., Alaa, R., & Elsebaey, S. (2024, April). Dantrolene and coenzyme Q10 as a suggested combination therapy to statin-induced myopathy in high fat diet rats: A possible interference with ROS/ TGF-β / Smad4 signaling pathway. *Toxicology and Applied Pharmacology*, *485*, 116900. DOI: 10.1016/j.taap.2024.116900 PMID: 38508403

Zhang, J., Wang, X., Vikash, V., Ye, Q., Wu, D., Liu, Y., & Dong, W. (2016). ROS and ROS-Mediated Cellular Signaling. *Oxidative Medicine and Cellular Longevity*, *2016*(1), 4350965. DOI: 10.1155/2016/4350965 PMID: 26998193

. Zhang K, Liu C, Yang K. Flavonoids Ameliorate Ferroptotic Cell Death in Cardiomyocytes through ROS Scavenging and Lipid Peroxide Reduction. FASEB J. 2022 May;36(S1):fasebj.2022.36.S1.0R522.

Zhang, Y., Murugesan, P., Huang, K., & Cai, H. (2020, March). NADPH oxidases and oxidase crosstalk in cardiovascular diseases: Novel therapeutic targets. *Nature Reviews. Cardiology*, *17*(3), 170–194. DOI: 10.1038/s41569-019-0260-8 PMID: 31591535

Zhao, C., Deng, H., & Chen, X. (2022, September). Harnessing immune response using reactive oxygen Species-Generating/Eliminating inorganic biomaterials for disease treatment. *Advanced Drug Delivery Reviews*, *188*, 114456. DOI: 10.1016/j.addr.2022.114456 PMID: 35843505

Zhao, R. Z., Jiang, S., Zhang, L., & Yu, Z. B. (2019). Mitochondrial electron transport chain, ROS generation and uncoupling. *International Journal of Molecular Medicine*, *44*(1), 3–15.

Zhou, X., An, B., Lin, Y., Ni, Y., Zhao, X., & Liang, X. (2023, September). Molecular mechanisms of ROS-modulated cancer chemoresistance and therapeutic strategies. *Biomedicine and Pharmacotherapy*, *165*, 115036. DOI: 10.1016/j.biopha.2023.115036 PMID: 37354814

Zhou, Y., Zhen, Y., Wang, G., & Liu, B. (2022, June 9). Deconvoluting the Complexity of Reactive Oxygen Species (ROS) in Neurodegenerative Diseases. *Frontiers in Neuroanatomy*, *16*, 910427. DOI: 10.3389/fnana.2022.910427 PMID: 35756499

Chapter 2
Antioxidant Defense Mechanisms:
Enzymatic and Non-Enzymatic

Vajra Vivek Bandodkar
https://orcid.org/0009-0008-5370-1085
St. Aloysius University (Deemed), India

Vishal Shridhar Moger
https://orcid.org/0009-0003-5175-7433
St. Aloysius University (Deemed), India

Pallavi Baliga
https://orcid.org/0000-0002-5485-9537
St. Aloysius University (Deemed), India

ABSTRACT

Reactive oxygen species (ROS) are highly reactive molecules generated as a by-product of normal metabolism and through exposure to external factors, such as radiation, pollutants, and certain medications. Overproduction of ROS leads to oxidative stress. Stress can damage vital biomolecules like DNA, proteins, and lipids, leading to chronic diseases. This chapter discusses the response of cells to oxidative stress. Roles and regulatory mechanisms of superoxide dismutase, catalase, the glutathione system, peroxiredoxin, and thioredoxin, in ROS defense have been discussed. The benefits of non-enzymatic antioxidants such as vitamins C and E, and minerals like selenium, and zinc have been highlighted. The synergistic interactions between enzymatic and non-enzymatic components have been detailed. This chapter highlights the potential of antioxidants to prevent chronic diseases, such as cardiovascular disorders, cancer, and neurodegeneration. The chapter concludes by discussing the intricacies of antioxidant supplementation, including challenges

DOI: 10.4018/979-8-3693-7919-6.ch002

and future directions.

1. INTRODUCTION TO OXIDATIVE STRESS

Oxidative stress is a state of imbalance between the generation of free radicals and antioxidant defense mechanisms of the body (Betteridge, 2000). Free radicals are atoms or molecules with unpaired electrons, and this characteristic makes them highly reactive (Shields et al. 2021). Due to their high reactivity, they can accept electrons from other compounds to become stable. As a result, the attacked molecule loses an electron and becomes a free radical, triggering a cascade of events that eventually harm the living cell. While free radicals can be of various types, this chapter will specifically focus on reactive oxygen species (ROS) and their implications in biological systems. Free radicals derived from oxygen are byproducts of metabolic reactions (Phaniendra et al. 2015). ROS refers to oxygen-containing reactive species like superoxide ($O_2^{\bullet-}$), hydrogen peroxide (H_2O_2), hydroxyl radical (OH■), singlet oxygen ($^1O_2$), peroxyl radical ($LOO^\bullet$), alkoxyl radical ($LO^\bullet$), lipid hydroperoxide (LOOH), peroxynitrite ($ONOO^-$), hypochlorous acid (HOCl), and ozone (O_3) (Li et al. 2016). Cellular ROS can be produced endogenously in the mitochondria, the intracellular site for oxidative phosphorylation. Other organelles associated with ROS production are the endoplasmic reticulum, where misfolded protein aggregation causes 'ER stress' and subsequently increases ROS, and the peroxisome, which contains several peroxisomal enzymes involved in the generation and metabolism of ROS (Hong et al. 2024; Phaniendra et al. 2015). Additionally, by producing ROS, metabolic enzymes like NADPH oxidases (NOX), nitric oxide synthases, lipoxygenases, xanthine oxidases, and cyclooxygenases can contribute to oxidative stress (Hong et al. 2024; Villalpando-Rodriguez & Gibson, 2021). Extracellular factors that generate ROS include UV radiation, industrial solvents, pollution, alcohol, tobacco smoke, ionizing radiations, chronic infections, inflammatory disorders, and nonsteroidal anti-inflammatory drugs (Bhattacharyya et al. 2014; Phaniendra et al. 2015). At moderate or low levels, ROS can be favorable by involving in several physiological activities like killing invading pathogens, wound healing, tissue repair, cell signaling pathways, mitogenic responses, and redox regulation (Bhattacharyya et al. 2014; Phaniendra et al. 2015). On the other hand, an accumulation of ROS in the cell can cause RNA and DNA damage, membrane phospholipid peroxidation, and protein oxidation, which can alter these biomolecules beyond repair and cease their function (Juan et al. 2021). Free radical-induced oxidative stress has been linked to several diseases (Figure 1) like cancers (lung, breast, prostate, colorectal, and bladder), cataract development, premature aging, neurodegenerative disorders (Parkinson's disease, Alzheimer's disease, and multiple sclerosis), rheumatoid arthritis, diabetes mellitus, cardiovascular diseases (atherosclerosis and hypertension), and respiratory diseases (asthma) (Phaniendra et al. 2015).

Understanding ROS sources and their effects helps us to appreciate the need to maintain a healthy balance of ROS production and antioxidant defenses. This understanding can help direct strategies to improve cellular health, potentially avert chronic illnesses, and maintain good health.

Figure 1. An overview of ROS sources (exogenous and endogenous), targets, and damage

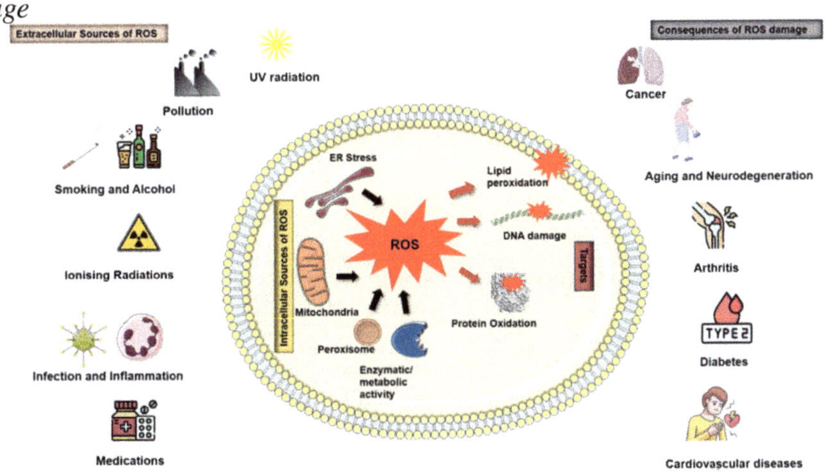

2. ENZYMATIC DEFENSES

Aerobic organisms exhibited two major antioxidant defense systems that minimize ROS-mediated damage due to oxygen-free radicals because of aerobic respiration and substrate oxidation. The first is enzymatic defense, and the second is a nonenzymatic antioxidant defense mechanism.

In general, cells control oxidative stress by essential antioxidant enzymes such as (i) superoxide dismutase: a class of oxidoreductase enzymes that contain metal ions in their active sites (Mn or Cu/Zn) and are responsible for converting superoxide anion into less toxic H_2O_2 (ii) Catalase breaks down H_2O_2 into a non-toxic form of O_2 and H_2O (iii) The glutathione system includes the glutathione peroxidase enzyme, responsible for the reduction of H_2O_2 into H_2O using glutathione as a cofactor that donates hydrogen, and the glutathione reductase enzyme involved in NADPH-dependent reduction of oxidized glutathione replenishing glutathione. Lastly, less used enzymes, (iv) Peroxiredoxins & Thioredoxins are less frequently used enzymes that help reduce H_2O_2 to H_2O and provide electrons for redox reactions.

2.1 Superoxide Dismutase

The SOD catalyzes the conversion of the two molecules of reactive oxygen free radical ($O_2^{\bullet-}$) into a less toxic form: hydrogen peroxide (H_2O_2) and molecular oxygen (O_2) (Figure 2), utilizing two equivalents of H^+ ions (Saxena et al. 2022). In addition, SODs play a critical role in inhibiting oxidative inactivation of nitric oxide, preventing peroxynitrite formation, which may lead to endothelial and mitochondrial dysfunction. There are three SOD isoforms in mammals: the cytoplasmic Cu/ZnSOD (SOD1), the mitochondrial MnSOD (SOD2), and the extracellular Cu/ZnSOD (SOD3), all of which require catalytic metals (Cu or Mn) for their activation.

Table 1. Classification of SOD

ISOFORM	CHARACTERISTICS	ASSOCIATED-METAL ION	LOCALIZATION
SOD 1 (Cu/Zn SOD)	32kDa (Homodimer)	Cu^{+2} (catalysis) Zn^{+2} (stability)	Cytoplasm, mitochondrial IMS, organelles such as nucleus, peroxisomes, lysosomes
SOD 2 (Mn-SOD)	96 kDa (Homotetramer)	Mn^{+3} (catalysis)	Mitochondrial matrix
SOD 3 (Cu/Zn SOD)/ (ecSOD)	135 kDa (Homotetramer)	Cu^{+2} (catalysis) Zn^{+2} (stability)	Extracellular matrix, extracellular fluids

Figure 2. The SOD catalyzes the conversion of the two molecules of reactive oxygen free radical ($O_2^{\bullet-}$) into a less toxic form: hydrogen peroxide (H_2O_2) and molecular oxygen (O_2), utilizing two equivalents of H^+ ions. Enzymatic activity of SOD involves alternate reduction and reoxidation of central catalytic metal (i.e., Cu or Mn) at the active site of the enzyme. Thus, Cu or Mn will be a key modulator of SOD activity of SOD1/SOD3 or SOD2, respectively

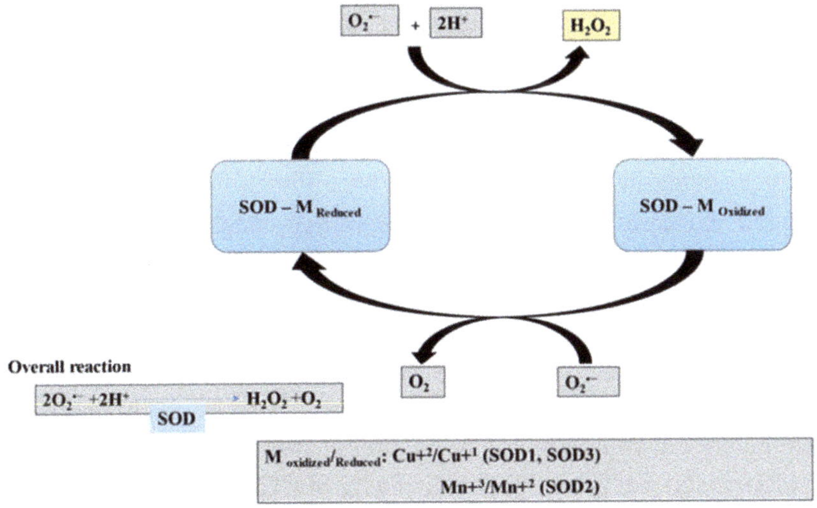

2.1.1. SOD 1

SOD1 (cytosolic Cu/ZnSOD) is the predominant intracellular SOD (Table 1). It is found mostly in the cytosol and to a lesser extent in the mitochondrial intermembrane space (IMS) as a 32 kDa homodimer (Okado-Matsumoto & Fridovich 2001; Sturtz et al. 2001). Additionally, it has been documented that SOD1 is found in peroxisomes, lysosomes, and nuclei. Copper (Cu) and zinc (Zn) are necessary for the SOD1 enzymatic activity. Zn is replaceable and contributes to correct protein folding and stability, while Cu is involved in catalytic activity and cannot be replaced by other metal ions (Fukai & Ushio-Fukai 2011).

According to Rakhit and Chakrabartty (2006), the amount of Cu bonded in the native Cu site determines the SOD activity of the remetallated derivatives. More significantly, the process by which SOD scavenges $O2^{\bullet-}$ requires the alternating reduction and reoxidation of Cu at the enzyme active site. Therefore, catalytically active Cu is needed for SOD1 activity to scavenge $O2^{\bullet-}$. The amount of intracellular free Cu is quite limited in a physiological setting. Accordingly, direct Cu transfer to cellular targets requires soluble Cu carrier proteins known as “Cu chaperones” (Itoh et al. 2009).

2.1.2 SOD2 (MnSOD)

SOD2, also known as mitochondrial manganese (Mn)-containing enzyme (Mn-SOD), is found in the mitochondrial matrix and is made up of a 96 kDa homotetramer (Luk et al. 2005). Comparable to SOD1 and SOD3 (Cu/ZnSODs), Mn in the active site of SOD2 catalyzes the disproportionation of $O2^{\bullet-}$ to oxygen and H_2O_2 (Miriyala et al. 2012). It is synthesized in the cytoplasm, and a signal peptide directs it to the mitochondria, where it participates in the dismutation of $O2^{\bullet-}$produced by the respiratory chain of enzymes.

While the active site of SOD2 differs from that of SOD1, it is comparable to the Fe SOD found in certain bacteria (Fukai & Ushio-Fukai 2011). Culotta, Yang, and O'Halloran's (2006) investigations on yeast genetics demonstrate that the metal insertion process of SOD2 differs significantly from SOD1, as predicted by the unique genomic and protein structure of SOD2.

The mortality of neonatal mice whose SOD2 gene is specifically disrupted indicates the critical role SOD2 plays in preserving mitochondrial function. According to Jang and Remmen (2009), such total SOD2 disruption results in dilated cardiomyopathy and neurodegeneration, which induce early postnatal mortality.

2.1.3 SOD3 (extracellular Cu/ZnSOD or ecSOD)

The predominant SOD in the vascular extracellular area is SOD3, a secretory extracellular Cu/Zn-containing SOD (ecSOD) (Table 1). SOD3 is a 135 kDa homotetramer comprising two disulfide-linked dimers in most species. SOD3 is primarily found in tissues (between 90 and 99 percent), with the remaining portion found in the extracellular matrix and on cell surfaces. SOD1 and SOD3 activity is highly susceptible to cyanide.

The protein enzyme SOD3 is made up of three functional domains. First, as mentioned by Sah, Agrahari, and Kim (2020), the amino-terminal residues [1–95] have an N-linked glycosylation site at Asn-89 that helps in separate cytosolic SOD1 from SOD3 and significantly improves the protein solubility. Secondly, the active site found in amino-terminal residues [96–193] exhibits approximately 50% homology with SOD1. Indeed, all the ligands to Cu and Zn and the arginine in the entrance to the active site in SOD1 can also be identified in this domain of SOD3. Third, the amino terminal residue [194-222] contain a C-terminal region corresponding to heparin-binding domain, which has a cluster of positively charged residues and is critical for binding to heparan sulfate proteoglycan. This area has a cluster of positively charged residues essential for binding to the heparan sulfate proteoglycan. The various sensitivities of these isoforms to cyanide (CN) and H_2O_2 were used to identify them all. Cu, Zn-SOD exhibits high sensitivity to H_2O_2 and CN. Whereas Fe-SOD is sensitive to H_2O_2 but not CN, Mn-SOD is insensitive to both. Furthermore, Cu and Zn-SOD are insensitive to chloroform-ethanol, although Mn-SOD and Fe-SOD were inhibited by it (Maurya & Namdeo 2022).

2.2 Catalase

The 244 kDa tetrameric enzyme catalase consists of four identical subunits, each measuring 59.7 kDa. About 527 amino acid residues make up each subunit, together with one haem group—iron (III) protoporphyrin—which is necessary for the catalytic activity of the unit and a firmly bound NADPH molecule (Kirkman & Gaetani 2007; Ko et al. 2000; Safo et al. 2001). A highly conserved "CAT fold" made up of an alpha-helical domain, a wrapping domain, a beta-barrel domain, and an N-terminal arm is seen in each monomer (Baker et al. 2023). Hydrogen peroxide (H_2O_2) is converted to water (H_2O) by catalase, which also releases an oxygen molecule (O_2). Consequently, controlling the abundance of H_2O_2, a relatively less toxic byproduct of dismutation of superoxide radical ($O_2^{\bullet -}$), catalyzed by enzyme superoxide dismutase. Although H_2O_2 can also function as a molecule for cellular signaling, it must be broken down because it can become hazardous if left to build up. The catalase enzyme is in organs like the kidney, liver, and peroxisomes. Conversely, this enzyme is produced at smaller levels in other organs. Most catalases are heme-containing enzymes, yet prokaryotes have a manganese catalase (Whittaker, 2012).

The reaction starts when a single H_2O_2 molecule oxidizes the ferricatalase haem group core Fe atom to create compound 1, an oxoferryl porphyrin cation radical (Alfonso-Prieto et al. 2009), and releases a single H_2O molecule. The next phase involves reducing compound I to its original state with the help of a second H_2O_2 molecule, producing two byproducts: an oxygen molecule and a water molecule (Figure 3).

Figure 3. One molecule of H_2O_2 oxidizes the central Fe atom in the haem group of ferricatalase to form compound 1 which is an oxoferryl porphyrin cation radical, releasing one molecule of H_2O. In the next a second molecule of H_2O_2 reduces compound I back to its initial state leading to the production of one molecule of oxygen and another molecule of water as by-products

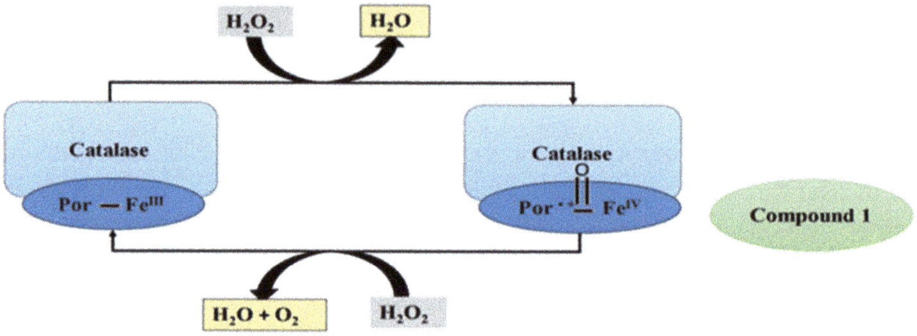

2.2.1 Regulation

Catalase expression and activity are tightly regulated at multiple levels, including transcription, post-transcription, and post-translation (Nishikawa et al. 2009), allowing for precise control of cellular redox homeostasis and oxidative stress responses. Transcriptional regulation is the primary method for controlling catalase abundance. Numerous transcription factors influence gene expression in response to environmental cues (Sirivarasai et al. 2015). During oxidative stress, the redox-sensitive transcription factor nuclear factor erythroid 2-related factor 2 (Nrf2) regulates the expression of antioxidant genes, including catalase (Bukke et al. 2022). When Nrf2 is activated, it moves to the nucleus and binds to antioxidant response elements in target gene promoter regions, enhancing transcriptional output. In addition to transcriptional regulation, post-translational modifications alter catalase activity in response to various physiological conditions. Enzymatic regulation such as phosphorylation, acetylation, and ubiquitination are significant mechanisms that regulate catalase activity depending on the context.

Furthermore, catalase dynamically shuttles between the cytosol and peroxisomal compartments, depending on its subcellular location, which affects its accessibility to substrates and regulatory factors. Changes in subcellular location can significantly affect catalase function and its capacity to defend against damage caused by oxidative stress (Bukke et al. 2022; Shen et al. 2023; Sirivarasai et al. 2015). Recent evidence points to the involvement of non-coding RNAs in the post-transcriptional regulation of catalase expression, including microRNAs and long non-coding RNAs. These regulatory RNAs provide another level of complexity to the regulatory landscape controlling catalase activity by altering mRNA stability and translation efficiency, which controls the amount of catalase enzyme (Carbonell & Gomes 2020; Haque et al. 2012).

2.3 Glutathione Peroxidase System

Enzyme glutathione reductase (GR) and glutathione peroxidase (GPx) are the two primary enzymes that make up the complex glutathione peroxidase system. GPx requires two molecules of GSH for the reduction of H_2O_2 to H_2O and the conversion of lipid peroxide (HOOR) to its corresponding alcohol (HOR). Two molecules of oxidized GSH form glutathione disulfide (GSSG) which is recycled by GR using the reducing power of NADPH (Figure 4). NADPH from the pentose phosphate pathway is used for the reaction. NADPH can also be formed directly in the mitochondria by $NAD(P)^+$ transhydrogenase, mitochondrial/cytosolic NADP-dependent isocitrate dehydrogenase, or cytosolic malate dehydrogenase (Rydström, 2006).

GR is a flavoprotein composed of 52 kD monomers. Apart from synthesis, GR activity provides a secondary source of GSH in the cytosol and some organelles, such as mitochondria. Although inhibiting GR depletes GSH and accumulates GSSG (Cereser et al. 2001), a full investigation of the GR and the cellular thiol redox system is lacking (Zhao et al. 2009). Enzyme inhibition has been linked to chemical and metal toxicity (Arning et al. 2008).

Glutathione peroxidase refers to a whole set of enzymes that are phylogenetically linked. GPx 1-4 are selenoproteins with selenocysteine (SeCys) in their catalytic center. Toppo et al. (2009) identified GPx6 as a human selenoprotein. Their significant antioxidant role has been demonstrated in a range of cell structures: GPx1 is found in the cytosol and mitochondria, GPx2 in the intestinal epithelium, and GPx3 in the plasma; all three work in the aqueous phase to reduce H_2O_2 and free fatty acid peroxides (Toppo et al. 2009). GPx4 primarily defends membranes by lowering phospholipid and cholesterol peroxides (Toppo et al. 2009; Flohé et al. 2011). Gpx5, which has cysteine instead of selenium in its active core, is an epididymal secretory enzyme. GPx6 is a human selenoprotein produced by the olfactory epithelium. GPx7 and GPx8 are both cysteines containing peroxidase with insignificant activity. GPx1, 2, 3, 4, 5, and 6 are homotetramers, which explains their specificity for hydrogen peroxide. GPx 4, 7, and 8 are monomers. This structure allows for the reaction with more complicated lipid hydroperoxides but has only been demonstrated for GPx4 (Flohé et al. 2011). The catalytic core of GPx was first identified as a triad consisting of SeCys or Cys, Gln, and Trp. It was later shown to form a tetrad with Asn (Vašková et al.2023). Individual GPx isoforms have different response mechanisms, and their activity requires GSH. The enzyme, hydroperoxide, and GSH do not form a ternary complex, but the process includes oxidation and reduction steps. Deprotonation occurs in the same manner during the oxidation process. The side chains of the Glu, Try, and Asp residues form a highly nucleophilic region in the active core of the enzyme, where the active site selenocysteine (RSeH) or cysteine (RSH) is oxidized upon peroxide binding. This process produces a selenenic acid (RSeOH) derivative. The selenenic acid is subsequently transformed to selenol (RseH) in two steps, beginning with a reaction with GSH to produce GS-SeR and water. A second GSH molecule reduces the GS-SeR intermediate to the selenol, releasing GS-SG as the byproduct (Deponte 2013; Flohé et al. 2011).

Figure 4. H_2O_2 leads to the oxidation of the active site selenocysteine (RSeH) or cysteine (RSH), resulting in the formation of a selenenic acid (RSeOH) derivative. The selenenic acid is then converted back to the selenol (RseH) by a two-step process that begins with a reaction with GSH to form the GS-SeR and water. A second GSH molecule reduces the GS-SeR intermediate to the selenol, releasing GS-SG as the byproduct. The glutathione disulfide is converted back to its active form by the enzyme glutathione reductase in the presence of NADPH which is attained from various sources such as the pentose phosphate pathway

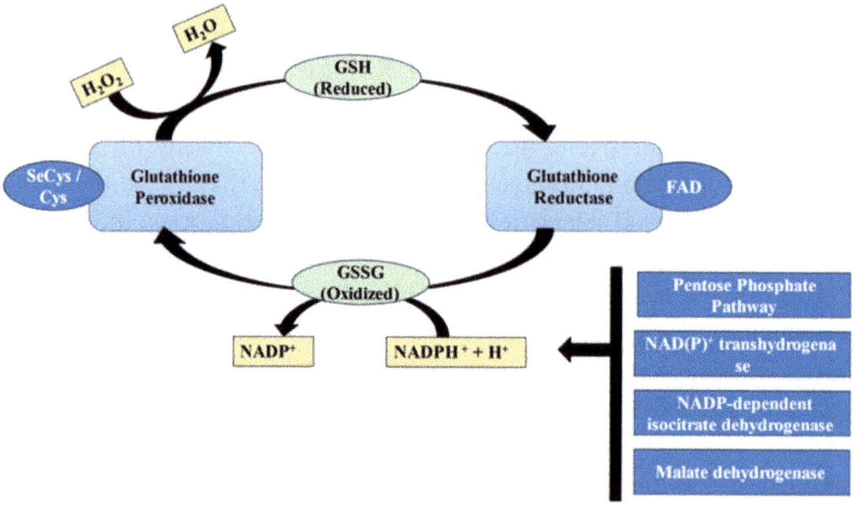

2.3.1 Regulation

Glutathione peroxidase system is regulated at different levels by various factors. The catalytic activity of GPx is directly linked to the presence of selenium incorporated into selenocysteine residue at its active site. A deficiency in dietary selenium can deplete GPx activity and levels, making individuals more susceptible to oxidative stress (Arthur et al. 2003). One important transcription factor essential for upregulating the expression of GPx in response to oxidative stress is nuclear factor erythroid 2-related factor 2 (Nrf2). Oxidative stress triggers Nrf2 activation, which increases gene transcription by binding to the antioxidant response element (ARE) in the GPx promoter (Ngo & Duennwald 2022). Other transcription factors that can control GPx expression include nuclear factor kappa B (NF-κB) and activator protein 1 (AP-1) (Priyadharshini et al. 2020). This is particularly true in response to inflammatory stimuli. The activity, stability, and localization of GPx can be impacted by phosphorylation, nitration, and ubiquitination (Vašková et al. 2023). Furthermore, several microRNAs can target GPx mRNA and cause translational inhibition or destruction (Matoušková et al. 2018). This offers a way to regulate GPx levels and stop overexpression, which might have unfavorable effects.

2.4 Peroxiredoxin and Thioredoxin System

2.4.1 Peroxiredoxin

Thiol-based peroxidases: Peroxiredoxins (Prdxs) are cysteine-dependent peroxidase enzymes that primarily reduce H_2O_2 to H_2O. However, they can also lower peroxynitrite, peroxynitrous acid, and lipid peroxides (hydroperoxide substrates) (Bryk, Griffin & Nathan 2000; Rhee et al. 2012). Their peroxidation functions overlap with GPx and catalase, although their catalytic efficiency is lower (~105 M−1 s−1) than GPx (~108 M−1 s−1) and catalase (~106 M−1 s−1) (Wood et al. 2003). The six mammalian peroxiredoxins are divided into three classes: conventional 2-Cys, atypical 2-Cys, and 1-Cys.

The largest class consists of conventional 2-Cys isoforms, including PRDX1, PRDX2, PRDX3 (mitochondria), and PRDX4 (ER). They are homodimers that contain two active sites (a CP and a CR). The disulfide bond between the two subunits is formed by the reaction of CP-SOH and the accessible CR of the other subunit. PRDX5 is the only unusual 2-cys peroxiredoxin found in mammals (Rhee & Woo 2011). The "atypical 2-Cys" Prdxs (Prdx5) are monomers that form intramolecular disulfide because both CR and CP are present in the same molecule; their reduction is achieved by Trx. The "atypical 2-Cys" Prdxs can form dimers regardless of redox status (Wood et al. 2003). PRDX6 is the only mammalian peroxiredoxin that belongs to the 1-Cys class. PRDX6 catalytic mechanism differs from the other peroxiredoxins, as no disulfide bond is formed due to the absence of accessible adjacent Cys residue and only the N terminal Cp is present (Rhee, Chae, & Kim 2005; Rhee & Woo 2011). Furthermore, PRDX6 relies on glutathione rather than reduction by thioredoxin, thioredoxin reductase, or NADPH (Rhee & Woo 2011).

2.4.2 Thioredoxin

The thioredoxin system consists of thioredoxin (Trx), thioredoxin reductases (TrxR), and NADPH (Figure 5). Trx is a disulfide-containing oxidoreductase that donates electrons to various redox processes while oxidizing itself. Trx is found in the cytoplasm, membranes, mitochondria, and extracellular spaces (Koharyova & Kolarova, 2008). The active site comprises the conserved sequence Cys-Gly-Pro-Cys. Oxidized Trx has a disulfide bond between the Cys residues (Trx-S-S) that is reduced by a flavoenzyme TrxR and NADPH to its active dithiol form, which is required for maintaining the redox balance of the cell by scavenging ROS and protecting proteins from oxidative damage by keeping them in their reduced form (Arner & Holmgren 2000).

Prdxs follow the same catalytic cycle, where the active site cysteine (peroxidative cysteine (CP) thiolate (CP-S−) attacks a hydroperoxide substrate and is oxidized to a CP-sulfenic acid (CP-SOH). The hydroperoxide substrate (including hydrogen peroxide, lipid peroxides, or peroxynitrite) is reduced to water or the corresponding alcohol. Some Prdxs have a second cysteine (CR), which interacts with CP-SOH to produce disulfide (CP-S-S-CR) and water (Nelson et al. 2018). CR can come from a neighboring monomer, the same monomer, glutathione, or a redox-relay binding partner (Bolduc et al. 2021). Thioredoxin, a tiny disulfide reductase, was initially found to reduce ribonucleotide reductase. It is the enzyme that catalyzes the synthesis of deoxyribonucleotides from ribonucleotides and is necessary for DNA replication (Stancill & Corbett 2021). Thioredoxin undergoes oxidation when it reduces the disulfide bond of peroxiredoxin, losing its catalytic activity. Thioredoxin reductase takes electrons from NADPH to reduce thioredoxin, allowing the antioxidant cycle to continue. If the peroxide level surpasses a particular threshold, peroxiredoxins undergo hyperoxidation, creating sulfinic or sulfonic acids, inhibiting the enzyme (Rhee & Woo 2011).

Figure 5. Peroxiredoxin and thioredoxin system

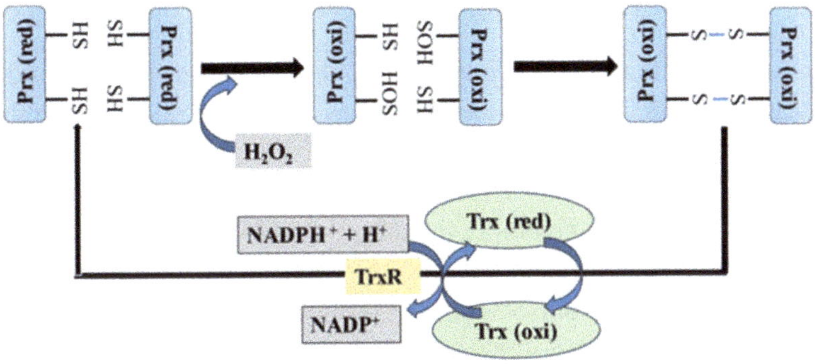

The active site cysteine (peroxidative cysteine (C_P) thiolate (C_P-S^-) attacks a hydroperoxide substrate and is oxidized to a C_P-sulfenic acid (C_P-SOH), the hydroperoxide substrate gets reduced to water or the corresponding alcohol. Some Prdxs contain a second, resolving cysteine (C_R), which reacts with C_P-SOH forming disulfide (C_P-S-S-C_R) and water. This disulfide is reduced to reform the thiolate by thioredoxin. In this process, Trx itself gets oxidized which is reduced by Thioredoxin Reductase (trxR) at the expense of $NADPH^+$ and H^+.

3. NON-ENZYMATIC DEFENSES

Nonenzymatic antioxidants impact human health and serve as food, medicine, and cosmetic preservation. Antioxidants that are not enzymatic can also stop the spread of free radicals. These include vitamins (C and E), minerals (selenium and zinc), phytochemicals, glutathione, and uric acid. While some antioxidants are liposoluble and are in cell membranes, others are water-soluble and occur in the cytosol or cytoplasmic matrix (Moussa et al. 2019).

3.1 Vitamins

3.1.1 Vitamin C

Potent antioxidant vitamin C can give up a hydrogen atom and create a comparatively stable ascorbate free radical. Vitamin C is a significant water-soluble antioxidant that neutralizes ROS and lowers oxidative stress. It contributes to the initial line of defense against oxidative damage to proteins and lipid membranes (Hamza, 2017). Vitamin C is a potent scavenger of oxygen and nitrogen oxide species, including singlet oxygen, hydrogen peroxide, hydroxyl radical, and superoxide radical ions. This characteristic of vitamin C is essential for procedures that shield cellular constituents from harm caused by free radicals (Hamza, 2017). Animals and plants contain vitamin C, a redox catalyst that facilitates the oxidation and reduction of monosaccharides. The capacity of vitamin C to transfer electrons from the second and third carbons accounts for its antioxidant properties (Hamza, 2017). Due to its low standard 1-electron reduction potential (282 mV), ability to convert DHA acid to ascorbic acid, and production of very stable semi-dehydroascorbic acid, ascorbic acid is an efficient electron donor. In the cytosol, thioredoxin reductase and cytochrome b reductase, which involve NADH and NADPH, respectively, convert semi-dehydroascorbate to ascorbate (Akbari et al. 2016). Intracellular substances including glutathione (GSH), NADH, and NADP are renewed by vitamin C. While most animals can generate vitamin C in their bodies and do not need it in their meals, humans must get it from their diet.

Vitamin C is required to convert proline residues to hydroxyproline and turn procollagen into collagen (Hamza, 2017). GSH-dependent dehydroascorbate reductase, prevalent in the testes, reacts with glutathione to retain it in its reduced form in other cells (Akbari et al. 2016). The highly reactive hydroxyl radical can be produced when H_2O_2 reacts with the superoxide anion produced during oxidative metabolism. Due to their strong affinity for H_2O_2, GSH peroxidases (GSH-px) and AA peroxidases (AA-px) can scavenge H_2O_2 (Hamza, 2017). The well-known antioxidant processes of ascorbic acid also include scavenging aqueous radicals and regeneration of α-tocopherol from the tocopheroxyl radical species (Akbari et al. 2016). Ascorbic acid was also considerably more effective than other plasma components in suppressing lipid peroxidation caused by the peroxyl radical initiator in experiments using human plasma lipids (Sies & Stahl, 1995).

3.1.2 Vitamin E

As fat oxidizes and free radical reactions spread, vitamin E, an antioxidant that breaks down chains, neutralizes reactive oxygen species. Activities of Vitamin C, B3, selenium and glutathione depend on vitamin E (Rizvi et al. 2014). The best protectant against oxidant-induced membrane damage is lipid-soluble vitamin E, which is concentrated in the hydrophobic inner location of the cell membrane (Birben et al. 2012). During cellular lipid peroxidation of polyunsaturated fatty acids (PFA) and low-density lipoprotein (LDL), it has been demonstrated to be the most potent membrane-bound antioxidant that cells use to scavenge reactive nitrogen and oxygen species, thereby disrupting oxidative damage to cell membrane phospholipids (Moussa et al. 2019). The major biological function of vitamin E is its antioxidant activity, which is explained by its reactivity with organic peroxyl radicals and linked to the chromane ring redox characteristics (Sies & Stahl, 1995). As the primary component of the anti-radical defense mechanism, vitamin E benefits from the ascorbate ion of vitamin C and their mutually beneficial antioxidation interactions. Since the body cannot produce vitamin E, it must be received through diet in modest doses as it is an exogenous vitamin (Moussa et al. 2019).

3.2 Minerals (Selenium and Zinc)

The healthy operation of an animal immune system and metabolism depends on a few critical micronutrients called trace elements. Among these, zinc (Zn) and selenium (Se) are especially crucial because of their antioxidant qualities. They protect DNA and cell membranes from oxidative damage brought on by oxygen radicals (ROS), which are produced during normal cellular metabolism (Ghorbani et al. 2024). Zinc and selenium play a specific role in the production and activity control of proteins and enzymes, such as glutathione peroxidase (GPX) and metallothioneins (MT), which are essential to the natural defense of the body against oxidative stress (Mocchegiani & Malavolta, 2019).

Small, cysteine-rich metallothionein binds several divalent cations and most likely serves as a zinc storage protein. Zinc is a strong inducer of the protein, and the promoter of the protein contains several closely linked nucleotide sequences known as metal-responsive elements (MRE), which give sensitivity to metals and can confer a similar sensitivity when inserted into the promoters of unrelated genes. Since zinc plays a crucial part in the so-called "Zn-finger" proteins, zinc induces metallothionein and regulates the expression of several other genes (Chesters, 2005a). Furthermore, animal growth, reproduction, and immune system function are among the physiological processes in which zinc (Zn) is crucial. The activity and expression of enzymes, mitogenic hormones, signaling pathways, and transcription-related genes are among the processes that zinc can impact (Chesters, 2005b). Animals with zinc deficiencies exhibit stunted growth and reduced appetite (Kendall et al. 2000).

Glutathione peroxidases (GPxs) are a family of selenoproteins that catalyze the reduction of hydroperoxides, including hydrogen peroxide, to shield cells from oxidative damage (Hall et al. 1998). There are four different GPx isoforms in mice and five human isoforms that depend on selenium. Nearly every cell has GPx1 in its cytoplasm, which catalyzes the reduction of free hydroperoxides. Like GPx1, GPx2 is also expressed in the gastrointestinal system and has a similar substrate specificity. The extracellular enzyme GPx3 decreases phospholipid hydroperoxides bonded to membranes and is present in plasma. GPx4 is expressed in several tissues and functions as a reductant agent for phospholipid hydroperoxide and hydrogen peroxide by utilizing thiols, such as homocysteine, cysteine, and 2-mercaptoethanol, in addition to GSH (Mocchegiani & Malavolta, 2019). The olfactory epithelium and embryonic tissues are the only places where the isoform GPx6 is expressed (Kryukov et al. 2003). Additionally, it exists as the epididymis isoenzyme GPx5, which is selenium-independent and found in both humans and mice (Hall et al. 1998). A lack of selenium can lead to several health problems, such as decreased immunity, cardiovascular consequences, concerns with reproduction and fertility, thyroid dysfunction, neurological symptoms, and abnormalities related to the musculoskeletal system (Bomer et al. 2020).

3.3 Phytochemicals

Phytochemicals, which are non-nutritive molecules generated from plants, are among the numerous dietary components crucial to the various processes carried out by the animal body (Abbas et al. 2014). Even though vitamins C and E are common antioxidants, fruits, vegetables, spices, and herbs may include substances such as flavonoids and phenolic acids with far higher antioxidant activity than the necessary vitamins (Lee et al. 2017).

3.3.1 Sources

Bioactive chemicals are abundant in traditional fruits such as guavas, sweetsops, persimmons, Chinese wampees, plums, pomegranates, Chinese dates, and berries (Manganaris et al. 2014). Some vegetables, including broccoli, cowpeas, Allium cepa, sweet potatoes, green soybeans, pepper, ginseng, and broccoli, have higher levels of total phenolics and antioxidant potential (Deng, Lin, et al. 2013). Among cereal grains, colored rice variants such as black, red, and purple rice are notable for having higher concentrations of phytochemicals that are antioxidants, including flavones and tannins (Deng, Xu, et al. 2013). Additionally, leftover fruit components like peels and seeds are a great source of phytochemicals like epicatechin, gallic acid, kaempferol, catechin, and cyanidin-3-glucoside (Deng et al. 2012). Fruits with elevated antioxidant potential *Eucalyptus robusta, Eurya nitida, Melastoma sanguineum, Melaleuca leucadendron, Lagerstroemia indica, Caryota mitis, Lagerstroemia speciosa, and Gordonia axillar (*Fu et al. 2010).

3.3.2 Benefits

3.3.2.1 Flavonoids

Flavonoids can directly react with a variety of ROS types. Flavonol quercetin exhibits a quenching ability for the non-radical ROS molecule singlet oxygen ($^1O_2$). The flavonoids kaempferol and cyanidin anthocyanidin may scavenge superoxide radicals (O_2^-) *in vitro*. The flavonoid quercetin scavenges hydroxyl radicals (•OH) generated by water radiolysis. The flavonols quercetin and rutin scavenge the linoleic acid hydroperoxide (LOO•) to prevent lipid peroxidation. The antioxidant or free radical scavenging properties of flavonoids are mostly attributed to the OH groups on their aromatic rings (Sakihama & Yamasaki, 2021).

3.3.2.2 Betalains

The skeleton of betalamic acid is a possible contributor to their antioxidant properties. The capacity of betalamic acids to scavenge free radicals may be enhanced by the phenolic hydroxy group in the *cyclo*-dopa moiety of betacyanin and the amino acid/amine component of betaxanthin. To detoxify hydrogen peroxide (H_2O_2), betalains can serve as electron donors for the peroxidase enzyme (Sakihama & Yamasaki, 2021).

3.3.2.3 Carotenoids

A class of phytonutrients generated by plants, algae, some bacteria and fungi called carotenoids are particularly liposoluble because of their long alkyl chain of unsaturated hydrocarbons (Stahl & Sies, 2003).

As the most prevalent lipid-soluble antioxidants, they prevent lipid peroxidation from damaging lipoproteins and cellular membranes and are highly effective in scavenging peroxyl radicals compared to other ROS (Mueller & Boehm, 2011).

3.3.2.4 Allyl sulfides & curcumin

In garlic, curcumin and allicin (diallyl thiosulfinate) are physiologically active substances with antioxidant qualities. Through a subsequent elimination reaction of the previous precursor, 2-propene sulfenic acid is the active form of allicin responsible for its antioxidant effect. The radical-scavenging process of Allicin includes the abstraction of an H atom from the sulfenic acid residue by a peroxyl radical. Curcumin is a liposoluble free radical scavenger with amazing chain-breaking power comparable to vitamin E and is also known as the bis-α, β-unsaturated β-diketone (Moussa et al. 2019).

3.3.2.5 Lycopene

Lycopene, a potent antioxidant that protects cells from free radical damage, is a non-provitamin A carotenoid found in abundance in tomatoes and their byproducts (Abdul-Lateef Ali & Kaid, 2014). Its ability to quench oxygen is twice as high as that of β-carotene and ten times higher than that of α-tocopherol. Lycopene controls redox-sensitive molecular targets, including Nrf2, and the activation of mitogen-activated protein kinases, which in turn affect the activity of other enzymes (Lee et al. 2017).

3.4 Glutathione (GSH)

Glutathione (GSH) is a tripeptide composed of L-γ-glutamyl-L-cysteinylglycine and serves several purposes in living things (Lv et al. 2019). Being a cysteine residue that carries an active thiol group, it functions as an antioxidant directly through interactions with electrophiles and reactive oxygen/nitrogen species (ROS and RNS, respectively) or indirectly as an enzyme cofactor (Duan & Chen, 2007).

3.4.1 Synthesis

Two ATP-consuming enzyme stages are involved in the synthesis of GSH from its component amino acids: the first is the synthesis of γ-glutamylcysteine from glutamate and cysteine, and the second is the generation of GSH from γ-glutamylcysteine and glycine. Glutamate–cysteine ligase (GCL) (EC 6.3.2.2, formerly γ-glutamylcysteine synthetase) catalyzes the rate-limiting first step of GSH biosynthesis. It is made up of two subunits: a light or modifier (GCLM, Mr. ~ 31 kDa) and a heavy or catalytic (GCLC, Mr. ~ 73 kDa), which are encoded by separate genes in humans, rodents, and fruit flies (Huang et al. 1993, 1995). GCL in bacteria and yeast, on the other hand, only consists of one polypeptide. GCLC demonstrates GSH-induced feedback inhibition as well as the entire catalytic activity of the isolated enzyme (Seelig et al. 1984). Even though GCLM lacks enzyme activity, it regulates glutamate and GSH levels by decreasing the Km of GCL and increasing the Ki of GSH (Huang et al. 1993).

γ-L-Glutamate + L-cysteine + ATP —→ γ-L-glutamyl-L-cysteine + ADP + Pi (Lushchak, 2012)

GSH synthetase, also known as GSH synthase (GS, EC 6.3.2.3), is the catalyst for the second phase of GSH production. GSH does not hinder the ability of GS to function because it consists of two subunits that are equal in size (Mr ~ 118 kDa) (Oppenheimer et al. 1979).

γ-L-Glutamyl-L-cysteine + glycine + ATP —→ GSH + ADP + Pi (Lushchak, 2012)

GCL is regarded as rate limiting since its byproduct, γ-glutamylcysteine, is present when GS is present at relatively low amounts (Lim & Luderer, 2018).

3.4.2 Role of GSH

The antioxidant function of GSH is mostly fulfilled by GSH peroxidase (GPx)-catalyzed reactions, which reduce hydrogen peroxide and lipid peroxide as GSH is oxidized to GSSG. A redox cycle is created when GSSG reductase reduces GSSG back to GSH at the expense of NADPH (Lu, 2009). GPx and GSH S-transferase are also capable of neutralizing organic peroxides. Although it is only found in peroxisomes, catalase can also reduce hydrogen peroxide. Because of this, GSH plays a crucial role in the mitochondrial defense against oxidative stress that is produced both physiologically and pathologically (Chen et al. 2024).

3.5 Uric Acid

About 66% of the oxygen-scavenging activity in blood serum is attributed to uric acid (UA), a hydrophilic antioxidant produced during the breakdown of purine nucleotides. Uric acid is the most abundant aqueous antioxidant in humans, with a blood level of about 3.5–7.5 mg/dL. It can be produced by both mammals and humans (Leonard et al. 2014; Waring, 2002). UA is a potent electron donor and a selective scavenger of peroxynitrite ($ONOO^-$); however, to fully scavenge such species, ascorbic acid, and thiols must be included in its cycle (Galano & Alvarez-Idaboy, 2011; Squadrito et al. 2000). Nitric oxide (·NO) and superoxide radicals ($O_2^{\cdot-}$) react to generate peroxynitrite, which has been linked to numerous diseases. In addition to scavenging peroxynitrite, UA reacts with hypochlorous acid, lipid peroxides, singlet oxygen, and hydroxyl radicals, converting them into harmless chemical species like urea and allantoin. Moreover, it has been linked to the scavenging of nitrogen dioxide ($NO_2^{\cdot}$) and carbonate ions ($CO_3^{\cdot-}$) (Settle, 2014)

4. ENZYMATIC AND NON-ENZYMATIC INTERPLAY

4.1 Recycling of Ascorbate

Ascorbate is a major antioxidant involved in the direct scavenging of free radicals. It is present in two different oxidized forms, monodehydroascorbate (MDHA), a one-electron oxidized form, and dehydroascorbate (DHA), a two-electron oxidized form of ascorbate. Ascorbate recycling is vital to combat oxidative stress and maintain the supply of the active vitamin in the body. There are two main enzymes involved in recycling oxidized ascorbate to its fully reduced active form, dehydroascorbate reductase and monodehydroascorbate reductase. DHAR catalyzes the reduction of DHA to the active form of ascorbate using glutathione (GSH) as a hydrogen donor. In contrast, (MDHAR) utilizes NADH/NADPH as an electron donor to reduce MDHA to ascorbate before the spontaneous oxidation of MDHA takes place to form DHA (Akbari et al. 2016; Hamza, 2017).

The enzyme GR plays a great role in catalyzing the reduction of glutathione disulfide (GSSG) back into two molecules of reduced glutathione (GSH), allowing glutathione to continue functioning as a hydrogen donor while reducing DHA. This reaction requires NADPH as an electron donor. GR uses the reducing power of NADPH to donate electrons to GSSG, breaking the disulfide bond and regenerating GSH (Couto et al. 2016). Studies show that TrxR, a selenium-dependent enzyme, is also involved in the conversion of DHA back to ascorbic acid using NADPH as a reducing agent apart from its primary function of regenerating peroxiredoxin (Nordberg & Arnér 2001).

4.2. Redox-active metals

Redox-active metals like copper, iron, and selenium participate in cycling reactions involving electron transfer between metals and as substrates to combat oxidative stress and maintain cellular homeostasis. These metals are a part of antioxidant enzymes and play a great role in their functioning.

Copper is a cofactor for Cu/Zn-SOD, an essential enzyme that catalyzes the dismutation of the superoxide anion ($O_2^{\bullet-}$), a highly reactive oxygen species, into less toxic hydrogen peroxide (H_2O_2) and molecular oxygen (O_2). This reaction neutralizes the damaging superoxide anion, which can contribute to lipid peroxidation, protein oxidation, and DNA damage if left unchecked (Maurya & Namdeo 2021).

Iron is a cofactor in the catalase enzyme, which catalyzes the decomposition of hydrogen peroxide (H_2O_2) into water (H_2O) and oxygen (O_2). This reaction leads to the degradation of accumulated hydrogen peroxide (Baker et al. 2023).

Selenium is a main component of enzymes that are directly involved in neutralizing ROS. Selenium is incorporated into the active site of glutathione peroxidase (GPx), which reduces hydrogen peroxide (H_2O_2) and lipid hydroperoxides into water (H_2O) and alcohols, respectively, using glutathione (GSH) as a substrate. Selenium is also a part of thioredoxin reductase (TrxR), an enzyme that regenerates the reduced form of thioredoxin, which plays a key role in maintaining the redox balance in cells. TrxR helps to combat oxidative stress by regenerating antioxidant molecules and maintaining cellular homeostasis (Flohé et al. 2011).

4.3. Betalains

Betalains are known to modulate the Nrf2/ARE signaling pathway, which plays a critical role in the defense against oxidative stress. Nrf2 (Nuclear factor erythroid 2–related factor 2) is a transcription factor that regulates the expression of various antioxidant and cytoprotective genes. Under normal conditions, Nrf2 is kept inactive in the cytoplasm by its inhibitor, Keap1. When oxidative stress or electrophiles are detected, Nrf2 is released from Keap1 and translocated to the nucleus, where it binds to Antioxidant Response Elements (ARE) present in DNA. ARE are specific sequences in the promoter regions of genes that encode antioxidant enzymes. When

Nrf2 binds to ARE, it stimulates the transcription of these antioxidant genes, thereby synthesizing enzymes such as glutathione peroxidase (GPx), superoxide dismutases (SOD), and catalase (CAT) involved in the detoxification and elimination of reactive oxygen species (ROS), enhancing the ability of the cell to combat oxidative stress (Wang et al. 2022; Nirmal et al.2024).

5. ANTIOXIDANTS IN HEALTH AND DISEASE

5.1 Cardiovascular disease

An increase in antioxidant levels minimizes the clinical manifestation of coronary artery disease (CAD) according to several large-scale epidemiological studies. People who consume large amounts of beta-carotene, vitamin C, and vitamin E through their diet are at a lower risk of developing CAD and may live longer (Yang et al. 2022). Results from cohort and case-control studies suggest that beta-carotene and the antioxidant vitamins C and E may be essential in the primary prevention of cardiovascular disease. Studies on antioxidants also revealed that they were most beneficial in the early stages of atherosclerosis, which is the condition that causes peripheral vascular disease, heart attacks, and strokes. Vitamin C intake of 45–113 mg/day has been linked to a lower risk of cardiovascular disease. 60% decreased risk of stroke could be related to vitamin C consumption of over 45 mg per day compared to those who consumed less than 28 mg per day (Osganian et al. 2003). Increased vitamin E intake was linked to a lower incidence of coronary heart disease in three large-scale prospective cohort studies on both men and women. Two-year research involving 87,245 nurses found that taking dietary supplements daily that include vitamin E at a level exceeding 100 IU reduced the incidence of coronary heart disease (Debreceni & Debreceni 2014).

5.2 Cancer

The National Cancer Institute (NCI) states that vitamin C may lower the chance of developing cancers of the rectum, pancreas, and cervix in addition to protecting against malignancies of the stomach, esophagus, and oral cavity. Vitamin C/ ascorbic acid may offer a defense against lung and breast cancer. Provitamin A/beta carotene may help lower the risk of cancer. The American Cancer Society claims that vitamin A strengthens the immune system, which may help prevent some malignancies. Free radicals that cause cell damage are obstructed by white blood cells (Antioxidants as Part of Your Cancer Diet | Stanford Health Care, 2024).

5.3 Neurodegeneration

It has been demonstrated that vitamin E exhibits exceptional anti-peroxyl radical activity and can mitigate the neurotoxic consequences of these radicals (Lloret et al. 2019). In Alzheimer's disease (AD), vitamin E dramatically lowers nitrosative and oxidative damage (Casati et al. 2020). Ascorbyl palmitate is another significant antioxidant therapeutic for AD (Frank & Gupta 2005). It has been demonstrated that diets high in carotenoids, flavonoids, polyphenols, and vitamins C, and E support standard Alzheimer's disease therapy (Burgener et al. 2008).

Vitamins C and E, in combination at high doses (3000 mg/day and 3200 IU/day), may decrease the progression of Parkinson's disease, according to pilot research (Fahn 1992). An endogenous antioxidant called coenzyme Q10 (CoQ10)/ubiquinone is a cofactor in the electron transport chain during the synthesis of ATP. The brain, heart, liver, and kidney are among the tissues with higher metabolism rates. Thus, they need a lot of this antioxidant. CoQ10 has been shown in numerous trials to have positive benefits for Parkinson's disease (Shults et al. 2002).

Similarly, numerous antioxidant compounds have demonstrated potential therapeutic benefits in treating Huntington's disease. Excellent antioxidant benefits are documented for coQ10, lycopene, vitamins C and E, melatonin, and naturally occurring substances such as grape seed extracts. These substances may also decrease the progression of the disease (Khan et al. 2018). Antioxidant drugs have also been shown to be effective in treating Amyotrophic Lateral Sclerosis in numerous investigations. It has been demonstrated that increasing plasma vitamin E levels by supplementation offers partial protection against the onset of amyotrophic lateral sclerosis (Wang et al. 2011).

5.4 Aging

Aging is the progressive loss of function combined with heightened susceptibility to illness. There are several theories on the aging process, even though the precise cause of aging is not entirely known. The oxidative stress theory of aging has received much attention lately. It contends that vascular inflammation results from an imbalance in the reduction-oxidation process in the body through a convoluted genetic pathway. Although many scientists think inflammation may be the root cause of age-related degeneration and disease, inflammation is the natural defense mechanism for preventing or healing damage from external substances. The literature suggests that antioxidants may slow down the aging process. Antioxidants can lessen inflammation, boost collagen, and shield the skin from sun damage (Park 2009). The following antioxidants may slow down the aging process:

The human body has the largest concentration of vitamin C, which is necessary for collagen production, whose formation is crucial for healthy, young skin (Miyazawa et al. 2022). Vitamin E (α-tocopherol) is a lipophilic compound found naturally in food (Tan et al. 2018). Red grapes and berries include resveratrol, which can increase collagen production and shield the skin from damage caused by free radicals. Additionally, it might lessen inflammation and redness (*How to Use Antioxidants in Your Skincare Routine for Anti-Aging*, 2022). Coenzyme Q10 (CoQ10) is an antioxidant that the body produces and is crucial for preventing cell damage and producing energy. Studies indicate that as people age, their levels of CoQ10 decrease (*The 14 Best Anti-Aging Supplements*, 2022). Niacinamide/ vitamin B3 has anti-inflammatory and skin-texturizing effects that reduce wrinkles and fine lines (*The Best Antioxidants for Your Skin & Their Benefits | Dermstore* 2022).

5.5 Supplementation challenges

Antioxidant consumption results in dose-related effects, whereby high dosages have detrimental effects and low amounts have beneficial ones (Bentley et al. 2015). The antioxidant defenses of the body may be maintained by antioxidant doses based on the Recommended Dietary Allowance (RDA), even for competitive endurance athletes. However, the adverse effects of antioxidant supplements arise when taking levels that are 5–17 times more than the RDA (Mason et al. 2020). For athletes and people who exercise frequently, taking large amounts of antioxidants may have unfavorable consequences (Gomez-Cabrera et al. 2012). High dosages of antioxidants over an extended period diminish physiological adaptations brought on by exercise, including muscular growth, enhanced antioxidant capacity, mitochondrial biogenesis, and higher insulin sensitivity. They also block redox-sensitive signaling pathways (Merry & Ristow 2016; Peternelj & Coombes 2011).

6. CONCLUSION AND FUTURE DIRECTIONS

The enzymatic and non-enzymatic defense systems work concertedly to protect the cells against the harmful effects of ROS. Synergism between these components is essential in preventing several chronic illnesses. Despite significant advancements in understanding the antioxidant defense system, several areas require attention. The effectiveness of antioxidant supplementation can be variable due to genetic differences, lifestyle factors, and dietary habits. Future research can focus on developing personalized antioxidant strategies based on individual needs and risk factors. The discovery of novel antioxidant compounds can help expand our arsenal against harmful ROS species. Developing a deeper understanding of the interaction between antioxidants and cellular processes such as metabolism and inflammation could help devise novel therapeutic strategies to combat chronic diseases resulting from free radicals. Developing efficient delivery systems can also enhance their effectiveness and reduce side effects. Researchers can develop more effective strategies for preventing and treating diseases associated with oxidative stress if these issues are addressed in future research.

REFERENCES

Abbas, Z. K., Saggu, S., Sakeran, M. I., Zidan, N., Rehman, H., & Ansari, A. A. (2014). Phytochemical, antioxidant, and mineral composition of hydroalcoholic extract of chicory (Cichorium intybus L.) leaves. *Saudi Journal of Biological Sciences, 22*(3), 322–326. DOI: 10.1016/j.sjbs.2014.11.015 PMID: 25972754

Akbari, A., Jelodar, G., Nazifi, S., & Sajedianfard, J. (2016). (in press). An Overview of the Characteristics and Function of Vitamin C in Various Tissues: Relying on its Antioxidant Function. *Zahedan Journal of Researches in Medical Sciences, In Press*(In Press). Advance online publication. DOI: 10.17795/zjrms-4037

Alfonso-Prieto, M., Biarnés, X., Vidossich, P., & Rovira, C. (2009). The molecular mechanism of the catalase reaction. *Journal of the American Chemical Society, 131*(33), 11751–11761. DOI: 10.1021/ja9018572 PMID: 19653683

Ali, N. A. L., Mohammed, A. B., & Allow, A. A. (2014). Effect of adding different levels of lycopene to the ration on some lipid profile traits of the laying hens ISA-Brown.

Antioxidants as Part of Your Cancer Diet | Stanford Health Care. (2024). Retrieved August 17, 2024, from https://stanfordhealthcare.org/medical-clinics/cancer-nutrition-services/reducing-cancer-risk/antioxidants.html

Arnér, E. S. J., & Holmgren, A. (2000). Physiological functions of thioredoxin and thioredoxin reductase. *European Journal of Biochemistry, 267*(20), 6102–6109. DOI: 10.1046/j.1432-1327.2000.01701.x PMID: 11012661

Arning, J., Dringen, R., Schmidt, M., Thiessen, A., Stolte, S., Matzke, M., Bottin-Weber, U., Caesar-Geertz, B., Jastorff, B., & Ranke, J. (2008). Structure-activity relationships for the impact of selected isothiazol-3-one biocides on glutathione metabolism and glutathione reductase of the human liver cell line Hep G2. *Toxicology, 246*(2–3), 203–212. DOI: 10.1016/j.tox.2008.01.011 PMID: 18313196

Arthur, J. R., McKenzie, R. C., & Beckett, G. J. (2003). Selenium in the immune system. *The Journal of Nutrition, 133*(5, Suppl 1), 1457S–1459S. Advance online publication. DOI: 10.1093/jn/133.5.1457S PMID: 12730442

Baker, A., Lin, C. C., Lett, C., Karpinska, B., Wright, M. H., & Foyer, C. H. (2023a). Catalase: A critical node in the regulation of cell fate. *Free Radical Biology & Medicine, 199*, 56–66. DOI: 10.1016/j.freeradbiomed.2023.02.009 PMID: 36775107

Bentley, D. J., Ackerman, J., Clifford, T., & Slattery, K. S. (2015). *Acute and Chronic 9 Effects of Antioxidant Supplementation on Exercise Performance.*

Betteridge, D. J. (2000). What is oxidative stress? *Metabolism: Clinical and Experimental, 49*(2, Suppl 1), 3–8. DOI: 10.1016/S0026-0495(00)80077-3 PMID: 10693912

Bhattacharyya, A., Chattopadhyay, R., Mitra, S., & Crowe, S. E. (2014). Oxidative stress: An essential factor in the pathogenesis of gastrointestinal mucosal diseases. *Physiological Reviews, 94*(2), 329–354. DOI: 10.1152/physrev.00040.2012 PMID: 24692350

Birben, E., Sahiner, U. M., Sackesen, C., Erzurum, S., & Kalayci, O. (2012). Oxidative stress and antioxidant defense. *The World Allergy Organization Journal, 5*(1), 9–19. DOI: 10.1097/WOX.0b013e3182439613 PMID: 23268465

Bolduc, J., Koruza, K., Luo, T., Malo Pueyo, J., Vo, T. N., Ezeriņa, D., & Messens, J. (2021). Peroxiredoxins wear many hats: Factors that fashion their peroxide-sensing personalities. *Redox Biology, 42*, 101959. DOI: 10.1016/j.redox.2021.101959 PMID: 33895094

Bomer, N., Grote Beverborg, N., Hoes, M. F., Streng, K. W., Vermeer, M., Dokter, M. M., IJmker, J., Anker, S. D., Cleland, J. G. F., Hillege, H. L., Lang, C. C., Ng, L. L., Samani, N. J., Tromp, J., van Veldhuisen, D. J., Touw, D. J., Voors, A. A., & van der Meer, P. (2020). Selenium and outcome in heart failure. *European Journal of Heart Failure, 22*(8), 1415–1423. DOI: 10.1002/ejhf.1644 PMID: 31808274

Bryk, R., Griffin, P., & Nathan, C. (2000). Peroxynitrite reductase activity of bacterial peroxiredoxins. *Nature, 407*(6801), 211–215. DOI: 10.1038/35025109 PMID: 11001062

Bukke, V. N., Moola, A., Serviddio, G., Vendemiale, G., & Bellanti, F. (2022). Nuclear factor erythroid 2-related factor 2-mediated signaling and metabolic associated fatty liver disease. *World Journal of Gastroenterology, 28*(48), 6909–6921. DOI: 10.3748/wjg.v28.i48.6909 PMID: 36632321

Burgener, S. C., Buettner, L. L., Coen Buckwalter, K., Beattie, E., Bossen, A. L., Fick, D. M., Fitzsimmons, S., Kolanowski, A., Richeson, N. E., Rose, K., Schreiner, A., Pringle Specht, J. K., Testad, I., Yu, F., & Mckenzie, S. (2008). Evidence supporting nutritional interventions for persons in early-stage Alzheimer's disease (AD). *The Journal of Nutrition, Health & Aging, 12*(1), 18–21. DOI: 10.1007/BF02982159 PMID: 18165840

Carbonell, T., & Gomes, A. (2020). MicroRNAs in the regulation of cellular redox status and its implications in myocardial ischemia-reperfusion injury. *Redox Biology, 36*, 101607. DOI: 10.1016/j.redox.2020.101607 PMID: 32593128

Casati, M., Boccardi, V., Ferri, E., Bertagnoli, L., Bastiani, P., Ciccone, S., Mansi, M., Scamosci, M., Rossi, P. D., Mecocci, P., & Arosio, B. (2020). Vitamin E and Alzheimer's disease: The mediating role of cellular aging. *Aging Clinical and Experimental Research, 32*(3), 459–464. DOI: 10.1007/s40520-019-01209-3 PMID: 31054115

Cereser, C., Boget, S., Parvaz, P., & Revol, A. (2001). Thiram-induced cytotoxicity is accompanied by rapid and drastic oxidation of reduced glutathione with consecutive lipid peroxidation and cell death. *Toxicology, 163*(2–3), 153–162. DOI: 10.1016/S0300-483X(01)00401-2 PMID: 11516525

Chen, T. H., Wang, H. C., Chang, C. J., & Lee, S. Y. (2024). Mitochondrial glutathione in cellular redox homeostasis and disease manifestation. In *International Journal of Molecular Sciences* (Vol. 25, Issue 2). Multidisciplinary Digital Publishing Institute (MDPI). DOI: 10.3390/ijms25021314

Chesters, J. K. (2005a). Trace element–gene interactions with particular reference to zinc. *The Proceedings of the Nutrition Society, 50*(2), 123–129. DOI: 10.1079/PNS19910023 PMID: 1749786

Couto, N., Wood, J., & Barber, J. (2016). The role of glutathione reductase and related enzymes on cellular redox homoeostasis network. *Free Radical Biology & Medicine, 95*, 27–42. DOI: 10.1016/j.freeradbiomed.2016.02.028 PMID: 26923386

Culotta, V. C., Yang, M., & O'Halloran, T. (2006). Activation of superoxide dismutases: Putting the metal to the pedal. *Biochimica et Biophysica Acta, 1763*(7), 747–758. DOI: 10.1016/j.bbamcr.2006.05.003 PMID: 16828895

Debreceni, B., & Debreceni, L. (2014). Role of vitamins in cardiovascular health and disease. *Research Reports in Clinical Cardiology, 5*, 283–295. DOI: 10.2147/RRCC.S44465

Deng, G. F., Lin, X., Xu, X. R., Gao, L. L., Xie, J. F., & Li, H. (2013). Antioxidant capacities and total phenolic contents of 56 vegetables. *Journal of Functional Foods, 5*(1), 260–266. DOI: 10.1016/j.jff.2012.10.015

Deng, G. F., Shen, C., Xu, X. R., Kuang, R. D., Guo, Y. J., Zeng, L. S., Gao, L. L., Lin, X., Xie, J. F., Xia, E. Q., Li, S., Wu, S., Chen, F., Ling, W. H., & Li, H. (2012). Potential of fruit wastes as natural resources of bioactive compounds. *International Journal of Molecular Sciences, 13*(7), 8308–8323. DOI: 10.3390/ijms13078308 PMID: 22942704

Deng, G. F., Xu, X. R., Zhang, Y., Li, D., Gan, R. Y., & Li, H. (2013). Phenolic compounds and bioactivities of pigmented rice. *Critical Reviews in Food Science and Nutrition*, *53*(3), 296–306. DOI: 10.1080/10408398.2010.529624 PMID: 23216001

Deponte, M. (2013). Glutathione catalysis and the reaction mechanisms of glutathione-dependent enzymes. *Biochimica et Biophysica Acta (BBA) -. Biochimica et Biophysica Acta. G, General Subjects*, *1830*(5), 3217–3266. DOI: 10.1016/j.bbagen.2012.09.018 PMID: 23036594

Duan, S., & Chen, C. (2007). *S-nitrosylation/Denitrosylation and Apoptosis of Immune Cells* (Vol. 4).

Fahn, S. (1992). A pilot trial of high-dose alpha-tocopherol and ascorbate in early Parkinson's disease. *Annals of Neurology*, *32 Suppl*(1 S), S128–S132. DOI: 10.1002/ana.410320722

Flohé, L., Toppo, S., Cozza, G., & Ursini, F. (2011). A comparison of thiol peroxidase mechanisms. *Antioxidants & Redox Signaling*, *15*(3), 763–780. DOI: 10.1089/ars.2010.3397 PMID: 20649470

Frank, B., & Gupta, S. (2005). A review of antioxidants and Alzheimer's disease. In *Annals of Clinical Psychiatry* (Vol. 17, Issue 4, pp. 269–286). DOI: 10.1080/10401230500296428

Fu, L., Xu, B. T., Xu, X. R., Qin, X. S., Gan, R. Y., & Li, H. (2010). Antioxidant capacities and total phenolic contents of 56 wild fruits from South China. *Molecules (Basel, Switzerland)*, *15*(12), 8602–8617. DOI: 10.3390/molecules15128602 PMID: 21116229

Fukai, T., & Ushio-Fukai, M. (2011). Superoxide dismutases: Role in redox signaling, vascular function, and diseases. *Antioxidants & Redox Signaling*, *15*(6), 1583–1606. DOI: 10.1089/ars.2011.3999 PMID: 21473702

Galano, A., & Alvarez-Idaboy, J. R. (2011). Glutathione: Mechanism and kinetics of its non-enzymatic defense action against free radicals. *RSC Advances*, *1*(9), 1763–1771. DOI: 10.1039/c1ra00474c

Ghorbani, A., Moeini, M. M., Souri, M., Hajarian, H., & Kachuee, R. (2024). Effect of dietary zinc, selenium and their combination on antioxidant parameters in serum and Semen of Sanjabi mature rams. *Journal of Trace Elements and Minerals*, *8*, 100118. DOI: 10.1016/j.jtemin.2024.100118

Gomez-Cabrera, M. C., Ristow, M., & Viña, J. (2012). Antioxidant supplements in exercise: Worse than useless? *American Journal of Physiology. Endocrinology and Metabolism, 302*(4), E476–E477. Advance online publication. DOI: 10.1152/ajpendo.00567.2011 PMID: 22307485

Hall, L., Williams, K., Perry, A. C. F., Frayne, J., & Jury, J. A. (1998). The majority of human glutathione peroxidase type 5 (GPX5) transcripts are incorrectly spliced: Implications for the role of GPX5 in the male reproductive tract. *The Biochemical Journal, 333*(1), 5–9. DOI: 10.1042/bj3330005 PMID: 9639555

Hamza, A. H. (Ed.). (2017). *Vitamin C*. InTech., DOI: 10.5772/66058

Haque, R., Chun, E., Howell, J. C., Sengupta, T., Chen, D., & Kim, H. (2012). MicroRNA-30b-mediated regulation of catalase expression in human ARPE-19 cells. *PLoS One*, *7*(8), e42542. DOI: 10.1371/journal.pone.0042542 PMID: 22880027

Hong, Y., Boiti, A., Vallone, D., & Foulkes, N. S. (2024). Reactive Oxygen Species Signaling and Oxidative Stress: Transcriptional Regulation and Evolution. *Antioxidants, 13*(3), 312. Advance online publication. DOI: 10.3390/antiox13030312 PMID: 38539845

How to Use Antioxidants in Your Skincare Routine for Anti-Aging. (2022). Retrieved August 18, 2024, from https://www.byrdie.com/antioxidant-skincare

Huang, C.-S., Anderson, M. E., & Meister, A. (1993). *THE JOURNAL OF BIOLOGICAL CHEMISTRY Amino acid sequence and function of the light subunit of rat kidney y-glutamylcysteine synthetase* (Vol. 268, Issue 27).

Huang, C.-S., He, W., Meister, A., & Anderson, M. E. (1995). The amino acid sequence of rat kidney glutathione synthetase (inborn errors of metabolism/glycoprotein/cloning/primary structure). [). https://www.pnas.org]. *Biochemistry, 92*, •••.

Huang, C. S., Sen Chang, L., Anderson, M. E., & Meister, A. (1993). Catalytic and regulatory properties of the heavy subunit of rat kidney γ-glutamylcysteine synthetase. *The Journal of Biological Chemistry, 268*(26), 19675–19680. DOI: 10.1016/S0021-9258(19)36569-X PMID: 8103521

Itoh, S., Ozumi, K., Kim, H. W., Nakagawa, O., McKinney, R. D., Folz, R. J., Zelko, I. N., Ushio-Fukai, M., & Fukai, T. (2009). Novel mechanism for regulation of extracellular SOD transcription and activity by copper: Role of antioxidant-1. *Free Radical Biology & Medicine, 46*(1), 95–104. DOI: 10.1016/j.freeradbiomed.2008.09.039 PMID: 18977292

Jang, Y. C., & van Remmen, H. (2009). The mitochondrial theory of aging: Insight from transgenic and knockout mouse models. *Experimental Gerontology*, *44*(4), 256–260. DOI: 10.1016/j.exger.2008.12.006 PMID: 19171187

Juan, C. A., de la Lastra, J. M., Plou, F. J., & Pérez-Lebeña, E. (2021). The chemistry of Reactive Oxygen Species (ROS) revisited: Outlining their role in biological macromolecules (DNA, lipids and proteins) and induced pathologies. *International Journal of Molecular Sciences*, *22*(9), 4642. Advance online publication. DOI: 10.3390/ijms22094642 PMID: 33924958

Kendall, N. R., Mcmullen, S., Green, A., & Rodway, R. G. (2000). The effect of a zinc, cobalt and selenium soluble glass bolus on trace element status and semen quality of ram lambs. In *Animal Reproduction Science* (Vol. 62). www.elsevier.co mrlocateranireprosci

Khan, F., Kumar Garg, V., Kumar Singh, A., & Tinku, T. (2018). Role of free radicals and certain antioxidants in the management of Huntington's disease: A review. *Journal of Analytical & Pharmaceutical Research*, *7*(4). Advance online publication. DOI: 10.15406/japlr.2018.07.00256

Kirkman, H. N., & Gaetani, G. F. (2007). Mammalian catalase: A venerable enzyme with new mysteries. *Trends in Biochemical Sciences*, *32*(1), 44–50. DOI: 10.1016/j.tibs.2006.11.003 PMID: 17158050

Ko, T. P., Safo, M. K., Musayev, F. N., di Salvo, M. L., Wang, C., Wu, S. H., & Abraham, D. J. (2000). Structure of human erythrocyte catalase. *Acta Crystallographica. Section D, Biological Crystallography*, *56*(Pt 2), 241–245. DOI: 10.1107/S0907444999015930 PMID: 10666617

Koháryová, M., & Kolarova, M. (2008). Oxidative stress and thioredoxin system. *General Physiology and Biophysics*. PMID: 18645221

Kryukov, G. V., Castellano, S., Novoselov, S. V., Lobanov, A. V., Zehtab, O., Guigó, R., & Gladyshev, V. N. (2003). Characterization of mammalian selenoproteomes. *Science*, *300*(5624), 1439–1443. DOI: 10.1126/science.1083516 PMID: 12775843

Lee, M. T., Lin, W. C., Yu, B., & Lee, T. T. (2017). Antioxidant capacity of phytochemicals and their potential effects on oxidative status in animals - A review. In *Asian-Australasian Journal of Animal Sciences* (Vol. 30, Issue 3, pp. 299–308). Asian-Australasian Association of Animal Production Societies. DOI: 10.5713/ajas.16.0438

Leonard, S. S., Blemings, K. P., & Shi, X. (2014). *Free radical scavenging, DNA protection, and inhibition of lipid peroxidation mediated by uric acid.* https://www.researchgate.net/publication/7905814

Li, R., Jia, Z., & Trush, M. A. (2016). Defining ROS in biology and medicine. *Reactive Oxygen Species (Apex, N.C.), 1*(1), 9–21. DOI: 10.20455/ros.2016.803 PMID: 29707643

Li, S., Tan, H. Y., Wang, N., Zhang, Z. J., Lao, L., Wong, C. W., & Feng, Y. (2015). The role of oxidative stress and antioxidants in liver diseases. *International Journal of Molecular Sciences, 16*(11), 26087–26124. DOI: 10.3390/ijms161125942 PMID: 26540040

Lim, J., & Luderer, U. (2018). Glutathione deficiency sensitizes cultured embryonic mouse ovaries to benzo[a]pyrene-induced germ cell apoptosis. *Toxicology and Applied Pharmacology, 352*, 38–45. DOI: 10.1016/j.taap.2018.05.024 PMID: 29800640

Lloret, A., Esteve, D., Monllor, P., Cervera-Ferri, A., & Lloret, A. (2019). The effectiveness of vitamin E treatment in alzheimer's disease. In *International Journal of Molecular Sciences* (Vol. 20, Issue 4). MDPI AG. DOI: 10.3390/ijms20040879

Lu, S. C. (2009). Regulation of glutathione synthesis. In *Molecular Aspects of Medicine* (Vol. 30, Issues 1–2, pp. 42–59). DOI: 10.1016/j.mam.2008.05.005

Luk, E., Yang, M., Jensen, L. T., Bourbonnais, Y., & Culotta, V. C. (2005). Manganese activation of superoxide dismutase 2 in the mitochondria of *Saccharomyces cerevisiae.The Journal of Biological Chemistry, 280*(24), 22715–22720. DOI: 10.1074/jbc.M504257200 PMID: 15851472

Lushchak, V. I. (2012). Glutathione Homeostasis and Functions: Potential targets for medical interventions. *Journal of Amino Acids, 2012*, 1–26. DOI: 10.1155/2012/736837 PMID: 22500213

Lv, H., Zhen, C., Liu, J., Yang, P., Hu, L., & Shang, P. (2019). Unraveling the potential role of glutathione in multiple forms of cell death in cancer therapy. In *Oxidative Medicine and Cellular Longevity* (Vol. 2019). Hindawi Limited., DOI: 10.1155/2019/3150145

Manganaris, G. A., Goulas, V., Vicente, A. R., & Terry, L. A. (2014). Berry antioxidants: Small fruits provide large benefits. In *Journal of the Science of Food and Agriculture* (Vol. 94, Issue 5, pp. 825–833). DOI: 10.1002/jsfa.6432

Mason, S. A., Trewin, A. J., Parker, L., & Wadley, G. D. (2020). Antioxidant supplements and endurance exercise: Current evidence and mechanistic insights. *Redox Biology, 35*, 101471. Advance online publication. DOI: 10.1016/j.redox.2020.101471 PMID: 32127289

Matoušková, P., Hanousková, B., & Skálová, L. (2018). MicroRNAs as potential regulators of glutathione peroxidases expression and their role in obesity and related pathologies. *International Journal of Molecular Sciences 2018, Vol. 19, Page 1199, 19*(4), 1199. DOI: 10.3390/ijms19041199

Maurya, R., & Namdeo, M. (2021). *Superoxide Dismutase: a key enzyme for the survival of intracellular pathogens in host.* Reactive Oxygen Species. [Working Title], DOI: 10.5772/intechopen.100322

Merry, T. L., & Ristow, M. (2016). Do antioxidant supplements interfere with skeletal muscle adaptation to exercise training? *The Journal of Physiology, 594*(18), 5135–5147. DOI: 10.1113/JP270654 PMID: 26638792

Miriyala, S., Spasojevic, I., Tovmasyan, A., Salvemini, D., & Vujaskovic, Z. (1822). st. Clair, D., & Batinic-Haberle, I. (2012). Manganese superoxide dismutase, MnSOD and its mimics. *Biochimica et Biophysica Acta (BBA) -. Molecular Basis of Disease*, (5), 794–814. DOI: 10.1016/j.bbadis.2011.12.002 PMID: 22198225

Miyazawa, T., Abe, C., Burdeos, G. C., Matsumoto, A., & Toda, M. (2022). Food antioxidants and aging: Theory, current evidence and perspectives. *Nutraceuticals, 2*(3), 181–204. DOI: 10.3390/nutraceuticals2030014

Mocchegiani, E., & Malavolta, M. (2019). Role of zinc and selenium in oxidative stress and immunosenescence: implications for healthy aging and longevity. In *Handbook of Immunosenescence* (pp. 2539–2573). Springer International Publishing., DOI: 10.1007/978-3-319-99375-1_66

Moussa, Z., Judeh, Z. M. A., & Ahmed, S. A. (2019). *Nonenzymatic exogenous and endogenous antioxidants.* www.intechopen.com

Mueller, L., & Boehm, V. (2011). Antioxidant activity of β-carotene compounds in different in vitro assays. *Molecules (Basel, Switzerland), 16*(2), 1055–1069. DOI: 10.3390/molecules16021055 PMID: 21350393

Nelson, K. J., Perkins, A., van Swearingen, A. E. D., Hartman, S., Brereton, A. E., Parsonage, D., Salsbury, F. R.Jr, Karplus, P. A., & Poole, L. B. (2018). Experimentally dissecting the origins of peroxiredoxin catalysis. *Antioxidants & Redox Signaling, 28*(7), 521–536. DOI: 10.1089/ars.2016.6922 PMID: 28375740

Ngo, V., & Duennwald, M. L. (2022). Nrf2 and Oxidative Stress: A General Overview of Mechanisms and Implications in Human Disease. *Antioxidants, 11*(12), 2345. Advance online publication. DOI: 10.3390/antiox11122345 PMID: 36552553

Nirmal, N. P., Medhe, S., Dahal, M., Koirala, P., Nirmal, S., Al-Asmari, F., & Xu, B. (2024). Betalains protect various body organs through antioxidant and anti-inflammatory pathways. *Food Science and Human Wellness, 13*(3), 1109–1117. DOI: 10.26599/FSHW.2022.9250093

Nishikawa, M., Hashida, M., & Takakura, Y. (2009). Catalase delivery for inhibiting ROS-mediated tissue injury and tumor metastasis. *Advanced Drug Delivery Reviews, 61*(4), 319–326. DOI: 10.1016/j.addr.2009.01.001 PMID: 19385054

Nordberg, J., & Arnér, E. S. J. (2001). Reactive oxygen species, antioxidants, and the mammalian thioredoxin system. *Free Radical Biology & Medicine, 31*(11), 1287–1312. DOI: 10.1016/S0891-5849(01)00724-9 PMID: 11728801

Nuttall, S. L., Kendall, M. J., & Martin, U. (1999). Mini-review QJM Antioxidant therapy for the prevention of cardiovascular disease. In *Q J Med* (Vol. 92).

Okado-Matsumoto, A., & Fridovich, I. (2001). Subcellular distribution of superoxide dismutases (SOD) in rat liver: Cu, Zn-SOD in mitochondria. *The Journal of Biological Chemistry, 276*(42), 38388–38393. DOI: 10.1074/jbc.M105395200 PMID: 11507097

Oppenheimer, L., Wellner, V. P., Griffith, O. W., & Meister, A. (1979). Glutathione Synthetase purification from rat kidney and mapping of the substrate binding sites. []. *The Journal of Biological Chemistry, 254*(12), 5184–5190. www.jbc.org. DOI: 10.1016/S0021-9258(18)50577-9 PMID: 447639

Osganian, S. K., Stampfer, M. J., Rimm, E., Spiegelman, D., Hu, F. B., Manson, J. A. E., & Willett, W. C. (2003). Vitamin C and risk of coronary heart disease in women. *Journal of the American College of Cardiology, 42*(2), 246–252. DOI: 10.1016/S0735-1097(03)00575-8 PMID: 12875759

Park, K. (2009). Biochemical Nutrition Commons Recommended Citation Recommended Citation Park, KwangLi. In *Inquiry Journal* (Vol. 13). https://scholars.unh.edu/inquiry_2009https://scholars.unh.edu/inquiry_2009/13

Peternelj, T. T., & Coombes, J. S. (2011). Antioxidant supplementation during exercise training: Beneficial or detrimental? In *Sports Medicine* (Vol. 41, Issue 12, pp. 1043–1069). DOI: 10.2165/11594400-000000000-00000

Phaniendra, A., Jestadi, D. B., & Periyasamy, L. (2015). Free radicals: Properties, sources, targets, and their implication in various diseases. *Indian Journal of Clinical Biochemistry : IJCB*, *30*(1), 11–26. DOI: 10.1007/s12291-014-0446-0 PMID: 25646037

Priyadharshini, L. C., Vishnupriya, S., Sakthivel, K. M., & Rasmi, R. R. (2020). Oxidative stress responsive transcription factors in cellular signaling transduction mechanisms. *Cellular Signalling*, *72*, 109670. DOI: 10.1016/j.cellsig.2020.109670 PMID: 32418887

Rakhit, R., & Chakrabartty, A. (2006). Structure, folding, and misfolding of Cu, Zn superoxide dismutase in amyotrophic lateral sclerosis. *Biochimica et Biophysica Acta (BBA) -. Biochimica et Biophysica Acta. Molecular Basis of Disease*, *1762*(11–12), 1025–1037. DOI: 10.1016/j.bbadis.2006.05.004 PMID: 16814528

Rayman, M. P. (2012). Selenium and human health. *Lancet*, *379*(9822), 1256–1268. DOI: 10.1016/S0140-6736(11)61452-9 PMID: 22381456

Rhee, S. G., & Woo, H. A. (2011). Multiple functions of peroxiredoxins: Peroxidases, sensors, and regulators of the intracellular messenger H_2O_2, and protein chaperones. *Antioxidants & Redox Signaling*, *15*(3), 781–794. DOI: 10.1089/ars.2010.3393 PMID: 20919930

Rhee, S. G., Woo, H. A., Kil, I. S., & Bae, S. H. (2012). Peroxiredoxin functions as a peroxidase and a regulator and sensor of local peroxides. *The Journal of Biological Chemistry*, *287*(7), 4403–4410. DOI: 10.1074/jbc.R111.283432 PMID: 22147704

Rizvi, S., Raza, S. T., Ahmed, F., Ahmad, A., Abbas, S., & Mahdi, F. (2014). *The Role of Vitamin E in Human Health and Some Diseases* (Vol. 14, Issue 2). Epub.

Rydström, J. (2006). Mitochondrial NADPH, transhydrogenase and disease. *Biochimica et Biophysica Acta*, *1757*(5–6), 721–726. DOI: 10.1016/j.bbabio.2006.03.010 PMID: 16730324

Safo, M. K., Musayev, F. N., Wu, S. H., Abraham, D. J., & Ko, T. P. (2001). Structure of tetragonal crystals of human erythrocyte catalase. *Acta Crystallographica. Section D, Biological Crystallography*, *57*(1), 1–7. DOI: 10.1107/S0907444900013767 PMID: 11134921

Sah, S. K., Agrahari, G., & Kim, T. Y. (2020). Insights into superoxide dismutase 3 in regulating biological and functional properties of mesenchymal stem cells. *Cell & Bioscience*, *10*(1), 22. Advance online publication. DOI: 10.1186/s13578-020-00386-3 PMID: 32128111

Sakihama, Y., & Yamasaki, H. (2021). *Phytochemical Antioxidants: Past, Present and Future*. www.intechopen.com

Seelig, G. F., Simondsens, R. P., & Meister, A. (1984). Reversible Dissociation of y-Glutamylcysteine Synthetase into Two Subunits*. *The Journal of Biological Chemistry*, *259*(15), 9345–9347. DOI: 10.1016/S0021-9258(17)42703-7 PMID: 6146611

Settle, T. (2014). The role of uric acid as an antioxidant in selected neurodegenerative disease pathogenesis: A Short Review. *Brain Disorders & Therapy*, *03*(03). Advance online publication. DOI: 10.4172/2168-975X.1000129

Shen, S., Zhang, C., Meng, Y., Cui, G., Wang, Y., Liu, X., & He, Q. (2023). Sensing of H2O2-induced oxidative stress by the UPF factor complex is crucial for activation of catalase-3 expression in Neurospora. *PLOS Genetics*, *19*(10), e1010985. DOI: 10.1371/journal.pgen.1010985 PMID: 37844074

Shields, H. J., Traa, A., & Van Raamsdonk, J. M. (2021). Beneficial and detrimental effects of reactive oxygen species on lifespan: A comprehensive review of comparative and experimental studies. *Frontiers in Cell and Developmental Biology*, *9*, 628157. DOI: 10.3389/fcell.2021.628157 PMID: 33644065

Shults, C. W., Oakes, D., Kieburtz, K., Flint Beal, M., Haas, R., Plumb, S., Juncos, J. L., Nutt, J., Shoulson, I., Carter, J., Kompoliti, K., Perlmutter, J. S., Reich, S., Stern, M., Watts, R. L., Kurlan, R., Molho, E., Harrison, M., & Lew, M. (2002). Effects of coenzyme Q 10 in early Parkinson's disease: Evidence of slowing of the functional decline. *Archives of Neurology*, *59*(10), 1541–1550. DOI: 10.1001/archneur.59.10.1541 PMID: 12374491

Sies, H., & Stahl, W. (1995). Vitamins E and C, β-carotene, and other carotenoids as antioxidants. *The American Journal of Clinical Nutrition*, *62*(6, SUPPL.), 1315S–1321S. Advance online publication. DOI: 10.1093/ajcn/62.6.1315S PMID: 7495226

Sirivarasai, J., Kaojarern, S., Chanprasertyothin, S., Panpunuan, P., Petchpoung, K., Tatsaneeyapant, A., Yoovathaworn, K., Sura, T., Kaojarern, S., & Sritara, P. (2015). Environmental lead exposure, catalase gene, and markers of antioxidant and oxidative stress relation to hypertension: An analysis based on the EGAT study. *BioMed Research International*, *2015*(1), 856319. DOI: 10.1155/2015/856319 PMID: 25793211

Squadrito, G. L., Cueto, R., Splenser, A. E., Valavanidis, A., Zhang, H., Uppu, R. M., & Pryor, W. A. (2000). Reaction of uric acid with peroxynitrite and implications for the mechanism of neuroprotection by uric acid. *Archives of Biochemistry and Biophysics*, *376*(2), 333–337. DOI: 10.1006/abbi.2000.1721 PMID: 10775420

Stahl, W., & Sies, H. (2003). Antioxidant activity of carotenoids. In *Molecular Aspects of Medicine* (Vol. 24, Issue 6, pp. 345–351). Elsevier Ltd. DOI: 10.1016/S0098-2997(03)00030-X

Stancill, J. S., & Corbett, J. A. (2021). The role of thioredoxin/peroxiredoxin in the β-cell defense against oxidative damage. *Frontiers in Endocrinology*, *12*, 718235. Advance online publication. DOI: 10.3389/fendo.2021.718235 PMID: 34557160

Sturtz, L. A., Diekert, K., Jensen, L. T., Lill, R., & Culotta, V. C. (2001). A fraction of yeast Cu, Zn-superoxide dismutase and its metallochaperone, CCS, localize to the intermembrane space of mitochondria. A physiological role for SOD1 in guarding against mitochondrial oxidative damage. *The Journal of Biological Chemistry*, *276*(41), 38084–38089. DOI: 10.1074/jbc.M105296200 PMID: 11500508

Sue, G. R., Ho, Z. C., & Kim, K. (2005). Peroxiredoxins: A historical overview and speculative preview of novel mechanisms and emerging concepts in cell signaling. *Free Radical Biology & Medicine*, *38*(12), 1543–1552. DOI: 10.1016/j.freeradbiomed.2005.02.026 PMID: 15917183

Tan, B. L., Norhaizan, M. E., Liew, W. P. P., & Rahman, H. S. (2018). Antioxidant and oxidative stress: A mutual interplay in age-related diseases. In *Frontiers in Pharmacology* (Vol. 9, Issue OCT). Frontiers Media S.A. DOI: 10.3389/fphar.2018.01162

The 14 Best Anti-Aging Supplements. (2022). Retrieved August 18, 2024, from https://www.healthline.com/nutrition/anti-aging-supplements

The Best Antioxidants for Your Skin & Their Benefits | Dermstore. (2022). Retrieved August 18, 2024, from https://www.dermstore.com/blog/antioxidants-in-skin-care/

Toppo, S., Flohé, L., Ursini, F., Vanin, S., & Maiorino, M. (2009). Catalytic mechanisms and specificities of glutathione peroxidases: Variations of a basic scheme. *Biochimica et Biophysica Acta*, *1790*(11), 1486–1500. DOI: 10.1016/j.bbagen.2009.04.007 PMID: 19376195

Vašková, J., Kočan, L., Vaško, L., & Perjési, P. (2023a). Glutathione-related enzymes and proteins: A review. *Molecules (Basel, Switzerland)*, *28*(3), 1447. Advance online publication. DOI: 10.3390/molecules28031447 PMID: 36771108

Villalpando-Rodriguez, G. E., & Gibson, S. B. (2021). Reactive Oxygen Species (ROS) regulates different types of cell death by acting as a rheostat. oxidative medicine and cellular longevity, 2021, 9912436. DOI: 10.1155/2021/9912436

Wang, H., O'Reilly, É. J., Weisskopf, M. G., Logroscino, G., McCullough, M. L., Schatzkiny, A., Kolonel, L. N., & Ascherio, A. (2011). Vitamin e intake and risk of amyotrophic lateral sclerosis: A pooled analysis of data from 5 prospective cohort studies. *American Journal of Epidemiology*, *173*(6), 595–602. DOI: 10.1093/aje/kwq416 PMID: 21335424

Wang, Y., Fernando, G. S. N., Sergeeva, N. N., Vagkidis, N., Chechik, V., Do, T., Marshall, L. J., & Boesch, C. (2022). Uptake and immunomodulatory properties of betanin, vulgaxanthin i and indicaxanthin towards Caco-2 intestinal cells. *Antioxidants*, *11*(8), 1627. DOI: 10.3390/antiox11081627 PMID: 36009345

Waring, W. S. (2002). Uric acid: An important antioxidant in acute ischaemic stroke, QJM. *International Journal of Medicine*, *95*(10), 691–693. https://academic.oup.com/qjmed/article/95/10/691/1538330 PMID: 12324642

Whittaker, J. W. (2012). Non-heme manganese catalase – The 'other' catalase. *Archives of Biochemistry and Biophysics*, *525*(2), 111–120. DOI: 10.1016/j.abb.2011.12.008 PMID: 22198285

Wood, Z. A., Schröder, E., Harris, J. R., & Poole, L. B. (2003). Structure, mechanism and regulation of peroxiredoxins. *Trends in Biochemical Sciences*, *28*(1), 32–40. DOI: 10.1016/S0968-0004(02)00003-8 PMID: 12517450

Yang, J., Zhang, Y., Na, X., & Zhao, A. (2022). β-Carotene supplementation and risk of cardiovascular disease: A systematic review and meta-analysis of randomized controlled trials. *Nutrients*, *14*(6), 1284. DOI: 10.3390/nu14061284 PMID: 35334942

Zhao, Y., Seefeldt, T., Chen, W., Wang, X., Matthees, D., Hu, Y., & Guan, X. (2009). Effects of glutathione reductase inhibition on cellular thiol redox state and related systems. *Archives of Biochemistry and Biophysics*, *485*(1), 56–62. DOI: 10.1016/j.abb.2009.03.001 PMID: 19272349

Chapter 3
Impact of Dietary Antioxidants on Redox Homeostasis and Its Effects on Health and Disease

Reyed M. Reyed
https://orcid.org/0000-0001-5027-3478
Genetic Engineering and Biotechnology Research Institute (GEBRI) & City of Scientific Research and Technological Applications, Egypt

P. K. Prabhakar
https://orcid.org/0000-0001-8130-1822
Lovely Professional University, India & Nagaland University, India

ABSTRACT

This chapter discusses the roles of reactive oxygen species (ROS) in cells, including their role in oxidative stress and redox balance. It explores how dietary antioxidants affect redox homeostasis, their sources, mechanisms of action, and antioxidant capacity. The chapter also discusses how antioxidants affect ROS-mediated signaling cascades and endogenous antioxidant defense mechanisms. It discusses the use of antioxidants in pharmacological approaches to treat chronic illnesses like cancer, neurological disorders, and cardiometabolic diseases. It also addresses the antioxidant paradox, where excessive antioxidant supplementation can disrupt redox equilibrium, emphasizing the importance of balancing pro- and antioxidative forces.

DOI: 10.4018/979-8-3693-7919-6.ch003

1. INTRODUCTION

1.1 Overview of Reactive Oxygen Species (ROS)

Free radical scavengers are highly reactive molecules containing oxygen radicals such as superoxide anions ($O2-$), hydroxyl radicals (•OH), and hydrogen peroxide ($H2O2$). These molecules act as proactive and reactive species in biological systems; they are related to cellular damage and oxidative stress, while at the same time, they are decisive factors in cellular signaling and metabolic regulation (Jomova et al., 2023). ROS generation is a consequence of normal cell functions, and it is provoked mainly within the mitochondria and other organelles and can be modulated by genetic and environmental factors, diet, etc. Several features of diet and nutrition affect the synthesis and scavenging of ROS. Natural foods are rich sources of vitamins and phytochemicals, and foods that are rich in such components include fruits, vegetables, and whole grains, which contain vitamin C, vitamin E, and flavonoids that help eliminate free radicals and maintain redox balance. On the other hand, unhealthy eating patterns involving high quantities of processed foods, sugars, and unhealthy fat greatly increase ROS formation and therefore oxidative stress, thus increasing cellular impairment and disease progression (Lobo et al., 2010). Furthermore, drinking water contaminated with additional heavy metals such as lead, cadmium, and mercury poses additional hazards. Heavy metals can produce ROS through different biochemical mechanisms and cause oxidative stress and damage in cells (Balali-Mood et al., 2021). Consuming unhealthy foods and being in contact with polluted water has a cumulative impact on ROS levels, which enhances the effect of oxidative stress in the body.

1.2 Effectiveness of Regulating Redox Balance

Oxidative homeostasis, on the other hand, involves the balance between the generation and neutralization of ROS. Oxidative stress is therefore a situation where the generation of ROS surpasses the body's capacity to control or neutralize these reactive species through antioxidant systems. This balance is essential for sustaining cellular processes and eliminating oxidative stress, which impairs lipids, proteins, and DNA (Pizzino, 2017). Abnormal redox status is implicated in many diseases, including cancer, neurodegenerative diseases, and metabolic diseases. The role of ROS in cancer development is worthy of attention. High concentrations of ROS can cause mutations in different DNA sequences in the cell nucleus, which, in turn, contribute to the development of cancer in different tissues (Baek et al., 2018). ROS reportedly participate in many signaling pathways that involve the control of cell division and survival and can also support tumor growth (Kumari, et al. 2018). Therefore, for the purpose of cancer prevention and therapy, identifying the sources and impacts of ROS is crucial. Sources of pollution, UV radiation, and dietary practices have a great impact on redox status. For example, increased ROS occur due to the effects of environmental toxins, which cause inflammation and cell damage (Ashique et al., 2024). First, mitochondria, which are the powerhouses of cells, play a crucial role in redox regulation, in addition to the human microbiota, which consists of trillions of microbes in the gut. Microbial metabolites, including short-chain fatty acids, can also alter oxidative stress and the body's antioxidant capacity (Fehér et al., 2022). Environmental factors and their effects on human microbial composition Again, the composition of the human microbiota depends on diet, exercise, and exposure to various chemicals in the environment. These factors shift either the microbial compositional and/or functional profile, which affects the production of species relevant to redox homeostasis. For example, the consumption of fiber allows favorable gut bacteria to drive the synthesis of SCFAs that offer antioxidants that alleviate oxidative stress (Cuciniello, et al 2023).

On the other hand, consuming processed foods in large quantities, such as sugars, refined white bread, and candy, may lead to dysbiosis, which is characterized by low levels of microbial diversity and increased levels of pathogenic species. An imbalance has been associated with obesity, diabetes, and inflammatory diseases such as inflammatory bowel diseases (Tristan et al., 2023). Such imbalances can indeed amplify oxidative stress, which impairs health and results in a cycle of misery. Drinking polluted water increases this problem only because heavy metals affect the composition and function of the gut microbiota in addition to their detrimental effects on health.

1.3 Effects on the Human Genome and Epigenome

ROS, the microbiome, and nutrition also interact through the genetic and epigenetic changes that occur within the mammalian gut. Glossary oxidative stress and DNA damage mutations may cause cancer and other genetic diseases (Shandilya et al., 2021). In addition, ROS can affect other subcellular pathways, including DNA methylation and histone modification, which involve reversible modification of the physical structure of DNA and the management of gene expression independently of its sequence (Huang et al., 2022). Drinking cow milk, especially milk produced from cows that have been administered hormones or antibiotics, is seen as a major competitor to oxidative stress due to possible residues present in the milk. According to the literature, some of the contents in milk could elicit ROS production and thus modulate health effects (Niu et al, 2023). From those sources and exposures, the term "passive grievous harm: Silent killers" alludes to the consequences of free radicals, oxidative stress and chronic diseases, including cancer. Stimuli that influence the microbiome are also capable of eliciting epigenetic shifts. For example, dietary factors such as folate, which is found in greens and lentils, play key roles in methylation, which controls gene functions. If these genes are missing in an individual's diet, then normal epigenetic functioning might be disrupted, increasing the likelihood of genetic disorders (Li et al., 2021).

Lifestyle components of this model include food and nutrition, which can inform dietary manipulation for the purpose of regulating the levels of oxidative stress and genomic stability. To some extent, individuals who consume a meal containing antioxidants and other anti-inflammatory nutrients support their redox balance and possibly decrease the risk of disease from oxidative damage and dysbiosis, according to Man et al., 2020.

Free radicals and oxidative stress occur in the form of reactive oxygen species (ROS), and redox homeostasis in cellular compartments plays a crucial role in cellular metabolism and disease. One can single out the role of dietary antioxidants, environmental factors, and the human microbiome as a number of factors that complicate these relationships. In dissecting how these elements interact, the authors effectively convey the importance of food and nutrition with respect to health and combating oxidative stress diseases Yoshikawa, &You, (2024).

2. ROLES OF REACTIVE OXYGEN SPECIES (ROS)

2.1 Types of ROS

Free radicals and ROS are unsaturated molecules with at least one unpaired electron, and an overproduction of these molecules can cause oxidative stress, leading to the destruction of cells and tissues. There are a number of ROS that are known to affect biological systems, and the different types have different properties in terms of their role in oxidation.

2.1.1 Free Radicals

Free radicals are molecules or atoms with unpaired electrons that can react with other molecules or atoms in nearby cells through the transfer of electrons (Tripathi, et al.,2022). In biological systems, free radicals act simultaneously as signaling substances and toxic elements if they are formed in large quantities (Rahaman et al., 2023).

2.1.2 Singlet Oxygen (O2)

Singly ionized oxygen is an oxygen molecule in which none of the electrons are unpaired, even though they are at a relatively high energy level. It is a strong oxidant, and it harms lipids and proteins as well as DNA by interacting with fatty acid double bonds. Singlet oxygen is usually generated in photochemical reactions when or through energy transfer from photoactivatable sensitizers in cells that have been irradiated with UV radiation (Murotomiet al., 2023).

2.1.3 Superoxide Radical (O2•⁻)

Superoxide radicals are highly formed ROS and are generated mainly from the leakage of electrons in the mitochondrial electron transport series. However, as the least oxidatively potent form of ROS, superoxide is a precursor to H_2O_2 and OH• through enzymatic or nonenzymatic reactions (Turrens, 2003).

2.1.4 Hydroxyl Radical (OH•)

Hydroxyl radicals are among the most reactive free radical species, and on the basis of the previous discussion, they can react with virtually all biomolecules, such as nucleic acids, lipids, and proteins. Hydroxyl radicals are formed by the Fenton process, in which hydrogen peroxide reacts with transition metal ions such as Fe or Cu, resulting in oxidative stress (Halliwell 2021).

2.1.5 Peroxyl Radical (ROO•)

ROO•s include peroxyl radicals, which are formed during lipid peroxidation, in which free radicals attack unsaturated fatty acids in cell membranes. These radicals can create sequence reactions that ultimately cause extensive loss of the cell membrane and reduced cell functionality. By comparison, peroxyl radicals are comparatively stable and long-lived forms of radicals, which means that they induce severe long-term oxidative stress (Su et al., 2019).

2.1.6 Lipid Peroxyl Radical (LOO•)

There is a special group of peroxyl radicals derived from lipids known as lipid peroxyl radicals that home to polyunsaturated fatty acid MS within the lipid bilayer of various cell structures. These radicals prolong the lipid peroxidation process and cause the loss of membrane integrity, disturbances in cellular signaling, and consequently, cell death (Ayala et al., 2014).

2.1.7 Various ROS

Free radicals are only a form of ROS; other forms include nonradical species such as hydrogen peroxide (H2O) and ozone (O2). These molecules can easily cross membranes and maintain oxidative stress in other compartments of the cell. Even though nonradical ROS are slightly less reactive than free radicals are, they act as precursors to other damaging species (Chaudhary et al., 2023).

2.1.8 The Hydroxyl Radical (OH•) Revisited

The hydroxyl radical occurs in its free form and is highly reactive; consequently, it can initiate many oxidation reactions in cells. Its formation is widespread, and it is associated with cellular iron and copper metabolism because the Fenton reaction is an important source of hydroxyl radicals in biological systems (Jomova et al., 2023). OH radicals are recognized to instigate strand breaks in DNA, cross-link proteins, and peroxidize membrane lipids (Ayala et al., 2014). The relationships among these different forms of ROS are important for maintaining redox homeostasis and thus controlling the effects of oxidative stress on cellular processes. Flowing free in the body, they cause damage to body tissues and cells, and antioxidant enzymes, such as superoxide dismutase (SOD) and glutathione peroxidase (GPx), offer protection against such species (Jomova et al., 2023). However, where the production of excess ROS exceeds the antioxidant capacity, free radical-induced oxidative injury results in diseases such as cancer and neurodegenerative disorders.

2.2 Biological ROS

For example, ROS are composed of free radicals, which include the superoxide anion radical (O2−), the hydroxyl radical (•OH), and nonradical species, including hydrogen peroxide (H2O2). These molecules are generated as intermediates in normal cellular processes, such as oxidative phosphorylation, mainly in the mitochondria (de Almeidaet al., 2022). Although they are generally considered detrimental because of their ability to induce oxidative damage, ROS are known to play functional roles in physiological processes such as signal transduction and cellular homeostasis (Checa, et al.). 2020). Naturally, in biological systems, ROS can react with different biomolecules, such as lipids, proteins, and nucleic acids, as shown in Figure 1. Such interactions are able to generate changes that can interfere with cellular housekeeping genes or initiate protective signaling cascades. Therefore, proper production and detoxification of ROS are critical for healthy cellular status and function (Xie et al., 2024).

2.3 Oxidative Stress and Cell Damage

In other words, oxidative stress occurs when the concentration of ROS exceeds the antioxidant potential of the body. This condition can lead to significant cellular damage, including the following:

- **Lipid peroxidation**: ROS can cause lipid peroxidation, leading to the generation of aldehyde products such as MDA, which can compromise cell membrane integrity and function (Su, 2019).
- **Protein oxidation**: ROS can alter the amino acid residues of proteins, resulting in diminished enzyme function and disrupted signal transduction (Andrés et al., 2022).
- **DNA damage**: ROS reportedly causes changes in bases, breaks in DNA strands, and cross-linking, resulting in mutations and genomic instability (Srinivas et al., 2019).

2.4 ROS in Redox Signaling Pathways

However, in addition to being able to cause cellular damage, ROS are important signaling molecules in multiple cellular processes. These proteins are involved in redox signaling, which is responsible for many biological processes, as highlighted in Table 1. The key roles that ROS play in these regulatory pathways are as follows:

- **Cell Proliferation:** The growth-promoting effects of ROS are mediated through the regulation of particular signal transduction pathways. Interestingly, some of these proteins engage with the mitogen-activated protein kinase (MAPK) pathway, which is essential in the processes of cell growth, differentiation, and survival (Zhang et al., 2016). In this way, ROS affect cell survival and apoptosis and can induce cellular signaling in response to environmental stimulus.
- **Inflammation:** ROS are also involved in inflammation through the activation of an essential transcription factor known as nuclear factor kappa B (NF-κB). NF-κB is involved in various functions, such as the immune response, inflammatory diseases, and, more specifically, chronic inflammation (Lingappan, 2018). Given the novel findings concerning ROS-mediated NF-κB activation, the molecular species in question serve as typical signaling and oxidative stress agents in this context.
- **Apoptosis:** At high levels, ROS causes the intrinsic apoptotic cascade, resulting in the apoptosis of the cell. This mechanism plays an important role in the proper functioning and balance of cells because it controls the rate of

production of new cells and clears damaged and nonfunctioning cells from the body. Significantly, Guo et al. (2024) demonstrated that ROS cause apoptosis through mitochondrial disruption and changes in proapoptotic proteins. These mechanisms highlight ROS as molecules that allow cells to transmit signals and cause oxidative stress depending on the environment of the cells and the concentration of these ROS.

Table 1. Roles of ROS in signaling pathways

Signaling Pathway	Role of ROS	References
MAPK Pathway	ROS activation of MAPK pathways promotes cell proliferation and survival.	(Zhang, et al., 2016)
NF-κB Activation	ROS activate NF-κB, leading to the expression of pro-inflammatory cytokines.	(Lingappan 2018).).
Apoptosis	Increased ROS levels can trigger intrinsic apoptotic pathways.	(Gao et al., 2024).

This table summarizes the functions of ROS in major signaling pathways, stressing their cytotoxic and signaling potentials at the same time

3. MECHANISMS OF ROS PRODUCTION

Free radicals, including reactive oxygen species (ROS), are generated during various biological activities and environmental processes, leading to oxidative stress and the disruption of cellular functions. To better understand the roles and dysfunctions of ROS, it is crucial to examine the mechanisms underlying their synthesis (Bhattacharyya et al., 2014). Table 2 outlines the primary mechanisms through which ROS are produced, highlighting key processes and their biological significance. These mechanisms include mitochondrial respiration, the immune response during inflammation, environmental factors such as UV radiation and pollution, cellular metabolism, and endoplasmic reticulum (ER) stress. Each pathway contributes to ROS production in unique ways, affecting cell signaling, metabolism, and, when imbalanced, promoting oxidative damage and disease.

Table 2. Key mechanisms of Reactive Oxygen Species (ROS) production

Mechanism	Description	Key Processes	Significance
Mitochondrial Respiration (Mukherjee, et al., 2024)	Major site of ROS production during aerobic respiration in eukaryotic cells.	- Electron transport chain (ETC) - Leakage at complexes I & III - Superoxide formation ($O2-$)	- Regulates apoptosis and metabolism. - Excess ROS leads to oxidative stress and cellular injury.
Inflammation & Immune Response (Bode et al., 2024)	ROS produced by neutrophils and macrophages during the respiratory burst as a defense mechanism.	- NADPH oxidase activation - Conversion of O2 to superoxide, H2O2, and hypochlorous acid	- Neutralizes pathogens - Chronic activation can lead to tissue damage and inflammation.
Environmental Factors (Shrestha, et al.,2024)	External stressors such as UV radiation, pollution, and heavy metals can induce ROS production.	- UV radiation excites biomolecules - Pollutants activate inflammatory processes - Heavy metals catalyze reactions	- Contributes to oxidative stress and various diseases.
Cellular Metabolism (Mu et al.,2024)	Various metabolic pathways produce ROS during substrate metabolism, particularly fatty acids and glucose.	- Fatty acid beta-oxidation - Lipid peroxides formation - Glycolysis and pentose phosphate pathway (NADPH)	- Disruption of metabolic balance can lead to ROS-related tissue damage.
5. Endoplasmic Reticulum Stress (Chen et al., 2023)	Disruption in ER function leads to misfolded proteins and subsequent ROS formation.	- Misfolded protein retention - Enhanced Ca2+ release - Activation of stress kinases	- Activates signaling cascades promoting apoptosis, linked to diseases like neurodegenerative disorders.

This table provides an overview of the primary pathways through which ROS are generated, including mitochondrial respiration, the immune response, environmental factors, cellular metabolism, and endoplasmic reticulum stress, highlighting their processes and biological importance. Each mechanism is described concisely, with a focus on both the processes involved and the significance of the ROS produced. This format helps in understanding the complexity and interconnectedness of ROS production mechanisms while providing a visually appealing and informative overview.

3.2 Pathological Conditions Associated With ROS-Related Symptoms

There are numerous diseases related to ROS, as illustrated in Figure 3, which means that these radical species take part in both normal and pathological processes. However, when they exist at high concentrations, ROS are toxic and aggressively participate in oxidation, which is an early event in various diseases. Both types of ROS play physiological roles in cellular homeostasis and can contribute to disease progression; thus, deciphering their roles in disorders is central for understanding and treating diseases.

1. **Cancer:** ROS also has the potential to cause DNA oxidation, which leads to mutations that cause carcinogenesis. This is especially manifested in lung, breast, and colorectal cancers, where increased ROS are pivotal in tumor generation, growth, and invasion (Katakwar et al., 2016). ROS may trigger oncogenes and repress tumor suppressor genes, which can cause uncontrolled cell proliferation to become an essential factor in tumors.
2. **Cardiovascular Diseases:** ROS are thus key players in the pathogenesis of cardiovascular diseases through endothelial dysfunction, atherosclerotic plaque formation, and myocardial infarction. Oxidative stress is the main trigger of atherosclerotic plaques because modifications of low-density lipoprotein (LDL) by oxidization lead to an inflammatory response in the vascular endothelium layer. This chronic low-grade inflammation results in plaque formation and vascular remodeling, a high risk of cardiovascular events, myocardial infarctions, etc. (Jin et al., 2024).
3. **Neurodegenerative Disorders:** Some of these diseases include Alzheimer's disease and Parkinson's disease and are strongly correlated with oxidative damage to neurons. ROS participate in the pathophysiological process of these disorders via the mechanisms of inflammation, mitochondrial disorders, and neuronal death. Oxidative stress contributes to the formation of additional toxic protein aggregates such as amyloid β in Alzheimer's disease as well as α-synuclein in Parkinson's disease and thus contributes to neurodegeneration (Houldsworth, 2024).
4. **Diabetes Mellitus:** In diabetes, the ROS level is clearly involved in the development of insulin resistance and dysfunction of pancreatic beta-cells. Hyperglycemia increases oxidative stress and impairs insulin signaling and insulin secretion, leading to further deterioration of glycemic control. A specific byproduct of cellular redox processes is reactive oxygen species (ROS), which are involved in the initiation and promotion of diabetes and its complications, such as neuropathy, retinopathy, and nephropathy (Banaszaket al., 2024).

5. **Chronic Respiratory Diseases:** ROS are implicated in the development of chronic respiratory illnesses such as COPD and asthma due to airway inflammation and hyperresponsiveness. Lung endothelial cells contain heme proteins that produce free radicals, causing oxidative stress to the lung tissue, airway remodeling, and distortion of normal gas exchange, all of which exacerbate symptoms and disease. Cigarette smoke and other imprecise agents are well-known producers of ROS in these diseases (Chang et al., 2024).
6. **Aging:** Excess accumulation of reactive oxygen species throughout the lifespan is connected with the process of aging and the development of age-related disorders, such as sarcopenia and frailty. According to Harman's free radical theory of aging, the accumulation of cellular damage resulting from ROS is the primary cause of the physiological decline observed during aging. Several studies support this theory by indicating that oxidative stress leads to impaired mitochondrial function, strand breakage, and limited stem cell division in elderly individuals (Maldonado et al., 2023).

While ROS play a role in normal cellular functions, overproduction of ROS can result in cellular injuries that contribute to the pathogenesis of certain diseases. ROS are a part of many signaling pathways, and when their balance is high, cell dysfunction and the onset of disease occur. The equilibrium between the generation of ROS and antioxidant functions is essential for the support of homeostasis. Hence, subsequent experiments should be directed toward identifying the pathways through which ROS are produced and controlled as new targets for the treatment of oxidative stress-associated diseases.

4. DIETARY ANTIOXIDANTS AND REDOX HOMEOSTASIS

Antioxidants are vital for regulating redox status within the body, which involves eliminating reactive oxygen species (ROS) to control the oxidation process and the resulting damage. These compounds are important for maintaining health and well-being and preventing damage to cells. Antioxidants originate from various food groups: conventional groups such as vitamins and polyphenols and modern groups that include postbiotics (Pizzino et al., 2017). This section includes detailed information about different antioxidants, how they can preserve cells, and what antioxidant defense the body possesses against the hazardous effects of oxidants. There are also two general categories of antioxidants derived from the diet: conventional (Rahamanet al., 2023) and postbiotic (Mabrouk, et al., 2024) antioxidants. Other conventional antioxidants include vitamins, vitamins C and E, polyphenols, flavonoids, and carotenoid substances. However, postbiotics are substances generated by the gut microbiota that have antioxidant activity. Table 3 shows the classification, source, and documented examples of conventional and postbiotic antioxidants. Postbiotic antioxidants are generated through bacterial fermentation during gut microbiota metabolism and are essential for maintaining and controlling systemic levels of oxidative stress and inflammation.

Table 3. Comparative overview of conventional and postbiotic antioxidants

Antioxidant	Source	Examples	Type	Mechanism of Action	References
Vitamin C	Citrus fruits, berries, broccoli	Ascorbic acid	Conventional	Neutralizes ROS by donating electrons	Gęgotek,, & Skrzydlewska, (2023)
Vitamin E	Nuts, seeds, vegetable oils	α-Tocopherol	Conventional	Protects cell membranes from lipid peroxidation	Sun et al ., 2024
Polyphenols	Green tea, dark chocolate, grapes	Catechins, Resveratrol	Conventional	Scavenges free radicals, regulates oxidative enzymes	Ballini et al., 2024
Carotenoids	Carrots, sweet potatoes, leafy greens	Beta-carotene, Lycopene	Conventional	Neutralizes singlet oxygen, protects DNA	González-Peña et al., 2023
Glutathione	Spinach, avocados, asparagus	Reduced glutathione	Conventional	Recycles other antioxidants, detoxifies free radicals	Roy, & Paira, (2024).
Short-chain fatty acids (SCFAs)	Fermented foods, gut bacteria	Butyrate, acetate	Postbiotic	Modulates gut health, reduces oxidative stress	Fusco et al., 2023
Polyphenol metabolites	Gut microbiota metabolizing polyphenols	Urolithins, phenolic acids	Postbiotic	Anti-inflammatory, antioxidant effects via microbiota metabolism	Ilie et al., 2024
Gut-derived Glutathione	Gut bacteria, fermented foods	Glutathione produced by microbiota	Postbiotic	Enhances antioxidant defense, supports cellular detoxification	Ottria et al., 2024

The body's first line of defense against oxidative stress is composed of the following:

- **Traditional antioxidants**: These compounds are obtained directly from dietary sources such as vitamins and plant-based substances. Reactive oxygen species (ROS) are primarily neutralized to avoid damaging cells and maintain the structural stability of molecules. Given their critical role in protecting cells from oxidative damage, these antioxidants are appropriately referred to as the "Supreme Guardians of Molecules," emphasizing their importance in preserving cell-based health and mental and physical health.
- **Postbiotic Antioxidants**: Generated by the activity of the gut microbiota, postbiotics such as SCFAs and polyphenol metabolites exert antioxidant effects indirectly by enhancing gut health, reducing inflammation, and modulating systemic oxidative stress.

This table highlights the functional differences between conventional antioxidants (which act directly on free radicals) and postbiotic antioxidants (which often work through gut-mediated mechanisms).

5. MECHANISMS OF ACTION

Free radical defensive elements are necessary for reducing OS through the neutralization of ROS before they champion damaging effects on cells. As shown in Table 4 below, these mechanisms of action reveal how antioxidants work to prevent oxidative damage in the body. The various protective mechanisms are illustrated in Figure 4, where different forms of antioxidants and their contributions to cell health and human well-being are shown. Since free radicals are usually linked to the deterioration of the cell and the impairment of functions, antioxidants are extremely important because they help neutralize these radicals. The specifics of antioxidant activity depend on the kind of antioxidant; however, the universal concept involves package transfer of electrons to free radicals, which makes them stable and prevents the destruction of several cellular structures, such as lipids, proteins, and DNA.

1. **Free Radical Scavenging:** Most antioxidants work within a chemical process by providing an electron to a free radical in an effort to stabilize it. For example, vitamin C is a water-soluble antioxidant that directly neutralizes ROS in the cytoplasm and extracellular space by simply removing an electron to form a stable compound. Vitamin E is a fat-soluble antioxidant that acts to counteract the effects of lipid peroxyl radicals on cell membranes via direct interactions in the lipid bilayer (Ungureanu, et al. 2024).
2. **Immobilizing Redox-Reactive Transition Mals:** Polyphenols and flavonoids are two examples of antioxidant classes that work by immobilizing or trapping redox-active transition metals such as iron and copper. These metals can then use Fenton reactions to produce extremely powerful hydroxyl radicals. Quercetin, a flavonoid found in apples and onions, chelates these metal ions, reducing their capacity to initiate harmful redox reactions (Fontes, et al., 2024).
3. **Enzymatic regulation:** Enzymatic antioxidants include superoxide dismutase, catalase, and glutathione peroxidase via the reduction of the abovementioned ROS to a less reactive species. For example, SOD converts superoxide radicals to hydrogen peroxide, which is further converted into water and hydrogen by catalase or GLUT peroxidase (Anwar, 2024).
4. **Redox** signaling modulates antioxidants associated with various biological functions, including cell division, programmed cell death, and inflammation. The redox cycling molecule glutathione, also commonly known as the mother

antioxidant, shifts between the reduced form (GSH) and the oxidized or disulfide form (GSSG). This balance is important for maintaining cellular homeostasis and protecting the cell from oxidative stress (Jomovaet al.,2023).

Table 4. Advanced mechanisms of action of antioxidants in case studies

Antioxidant	Mechanism of Action	Example	Case Study & Outcome
Vitamin C Traber, 2013	Direct scavenging of ROS by donating electrons, aids in regeneration of Vitamin E to restore antioxidant function	Ascorbic acid	Study: Vitamin C supplementation in patients with coronary artery disease. Outcome: Improved endothelial function and reduced oxidative markers.
Vitamin E Lloret, et al., 2019	Neutralizes lipid peroxyl radicals, protecting cell membranes from lipid peroxidation	α-Tocopherol	Study: Vitamin E in neurodegenerative diseases. Outcome: Delayed progression of Alzheimer's disease in patients due to reduced oxidative stress.
Polyphenols Krawczyk et al 2023	Metal chelation, scavenging of ROS, inhibition of oxidative enzymes	Catechins, Resveratrol	Study: Resveratrol's impact on prediabetic individuals. Outcome: Improved insulin sensitivity and decreased oxidative stress.
Glutathione Lushchak et al., 2012	Enzymatic neutralization of ROS via the GSH/GSSG redox cycle, vital in detoxification pathways	Reduced glutathione	Study: Glutathione supplementation in patients with chronic liver disease. Outcome: Reduced liver inflammation and oxidative stress.
Postbiotics (SCFAs) Fusco et al ., 2023	Modulate redox signaling pathways, exert anti-inflammatory effects, enhance gut health	Butyrate, acetate	Study: SCFA supplementation in inflammatory bowel disease (IBD). Outcome: Reduced intestinal inflammation and oxidative damage.

This table summarizes the various methods of action of several antioxidants, as well as clinical data supporting their treatment usefulness in treating oxidative stress-related illnesses.

6. ANTIOXIDANT EFFECTIVENESS AND CYTOPROTECTIVE PROPERTIES

The ability of antioxidants to shield cells from oxidative stress is essential for maintaining cellular homeostasis and preventing disease development. Their capacity to inhibit redox-dependent signaling mechanisms, mitigate reactive oxygen species (ROS), and affect gene expressions is primarily responsible for their cytoprotective properties. Chronic oxidative stress is related to a wide range of pathological illnesses, including diabetes, cancer, heart disease, and dementia-related diseases such as cognitive impairment and Parkinson's disease (Kozlovet al., 2024). Overcoming free radicals directly, chelating metal ions, and bolstering the body's own antioxidant defenses are some of the ways that antioxidants work to protect the body.

Glutathione, also known as the "prime regulator of redox balance," is a tripeptide made up of L-γ-glutamyl-L-cysteinyl-glycine. It is essential for cellular defense, detoxification, and immunological modulation and maintaining redox equilibrium. Its unusual capacity to exist in both the reduced (GSH) and oxidized (GSSG) states makes it an important biomarker for measuring oxidative stress, with the GSH/GSSG ratio offering valuable insights into the redox status of cells. The dual function of glutathione is distinguished. It not only directly neutralizes free radicals but also acts as an important cofactor for enzymes such as glutathione peroxidase, thereby considerably improving the body's total antioxidant defense system.

In addition, glutathione is essential for the regeneration of other powerful antioxidants, such as vitamins C and E, which improves their effectiveness in countering oxidative stress and maintaining cellular integrity. This multimodal activity highlights its importance as a pillar in the maintenance of redox equilibrium and cellular health.

7. SYNERGISTIC ANTIOXIDANTS: BRIDGING NATURAL, SYNTHETIC, AND POSTBIOTIC FRONTIERS

As the world continues to seek new antioxidant nutrients, the search remains central from biomedical and nutritional points of view as a means of addressing the effects of oxidative stress, an important aspect associated with most diseases. This approach encompasses broad categories of natural compounds, synthetic agents, and the newest category of postbiotic antioxidants, which all play a role in the modulation of oxidative stress (Table 5).

7.1 Astaxanthin

Astaxanthin, a potent marine antioxidant, is a widely known carotenoid sourced from sea organisms, including algae and krill; the antioxidant capacity of astaxanthin is 550 times greater than that of vitamin C and 67 times greater than that of vitamin E. Due to its special molecular configuration, penetration within the cellular membranes is enhanced, and thus, shielding of both the inner and outer layers from oxidative damage is provided. Astaxanthin has multiple benefits in clinical use; it is effective in preventing skin damage caused by ultraviolet radiation, promoting cardiovascular health and increasing muscle stamina. Most remarkably, it can cross the blood–brain barrier and may be used to treat neurodegenerative diseases such as Alzheimer's disease and Parkinson's disease because of its anti-inflammatory and antioxidant effects (Utomo et al., 2024).

7.2 Edaravone: A Synthetic Antioxidant for Neuroprotection

Among synthetic antioxidants, edaravone deserves special attention because it is an effective drug for the treatment of amyotrophic lateral sclerosis (ALS), which is a serious neurodegenerative disease characterized by the destruction of motor neurons. There is increasing evidence that oxidative stress is implicated in ALS because the overproduction of ROS causes neuronal apoptosis. In this way, edaravone reduces the impact of free radicals on neurons and thus prevents their damage. Clinical trials have proven its effectiveness, where ALS degeneration and motor function decline rates have slowed considerably and oxidative stress markers have decreased (Yamashita, et al., 2024). The findings of edaravone also extend beyond ALS and could be useful in other neurodegenerative diseases that involve oxidative stress.

7.3 Honey: A Rich Source of Natural Antioxidants

Honey, which contains high concentrations of bioactive compounds such as polyphenols, flavonoids, and organic acids, has long been used for its therapeutic effects. Collectively, these constituents have strong antioxidant and anti-inflammatory properties. Free radicals are neutralized by honey, which promotes tissue repair, especially in wound care, such as burns or ulcers (Al-Kafaweenet al., 2023). This serves the purpose of acting as an antioxidant and reducing inflammation, making it play two essential roles in conventional and contemporary medicine.

7.4 Postbiotic Antioxidants: The New Frontier

Innovative studies have shed light on the actual nature of postbiotic antioxidants or the bioactive (organic) metabolites released by the gut microbiota, which play a role in the control of oxidative stress and inflammation. Butyrate and propionate, short-chain fatty acids, and polyphenol metabolites are key regulatory agents of the body's redox status. SCFAs, formed from the fermentation of dietary fiber, not only supply energy to colonic epithelial cells but also increase the intrinsic anti-oxidant potential and decrease inflammation in the body. The metabolites of plant polyphenols, which include urolithins, have been shown to combat free radicals and reduce inflammation (Gurunathan, et al., 2023). Since the inclusion of postbiotic antioxidants in functional foods has not previously been considered a method for enhancing gut health or reducing systemic oxidative stress, this innovation can be a game changer in the functional food industry. Unlike usual antioxidants, postbiotics are generated internally, which results in mutually beneficial interactions among the diet, microbiota, and host. Postbiotics obtained from functional food products can absorb prebiotics and dietary fibers in the human body to increase the body's antioxidant activity.

The discussion of antioxidants, such as astaxanthin or synthetic compounds such as edaravone, as well as potential postbiotic agents suggests a broad and complex approach to prevent oxidative stress, which is considered to have a major impact on different chronic diseases. The different categories of antioxidants exhibit unique actions that are crucial for maintaining the shielding and balance of cells. Astaxanthin, a natural antioxidant, has increased ROS-scavenging capacity and increased cellular protective abilities through the regulation of inflammation and mitochondrial biogenesis. Research has revealed that astaxanthin not only is beneficial for preserving the nervous system but also has potential for the skin and cardiovascular system, making it a potential therapeutic agent in healthcare (Utomo et al., 2024). Some synthetic antioxidants, such as edaravone, provide specific remedies for pathological disorders, such as amyotrophic lateral sclerosis (ALS). As a synthetic compound that successfully decreases oxidative damage and enhances neuronal function, edaravone is an excellent example of the therapeutic use of synthetic compounds in medicine. Synthetic antioxidants have been shown to prolong ALS survival through phase 3 clinical trials, and the use of synthetic antioxidants has led to more extensive development to address oxidative stress illnesses comprehensively (Yamashita, et al., 2024). Modern trends in nutritional science are represented by newly discovered postbiotic antioxidants synthesized by the gut microbiota. These products, such as SCFAs and polyphenol metabolites, strengthen endogenous antioxidant networks and connect the gut and systemic oxidation. These postbiotics constitute another potential path to explore functional foods for enhancing health performance on the basis of the mutual relationship between food and the microbiota (Gurunathan, et al., 2023. The more we learn about these various antioxidant stances, the clearer it is that a complex approach involving the use of both natural and synthetic antioxidants along with the optimization of the antioxidant properties of postbiotic compounds will be indispensable in the fight against oxidative stress and associated diseases. More research should be encouraged to further understand the relationships of such compounds in use today as well as the potential health benefits and potential hazardous effects of such compounds. Such efforts could revolutionize approaches to disease prevention and management, thereby changing the entire paradigm regarding nutrition, health, and aging under conditions of oxidative stress.

Table 5. Overview of antioxidants: Sources, mechanisms, clinical outcomes, and pharmacokinetics

antioxidant	Source/ Condition Studied	Mechanism of Action	Clinical Outcomes	Physicochemical Characteristics	Pharmacokinetic Profile	Natural/ Synthesized
Astaxanthin (Ambati et al., 2014)	Marine algae, cardiovascular diseases	Scavenges ROS, anti-inflammatory	Reduced skin aging, improved cardiovascular function	Lipophilic, high molecular weight	Bioavailability enhanced with fats; peak plasma levels within 3-4 hours	Natural
Glutathione Miller & Adams, 2020	Endogenous, neurodegenerative diseases, cancer	Detoxification, immune regulation, regenerates other antioxidants	Improved cognitive function, reduced oxidative damage	Tripeptide (Gly-Cys-Glu), water-soluble	Rapidly metabolized; short half-life (~30 minutes)	Natural
Edaravone Smith et al., 2020	Synthetic, used in ALS treatment	Scavenges free radicals in neurons	Slowed ALS progression, improved motor function	Small molecular weight, hydrophilic	Peak plasma concentration within 1 hour; half-life ~4 hours	Synthetic
Honey Brown et al., 2020	Bee-derived, wound healing, inflammation	Contains antioxidant polyphenols that reduce oxidative stress	Accelerated wound healing, reduced inflammation	Viscous liquid, contains various sugars and polyphenols	Bioavailability varies; effects seen within 30 minutes	Natural
Coenzyme Q10 (Water-Soluble) Mantle, et al.,2023	Synthetic, cardiovascular diseases, fatigue	Electron transport, antioxidant	Enhanced mitochondrial function, reduced oxidative stress	Lipophilic, crystalline structure	Peak plasma levels achieved within 6-8 hours; half-life ~33 hours	Synthetic
L-Carnitine (Water-Soluble) Koozehchianet al., 2018	Synthetic, cardiovascular diseases, fatigue	Fatty acid transport, mitochondrial energy production	Improved endurance, reduced fatigue in athletes	Water-soluble, stable under physiological conditions	Peak plasma levels within 2-3 hours; half-life ~15-20 hours	Synthetic
Coenzyme Q10 and L-Carnitine (Combination) Zozinaet al., 2018	Synthetic, cardiovascular diseases, fatigue	Coenzyme Q10: Electron transport, antioxidant; L-Carnitine: Fatty acid transport	Enhanced mitochondrial function, improved endurance, cardiovascular health	Combined properties of both compounds	Similar to individual components; synergistic effects on metabolism	Synthetic

This table presents a comprehensive overview of various antioxidants, detailing their sources, mechanisms of action, clinical outcomes, physicochemical characteristics, pharmacokinetic profiles, and recommended daily dosages. The integration of both natural and synthetic antioxidants underscores their roles in managing oxidative stress and related health conditions. Each entry provides insights into how these compounds contribute to cellular protection and overall health, reflecting the growing importance of antioxidants in biomedical and nutritional research.

Notes:

- **Physicochemical Characteristics**: Discuss the attributes that are ratios to their roles and endurance on the basis of physical and fundamental substances.

- **Pharmacokinetic** profile: The absorption, distribution, metabolism, and elimination profiles of each antioxidant are described.

Table 5 lists current and potential sources of novel antioxidants along with their fundamental properties and functions in human health and disease prevention.

Understanding how these antioxidants function in the human body and how to use them effectively to treat issues is crucial.

Key Components of Table

1. **Antioxidants:** Since each compound was enumerated here, astaxanthin, edaravone, and honey-derived polyphenols have their own characteristic antioxidant effects. These substances are thought to quench ROS, thus reducing oxidative stress.
2. **Source:** Their sources include natural resources such as marine algae, honey, and many other synthetic compounds. This diversity shows how antioxidants can be accessed in different ways, including through food sources or the intake of supplements.
3. **Mechanism of action**: The table given below explains the different activities of each antioxidant. For example, astaxanthin is characterized by increased permeability of the blood–brain barrier, which provides neuroprotection, and edaravone is useful in preventing oxidation in neurodegenerative diseases such as ALS.
4. **Recommended Dosage**: Including the dosage range per day is imperative, as this helps consumers and healthcare professionals utilize these antioxidants in the right manner. In this way, their benefits can be optimized for use while simultaneously reducing their side effects as much as possible.
5. **Natural vs. synthetic**: The table differentiates between the antioxidants that are endogenous in nature and those that are exogenous in nature.

In brief, conventional and postbiotic antioxidants are essential for maintaining redox balance and protecting cells and tissues against free radical-induced injury. Specifically, it is important to mention agents such as coenzyme Q10 and L-carnitine; both possess distinctive modes of action and constitute a major part of the cell protection process. CoQ 10 is important in the transport of electrons and mitochondrial performance. L-carnitine, on the other hand, aids in fatty acid transport and energy efficiency, hence increasing endurance and decreasing fatigue levels. The former works can complement

Each other and provide an enhanced system for enhancing the cardiovascular system and metabolic processes in the body. New sources of antioxidants, such as edaravone and astaxanthin, also have immense therapeutic potential in the treatment of ailments, which are correlated with oxidative stress. This is why further investigation of their processes and functions is highly desirable and essential in the future. In expanding on the varied functions of these antioxidants, we allow for the development of novel approaches to dietary management as well as medical treatments meant to improve health or prevent diseases associated with oxidative stress. This table lists different antioxidants, with information on their sources, actions, clinical effects, physical–chemical properties, pharmacokinetics, suggested daily intake, and affected organs. The goal is to improve the understanding of the various tasks these compounds perform in relation to human health and therapeutic approaches and whether they are natural or synthetic.

8. THE ROLE OF ANTIOXIDANTS IN MODULATING ROS-MEDIATED SIGNALING

There is crucial involvement of ROS in signaling across networks in the steady state as antioxidants. Free radicals are typically referred to as ROS and are relevant in signal transduction, metabolism, and other cellular functions, as well as in the development of diseases. Antioxidants counteract ROS and their damaging effects on cells and help prevent diseases associated with oxidative stress. These processes are modulated by conventional and postbiotic antioxidants, and recently categorized antioxidants, such as astaxanthin and edaravone, have been shown to have therapeutic effects.

There is a major agreement that biotic and abiotic free radicals are detrimental to human health and are mostly characterized by oxidative stress (Tab. 6). Free radicals are highly reactive molecules with uneven electrons that are capable of destroying tissues in the body. It has been classified into biogenic and abiotic agents, the two of which originate from different places.

Case Study 1: Astaxanthin is an antioxidant found in marine algae that is effective in quenching singlet oxygen and preventing skin damage due to oxidative stress caused by UV irradiation. A study conducted by Zhou et al. (2021) revealed that the place where astaxanthin was administered helped lessen skin damage and even increase skin firmness in people who underwent UV irradiation; the role of astaxanthin as a potent antioxidant and free radical in the context of the environment manifested.

Case Study 2: Superoxide radicals in Alzheimer's disease Oxygen free radicals ($O2\bullet-$) generated during the process of the mitochondrial respiratory process are associated with neurodegenerative diseases such as Alzheimer's disease. Chidambaram et al. (2024) reported that antioxidants such as coenzyme Q10 can control oxidative stress in neurons through high superoxide production. This antioxidant enhances mitochondrial activity and decreases oxidative damage, which leads to the development of neurodegenerative disorders.

Postbiotic antioxidants: The effects of postbiotics on gut health indicate that they have antioxidant properties caused by the removal of oxidative stress. Butyrate is a postbiotic antioxidant that is a short-chain fatty acid.

Case Study 3: Butyrate in gut health Butyrate supplementation restored gut epithelial integrity and decreased IBD-associated inflammation through the modulation of NF-κB signaling pathways in patients with IBD (Nozu, et al., 2019). Notably, butyrate has antioxidant qualities that allow it to effectively reduce oxidative stress and enhance health.

Table 6. Examples of biotic- and abiotic-free radicals and associated antioxidant defenses

Free Radical Type	Radical	Source	Antioxidant Type	Antioxidant Example	Clinical Effect	Case Study/ Reference
Biotic	Superoxide Radical	Mitochondrial respiration	Conventional	Coenzyme Q10	Protection from oxidative damage in neurons	Alzheimer's Disease (Turrens., 2003)
Biotic	Hydroxyl Radical	Fenton reaction	Conventional	Glutathione	Detoxification, cellular repair	Liver Health (Roy, & Paira, (2024).)
Abiotic	Peroxyl Radical	Lipid oxidation	Conventional	Vitamin E	Cardiovascular protection	Heart Health (Lloret, et al., 2019)
Abiotic	Nitrogen Dioxide	Combustion processes	Conventional	Vitamin C	Reduced lung damage in air pollution	Lung Health (Nizet & Ghosh, 2017)
Biotic	Various ROS	Probiotic metabolism (postbiotic)	Postbiotic	Butyrate	Improved gut health, anti-inflammatory	IBD (Zhang et al., 2020)
Synthetic	Hydroxyl Radical	Stroke-induced ROS	Synthetic	Edaravone	Neuroprotection in ischemic stroke	Stroke (Mizuno et al., 2018)
Conventional	Hydroxyl Radical	Tumor-related ROS	Conventional	Curcumin	Reduced tumor growth in cancer patients	Cancer (Zhu et al., 2024)

This table shows examples of biotic and abiotic free radicals, along with the corresponding antioxidant defenses that combat them. Free radicals are categorized on the basis of their origin—biotic radicals are produced through metabolic processes in living organisms, whereas abiotic radicals arise from environmental factors. The table lists various antioxidants, specifying their types (conventional, postbiotic, or synthetic) and providing examples such as coenzyme Q10, glutathione, vitamin E, and curcumin. Each entry also highlights the clinical effects observed in relevant case studies, such as the neuroprotective role of coenzyme Q10 in Alzheimer's disease and the detoxifying effects of glutathione on liver health. Overall, the table underscores the importance of antioxidants in mitigating oxidative damage and promoting health across different physiological contexts.

9. ENGAGEMENT WITH SIGNALING CASCADES

In this way, antioxidants modulate pivotal molecular targets, including signals for inflammation, cell survival, and proliferation, through interactions with ROS. This mechanism controls multiple processes, such as nuclear factor kappa B (NF-κB), mitogen-activated protein kinase (MAPK), and phosphoinositide 3-kinase (PI3K)/ Akt (Morgan, M. J., & Liu, Z. G. 2011).

Case Study 4: Edaravone in neuroprotection. The synthetic antioxidant edaravone has been found to be useful in the management of acute ischemic stroke through the neutralization of hydroxyl radicals. It defends neurons by limiting oxidative substance levels and moderating pathways that are inclined to damage neurons. Yamashita & Abe (2024) executed a clinical trial showing that edaravone enhanced poor neurological recovery in stroke sufferers, thereby indicating its efficacy in curtailing ROS signaling.

Case Study 5: Curcumin in Cancer Therapy The literature produced by medical practitioners indicates a strong relationship between curcumin, an antioxidant within turmeric, and the suppression of NF-κB signaling and inflammation, as well as the regulation of cancer cell division. According to Islam et al. (2024), curcumin consumption decreases NF-κB activation and tumor progression in patients with colorectal cancer, suggesting that curcumin could be a choice for anticancer medications.

Molecules with antioxidant properties, including astaxanthin, glutathione, coenzyme Q10, curcumin, vitamin C, and vitamin E, play essential regulatory roles in ROS signaling pathways, while postbiotics such as butyrate also have critical functions in signaling. The antioxidant ability of free radicals is evident in clinical aspects such as Alzheimer's disease (coenzyme Q10), stroke (edaravone), cancer (curcumin), and skin protection (astaxanthin). The antioxidative characteristics of these antioxidants suggest that they also play a therapeutic role in moderating oxidative stress-related diseases and improving overall human wellness.

Table 7. Types of antioxidants, their sources, signaling pathways, and clinical effects

Antioxidant Type	Antioxidant	Source	Free Radical	Signaling Pathway	Clinical Effect	Case Study/ Reference
Conventional	Astaxanthin	Marine algae	Singlet oxygen	MAPK, PI3K/ Akt	Skin protection, reduced UV damage	Skin Health (Lee et al., 2010)
Conventional	Coenzyme Q10	Endogenous/ Supplements	Superoxide Radical	Mitochondrial	Neuroprotection in Alzheimer's disease	Alzheimer's Disease (Turrens et al., 2003)
Conventional	Glutathione	Endogenous	Hydroxyl Radical	MAPK (ERK)	Liver detoxification, cellular repair	Liver Health (Chen et al., 2016)
Conventional	Vitamin C	Citrus fruits	Peroxyl Radical	NF-κB, PI3K	Reduced inflammation, skin health	Skin & Immune Health (Jomova et al ., 2023)
Conventional	Vitamin E	Nuts, seeds	Lipid Peroxyl Radical	NF-κB	Cardiovascular protection	Heart Health (Lloret, et al., 2019
Postbiotic	Butyrate	Probiotic metabolism	Various ROS	NF-κB, PI3K/ Akt	Improved gut health, anti-inflammatory	IBD (Nozu, et al., 2019)
Synthetic	Edaravone	Synthetic	Hydroxyl Radical	NF-κB, MAPK	Neuroprotection in stroke	Stroke (Mizuno et al., 2018)
Conventional	Curcumin	Turmeric	Hydroxyl Radical	NF-κB, MAPK	Reduced inflammation, tumor inhibition	Cancer (Islam et al. 2024)

This table provides a detailed overview of different antioxidants classified into conventional, postbiotic, and synthetic categories. Each entry outlines the source of the antioxidant, the specific free radical it targets, the signaling pathways involved, and the clinical effects demonstrated in case studies. Collectively, the table highlights the various functions of antioxidants in safeguarding cellular integrity and enhancing health under a range of conditions.

10. ANTIOXIDANT DEFENSE

10.1 Intracellular or endogenous antioxidants:

Endogenous antioxidants are important for competing with ROS to prevent the degradation of cellular components. Among all natural antioxidants, the main ones produced include superoxide dismutase (SOD), glutathione peroxidase (GPx), and catalase.

- **Superoxide Dismutase (SOD):** SOD is a metalloenzyme that reduces superoxide radicals (O•–), which are among the most reactive ROS, and converts them into hydrogen peroxide (HO), which is less reactive in nature. HO is then neutralized by other antioxidant enzymes, such as catalase or GSH-Px. SOD is crucial for preventing the formation of cellular macromolecules, and especially in the mitochondrial respiratory process, a high percentage of ROS are formed. Research has also shown the importance of SOD in limiting the incidence of neurodegenerative diseases because of its effects on oxidation (Park, & Son 2024).

- **Glutathione peroxidase (GPx):** GPx is a selenium-containing enzyme that catalyzes the reduction of hydrogen peroxide to water and lipid peroxides to lipid alcohol. This process helps preserve the health of cell membranes through protection from lipid peroxidation, a process of oxidative damage to lipids in the membrane. GPx is especially highly expressed in tissues prone to oxidative stress, including the liver and kidneys. It is also involved in the detoxication of the endogenous hydrogen peroxide generated during cellular metabolism (Toppo et al., 2008).
- **Catalase:** Catalase is another important antioxidant enzyme that breaks HO into water and oxygen to prevent buildup that would otherwise produce destructive OH• radicals. Catalase is located mostly in peroxisomes, and it is among the most rapidly turned-over enzymes; this explains why it is very effective in degrading hydrogen peroxide (Aebi, 1984).

Table 8. Intercellular antioxidant enzymes and their functions

Enzyme	Function	References
Superoxide Dismutase	Converts O_2•− to H_2O_2	Park, & Son 2024
Glutathione Peroxidase	Reduces H_2O_2 and lipid peroxides	Toppo et al., 2008
Catalase	Decomposes H_2O_2 into water and oxygen	Aebi, 1984

10.2 Extracellular Antioxidants

Extracellular or exogenous sources of antioxidants include vitamins C and E, flavonoids, and polyphenols available in diets, which help increase antioxidant levels and replace exhausted endogenous antioxidants. Among these mediator species, nitric oxide (NO) is unique in that it also functions as a signaling molecule as well as an antioxidant. NO, which is produced liberally by NOS, is a vital component that maintains vascular function in a homeostatic manner. NO also efficiently clears free radicals such as superoxide radicals by generating peroxynitrite (Andrianov et al., 2024). This reduces oxidative damage since Caminiti et al. (2024) reported that NO rescues endothelial cells from oxidative stress. Furthermore, nitric oxide enhances the synthesis of the endogenous antioxidant SOD and amplifies the activity of antioxidant enzymes to counteract the effects of oxidative stress. Bartz et al. (2015) noted that NO donors upregulate SOD in cardiac tissue and that this upregulation offers critical protection against ischemia–reperfusion injury. All these interactions support the hypothesis that integrated exogenous and endogenous antioxidant systems act together to moderate redox balance and guard against oxidative insult.

Role of nitric oxide in signaling pathways Nitric oxide plays a role in managing inflammation by altering the signals in signaling pathways. Conversely, NO abrogates ROS levels to prevent the activation of inflammatory transcription factors, including NF-κB. Mao et al. (2013) reported that NO suppressed NF-κB activation and subsequent cytokine release by macrophages.

The gut microbiota and its relationship to antioxidant production: New research has focused on the NO-producing capacity of the gut microbiota and other natural antioxidants. The microbiota can convert the nitrates found in foods into nitrites and then into NO, which increases vascular and general health. For example, Rossman et al. (2021) reported that the endothelial function of human subjects was enhanced after the administration of dietary nitrate through an increase in NO.

Table 9. Dietary supplements that increase nitric oxide production.

Supplement	Mechanism of Action	Clinical Outcome	References
L-arginine	Precursor for nitric oxide synthesis	Improved exercise performance and recovery	Kiani, et al. (2022)
L-citrulline	Converted to L-arginine, enhancing NO production	Increased blood flow and reduced fatigue	Theodorou, (2021)
Beetroot Extract	High in dietary nitrates, converted to nitrites in the body	Enhanced cardiovascular function	Stanaway, et al. (2019)
Nitrate-Rich Greens	Source of nitrates that boost NO levels	Improved endothelial function	Sweazea et al. (2015)
Coenzyme Q10	Supports mitochondrial function, indirectly enhances NO	Improved endothelial function	Samimi et al. (2007)
Pycnogenol	Enhances endothelial NO synthase activity	Improved erectile function and blood flow	Weichmann, et al. (2024)
Astaxanthin	Reduces oxidative stress, supports NO production	Enhanced physical performance	Waldman. (2024)
Garlic Extract	Increases NO synthesis via sulfur compounds	Improved cardiovascular health	Li, et al. (2022)
Alpha Lipoic Acid	Antioxidant properties that enhance NO bioavailability	Improved insulin sensitivity and blood flow	Capece, et al. (2022)

This table summarizes various dietary supplements that have been shown to increase nitric oxide (NO) production in the body. Each supplement is listed along with its mechanism of action, clinical outcomes observed in studies, and relevant references. The supplements include amino acids, plant extracts, and antioxidants, each of which contributes to improved exercise performance, cardiovascular health, and overall physiological function by increasing NO levels.

10.3 Bifunctional Antioxidants

Multiple targeted antioxidants act both as ROS and free radical quenchers and as activators of intracellular antioxidant systems, especially the Nrf2-ARE pathway. This dual activity makes them suitable candidates for therapeutics against oxidative stress-related pathologies such as neurodegenerative disorders, cancer, and metabolic diseases (Table 10). They not only are simple radical neutralizers but also participate in redox homeostasis regulation, in the modulation of inflammation, and in the improvement of resistance to cellular oxidant stress.

Key Bifunctional Antioxidants

1. **Esculetin (6,7-dihydroxycoumarin):** An isolated coumarin that has shown pronounced ROS-reducing activity in addition to affecting several antioxidant enzymes, such as GPx and SOD. They also reported that esculetin suppresses the production of proinflammatory cytokines through the downregulation of the NF-κB and MAPK pathways and concluded that esculetin can be used as a therapeutic agent for neurodegenerative and cardiovascular diseases (Garg, et al., 2022). New research has demonstrated its capacity to improve mitochondrial function, further decreasing oxidative stress levels.
2. **Alpha-lipoic acid (ALA):** ALA is a native antioxidant and can also be used as a supplement. It is involved in mitochondrial bioenergetics and the recycling of other antioxidants, such as ascorbate and tocopherol. ALA has been investigated in DN, where the compound enhances mitochondrial functionality and suppresses the production of ROS by upregulating the Nrf2 pathway (Ramos, et al., 2023). The ability of ALA to chelate heavy metals provides another therapeutic perspective that earns it a place in treatment.
3. **Chalcones:** These flavonoids possess direct antioxidant properties by themselves, which also stimulate the expression of endogenous antioxidant enzymes comparable to those of catalase and SOD. Previous studies have indicated that chalcones have the ability to control apoptosis and autophagy in cancer cells and affect the PI3K/Akt and MAP kinase pathways; hence, the anticarcinogenic nature of these compounds is highlighted by Chen et al. (2020).

4. **Sulforaphane:** As one of the abundant isothiocyanates present in cruciferous vegetables, sulforaphane effectively activates the Nrf2-ARE pathway. It also stimulates the production of the body's own antioxidants, such as glutathione, while eliminating free radicals directly. Sulforaphane also has both anti-inflammatory and antioxidant effects in animal models of neurodegenerative diseases, including Alzheimer's disease and Parkinson's disease (Dinkova-Kostova, et al., 2017).

Modulation of the Gut Microbiota

Thus, the gut microbiota has a predictive influence on the bioavailability and efficacy of bifunctional antioxidants in humans. For example, some circuits in the gut can transform precursor bioelements, which comprise glucosinolates from sulforaphane, into forms with improved antioxidant activity Abukhabta, et al., 2021). The reduced bioavailability of these antioxidants due to dysbiosis, or an imbalance in the composition of the gut microbiota, limits their effectiveness in therapy. However, compounds such as ALA can alter the structure and performance of the gut microbiota in a way that favors the growth of beneficial bacteria such as Lactobacillus and Bifidobacterium, which strengthen the gut barrier and lower inflammation in the body.

Effectiveness and Mechanism of Action

The therapeutic usefulness of bifunctional antioxidants therefore depends on their availability, metabolism among body cells and tissues, and ability to modulate antioxidant activities within the cells. Lipoic acid, on the other hand, has a half-life in the plasma for a short time, and this can be overcome by formulation methods such as sustained-release tablets. Various studies have demonstrated that both esculetin and sulforaphane have dose-dependent protective effects in in vivo and in vitro experiments; as the concentration increases, neuroprotective and anti-inflammatory properties are even further amplified (Antioxidative activity and mechanisms of esculetin and sulforaphane: an update Pruccoli, 2020; Houghton C. A. 2019).

Table 10. Bifunctional antioxidants: Overview, clinical applications, and dosing

	Mechanism of Action	Clinical Applications	Recommended Dosage	Time of Administration
Esculetin Pruccoli et al. (2020).).	- Scavenges free radicals - Inhibits lipid peroxidation - Modulates antioxidant enzyme activity	- Neuroprotection - Anti-inflammatory effects	- 10-50 mg/kg/day in animal models	- Morning or with meals
Alpha-Lipoic Acid Hsieh., et al. (2023	- Directly scavenges ROS - Regenerates other antioxidants (e.g., vitamins C and E)	- Diabetic neuropathy - Cardiovascular protection	- 600 mg/day (oral supplementation)	- Once daily, preferably with meals
Chalcones Silva, et al. (2023).	- Scavenges ROS - Induces the expression of antioxidant enzymes - Inhibits lipid peroxidation	- Cancer prevention - Anti-inflammatory	- 100 mg/day (in preclinical studies)	- With meals
Sulforaphane Egner,., et al. (2012).	- Activates the Nrf2 pathway - Increases production of endogenous antioxidants like glutathione	- Cancer prevention - Neuroprotection	- 20-40 mg/day (broccoli extract equivalent)	- With meals
Pterostilbene Cui et al., (2024)	- Direct scavenger of ROS - Modulates antioxidant gene expression - Inhibits oxidative DNA damage	- Cardiovascular health - Anti-aging	- 50-150 mg/ day (oral supplementation)	- With meals
Coenzyme Q10 (Ubiquinone) Gutierrez-Mariscal,., et al. (2020).	- Electron carrier in the mitochondrial respiratory chain - Regenerates Vitamin E - Reduces oxidative damage	- Cardiovascular support - Mitochondrial function	- 100-200 mg/ day (oral supplementation)	- Preferably with meals

1. Although esculetin per se is not marketed as a dietary supplement, it can be purchased from sources such as nutraceuticals and phytopharmaceuticals.
2. Some of the brand names of alpha-lipoic acid include thioctic acid, ALA 600, and thiogamma because of its antioxidant and neuroprotective properties.
3. Rather uncommon in commercial supplements, chalcones appear in particular formulations created for cancer prevention and inflammation.
4. Sulforaphane supplements extracted from broccoli are available in major brands, including BroccoMax and Avmacol.
5. Pterostilbene is commercially distributed, such as pTeroPure or PteroMax, with a focus on its benefits for cardiovascular and antiaging health.
6. Mostly for cardiovascular and mitochondrial health, coenzyme Q10 (ubiquinone) is available under names that include Q-Sorb, ubiquinol, and CoQ10 Max.

Every antioxidant has defined clinical uses and market formulations that help facilitate its therapeutic application.

11. OXYSHIELD TRIAD: OUTSTANDING ANTIOXIDANT CHAMPION FOR EXTRAORDINARY HEALTH ASSURANCE

The OxyShield Triad comprises magnesium (Mg), zinc (Zn), and copper (Cu), a combination that increases the body's ability to fight free radicals and enhances physiological functions. The micronutrients that are deemed vital within these centrality areas include the following: tract integrity Nutrients shield against the plague of oxidative stress and continue health opposition. These three factors provide mutually enhancing effects that enhance the body's antioxidant potential and resist oxidative stress to help promote optimal health in the long term.

11.1 Key Micronutrients in the OxyShield Triad

Table 11 presents information on three micronutrients, which are parts of the OxyShield Triad. These micronutrient formulations, their dosages, trade names, dietary sources, and their antioxidative effects are presented in the table below. Additionally, their clinical uses and the impact of certain coadministered drugs on their absorption and efficacy are discussed. It is crucial to best comprehend these micronutrients to determine their optimal antioxidant consumption and well-being.

Table 11. The OxyShield Triad micronutrients are included in this table along with their respective dosages, dietary sources, antioxidative effects, clinical applications, and medication interactions.

Micronutrient	Dosage	Dietary Sources	Antioxidative Effects	Clinical Uses	Drug Interactions	References
Magnesium (Mg)	300 mg/ day	Nuts, Seeds, Leafy Greens	Helps reduce oxidative stress and maintains cellular function	Supports muscle function, bone health, and nerve function	Can be impacted by proton pump inhibitors, certain antibiotics	Sagar et al., (2024). de Oliveira et al.,(2024)
Zinc (Zn)	11 mg/ day	Meat, Shellfish, Legumes	Plays a role in the enzymatic function to combat oxidative stress	Boosts immune system, supports wound healing and metabolism	Absorption reduced by calcium supplements and some antibiotics	Matuszczak et al., (2024) Vargas-Uricoechea et al.,2024
Copper (Cu)	0.9 mg/ day	Organ Meats, Whole Grains	Involved in antioxidant defense and supports superoxide dismutase	Important for iron metabolism and connective tissue formation	Iron supplements may reduce absorption, and certain medications	Fan et al., 2024 Liu et al., (2024)

12. VITAGUARD ENSEMBLE

The VitaGuard Ensemble, also referred to as the "ACED" formula, represents a groundbreaking blend of four essential vitamins: A, C, E, and D. These vitamins act in concert to increase the body's antioxidant protection and support essential physiological functions that are vital to life. The incorporation of these vitamins guarantees coverage against oxidative stress and helps maintain the best physiological fitness. The IOM (US) Panel on Dietary Antioxidants and Related Compounds has noted that vitamins C and E are known for their ability to counteract reactive oxygen species, therefore preventing cell damage. As a source of carotenoids, vitamin A has roles in immunity and vision; vitamin D plays roles in the regulation of calcium and bone formation. The complementary actions of these vitamins improve not only individual antioxidant roles but also organelle function within all the systems.

This combination of vitamins provides good protection against oxidative stress and increases general health, which is supported by current nutritional insights.

- **Vitamin A:** Also called the "vision guardian," it is essential for the eyes, skin, and immune system to be healthy. It also functions as an antioxidant and assists in defending cellular organelles from the effects of free radicals.
- **Vitamin C:** First, it is a fat-soluble vitamin that works as a direct antioxidant and free radical fighter; in addition, it is a recycling agent because it regenerates other antioxidants present in the human body. Vitamin C is crucial for the formation of collagen and is related to skin integrity, tissue repair, and immune actions against pathogens.
- **Vitamin E:** Prolactin, a fat-soluble antioxidant, plays a very important role in skin health and has the ability to protect cell membranes from free radicals. This compound increases antioxidant defense against oxidative stress in the body and works independently to increase immunity via vitamins A and C.
- **Vitamin D:** In many ways, vitamin D is not similar to other vitamins; instead, it acts as a hormone. They are involved in calcium metabolism and affect the immune system; they are important in the bone and anti-inflammatory systems.
- **The Synergistic Power of ACED:** The VitaGuard ACED formula focuses on the impact of using these vitamins together to strengthen antioxidant functions. We already know that vitamin A helps to stabilize cell membranes, whereas vitamin C, with the rebirth of vitamin E, maintains its antioxidant properties. Vitamin D supports this synergy by modulating immune reactions and decreasing oxidative stress and inflammation. These vitamins constitute the entire antioxidant system necessary for health and the prevention of oxidative effects.

Table 12. Vita-Guard ensemble ACED overview

Vita ACED	Common Formulations	Typical Dosage (RDA)	Dietary Sources	Antioxidant Role	Clinical Applications	Drug Interaction
Vitamin A Didier et al., 2023	Retinyl palmitate, beta-carotene	900 mcg (men), 700 mcg (women)	Carrots, sweet potatoes, spinach, liver	Vitamin A acts primarily as a fat-soluble antioxidant, neutralizing free radicals in the lipid membrane. It also enhances immune responses and supports vision.	Essential for maintaining epithelial tissues, immune function, and eye health. Prevents night blindness and helps maintain healthy skin and mucous membranes	Can interact with anticoagulants (e.g., warfarin), leading to an increased risk of bleeding due to impaired clotting mechanisms
Vitamin C Carr, & Maggini. (2017).	Ascorbic acid, sodium ascorbate	90 mg (men), 75 mg (women)	Citrus fruits (oranges, lemons), strawberries, bell peppers, broccoli	Water-soluble antioxidant, scavenges free radicals, regenerates other antioxidants (like vitamin E), and supports collagen synthesis and immune function	Effective in reducing cold duration, enhancing immune function, and promoting skin health. It aids in the absorption of nonheme iron.	May enhance the absorption of certain iron supplements, potentially leading to iron overload if combined with high doses of iron supplements
Vitamin E Rizvi et al., 2014	Alpha-tocopherol, gamma-tocopherol, tocotrienols	15 mg (both men and women)	Nuts (almonds, hazelnuts), seeds, spinach, broccoli	Fat-soluble antioxidant protecting cell membranes from lipid peroxidation, works synergistically with vitamin C	Shown to support cardiovascular health by preventing oxidative damage to LDL cholesterol, improves skin health, and has anti-inflammatory properties	May interact with anticoagulants like warfarin and statins, increasing bleeding risks and interfering with cholesterol-lowering therapies.
Vitamin D https://ods.od.nih.gov/factsheets/VitaminD-HealthProfessional/	Ergocalciferol (D2), cholecalciferol (D3)	600 IU (under 70 years), 800 IU (70+)	Fatty fish (salmon, mackerel), fortified dairy, sunlight exposure	Regulates immune function, enhances calcium and phosphorus absorption, and has anti-inflammatory and immunomodulatory effects	Crucial for bone health, preventing osteoporosis, and boosting immune responses. It also reduces the risk of autoimmune diseases and chronic inflammation	May interact with corticosteroids, weight-loss drugs, and cholesterol-lowering medications. Prolonged corticosteroid use may inhibit calcium absorption, leading to bone loss

The table highlights the essential roles of vitamins A, C, E, and D in supporting overall health:

- **Vitamin A:** Vitamin A improves night vision, skin and immune status and has antioxidant properties. It is present in foods such as carrots, spinach, and the liver. It can react with other prescriptions that include blood thinners such as warfarin, which may lead to bleeding.

- **Vitamin C:** A nutrient that significantly prevents the body from oxidative harm and increases the body's ability to absorb iron while boosting the immune system. It is found in oranges, lemons, grapes, strawberries, bell peppers, etc. Taking high doses may result in increased interaction with iron when it is taken with iron supplements.
- **Vitamin E** serves to shield cell membranes against damage caused by oxidation and has positive effects on the cardiovascular system. It is found in nuts and seeds and may further thin the blood, especially when it is consumed with blood-thinning medications.
- **Vitamin D:** Promotion of calcium uptake in the body, bone formation, and diet inflammatory status. Like vitamin D, which is obtained from fatty fish and some dairy foods, it is more easily absorbed if it is taken in conjunction with fat. It may also react with other drugs, such as corticosteroids, that are known to cause a decrease in calcium uptake and ultimately bone loss.

Quite simply, each vitamin is a nutrient that the human body needs to stay healthy, and it is always beneficial to remember dosages and possible interactions with other medicines.

13. IMPACT OF DIETARY ANTIOXIDANT SUPPLEMENTS ON THE HUMAN GUT MICROBIOTA: IMPLICATIONS FOR HEALTH AND DISEASE PATHWAYS

The human gut microbiota, a vital determinant of health and disease, can be influenced by dietary antioxidant supplements. The gut microbiota is a delicate symbiotic community of microorganisms involved in numerous significant processes, including carbohydrate breakdown, vitamin production, and immunomodulatory actions. The proper distribution of gut microbes is said to increase the health of an individual, whereas an unbalanced population of microbes is said to cause obesity, type 2 diabetes, and IBD, among other diseases (Martínez, et al., 2024).

13.1 Modes of Action of Antioxidants

Antioxidants are substances that have the ability to counteract free radicals and thereby eliminate the oxidative pressure on cells and tissues. The preventive/therapeutic effects of dietary antioxidant supplements include L-carnitine, vitamins C and E, polyphenols, astaxanthin, curcumin, and the synthetic antioxidant edaravone (Table 13).

13.2 Alterations in the Composition and Function of the Gut Microbiota

1. **Modulation of Microbial Composition:** A study revealed that the consumption of antioxidants can increase the growth of healthy intestinal bacterial flora while simultaneously reducing the number of pathogenic organisms. For example, L-carnitine, which is easily obtained from animal sources, has been found to increase the number of authoritative bacterial flora. Polyphenols, which are part of the fruit and vegetable diet, help in the selection of healthy bacteria such as Lactobacillus and Bifidobacterium (David et al., 2024).
2. **Enhancement of Gut Barrier Integrity:** Antioxidants assist in preserving the integrity of the intestinal barrier and prevent the penetration of toxic substances into the bloodstream, preventing oxidative stress. Curcumin and astaxanthin, for example, can protect tissues and cells in the GIT against oxidative stress and inflammation (Gupta et al., 2013; Lee et al., 2022).
3. **Anti-inflammatory properties:** The ability of antioxidants to decrease inflammation is one of the ways in which they play a significant role in improving gut health. For example, edaravone slows oxidative injury in certain neurodegenerative diseases; therefore, it is logical to assume that it may also be helpful for gut health (Morén, et al., 2022).
4. **Prebiotic Functions:** Some antioxidants, such as honey and butyrate, contribute as prebiotics to the growth of favorable gut microbiota. According to Becerril-Sánchez et al. (2021), honey is made of phenolic compounds, which may be used to increase the functional and evolutionary capacity of microbes. Butyrate, another SCFA produced from the fermentation of fiber, possesses anti-inflammatory properties and plays a role in maintaining gut integrity (Du et al., 2024).
5. **Clinical Applications and Future Prospects:** The relationship between dietary antioxidant supplements and the gut microbiota has significant clinical implications.
    - **Metabolic health:** Several studies have indicated that diets with high antioxidant contents may improve metabolic performance through changes in the gut microbiome. Hence, to maintain good metabolic performance, there should be a balance in the microbiome, which may help to 'prevent insulin resistance—a state preceding type 2 diabetes (Qi, et al., 2023).

- **Aging and longevity:** Aging people experience deteriorated oxidation levels and microbiota imbalance. Because increased oxidative stress is counterproductive to healthy aging and longevity, taking antioxidant supplements can help reset the microbiota (Yang et al., 2024).
- **Diseaseprevention:** Increasing evidence indicates that diets rich in antioxidants might reduce the likelihood of chronic diseases. For example, vitamins C and E have established links between a reduced risk of cardiovascular diseases and certain cancers (Didier et al. 2023).

Hence, carnitine, astaxanthin, and honey are antioxidant dietary supplements that alter the composition and structure of the gut microbiota, which influences health and disease. The absence of inflammation, a healthy gut barrier, and a balanced microbiome, all of which are compelled by traces of antioxidants, becomes imperative for preventing chronic ailments and improving well-being. Research into the interactions between dietary antioxidants and the gut microbiota will provide a better understanding of their potential advantages for human health.

Table 13. Impact of dietary antioxidant supplements on the human gut microbiota

Antioxidant Type	Example	Effect on Gut Microbiota	Clinical Outcome	Case Study/ Reference	Clinical Case Studies
Vitamin C	Citrus fruits	Increases beneficial bacteria	Reduced inflammation	Miles, & Calder, (2021)	Miles,& Calder (2021) demonstrated that vitamin C supplementation increased beneficial gut bacteria, reducing systemic inflammation in patients with metabolic syndrome.
Vitamin E	Nuts, seeds	Protects against oxidative stress	Cardiovascular health	Liu et al. (2023)	Liu et al. (2023) found that dietary vitamin E improved gut microbiota diversity, enhancing cardiovascular health in older adults.
Polyphenols	Green tea	Promotes Lactobacillus growth	Enhanced metabolic health	Liu et al., (2023)	Liu et al., (2023) reported that green tea polyphenols promoted Lactobacillus growth, leading to improved metabolic parameters in overweight individuals.
Curcumin	Turmeric	Reduces gut inflammation	Alleviates IBD symptoms	Lin et al. (2023)	Lin et al. (2023) found that curcumin reduced gut inflammation and alleviated symptoms in IBD patients.
Coenzyme Q10	Supplements	Supports gut barrier function	Neuroprotection	Bagheri, et al. (2023)	Bagheri, et al. (2023) showed that coenzyme Q10 supports gut barrier integrity, crucial for neuroprotection in neurodegenerative disease patients.
Astaxanthin	Marine algae	Enhances gut microbiota balance	Skin protection, reduced UV damage	Bjørklund, et al. (2010)	Lee et al. (2010) indicated that astaxanthin balances gut microbiota, contributing to improved skin health and reduced UV damage.
Edaravone	Synthetic	Protects against oxidative damage	Neuroprotection in ischemic stroke	Watanabe et al. (2018)	Watanabe et al. (2018) documented edaravone's neuroprotective effects in ischemic stroke patients, highlighting its role in reducing oxidative stress and supporting overall gut health.

continued on following page

Table 13. Continued

Antioxidant Type	Example	Effect on Gut Microbiota	Clinical Outcome	Case Study/ Reference	Clinical Case Studies
Carnitine	Animal products	Supports beneficial gut bacteria	Improved metabolic health	Heianza, et al. (2019)	Heianza, et al. (2019) found that carnitine supports beneficial gut bacteria, improving metabolic markers in prediabetic individuals.
Honey	Natural sweetener	Enhances beneficial bacteria	Improved gut health	Schell, (2022)	Nizet & Ghosh (2017) reported that honey enhances the diversity of beneficial gut bacteria, improving gut health in gastrointestinal disorder patients.
Butyrate	Probiotic metabolism	Supports anti-inflammatory processes	Improved gut health	Recharla, et al. (2023)	Recharla, et al. (2020) highlighted that butyrate, produced from dietary fiber fermentation, supports anti-inflammatory processes and improves gut health in IBD patients.

These case studies underscore the profound influence of dietary antioxidants on the gut microbiota and overall health, emphasizing their crucial role in clinical nutrition and disease prevention. Antioxidant supplements such as carnitine, astaxanthin, and honey have been shown to significantly modulate the composition and function of the gut microbiome, thereby impacting key health and disease pathways. By promoting a balanced microbiome, strengthening gut barrier integrity, and reducing inflammation, these antioxidants are pivotal in preventing chronic diseases and enhancing overall well-being.

Continued research into the dynamic relationship between dietary antioxidants and the gut microbiota will expand our understanding of their potential health benefits. These findings further highlight the essential role of antioxidants in optimizing gut health and their ability to positively influence a wide range of health outcomes.

14. NANO-ANTIOXIDANTS: A REVOLUTIONARY APPROACH TO TRADITIONAL ANTIOXIDANT THERAPY

14.1 Cancer Therapy

Recent research has demonstrated the potential of nanoantioxidants in enhancing the effectiveness of cancer therapies:

- **Nano-Curcumin in Breast Cancer:** Using breast cancer cell lines, Gao et al. (2019) examined the effects of nanocurcumin formulations. A comparison of nanocurcumin with traditional curcumin revealed that the latter greatly increased the apoptotic effects on cancer cells. This was explained by increased cellular absorption and bioavailability, which made it possible to induce oxidative stress in tumor cells and scavenge reactive oxygen species (ROS) more successfully. As evidenced by the researchers' findings, the nanoformulation has therapeutic potential for the targeted treatment of cancer because it not only increases anticancer efficacy but also decreases cytotoxicity to normal cells.

- **Targeted drug delivery with nanoparticles:** Zhang et al. (2020) investigated the use of antioxidant-loaded silica-based nanoparticles, such as quercetin and resveratrol, for targeted delivery in pancreatic cancer. The results of the investigation demonstrated that these nanoparticles considerably decreased tumor growth in comparison to that of the controls and increased treatment-trained mouse survival. By lowering oxidative stress in the tumor microenvironment, the antioxidant impact and targeted administration increase the efficacy of chemotherapy.

14.2 Neuroprotection

The role of nanoantioxidants in neurodegenerative diseases has garnered significant interest:

- **Nanorresveratrol for Alzheimer's Disease: The neuroprotective effects of resveratrol-loaded nanoparticles** were investigated by Sharma et al. (2020) in an Alzheimer's disease model. These results showed that the formulation of nanoresveratrol enhanced cognitive performance and decreased brain amyloid-beta plaque buildup. This study focused on how nanoresveratrol can provide stronger neuroprotective and antioxidant benefits since it can pass through the blood–brain barrier more efficiently than traditional formulations can.
- **Nanoparticles in Parkinson's Disease:** In recent research, antioxidant-encapsulated iron oxide nanoparticles were studied in a Parkinson's disease model by Khameneh et al. (2021). The findings revealed that indicators of oxidative stress and neuroinflammation were significantly decreased. The ability of these nanoantioxidants to prevent the degeneration of dopaminergic neurons suggests that they may be used as a treatment approach for neurodegenerative diseases.

14.3 Cardiovascular Health

Nanoantioxidants have also been well documented for their beneficial cardiovascular properties.

- **CoQ10 Nanoparticles in Heart Disease:** Beydemir et al. (2021) conducted a clinical trial in which CoQ10 nanoparticles were used to treat patients with heart failure. This study revealed substantial improvements in heart function and considerable decreases in oxidative stress indicators. Compared with patients on regular CoQ10 supplements, patients on the nanoparticle formulation demonstrated improved exercise ability and overall heart health.
- **Omega-3 fatty acid nanoparticles:** Patients with metabolic syndrome may benefit from omega-3 fatty acid nanoparticle treatment, according to Calder's (2021) research. These findings indicate that these nanoparticles successfully lowered systemic inflammation and improved lipid profiles, which decreased the risk of cardiovascular events. This study demonstrated how the nanoformulation improved the distribution and absorption of omega-3 fatty acids, enhancing their antioxidant effects.

14.4 The Antioxidant Paradox

Examination of Excessive Antioxidant Supplementation

Despite being highly regarded for their health perks, researchers warn that high doses could induce contradictory impacts. Although antioxidants are important for addressing oxidative stress, large amounts of antioxidants can disrupt the delicate relationship between antioxidative and oxidative reactions in the body. Such evidence suggests that large amounts are advantageous since high doses of specific antioxidants may lead to negative results. The 2012 analysis by Bjelakovic et al. revealed an alarming association between the use of potent antioxidant supplements such as beta-carotene and vitamin E and increased mortality in certain communities. These results have ignited intense arguments concerning the potential dangers of nonspecific antioxidant use for people with smoking histories or chronic health issues. Excess beta-carotene found in smokers is associated with an increased incidence of lung cancer, indicating that high antioxidant intake might actually create oxidative injury rather than help protect against it. In 2023, Didier et al. examined the relationship between antioxidant supplementation and cancer risk and concluded that diets high in antioxidants lower cancer risk but that supplemental antioxidants provide no similar defensive benefits. These variations could be due to the body's elaborate mechanisms aimed at preserving the level of oxidation. On the basis of findings from the study. Antioxidant supplements might block the effectiveness of native antioxidants, as they share the same pathways. Balancing the activities of pro-oxidative and antioxidative elements is important for proper performance of the body's antioxidant defenses. Kozlov et al. (2024) reported that an imbalance of pro-oxidants causes oxidative stress and damages cells while increasing the risk of neurological and cardiovascular diseases. ROS are important for important physiological activities, including signaling between cells and immune responses. Huang et al. (2021) suggested that ROS play two essential roles: at lower doses, they act as communication molecules, whereas elevated amounts can cause severe damage. The concept of "hormesis" applies here: minimal oxidative stress can ignite the body's defense mechanisms against antioxidants and increase its capacity to cope with future oxidative issues. According to Kawamura, & Muraoka (2018), exercise-induced low-level oxidative stress can enhance muscle adaptations. Too much antioxidant intake could deter favorable reactions triggered by ROS. According to Merry & Ristow (2016. (2009), taking antioxidant supplements might cancel favorable results from exercise-related oxidative stress and lessen improvements in insulin sensitivity in fit people. Although antioxidants are crucial for overall health, their association is complicated, requiring an attentive strategy to gain maximum benefits while respecting the body's intrinsic abilities. Understanding the relationship between pro-oxidants and antioxidants is crucial for creating successful health strategies that increase well-being and reduce risks from excessive antioxidant intake (Sotler, et al.,2019).

15. CONCLUSION

Overall, this review of new antioxidant studies emphasizes the participation of radicals such as ROS and dietary antioxidants in health maintenance. With respect to targeted antioxidant therapies, specific examples include the OxyShield Triad, which is composed of magnesium (Mg), zinc (Zn), and copper (Cu), whereas the Vita-Guard Ensemble ACED presents the same applicability. These formulations supplement the body's protection against oxidative stress, as they perform other basic functions. ROS are inherently generated products of cellular metabolism that are involved in signaling and maintaining homeostasis. However, overproduction of ROS results in oxidative stress, which is associated with chronic diseases such as cancer, cardiac diseases, and neurodegenerative diseases. Consequently, the 3 balances between ROS and antioxidants are important since dietary antioxidants help to annul the above species to minimize oxidative injury. New data support the role of antioxidants in modulating ROS signaling cascades. Vitamin C, vitamin E, and polyphenols, as well as other bioactive molecules, can interact with signaling pathways, influencing the gene expression of inflammatory mediators and regulating cell survival. For example, flavonoid compounds stimulate the enzyme expression of transcription factors controlling the antioxidant response; curcumin modifies the gut microbiota to magnify the anti-inflammatory action. This awareness of antioxidant therapeutic uses is crucial in the development of specific treatment plans for various disorders caused by oxidative stress. The development of nanoantioxidants ushers in a revolution in conventional antioxidant treatment strategies. The feature that sets nanoantioxidants apart from conventional antioxidants is the problem of bioavailability, which could be avoided due to the nanotechnological approach. These characteristics enable them to scavenge ROS more effectively and/or to better regulate cellular responses at the molecular level. The current findings show that nanoantioxidants may have antidiabetic and anti-inflammatory effects on diabetic complications, neurodegenerative disorders, and ischemic-reperfusion diseases. However, blending natural antioxidants with biomaterials is becoming a new frontier in the development of tissue replacement in regenerative medicine.

In this approach, natural antioxidant compounds are incorporated into biomaterial scaffolds intended for tissue engineering applications, which may improve cell survival and proliferation and minimize oxidative stress in implanted tissues (Perez-Araluce et al., 2024). However, problems persist in controlling the release kinetics of antioxidants and their physical, chemical, and efficacy in vivo (Benbettaïeb, et al., 2018). In this manner, the OxyShield Triad dynamically adjusts the production of species such as ROS to improve the health status of the cells and decrease their vulnerability to chronic diseases. The Vita-Guard Ensemble ACED focuses on a more extensive strategy for antioxidant protection while ensuring that people obtain as many protective nutrients as possible. Together, these developments present fundamental directions for enhancing treatments in clinical practice. In overcoming oxidative stress difficulties and improving general wellbeing, OxyShield and Vita-Guard improve patient treatment and quality of life. Given that the understanding of antioxidants is gradually improving, their use in interventions that affect health will be useful in addressing oxidative stress-associated diseases.

REFERENCES

Abukhabta, S., Khalil Ghawi, S., Karatzas, K. A., Charalampopoulos, D., McDougall, G., Allwood, J. W., Verrall, S., Lavery, S., Latimer, C., Pourshahidi, L. K., Lawther, R., O'Connor, G., Rowland, I., & Gill, C. I. R. (2021). Sulforaphane-enriched extracts from glucoraphanin-rich broccoli exert antimicrobial activity against gut pathogens in vitro and innovative cooking methods increase in vivo intestinal delivery of sulforaphane. *European Journal of Nutrition*, *60*(3), 1263–1276. DOI: 10.1007/s00394-020-02322-0 PMID: 32651764

Adıgüzel, E., & Ülger, T. G. (2024). A marine-derived antioxidant astaxanthin as a potential neuroprotective and neurotherapeutic agent: A review of its efficacy on neurodegenerative conditions. *European Journal of Pharmacology*, *977*, 176706. DOI: 10.1016/j.ejphar.2024.176706 PMID: 38843946

Aebi, H. (1984). Catalase in vitro. *Methods in Enzymology*, *105*, 121–126. DOI: 10.1016/S0076-6879(84)05016-3 PMID: 6727660

Ahn, J., & Hayes, R. B. (2021). Environmental Influences on the Human Microbiome and Implications for Noncommunicable Disease. *Annual Review of Public Health*, *42*(1), 277–292. DOI: 10.1146/annurev-publhealth-012420-105020 PMID: 33798404

Al-Kafaween, M. A., Alwahsh, M., Mohd Hilmi, A. B., & Abulebdah, D. H. (2023). Physicochemical Characteristics and Bioactive Compounds of Different Types of Honey and Their Biological and Therapeutic Properties: A Comprehensive Review. *Antibiotics (Basel, Switzerland)*, *12*(2), 337. DOI: 10.3390/antibiotics12020337 PMID: 36830249

Ambati, R. R., Phang, S. M., Ravi, S., & Aswathanarayana, R. G. (2014). Astaxanthin: Sources, extraction, stability, biological activities and its commercial applications—a review. *Marine Drugs*, *12*(1), 128–152. DOI: 10.3390/md12010128 PMID: 24402174

Andrés, C. M. C., Pérez de la Lastra, J. M., Andrés Juan, C., Plou, F. J., & Pérez-Lebeña, E. (2022). Impact of Reactive Species on Amino Acids-Biological Relevance in Proteins and Induced Pathologies. *International Journal of Molecular Sciences*, *23*(22), 14049. DOI: 10.3390/ijms232214049 PMID: 36430532

Andrianov, V. V., Schepetkin, I. A., Bazan, L. V., Gainutdinov, K. L., Kovrizhina, A. R., Atochin, D. N., & Khlebnikov, A. I. (2024). Evaluation of Nitric Oxide-Donating Properties of 11H-indeno[1,2-b]quinoxalin-11-one Oxime (IQ-1) by Electron Paramagnetic Resonance Spectroscopy. *Molecules (Basel, Switzerland)*, *29*(16), 3820. DOI: 10.3390/molecules29163820 PMID: 39202899

Anwar, S., Alrumaihi, F., Sarwar, T., Babiker, A. Y., Khan, A. A., Prabhu, S. V., & Rahmani, A. H. (2024). Exploring Therapeutic Potential of Catalase: Strategies in Disease Prevention and Management. *Biomolecules*, *14*(6), 697. DOI: 10.3390/biom14060697 PMID: 38927099

Ashique, S., Mishra, N., Mantry, S., Garg, A., Kumar, N., Gupta, M., Kar, S. K., Islam, A., Mohanto, S., & Subramaniyan, V. (2025, January). (2024). Crosstalk between ROS-inflammatory gene expression axis in the progression of lung disorders. *Naunyn-Schmiedeberg's Archives of Pharmacology*, *398*(1), 417–448. Advance online publication. DOI: 10.1007/s00210-024-03392-1 PMID: 39196392

Ayala, A., Muñoz, M. F., & Argüelles, S. (2014). Lipid peroxidation: Production, metabolism, and signaling mechanisms of malondialdehyde and 4-hydroxy-2-nonenal. *Oxidative Medicine and Cellular Longevity*, *360438*, 1–31. Advance online publication. DOI: 10.1155/2014/360438 PMID: 24999379

Baek, S. H., Cho, Y., Lee, J., Choi, B. Y., Choi, Y., Park, J. S., Kim, H., Sul, J., Kim, E., Park, J. H., & Jo, D. G. (2018). Intracellular and Mitochondrial Reactive Oxygen Species Measurement in Primary Cultured Neurons. *Bio-Protocol*, *8*(11), e2871. DOI: 10.21769/BioProtoc.2871 PMID: 34285985

Bagheri, S., Haddadi, R., Saki, S., Kourosh-Arami, M., Rashno, M., Mojaver, A., & Komaki, A. (2023). Neuroprotective effects of coenzyme Q10 on neurological diseases: A review article. *Frontiers in Neuroscience*, *17*, 1188839. DOI: 10.3389/fnins.2023.1188839 PMID: 37424991

Balali-Mood, M., Naseri, K., Tahergorabi, Z., Khazdair, M. R., & Sadeghi, M. (2021). Toxic Mechanisms of Five Heavy Metals: Mercury, Lead, Chromium, Cadmium, and Arsenic. *Frontiers in Pharmacology*, *12*, 643972. DOI: 10.3389/fphar.2021.643972 PMID: 33927623

Ballini, A., Zhurakivska, K., Troiano, G., Lo Muzio, L., Caponio, V. C. A., Spirito, F., Porro, R., Rella, M., Cantore, S., Arrigoni, R., & Dioguardi, M. (2024). Dietary Polyphenols against Oxidative Stress in Head and Neck Cancer: What's New, What's Next. *Journal of Cancer*, *15*(2), 293–308. DOI: 10.7150/jca.90545 PMID: 38169656

Banaszak, M., Górna, I., Woźniak, D., Przysławski, J., & Drzymała-Czyż, S. (2024). The Impact of Curcumin, Resveratrol, and Cinnamon on Modulating Oxidative Stress and Antioxidant Activity in Type 2 Diabetes: Moving beyond an Anti-Hyperglycemic Evaluation. *Antioxidants*, *13*(5), 510. DOI: 10.3390/antiox13050510 PMID: 38790615

Bartz, R. R., Suliman, H. B., & Piantadosi, C. A. (2015). Redox mechanisms of cardiomyocyte mitochondrial protection. *Frontiers in Physiology*, *6*, 291. DOI: 10.3389/fphys.2015.00291 PMID: 26578967

Becerril-Sánchez, A. L., Quintero-Salazar, B., Dublán-García, O., & Escalona-Buendía, H. B. (2021). Phenolic Compounds in Honey and Their Relationship with Antioxidant Activity, Botanical Origin, and Color. *Antioxidants, 10*(11), 1700. DOI: 10.3390/antiox10111700 PMID: 34829570

Benbettaïeb, N., Tanner, C., Cayot, P., Karbowiak, T., & Debeaufort, F. (2018). Impact of functional properties and release kinetics on antioxidant activity of biopolymer active films and coatings. *Food Chemistry, 242*, 369–377. DOI: 10.1016/j.foodchem.2017.09.065 PMID: 29037702

Bhattacharyya, A., Chattopadhyay, R., Mitra, S., & Crowe, S. E. (2014). Oxidative stress: An essential factor in the pathogenesis of gastrointestinal mucosal diseases. *Physiological Reviews, 94*(2), 329–354. DOI: 10.1152/physrev.00040.2012 PMID: 24692350

Bjelakovic, G., Nikolova, D., Gluud, L. L., Simonetti, R. G., & Gluud, C. (2012). Antioxidant supplements for prevention of mortality in healthy participants and patients with various diseases. *Cochrane Database of Systematic Reviews, 2012*(3), CD007176. Advance online publication. DOI: 10.1002/14651858.CD007176.pub2 PMID: 22419320

Bjørklund, G., Gasmi, A., Lenchyk, L., Shanaida, M., Zafar, S., Mujawdiya, P. K., Lysiuk, R., Antonyak, H., Noor, S., Akram, M., Smetanina, K., Piscopo, S., Upyr, T., & Peana, M. (2022). The Role of Astaxanthin as a Nutraceutical in Health and Age-Related Conditions. *Molecules (Basel, Switzerland), 27*(21), 7167. DOI: 10.3390/molecules27217167 PMID: 36363994

Bode, K., Hauri-Hohl, M., Jaquet, V., & Weyd, H. (2023). Unlocking the power of NOX2: A comprehensive review on its role in immune regulation. *Redox Biology, 64*, 102795. DOI: 10.1016/j.redox.2023.102795 PMID: 37379662

Bondy, S. C. (2023). The Hormesis Concept: Strengths and Shortcomings. *Biomolecules, 13*(10), 1512. DOI: 10.3390/biom13101512 PMID: 37892194

Caminiti, R., Carresi, C., Mollace, R., Macrì, R., Scarano, F., Oppedisano, F., Maiuolo, J., Serra, M., Ruga, S., Nucera, S., Tavernese, A., Gliozzi, M., Musolino, V., Palma, E., Muscoli, C., Rubattu, S., Volterrani, M., Federici, M., Volpe, M., & Mollace, V. (2024). The potential effect of natural antioxidants on endothelial dysfunction associated with arterial hypertension. *Frontiers in Cardiovascular Medicine, 11*, 1345218. DOI: 10.3389/fcvm.2024.1345218 PMID: 38370153

Capece, U., Moffa, S., Improta, I., Di Giuseppe, G., Nista, E. C., Cefalo, C. M. A., Cinti, F., Pontecorvi, A., Gasbarrini, A., Giaccari, A., & Mezza, T. (2022). Alpha-Lipoic Acid and Glucose Metabolism: A Comprehensive Update on Biochemical and Therapeutic Features. *Nutrients, 15*(1), 18. DOI: 10.3390/nu15010018 PMID: 36615676

Carr, A. C., & Maggini, S. (2017). Vitamin C and Immune Function. *Nutrients, 9*(11), 1211. DOI: 10.3390/nu9111211 PMID: 29099763

Chang, W., Wang, M., Zhu, W., Dai, T., Han, Z., Sun, N., & Wang, D. (2024). Effect of allyl isothiocyanate on 4-HNE induced glucocorticoid resistance in COPD and the underlying mechanism. *Heliyon, 10*(17), e37275. DOI: 10.1016/j.heliyon.2024.e37275 PMID: 39296102

Chaudhary, P., Janmeda, P., Docea, A. O., Yeskaliyeva, B., Abdull Razis, A. F., Modu, B., Calina, D., & Sharifi-Rad, J. (2023). Oxidative stress, free radicals and antioxidants: Potential crosstalk in the pathophysiology of human diseases. *Frontiers in Chemistry, 11*, 1158198. DOI: 10.3389/fchem.2023.1158198 PMID: 37234200

Checa, J., & Aran, J. M. (2020). Reactive Oxygen Species: Drivers of Physiological and Pathological Processes. *Journal of Inflammation Research, 13*, 1057–1073. DOI: 10.2147/JIR.S275595 PMID: 33293849

Chen, X., Shi, C., He, M., Xiong, S., & Xia, X. (2023). Endoplasmic reticulum stress: Molecular mechanism and therapeutic targets. *Signal Transduction and Targeted Therapy, 8*(1), 352. DOI: 10.1038/s41392-023-01570-w PMID: 37709773

Chen, Y. F., Wu, S. N., Gao, J. M., Liao, Z. Y., Tseng, Y. T., Fülöp, F., Chang, F. R., & Lo, Y. C. (2020). The Antioxidant, Anti-Inflammatory, and Neuroprotective Properties of the Synthetic Chalcone Derivative AN07. *Molecules (Basel, Switzerland), 25*(12), 2907. DOI: 10.3390/molecules25122907 PMID: 32599797

Chidambaram, S. B., Anand, N., Varma, S. R., Ramamurthy, S., Vichitra, C., Sharma, A., Mahalakshmi, A. M., & Essa, M. M. (2024). Superoxide dismutase and neurological disorders. *IBRO Neuroscience Reports, 16*, 373–394. DOI: 10.1016/j.ibneur.2023.11.007 PMID: 39007083

Cuciniello, R., Di Meo, F., Filosa, S., Crispi, S., & Bergamo, P. (2023). The Antioxidant Effect of Dietary Bioactives Arises from the Interplay between the Physiology of the Host and the Gut Microbiota: Involvement of Short-Chain Fatty Acids. *Antioxidants, 12*(5), 1073. DOI: 10.3390/antiox12051073 PMID: 37237938

Cui, C., Zheng, J., Zhang, H., & Xing, Z. (2024). Pterostilbene ameliorates oxidative stress and neuronal apoptosis after intracerebral hemorrhage via the sirtuin 1-mediated Nrf2 pathway in vivo and in vitro. Journal of stroke and cerebrovascular diseases: the official journal of National Stroke Association, 33(11), 107950. Advance online publication. https://doi.org/DOI: 10.1016/j.jstrokecerebrovasdis.2024.107950

David, A., & Lev-Ari, S. (2024). Targeting the Gut Microbiome to Improve Immunotherapy Outcomes: A Review. *Integrative Cancer Therapies*, *23*, 15347354241269870. DOI: 10.1177/15347354241269870 PMID: 39223798

de Almeida, A. J. P. O., de Oliveira, J. C. P. L., da Silva Pontes, L. V., de Souza Júnior, J. F., Gonçalves, T. A. F., Dantas, S. H., de Almeida Feitosa, M. S., Silva, A. O., & de Medeiros, I. A. (2022). ROS: Basic Concepts, Sources, Cellular Signaling, and its Implications in Aging Pathways. *Oxidative Medicine and Cellular Longevity*, *1225578*, 1–23. Advance online publication. DOI: 10.1155/2022/1225578 PMID: 36312897

de Oliveira, A. R. S., Cruz, K. J. C., Morais, J. B. S., Dos Santos, L. R., de Sousa Melo, S. R., Fontenelle, L. C., Severo, J. S., Beserra, J. B., de Sousa, T. G. V., de Freitas, S. T., de Oliveira, E. H. S., Maia, C. S. C., de Matos Neto, E. M., de Oliveira, F. E., Henriques, G. S., & Marreiro, D. D. N. (2024). Magnesium, selenium and zinc deficiency compromises antioxidant defense in women with obesity. *Biometals: an international journal on the role of metal ions in biology, biochemistry, and medicine*, . Advance online publication. https://doi.org/DOI: 10.1007/s10534-024-00625-x

Didier, A. J., Stiene, J., Fang, L., Watkins, D., Dworkin, L. D., & Creeden, J. F. (2023). Antioxidant and Anti-Tumor Effects of Dietary Vitamins A, C, and E. *Antioxidants*, *12*(3), 632. DOI: 10.3390/antiox12030632 PMID: 36978880

Didier, A. J., Stiene, J., Fang, L., Watkins, D., Dworkin, L. D., & Creeden, J. F. (2023). Antioxidant and Anti-Tumor Effects of Dietary Vitamins A, C, and E. *Antioxidants*, *12*(3), 632. DOI: 10.3390/antiox12030632 PMID: 36978880

Dinkova-Kostova, A. T., Fahey, J. W., Kostov, R. V., & Kensler, T. W. (2017). KEAP1 and Done? Targeting the NRF2 Pathway with Sulforaphane. Trends in food science & technology, 69(Pt B), 257–269. https://doi.org/DOI: 10.1016/j.tifs.2017.02.002

Du, Y., He, C., An, Y., Huang, Y., Zhang, H., Fu, W., Wang, M., Shan, Z., Xie, J., Yang, Y., & Zhao, B. (2024). The Role of Short Chain Fatty Acids in Inflammation and Body Health. *International Journal of Molecular Sciences*, *25*(13), 7379. DOI: 10.3390/ijms25137379 PMID: 39000498

Egner, P. A., Chen, J. G., Wang, J. B., Wu, Y., Sun, Y., Lu, J. H., Zhu, J., Zhang, Y. H., Chen, Y. S., Friesen, M. D., Jacobson, L. P., Muñoz, A., Ng, D., Qian, G. S., Zhu, Y. R., Chen, T. Y., Botting, N. P., Zhang, Q., Fahey, J. W., & Kensler, T. W. (2011). Bioavailability of Sulforaphane from two broccoli sprout beverages: Results of a short-term, crossover clinical trial in Qidong, China. *Cancer Prevention Research (Philadelphia, Pa.)*, *4*(3), 384–395. DOI: 10.1158/1940-6207.CAPR-10-0296 PMID: 21372038

Fan, H., Wang, K., Zhao, X., Song, B., Yao, T., Liu, T., Gao, G., Lu, W., & Liu, C. (2024). Emerging insights into cuproptosis and copper metabolism: Implications for age-related diseases and potential therapeutic strategies. *Frontiers in Aging Neuroscience*, *16*, 1335122. DOI: 10.3389/fnagi.2024.1335122 PMID: 38715962

Fehér, J., Élő, Á., István, L., Nagy, Z. Z., Radák, Z., Scuderi, G., Artico, M., & Kovács, I. (2022). Microbiota mitochondria disorders as hubs for early age-related macular degeneration. *GeroScience*, *44*(6), 2623–2653. DOI: 10.1007/s11357-022-00620-5 PMID: 35978068

Fontes, A., Jauch, A. T., Sailer, J., Engler, J., Azul, A. M., & Zischka, H. (2024). Metabolic Derangement of Essential Transition Metals and Potential Antioxidant Therapies. *International Journal of Molecular Sciences*, *25*(14), 7880. DOI: 10.3390/ijms25147880 PMID: 39063122

Fusco, W., Lorenzo, M. B., Cintoni, M., Porcari, S., Rinninella, E., Kaitsas, F., Lener, E., Mele, M. C., Gasbarrini, A., Collado, M. C., Cammarota, G., & Ianiro, G. (2023). Short-Chain Fatty-Acid-Producing Bacteria: Key Components of the Human Gut Microbiota. *Nutrients*, *15*(9), 2211. DOI: 10.3390/nu15092211 PMID: 37432351

Gao, W., Liu, Y. F., Zhang, Y. X., Wang, Y., Jin, Y. Q., Yuan, H., Liang, X. Y., Ji, X. Y., Jiang, Q. Y., & Wu, D. D. (2024). The potential role of hydrogen sulfide in cancer cell apoptosis. *Cell Death Discovery*, *10*(1), 114. DOI: 10.1038/s41420-024-01868-w PMID: 38448410

Garg, S. S., Gupta, J., Sahu, D., & Liu, C. J. (2022). Pharmacological and Therapeutic Applications of Esculetin. *International Journal of Molecular Sciences*, *23*(20), 12643. DOI: 10.3390/ijms232012643 PMID: 36293500

Gęgotek, A., & Skrzydlewska, E. (2023). Ascorbic acid as antioxidant. *Vitamins and Hormones*, *121*, 247–270. DOI: 10.1016/bs.vh.2022.10.008 PMID: 36707136

González-Peña, M. A., Ortega-Regules, A. E., Anaya de Parrodi, C., & Lozada-Ramírez, J. D. (2023). Chemistry, Occurrence, Properties, Applications, and Encapsulation of Carotenoids-A Review. *Plants*, *12*(2), 313. DOI: 10.3390/plants12020313 PMID: 36679026

Gupta, S. C., Patchva, S., & Aggarwal, B. B. (2013). Therapeutic roles of curcumin: Lessons learned from clinical trials. *The AAPS Journal*, *15*(1), 195–218. DOI: 10.1208/s12248-012-9432-8 PMID: 23143785

Gurunathan, S., Thangaraj, P., & Kim, J. H. (2023). Postbiotics: Functional Food Materials and Therapeutic Agents for Cancer, Diabetes, and Inflammatory Diseases. *Foods*, *13*(1), 89. DOI: 10.3390/foods13010089 PMID: 38201117

Gutierrez-Mariscal, F. M., Arenas-de Larriva, A. P., Limia-Perez, L., Romero-Cabrera, J. L., Yubero-Serrano, E. M., & López-Miranda, J. (2020). Coenzyme Q10 Supplementation for the Reduction of Oxidative Stress: Clinical Implications in the Treatment of Chronic Diseases. *International Journal of Molecular Sciences*, *21*(21), 7870. DOI: 10.3390/ijms21217870 PMID: 33114148

Halliwell, B., Adhikary, A., Dingfelder, M., & Dizdaroglu, M. (2021). Hydroxyl radical is a significant player in oxidative DNA damage in vivo. *Chemical Society Reviews*, *50*(15), 8355–8360. DOI: 10.1039/D1CS00044F PMID: 34128512

Heianza, Y., Sun, D., Li, X., DiDonato, J. A., Bray, G. A., Sacks, F. M., & Qi, L. (2019). Gut microbiota metabolites, amino acid metabolites and improvements in insulin sensitivity and glucose metabolism: The POUNDS Lost trial. *Gut*, *68*(2), 263–270. DOI: 10.1136/gutjnl-2018-316155 PMID: 29860242

Houghton, C. A. (2019). Sulforaphane: Its "Coming of Age" as a Clinically Relevant Nutraceutical in the Prevention and Treatment of Chronic Disease. *Oxidative Medicine and Cellular Longevity*, *2716870*, 1–27. Advance online publication. DOI: 10.1155/2019/2716870 PMID: 31737167

Houldsworth, A. (2024). Role of oxidative stress in neurodegenerative disorders: A review of reactive oxygen species and prevention by antioxidants. *Brain Communications*, *6*(1), fcad356. Advance online publication. DOI: 10.1093/braincomms/fcad356 PMID: 38214013

Hsieh, R. Y., Huang, I. C., Chen, C., & Sung, J. Y. (2023). Effects of Oral Alpha-Lipoic Acid Treatment on Diabetic Polyneuropathy: A Meta-Analysis and Systematic Review. *Nutrients*, *15*(16), 3634. DOI: 10.3390/nu15163634 PMID: 37630823

Huang, M., Wu, Q., & Jiang, Z. H. (2022). Epigenetic Alterations under Oxidative Stress in Stem Cells. *Oxidative Medicine and Cellular Longevity*, *6439097*, 1–11. Advance online publication. DOI: 10.1155/2022/6439097 PMID: 36071870

Huang, R., Chen, H., Liang, J., Li, Y., Yang, J., Luo, C., Tang, Y., Ding, Y., Liu, X., Yuan, Q., Yu, H., Ye, Y., Xu, W., & Xie, X. (2021). Dual Role of Reactive Oxygen Species and their Application in Cancer Therapy. *Journal of Cancer*, *12*(18), 5543–5561. DOI: 10.7150/jca.54699 PMID: 34405016

Ilie, C. I., Spoiala, A., Chircov, C., Dolete, G., Oprea, O. C., Vasile, B. S., Crainiceanu, S. A., Nicoara, A. I., Marinas, I. C., Stan, M. S., Ditu, L. M., Ficai, A., & Oprea, E. (2024). Antioxidant, Antitumoral, Antimicrobial, and Prebiotic Activity of Magnetite Nanoparticles Loaded with Bee Pollen/Bee Bread Extracts and 5-Fluorouracil. *Antioxidants*, *13*(8), 895. DOI: 10.3390/antiox13080895 PMID: 39199141

Islam, M. R., Rauf, A., Akash, S., Trisha, S. I., Nasim, A. H., Akter, M., Dhar, P. S., Ogaly, H. A., Hemeg, H. A., Wilairatana, P., & Thiruvengadam, M. (2024). Targeted therapies of curcumin focus on its therapeutic benefits in cancers and human health: Molecular signaling pathway-based approaches and future perspectives. Biomedicine & pharmacotherapy = Biomedecine & pharmacotherapie, 170, 116034. https://doi .org/DOI: 10.1016/j.biopha.2023.116034

Jin, S., & Kang, P. M. (2024). A Systematic Review on Advances in Management of Oxidative Stress-Associated Cardiovascular Diseases. *Antioxidants*, *13*(8), 923. DOI: 10.3390/antiox13080923 PMID: 39199169

Jomova, K., Raptova, R., Alomar, S. Y., Alwasel, S. H., Nepovimova, E., Kuca, K., & Valko, M. (2023). Reactive oxygen species, toxicity, oxidative stress, and antioxidants: Chronic diseases and aging. *Archives of Toxicology*, *97*(10), 2499–2574. DOI: 10.1007/s00204-023-03562-9 PMID: 37597078

Katakwar, P., Metgud, R., Naik, S., & Mittal, R. (2016). Oxidative stress marker in oral cancer: A review. *Journal of Cancer Research and Therapeutics*, *12*(2), 438–446. DOI: 10.4103/0973-1482.151935 PMID: 27461591

Kawamura, T., & Muraoka, I. (2018). Exercise-Induced Oxidative Stress and the Effects of Antioxidant Intake from a Physiological Viewpoint. *Antioxidants*, *7*(9), 119. DOI: 10.3390/antiox7090119 PMID: 30189660

Kiani, A. K., Bonetti, G., Medori, M. C., Caruso, P., Manganotti, P., Fioretti, F., Nodari, S., Connelly, S. T., & Bertelli, M. (2022). Dietary supplements for improving nitric-oxide synthesis. *Journal of Preventive Medicine and Hygiene*, *63*(2, Suppl 3), E239–E245. DOI: 10.15167/2421-4248/jpmh2022.63.2S3.2766 PMID: 36479475

Koozehchian, M. S., Daneshfar, A., Fallah, E., Agha-Alinejad, H., Samadi, M., Kaviani, M., Kaveh, B. M., Jung, Y. P., Sablouei, M. H., Moradi, N., Earnest, C. P., Chandler, T. J., & Kreider, R. B. (2018). Effects of nine weeks L-Carnitine supplementation on exercise performance, anaerobic power, and exercise-induced oxidative stress in resistance-trained males. *Journal of Exercise Nutrition & Biochemistry*, *22*(4), 7–19. DOI: 10.20463/jenb.2018.0026 PMID: 30661327

Kozlov, A. V., Javadov, S., & Sommer, N. (2024). Cellular ROS and Antioxidants: Physiological and Pathological Role. *Antioxidants*, *13*(5), 602. DOI: 10.3390/antiox13050602 PMID: 38790707

Krawczyk, M., Burzynska-Pedziwiatr, I., Wozniak, L. A., & Bukowiecka-Matusiak, M. (2023). Impact of Polyphenols on Inflammatory and Oxidative Stress Factors in Diabetes Mellitus: Nutritional Antioxidants and Their Application in Improving Antidiabetic Therapy. *Biomolecules*, *13*(9), 1402. DOI: 10.3390/biom13091402 PMID: 37759802

Kumari, S., Badana, A. K., G, M. M., G, S., & Malla, R. R. (2018). Reactive Oxygen Species: A Key Constituent in Cancer Survival. *Biomarker Insights*, *13*, 1177271918755391. DOI: 10.1177/1177271918755391 PMID: 29449774

Lee, J., Kim, M. H., & Kim, H. (2022). Anti-Oxidant and Anti-Inflammatory Effects of Astaxanthin on Gastrointestinal Diseases. *International Journal of Molecular Sciences*, *23*(24), 15471. DOI: 10.3390/ijms232415471 PMID: 36555112

Li, D., Li, Y., Yang, S., Lu, J., Jin, X., & Wu, M. (2022). Diet-gut microbiota-epigenetics in metabolic diseases: From mechanisms to therapeutics. Biomedicine & pharmacotherapy = Biomedecine & pharmacotherapie, 153, 113290. https://doi .org/DOI: 10.1016/j.biopha.2022.113290

Li, M., Yun, W., Wang, G., Li, A., Gao, J., & He, Q. (2022). Roles and mechanisms of garlic and its extracts on atherosclerosis: A review. *Frontiers in Pharmacology*, *13*, 954938. DOI: 10.3389/fphar.2022.954938 PMID: 36263122

Lin, Y., Liu, H., Bu, L., Chen, C., & Ye, X. (2022). Review of the Effects and Mechanism of Curcumin in the Treatment of Inflammatory Bowel Disease. *Frontiers in Pharmacology*, *13*, 908077. DOI: 10.3389/fphar.2022.908077 PMID: 35795556

Lingappan, K. (2018). NF-κB in Oxidative Stress. *Current Opinion in Toxicology*, *7*, 81–86. DOI: 10.1016/j.cotox.2017.11.002 PMID: 29862377

Liu, S., Cheng, L., Liu, Y., Zhan, S., Wu, Z., & Zhang, X. (2023). Relationship between Dietary Polyphenols and Gut Microbiota: New Clues to Improve Cognitive Disorders, Mood Disorders and Circadian Rhythms. *Foods*, *12*(6), 1309. DOI: 10.3390/foods12061309 PMID: 36981235

Liu, S., He, Y., Zhang, Y., Zhang, Z., Huang, K., Deng, L., Liao, B., Zhong, Y., & Feng, J. (2023). Targeting gut microbiota in aging-related cardiovascular dysfunction: Focus on the mechanisms. *Gut Microbes*, *15*(2), 2290331. DOI: 10.1080/19490976.2023.2290331 PMID: 38073096

Liu, Y., Zhang, W., Liu, Z., Zheng, A., Liang, B., Li, H., & Meng, Q. (2024). Serum copper assessment in patients with polycystic ovary syndrome and tubal infertility: A retrospective 5-year study. *Food Science & Nutrition*, *12*(8), 5979–5989. DOI: 10.1002/fsn3.4258 PMID: 39139929

Lloret, A., Esteve, D., Monllor, P., Cervera-Ferri, A., & Lloret, A. (2019). The Effectiveness of Vitamin E Treatment in Alzheimer's Disease. *International Journal of Molecular Sciences*, *20*(4), 879. DOI: 10.3390/ijms20040879 PMID: 30781638

Lloret, A., Esteve, D., Monllor, P., Cervera-Ferri, A., & Lloret, A. (2019). The Effectiveness of Vitamin E Treatment in Alzheimer's Disease. *International Journal of Molecular Sciences*, *20*(4), 879. DOI: 10.3390/ijms20040879 PMID: 30781638

Lobo, V., Patil, A., Phatak, A., & Chandra, N. (2010). Free radicals, antioxidants and functional foods: Impact on human health. *Pharmacognosy Reviews*, *4*(8), 118–126. DOI: 10.4103/0973-7847.70902 PMID: 22228951

Lushchak, V. I. (2012). Glutathione homeostasis and functions: Potential targets for medical interventions. *Journal of Amino Acids*, *736837*, 1–26. Advance online publication. DOI: 10.1155/2012/736837 PMID: 22500213

Mabrouk, A., Sharaf, O., Ibrahim, G., & Mehaya, F. (2024). Comparative studies of the antimicrobial, antioxidant activities, vitamins, organic acid and short-chain fatty acids of postbiotics derived from five probiotic strains. *Egyptian Journal of Chemistry*, *0*(0), 0. Advance online publication. DOI: 10.21608/ejchem.2024.301125.9935

Maldonado, E., Morales-Pison, S., Urbina, F., & Solari, A. (2023). Aging Hallmarks and the Role of Oxidative Stress. *Antioxidants*, *12*(3), 651. DOI: 10.3390/antiox12030651 PMID: 36978899

Man, A. W. C., Li, H., & Xia, N. (2020). Impact of Lifestyles (Diet and Exercise) on Vascular Health: Oxidative Stress and Endothelial Function. *Oxidative Medicine and Cellular Longevity*, *1496462*, 1–22. Advance online publication. DOI: 10.1155/2020/1496462 PMID: 33062134

Mantle, D., Lopez-Lluch, G., & Hargreaves, I. P. (2023). Coenzyme Q10 Metabolism: A Review of Unresolved Issues. *International Journal of Molecular Sciences*, *24*(3), 2585. DOI: 10.3390/ijms24032585 PMID: 36768907

Mao, K., Chen, S., Chen, M., Ma, Y., Wang, Y., Huang, B., He, Z., Zeng, Y., Hu, Y., Sun, S., Li, J., Wu, X., Wang, X., Strober, W., Chen, C., Meng, G., & Sun, B. (2013). Nitric oxide suppresses NLRP3 inflammasome activation and protects against LPS-induced septic shock. *Cell Research*, *23*(2), 201–212. DOI: 10.1038/cr.2013.6 PMID: 23318584

Martínez, A., Velázquez, L., Díaz, R., Huaiquipán, R., Pérez, I., Muñoz, A., Valdés, M., Sepúlveda, N., Paz, E., & Quiñones, J. (2024). Impact of Novel Foods on the Human Gut Microbiome: Current Status. *Microorganisms*, *12*(9), 1750. DOI: 10.3390/microorganisms12091750 PMID: 39338424

Matuszczak, M., Kiljańczyk, A., Marciniak, W., Derkacz, R., Stempa, K., Baszuk, P., Bryśkiewicz, M., Sun, P., Cheriyan, A., Cybulski, C., Dębniak, T., Gronwald, J., Huzarski, T., Lener, M. R., Jakubowska, A., Szwiec, M., Stawicka-Niełacna, M., Godlewski, D., Prusaczyk, A., & Lubiński, J. (2024). Zinc and Its Antioxidant Properties: The Potential Use of Blood Zinc Levels as a Marker of Cancer Risk in BRCA1 Mutation Carriers. *Antioxidants*, *13*(5), 609. DOI: 10.3390/antiox13050609 PMID: 38790714

Merry, T. L., & Ristow, M. (2016). Do antioxidant supplements interfere with skeletal muscle adaptation to exercise training? *The Journal of Physiology*, *594*(18), 5135–5147. DOI: 10.1113/JP270654 PMID: 26638792

Miles, E. A., & Calder, P. C. (2021). Effects of Citrus Fruit Juices and Their Bioactive Components on Inflammation and Immunity: A Narrative Review. *Frontiers in Immunology*, *12*, 712608. DOI: 10.3389/fimmu.2021.712608 PMID: 34249019

Morén, C., deSouza, R. M., Giraldo, D. M., & Uff, C. (2022). Antioxidant Therapeutic Strategies in Neurodegenerative Diseases. *International Journal of Molecular Sciences*, *23*(16), 9328. DOI: 10.3390/ijms23169328 PMID: 36012599

Morgan, M. J., & Liu, Z. G. (2011). Crosstalk of reactive oxygen species and NF-κB signaling. *Cell Research*, *21*(1), 103–115. DOI: 10.1038/cr.2010.178 PMID: 21187859

Mu, B., Zeng, Y., Luo, L., & Wang, K. (2024). Oxidative stress-mediated protein sulfenylation in human diseases: Past, present, and future. *Redox Biology*, *76*, 103332. Advance online publication. DOI: 10.1016/j.redox.2024.103332 PMID: 39217848

Mukherjee, A., Ghosh, K. K., Chakrabortty, S., Gulyás, B., Padmanabhan, P., & Ball, W. B. (2024). Mitochondrial Reactive Oxygen Species in Infection and Immunity. *Biomolecules, 14*(6), 670. DOI: 10.3390/biom14060670 PMID: 38927073

Murotomi, K., Umeno, A., Shichiri, M., Tanito, M., & Yoshida, Y. (2023). Significance of Singlet Oxygen Molecule in Pathologies. *International Journal of Molecular Sciences, 24*(3), 2739. DOI: 10.3390/ijms24032739 PMID: 36769060

Niu, C., Yan, M., Yao, Z., & Dou, J. (2023). Antibiotic residues in milk and dairy products in China: Occurrence and human health concerns. *Environmental Science and Pollution Research International, 30*(53), 113138–113150. DOI: 10.1007/s11356-023-30312-2 PMID: 37851252

Nozu, T., Miyagishi, S., Nozu, R., Takakusaki, K., & Okumura, T. (2019). Butyrate inhibits visceral allodynia and colonic hyperpermeability in rat models of irritable bowel syndrome. *Scientific Reports, 9*(1), 19603. DOI: 10.1038/s41598-019-56132-4 PMID: 31862976

Ottria, R., Xynomilakis, O., Casati, S., & Ciuffreda, P. (2024). Pre- to Postbiotics: The Beneficial Roles of Pediatric Dysbiosis Associated with Inflammatory Bowel Diseases. *Microorganisms, 12*(8), 1582. DOI: 10.3390/microorganisms12081582 PMID: 39203424

Park, J., & Son, H. (2024). Antioxidant Systems of Plant Pathogenic Fungi: Functions in Oxidative Stress Response and Their Regulatory Mechanisms. *The Plant Pathology Journal, 40*(3), 235–250. DOI: 10.5423/PPJ.RW.01.2024.0001 PMID: 38835295

Perez-Araluce, M., Jüngst, T., Sanmartin, C., Prosper, F., Plano, D., & Mazo, M. M. (2024). Biomaterials-Based Antioxidant Strategies for the Treatment of Oxidative Stress Diseases. *Biomimetics (Basel, Switzerland), 9*(1), 23. DOI: 10.3390/biomimetics9010023 PMID: 38248597

Pizzino, G., Irrera, N., Cucinotta, M., Pallio, G., Mannino, F., Arcoraci, V., Squadrito, F., Altavilla, D., & Bitto, A. (2017). Oxidative Stress: Harms and Benefits for Human Health. *Oxidative Medicine and Cellular Longevity, 8416763*(1), 8416763. Advance online publication. DOI: 10.1155/2017/8416763 PMID: 28819546

Pratelli, G., Tamburini, B., Carlisi, D., De Blasio, A., D'Anneo, A., Emanuele, S., Notaro, A., Affranchi, F., Giuliano, M., Seidita, A., Lauricella, M., & Di Liberto, D. (2023). Foodomics-Based Approaches Shed Light on the Potential Protective Effects of Polyphenols in Inflammatory Bowel Disease. *International Journal of Molecular Sciences, 24*(19), 14619. DOI: 10.3390/ijms241914619 PMID: 37834065

Pruccoli, L., Morroni, F., Sita, G., Hrelia, P., & Tarozzi, A. (2020). Esculetin as a Bifunctional Antioxidant Prevents and Counteracts the Oxidative Stress and Neuronal Death Induced by Amyloid Protein in SH-SY5Y Cells. *Antioxidants*, *9*(6), 551. DOI: 10.3390/antiox9060551 PMID: 32630394

Qi, Y., & Wang, X. (2023). The Role of Gut Microbiota in High-Fat-Diet-Induced Diabetes: Lessons from Animal Models and Humans. *Nutrients*, *15*(4), 922. DOI: 10.3390/nu15040922 PMID: 36839280

Rahaman, M. M., Hossain, R., Herrera-Bravo, J., Islam, M. T., Atolani, O., Adeyemi, O. S., Owolodun, O. A., Kambizi, L., Daştan, S. D., Calina, D., & Sharifi-Rad, J. (2023). Natural antioxidants from some fruits, seeds, foods, natural products, and associated health benefits: An update. *Food Science & Nutrition*, *11*(4), 1657–1670. DOI: 10.1002/fsn3.3217 PMID: 37051367

Ramos, C. S. C. B., Silva, V. A. P. D., Corrêa, L. B. N. S., Abboud, R. S., Boaventura, G. T., & Chagas, M. A. (2023). Long-Term Alpha-Lipoic Acid (ALA) Antioxidant Therapy Reduces Damage in the Cardiovascular System of Streptozotocin-Induced Diabetic Rats. *Journal of Dietary Supplements*, *20*(3), 444–458. DOI: 10.1080/19390211.2021.2004290 PMID: 34806528

Recharla, N., Geesala, R., & Shi, X. Z. (2023). Gut Microbial Metabolite Butyrate and Its Therapeutic Role in Inflammatory Bowel Disease: A Literature Review. *Nutrients*, *15*(10), 2275. DOI: 10.3390/nu15102275 PMID: 37242159

Rizvi, S., Raza, S. T., Ahmed, F., Ahmad, A., Abbas, S., & Mahdi, F. (2014). The role of vitamin e in human health and some diseases. *Sultan Qaboos University Medical Journal*, *14*(2), e157–e165. PMID: 24790736

Rossman, M. J., Gioscia-Ryan, R. A., Santos-Parker, J. R., Ziemba, B. P., Lubieniecki, K. L., Johnson, L. C., Poliektov, N. E., Bispham, N. Z., Woodward, K. A., Nagy, E. E., Bryan, N. S., Reisz, J. A., D'Alessandro, A., Chonchol, M., Sindler, A. L., & Seals, D. R. (2021). Inorganic Nitrite Supplementation Improves Endothelial Function With Aging: Translational Evidence for Suppression of Mitochondria-Derived Oxidative Stress. Hypertension (Dallas, Tex.: 1979), 77(4), 1212–1222. https://doi.org/DOI: 10.1161/HYPERTENSIONAHA.120.16175

Roy, N., & Paira, P. (2024). Glutathione Depletion and Stalwart Anticancer Activity of Metallotherapeutics Inducing Programmed Cell Death: Opening a New Window for Cancer Therapy. *ACS Omega*, *9*(19), 20670–20701. DOI: 10.1021/acsomega.3c08890 PMID: 38764686

Sagar, A. N., Kalburgi, V., Vagha, J. D., Taksande, A., Meshram, R. J., & Lohiya, S. (2024). A Comprehensive Review of the Role of Magnesium in Critical Care Pediatrics: Mechanisms, Clinical Impact, and Therapeutic Strategies. *Cureus, 16*(8), e66643. DOI: 10.7759/cureus.66643 PMID: 39258079

Samimi, F., Namiranian, N., Sharifi-Rigi, A., Siri, M., Abazari, O., & Dastghaib, S. (2024). Coenzyme Q10: A Key Antioxidant in the Management of Diabetes-Induced Cardiovascular Complications-An Overview of Mechanisms and Clinical Evidence. *International Journal of Endocrinology, 2247748*, 1–13. Advance online publication. DOI: 10.1155/2024/2247748 PMID: 38524871

Schell, K. R., Fernandes, K. E., Shanahan, E., Wilson, I., Blair, S. E., Carter, D. A., & Cokcetin, N. N. (2022). The Potential of Honey as a Prebiotic Food to Re-engineer the Gut Microbiome Toward a Healthy State. *Frontiers in Nutrition, 9*, 957932. DOI: 10.3389/fnut.2022.957932 PMID: 35967810

Shandilya, S., Kumar, S., Kumar Jha, N., Kumar Kesari, K., & Ruokolainen, J. (2021). Interplay of gut microbiota and oxidative stress: Perspective on neurodegeneration and neuroprotection. *Journal of Advanced Research, 38*, 223–244. DOI: 10.1016/j.jare.2021.09.005 PMID: 35572407

Shrestha, J., Limbu, K. R., Chhetri, R. B., Paudel, K. R., Hansbro, P. M., Oh, Y. S., Baek, D. J., Ki, S. H., & Park, E. Y. (2024). Antioxidant genes in cancer and metabolic diseases: Focusing on Nrf2, Sestrin, and heme oxygenase 1. *International Journal of Biological Sciences, 20*(12), 4888–4907. DOI: 10.7150/ijbs.98846 PMID: 39309448

Silva, J. R., Souza, A. O., Anjos, I. C., Pádua, G. M. S., Vieira, L. C. C., & Gai, B. M. (2023). Evaluation of chalcones derivatives in lipid peroxidation reduction induced by Fe2+/EDTA in vitro. Research. *Social Development, 12*(1), e16012139541. DOI: 10.33448/rsd-v12i1.39541

Sotler, R., Poljšak, B., Dahmane, R., Jukić, T., Pavan Jukić, D., Rotim, C., Trebše, P., & Starc, A. (2019). prooxidant activities of antioxidants and their impact on health. *Acta Clinica Croatica, 58*(4), 726–736. DOI: 10.20471/acc.2019.58.04.20 PMID: 32595258

Srinivas, U. S., Tan, B. W. Q., Vellayappan, B. A., & Jeyasekharan, A. D. (2019). ROS and the DNA damage response in cancer. *Redox Biology, 25*, 101084. DOI: 10.1016/j.redox.2018.101084 PMID: 30612957

Stanaway, L., Rutherfurd-Markwick, K., Page, R., Wong, M., Jirangrat, W., Teh, K. H., & Ali, A. (2019). Acute Supplementation with Nitrate-Rich Beetroot Juice Causes a Greater Increase in Plasma Nitrite and Reduction in Blood Pressure of Older Compared to Younger Adults. *Nutrients*, *11*(7), 1683. DOI: 10.3390/nu11071683 PMID: 31336633

Su, L. J., Zhang, J. H., Gomez, H., Murugan, R., Hong, X., Xu, D., Jiang, F., & Peng, Z. Y. (2019). Reactive Oxygen Species-Induced Lipid Peroxidation in Apoptosis, Autophagy, and Ferroptosis. *Oxidative Medicine and Cellular Longevity*, *5080843*, 1–13. Advance online publication. DOI: 10.1155/2019/5080843 PMID: 31737171

Sun, W., Taniguchi, T., Clyma, K. R., Jacob, T. S., & Jernigan, V. B. B. (2024). Impact of Nutrition Interventions on Vitamin and Mineral Intake among Native American Children and Parents: Insights from Food Resource Equity for Sustainable Health (FRESH) Study. *International Journal of Environmental Research and Public Health*, *21*(8), 1014. DOI: 10.3390/ijerph21081014 PMID: 39200625

Sweazea, K. L., Johnston, C. S., Miller, B., & Gumpricht, E. (2018). Nitrate-Rich Fruit and Vegetable Supplement Reduces Blood Pressure in Normotensive Healthy Young Males without Significantly Altering Flow-Mediated Vasodilation: A Randomized, Double-Blinded, Controlled Trial. *Journal of Nutrition and Metabolism*, *1729653*, 1–10. Advance online publication. DOI: 10.1155/2018/1729653 PMID: 30305961

Theodorou, A. A., Zinelis, P. T., Malliou, V. J., Chatzinikolaou, P. N., Margaritelis, N. V., Mandalidis, D., Geladas, N. D., & Paschalis, V. (2021). Acute L-Citrulline Supplementation Increases Nitric Oxide Bioavailability but Not Inspiratory Muscle Oxygenation and Respiratory Performance. *Nutrients*, *13*(10), 3311. DOI: 10.3390/nu13103311 PMID: 34684312

Toppo, S., Vanin, S., Bosello, V., & Tosatto, S. C. (2008). Evolutionary and structural insights into the multifaceted glutathione peroxidase (Gpx) superfamily. *Antioxidants & Redox Signaling*, *10*(9), 1501–1514. DOI: 10.1089/ars.2008.2057 PMID: 18498225

Traber, M. G., & Stevens, J. F. (2011). Vitamins C and E: Beneficial effects from a mechanistic perspective. *Free Radical Biology & Medicine*, *51*(5), 1000–1013. DOI: 10.1016/j.freeradbiomed.2011.05.017 PMID: 21664268

Tripathi, R., Gupta, R., Sahu, M., Srivastava, D., Das, A., Ambasta, R. K., & Kumar, P. (2022). Free radical biology in neurological manifestations: Mechanisms to therapeutics interventions. *Environmental Science and Pollution Research International*, *29*(41), 62160–62207. DOI: 10.1007/s11356-021-16693-2 PMID: 34617231

Tristan Asensi, M., Napoletano, A., Sofi, F., & Dinu, M. (2023). Low-Grade Inflammation and Ultra-Processed Foods Consumption: A Review. *Nutrients, 15*(6), 1546. DOI: 10.3390/nu15061546 PMID: 36986276

Turrens, J. F. (2003). Mitochondrial formation of reactive oxygen species. *The Journal of Physiology, 552*(Pt 2), 335–344. DOI: 10.1113/jphysiol.2003.049478 PMID: 14561818

Ungureanu, L. B., Ghiciuc, C. M., Amalinei, C., Ungureanu, C., Petrovici, C. G., & Stănescu, R. . (2024). Antioxidants as Protection against Reactive Oxygen Stress Induced by Formaldehyde (FA) Exposure: A Systematic Review. *Biomedicines, 12*(8), 1820. DOI: 10.3390/biomedicines12081820 PMID: 39200284

Utomo, N. P., Pinzon, R. T., Latumahina, P. K., & Damayanti, K. R. S. (2024). Astaxanthin and improvement of dementia: A systematic review of current clinical trials. *Cerebral Circulation - Cognition and Behavior, 7*, 100226. DOI: 10.1016/j.cccb.2024.100226 PMID: 39036318

Vargas-Uricoechea, H., Urrego-Noguera, K., Vargas-Sierra, H., & Pinzón-Fernández, M. (2024). Zinc and Ferritin Levels and Their Associations with Functional Disorders and/or Thyroid Autoimmunity: A Population-Based Case–Control Study. *International Journal of Molecular Sciences, 25*(18), 10217. DOI: 10.3390/ijms251810217 PMID: 39337701

Vitamin D - Health Professional Fact Sheet https://ods.od.nih.gov/factsheets/VitaminD-HealthProfessional/

Waldman, H. (2024). Astaxanthin Supplementation as a Potential Strategy for Enhancing Mitochondrial Adaptations in the Endurance Athlete: An Invited Review. *Nutrients, 16*(11), 1750. DOI: 10.3390/nu16111750 PMID: 38892683

Watanabe, K., Tanaka, M., Yuki, S., Hirai, M., & Yamamoto, Y. (2018). How is edaravone effective against acute ischemic stroke and amyotrophic lateral sclerosis? *Journal of Clinical Biochemistry and Nutrition, 62*(1), 20–38. DOI: 10.3164/jcbn.17-62 PMID: 29371752

Weichmann, F., & Rohdewald, P. (2024). Pycnogenol® French maritime pine bark extract in randomized, double-blind, placebo-controlled human clinical studies. *Frontiers in Nutrition, 11*, 1389374. DOI: 10.3389/fnut.2024.1389374 PMID: 38757130

Xie, W., Koppula, S., Kale, M. B., Ali, L. S., Wankhede, N. L., Umare, M. D., Upaganlawar, A. B., Abdeen, A., Ebrahim, E. E., El-Sherbiny, M., Behl, T., Shen, B., & Singla, R. K. (2024). Unraveling the nexus of age, epilepsy, and mitochondria: Exploring the dynamics of cellular energy and excitability. *Frontiers in Pharmacology, 15*, 1469053. DOI: 10.3389/fphar.2024.1469053 PMID: 39309002

Yamashita, T., & Abe, K. (2024). Update on Antioxidant Therapy with Edaravone: Expanding Applications in Neurodegenerative Diseases. *International Journal of Molecular Sciences, 25*(5), 2945. DOI: 10.3390/ijms25052945 PMID: 38474192

Yan, S., Lu, J., Chen, B., Yuan, L., Chen, L., Ju, L., Cai, W., & Wu, J. (2024). The Multifaceted Role of Alpha-Lipoic Acid in Cancer Prevention, Occurrence, and Treatment. *Antioxidants, 13*(8), 897. DOI: 10.3390/antiox13080897 PMID: 39199143

Yang, J., Luo, J., Tian, X., Zhao, Y., Li, Y., & Wu, X. (2024). Progress in Understanding Oxidative Stress, Aging, and Aging-Related Diseases. *Antioxidants, 13*(4), 394. DOI: 10.3390/antiox13040394 PMID: 38671842

Yoshikawa, T., & You, F. (2024). Oxidative Stress and Bio-Regulation. *International Journal of Molecular Sciences, 25*(6), 3360. DOI: 10.3390/ijms25063360 PMID: 38542335

Zhang, J., Wang, X., Vikash, V., Ye, Q., Wu, D., Liu, Y., & Dong, W. (2016). ROS and ROS-Mediated Cellular Signaling. *Oxidative Medicine and Cellular Longevity, 4350965*(1), 4350965. Advance online publication. DOI: 10.1155/2016/4350965 PMID: 26998193

Zhou, X., Cao, Q., Orfila, C., Zhao, J., & Zhang, L. (2021). Systematic Review and Meta-Analysis on the Effects of Astaxanthin on Human Skin Aging. *Nutrients, 13*(9), 2917. DOI: 10.3390/nu13092917 PMID: 34578794

Zozina, V. I., Covantev, S., Goroshko, O. A., Krasnykh, L. M., & Kukes, V. G. (2018). Coenzyme Q10 in Cardiovascular and Metabolic Diseases: Current State of the Problem. *Current Cardiology Reviews, 14*(3), 164–174. DOI: 10.2174/1573403X14666180416115428 PMID: 29663894

Chapter 4
Environmental Exposure and Oxidative Stress:
Effects on Human Health

Ansari Rashid
https://orcid.org/0009-0000-0089-4153
Department of Pharmacology, Sanjay Education Society's Institute of Pharmacy, Dhule, India

Pawan Gupta
https://orcid.org/0000-0001-7201-1356
Department of Pharmaceutical Chemistry, Shri Vile Parle Kelavani Mandal's Institute of Pharmacy, Dhule, India

Yogesh Chaudhari
Department of Pharmaceutical Chemistry, Shri Vile Parle Kelavani Mandal's Institute of Pharmacy, Dhule, India

Thangaraj Devadoss
https://orcid.org/0000-0002-2228-9357
Department of Pharmaceutical Chemistry, Karpagam Academy of Higher Education, Coimbatore, India

Rahul Kumar Garg
https://orcid.org/0000-0003-3299-0988
Asian College of Pharmacy, Udaipur, India

Rewati Wanjari
https://orcid.org/0009-0008-6574-3047
Department of Pharmaceutical Chemistry, Shri Vile Parle Kelavani Mandal's Institute of Pharmacy, Dhule, India

ABSTRACT

Exposure to environmental pollutants/toxins has progressively been identified as a major cause of different health concerns. Air pollution, occupational and radiation exposures, and agricultural chemicals are significant environmental factors contributing to oxidative stress (OS) and adverse health outcomes. Air pollution (particulate matter/toxic gases) is worsening respiratory and cardiovascular diseases. The occupational exposures to harmful substances can create chronic oxidative conditions which affects overall health. Radiation exposures (medical procedures /

DOI: 10.4018/979-8-3693-7919-6.ch004

environmental sources) and agricultural chemicals cause the cellular damage, increasing the risk of cancer, affecting endocrine and neurological functions. Overall impact of these exposures is required rigorous monitoring and regulation to mitigate health risks. Addressing these environmental challenges is essential in protecting health and promoting public awareness to reduce harmful exposures and improve overall well-being.

1. INTRODUCTION:

Oxidative stress, a phenomenon of considerable importance within the realm of cellular biology, manifests within human cells primarily because of an imbalance that occurs between the production of reactive species and the capacity of antioxidant defenses to neutralize these potentially harmful entities, thereby contributing significantly to the pathogenesis of a myriad of diseases that afflict the population. The primary sources of this oxidative stress can be traced back to a variety of environmental factors, lifestyle choices, and intrinsic metabolic processes, which collectively culminate in chronic oxidative stress and, as a result, lead to extensive cellular damage that may compromise overall health and well-being. The exposure to a plethora of harmful chemicals, the consumption of a nutritionally poor diet, coupled with a sedentary lifestyle, can significantly elevate the levels of reactive species in the body, thus leading to an exacerbation of oxidative stress, as per the reports published (J Biochem Mol Toxicol, 2023). Within the complex framework of normal metabolic processes, free radicals, notably including reactive oxygen species (ROS), are generated. However, it is critical to recognize that these molecules can become deleterious to cellular integrity when they are produced in excessive quantities (Elsevier eBooks, 2023). Increased oxidative stress can affects the normal physiology of body by affecting the essential biomolecules such as DNA, proteins, and lipids, which in turn disrupts various critical cellular functions and intricate signaling pathways within the organism, potentially leading to the initiation and progression of a multitude of diseases including, but not limited to, cancer, diabetes, and an array of neurodegenerative disorders, as evidenced by the findings (Curr Mol Med, 2023) and supported by the research titled "Free Radicals, Oxidative Stress-Related Diseases and Antioxidant Supplementation" published in 2022. The observation of increased concentrations of specific markers, such as malondialdehyde (MDA), serves as a significant indicator of oxidative damage and has been shown to correlate with the presence and severity of various pathological conditions (J Cell Physiol, 2023). While it is well-established that oxidative stress constitutes a major contributing factor in the development and exacerbation of numerous diseases. It is also important to acknowledge that certain studies have proposed that low levels of

ROS may, in fact, serve beneficial roles in the processes of cellular signaling and the maintenance of homeostasis, thereby illustrating the intricate and multifaceted relationship that exists between oxidative stress and overall health (Arch Toxicol, 2023). In this chapter, various environmental factors affecting on progression of oxidative stress (OS) and its effects on human health is discussed. The OS in human beings predominantly originates from a variety of environmental pollutants which include, but are not limited to, heavy metals, fine particulate matter (PM), and an array of toxic chemicals that are prevalent in our surroundings. The presence of these harmful pollutants in the environment plays a crucial role in the generation of free radicals, which in turn instigates significant cellular damage and is intricately linked to the development of chronic diseases that can severely compromise health and well-being. To combat this pressing issue, a number of mitigation strategies have been proposed that primarily focus on enhancing the body's antioxidant defenses, as well as reducing overall exposure to these deleterious substances that contribute to OS.

Environmental Sources of Oxidative Stress

Air pollution, particularly in the form of PM2.5 that emanates from industrial emissions and the exhaust produced by vehicles, has been shown to significantly contribute to the exacerbation of OS as well as increased inflammation in the human body. This has been associated with a range of diseases including cardiovascular and pulmonary disorders (SciSpace, 2023) (Int J Sci Res Arch, 2024). Furthermore, chemical pollutants [toxic heavy metals such as lead and mercury; organic pollutants like pesticides] persist in the environment for extended periods and have been documented to further intensify oxidative stress, thereby posing a continuous threat to human health (Front Physiol, 2022). The various mitigation strategies are adopted nowadays to counteract OS, one of the most promising approaches involve the strategic intake of antioxidants, where nutritional interventions utilize phytochemicals and various antioxidants. These effectively neutralize free radicals, significantly reducing the extent of oxidative damage that is inflicted upon the cells in the body (J Pharm Pharmacol, 2021). Additionally, the implementation of comprehensive policy and regulatory frameworks aimed at enforcing stricter environmental regulations could play a pivotal role in the reduction of pollutant levels in the environment. This can contribute to a corresponding decrease in OS levels within various populations (Front Physiol, 2022). These strategies possessed substantial potential in addressing the issue of OS. Still significant challenges remain in achieving a comprehensive understanding of the intricate and complex interactions that exist between the myriad pollutants and their cumulative effects on human health, which warrants further research and investigation in this critical area.

Mechanisms of oxidative stress induction by environmental factors

The process of inducing OS as a reaction to environmental toxins is characterized by intricate and multifaceted molecular mechanisms. These exert a profound influence on a wide array of cellular processes, leading ultimately to detrimental outcomes within the affected cells. Among the primary factors that contribute to this phenomenon are mitochondrial dysfunction, the generation of ROS, and various epigenetic modifications, all of which together culminate in significant cellular damage and dysfunction, can lead to compromise the integrity of cellular operations. Mitochondrial dysfunction is coupled with the subsequent generation of ROS. The exposure to environmental pollutants has been demonstrated to inflict considerable damage to mitochondrial structures and functions. This turns to upsurge of ROS production. The elevation of ROS is recognized as a fundamental catalyst of OS (Toxicol Res, 2023). For example, it has illustrated that exposure to the environmental contaminant 2,3,7,8-tetrachlorodibenzo-p-dioxin (TCDD) can lead to a marked increase in ROS levels. This phenomenon has been empirically correlated with the incidence of DNA damage *via* OS-related pathways (SciSpace, 2024). In addition to these mechanisms, environmental factors can instigate epigenetic alterations that modify gene expression patterns associated with the cellular responses to OS. Notably, nutritional influences have been identified as a substantial factor which can effectively modulate these epigenetic responses, thereby impacting the overall capacity of cells to cope with OS (Antioxidants, 2023). Turning our attention to the plant kingdom, it becomes evident that a variety of environmental stressors can lead to the accumulation of ROS. These subsequently triggers intricate signaling pathways which culminate in OS. In severe cases, it may even lead to cell apoptosis and death throughout all over the body. This observation underscores the dualistic nature of ROS, which can exert both deleterious and advantageous effects depending on their concentration and context (Springer, 2021). Furthermore, in the realm of nematodes, OS emerges as a critical initial response to exposure to environmental toxicants, which serves to activate various defense mechanisms aimed at promoting cellular damage associated with such stressors (Springer, 2019).

1.1. Epidemiological evidence linking air pollution to oxidative damage

Epidemiological investigations provide robust evidence that establishes a significant correlation between the adverse effects of air pollution and the induction of oxidative damage within human biological systems, particularly through the intricate processes involved in the generation of ROS. These are known to be pivotal contributors to a myriad of health complications that afflict various populations worldwide. This noteworthy relationship is consistently observable across a diverse array of demographic groups and a multitude of health conditions, highlighting the universal impact of air pollution on human health. The phenomenon of air pollution, and more specifically, the presence of particulate matter, has been shown to provoke OS through the excessive production of ROS. This leads to a significant threat to the integrity of vital cellular components such as DNA, proteins, and lipids as per research report (SciSpace, 2024). Upon inhalation, these pollutants incite inflammatory responses and activate various molecular pathways that ultimately culminate in programmed cell death. Thus, exacerbating the prevalence of chronic respiratory diseases, as supported by findings (Int J Mol Sci, 2023). A plethora of studies have indicated that individuals who are already affected with pre-existing airway diseases tend to experience a markedly heightened level of OS as a direct response to exposure to pollutants such as nitrogen dioxide (NO_2) and fine particulate matter PM2.5 (Sci Total Environ, 2023). As per report, children represent a particularly vulnerable demographic, primarily due to their developing respiratory systems and inherently higher metabolic rates, which collectively contribute to a significantly increased susceptibility to oxidative damage resulting from exposure to air pollution (Int J Mol Sci, 2023). Implications for public health, the results presented in the research clearly indicated how important it is to implement appropriate measures of air quality management together with intensive public health programs fighting against the unfavorable impacts of the escalated level of OS as well as other health consequences that stem from this phenomenon as intensely as possible (Sci Total Environ, 2023). Viewing collectively, some scholars argue that the complexity of individual responses to air pollution, coupled with the multifactorial nature of oxidative stress. Necessitates a sustained and concerted effort in research to fully elucidate these complex interactions and their far-reaching implications for public health.

Impact on oxidative stress markers

The influence on the biomarkers which are associated with OS is profoundly significant, as the exposure to a variety of pollutants has been demonstrated to result in an elevation of OS (**as discussed above**), which in turn has been correlated with a myriad of health complications that affect human populations. This intricate relationship is further emphasized by an array of scientific studies that delve into the underlying mechanisms and specific biomarkers involved in this complex interplay between environmental exposures and biological responses. The inhalation and absorption of environmental pollutants, particularly fine particulate matter such as PM2.5 and toxic heavy metals like mercury, have been empirically established to significantly raise the levels of OS biomarkers, including but not limited to MDA and glutathione S-transferase (GST), particularly in vulnerable populations such as children (J Environ Prot, 2018). Additionally, various biomarkers, encompassing ROS and indicators of oxidative DNA damage, serve as crucial measures of OS that are provoked by exposure to environmental toxins and hazardous substances (Trends Anal Chem, 2023). As exposure to an air pollutant leads to a marked increase in the, a phenomenon that can further exacerbate various health conditions, including type 2 diabetes, through intricate mechanisms that involve inflammatory processes and dysfunction of the endothelial lining (Antioxidants, 2021). A thorough and comprehensive examination of urinary biomarkers such as urea and uric acid has unveiled that exposures to environmental chemicals exhibit a significant correlation with both oxidative and nitrative stress, thereby indicating a potential pathway that could lead to the manifestation of various diseases such as diabetes, neurodegenerative diseases, etc. (Anal Bioanal Chem, 2022). While the preponderance of evidence robustly supports the existence of a connection between environmental factors and the biomarkers of oxidative stress, it is imperative to recognize that individual susceptibility, as well as genetic predispositions, may also significantly influence the degree to which these environmental exposures impact health outcomes (Redox Biol, 2020).

2. OCCUPATIONAL EXPOSURES:

In this section the various industrial wastes are discussed which affect the OS environment in the body.

2.1. Industrial chemicals and solvents

The toxicity of certain environmental contaminants has been linked to the generation of ROS (J Neurochem, 2010). Four industrially significant chlorinated organic solvents were examined for their cytotoxic properties in vitro studies: dichloromethane (DCM), 1,2-dichloroethane (DCE), trichloroethylene (TCE), and tetrachloroethylene (PERC). After 72-hour exposure of Junket cells to each given solvent, the parameters were measured like intracellular free calcium concentration ([Ca^{2+}]), caspase-3 activity, cell proliferation, and the production of ROS. The results described that concentration-dependent rise in intracellular free [Ca^{2+}] and the production of ROS after being exposed to each of the solvents. However, there was a corresponding drop in cell proliferation. The potency of solvent was PERC > TCE > DCM > DCE. TCE and PERC boosted caspase-3 activity in a concentration-dependent manner, but DCM and DCE had no discernible effect. Changes in intracellular free [Ca^{2+}] and caspase-3 activity were demonstrated to be independent of ROS production with n-acetyl-L cysteine pretreatment. Nonetheless, a causal relationship existed between elevated ROS production and the observed reduction in cell growth (Int J Toxicol, 2013). Along with, impact of different solvents like C6–10 aromatic, aliphatic, and alicyclic on the generation of ROS in rat cerebellar granule cell cultures was studied in 2002. By tracking the oxidation of 2',7'-dichlorofluorescin (DCFH) to the fluorescent chemical 2',7'-dichlorofluorescein (DCF), ROS production was measured. In *in-vitro* studies, rat cerebellar granule cells generate ROS when exposed to aromatic solvents with $C > 7$ and aliphatic and alicyclic solvents with $C \geqslant 7$. As the concentration of the solvent increased, then response was increased. The octane-water partition coefficients of the chemicals in each homologous group appeared to be related to their potency. In general, it was found that aliphatic and alicyclic chemicals were more effective in causing ROS production than aromatic solvents (Int J Mol Sci, 2021).

2.2. Radiation exposure (ionizing and non-ionizing)

Superoxide ($O_2^{\bullet -}$) is one of the ROS that produces ionizing radiation, which damages the biological system. Different antioxidant systems have been created by the human body to protect against high ROS levels (27). DNA double-strand breaks are the most severe kind of DNA damage caused by ionizing radiation (IR) exposure. These breaks can result in genomic instability, cancer, and cell death. Cells have therefore evolved defenses against DNA damage and mechanisms for their repair. Also, endogenous ROS builds up in the irradiated cells as a result of IR. ROS-mediated, IR-induced biological effects are comparatively more substantial than DNA damage-mediated biological effects after exposure to low-dose irradiation. There is growing evidence that the alteration of mitochondria in irradiated cells is associated with an increase in endogenous ROS. When mammalian cells are exposed to infrared radiation, their mitochondria produce more endogenous ROS, which may cause mitochondrial malfunction. Damaged mitochondria are eliminated by mitophagy because their remnants may produce or leak additional ROS within the cell. This route is involved in maintaining mitochondrial integrity, and its disturbance may result in ageing and a number of diseases, including neurodegeneration (J Radiat Res (Tokyo), 2018). A pilot study was performed on 60 medical workers out of which there were forty age- and gender-matched unexposed controls and twenty occupationally exposed to IR (all radiologic technologists), split into three subgroups: conventional radiography (CR), computerized tomography (CT), and interventional radiography (IR). As a measure of redox state, blood levels of $O_2^{\bullet -}$ and MDA were examined, together with the activity of the antioxidant enzymes catalase and superoxide dismutase (SOD). Measurements of reduced and oxidized glutathione (GSH and GSSG, respectively) were also used to determine the redox status. It observed that the blood of IR and CT-exposed patients had considerably higher levels of $O_2^{\bullet -}$ and MDA, as well as SOD activity, compared to both CR-exposed people and control persons. The levels of GSH, catalase, and the ratio of GSH/GSSG (glutathione and glutathione disulfide) did not show any significant difference. However, there is a statistically significant difference observed between exposed workers and control subjects (Redox Rep, 2016).

2.3. Agricultural chemicals

Pesticide exposure appears to be linked to much higher ROS and OS production than what the intrinsic antioxidant system of the cell can lower to normal physiological levels (29). Pesticides' high lipid solubility and low molecular weight enhance their absorption and level of toxicity. Pesticides containing organophosphate (OP) and carbamate work by inhibiting carboxyl ester hydrolases, specifically acetylcholinesterase, which causes a buildup of acetylcholine (30). Furthermore, research has indicated that suppression of the acetylcholinesterase enzyme may be linked to an increase in ROS in agricultural workers who are exposed to OP pesticides and bipyridyl herbicides, such as paraquat (PQ). Both the chemicals (OP and PQ) raises lipid peroxidation and fall in antioxidant capacity, which causes OS (31,32). Many different biological reactions are implicated in the toxicological effects of pesticides, which are brought on by OS such as gene expression variations, selective signal transduction activation, and/or inhibition take place in a variety of cell signaling pathways. A suitable cell line has been used to study the function of particular pesticide-mediated OS in the induction of various cell signaling pathways, both in vitro and in vivo studies (React Oxyg Species, 2017).

A multitude of cell signaling pathways linked to apoptosis are often stimulated by an increased amount of OS, which is caused by the formation of ROS. The Keap1/Nrf2/ARE pathway is a key defensive mechanism of cells against OS.

Basic leucine zipper transcription factor Nrf2 separates from Keap1 and moves into the nucleus when the amount of ROS in the cytoplasm rises above a particular threshold. It then dimerises small Maf-binding proteins and attaches itself to antioxidant response elements (AREs) to trigger the production of genes that are dependent on AREs. The following phase involves the activation of antioxidative enzymes, which shield the cells from an excess of ROS generation. It should be noted that apoptosis, cell division, cell signaling, and other oxidative stress-mediated activities are all significantly influenced by Ca^{2+} ions (Biochim Biophys Acta BBA-Biomembr, 2015) (Mol Pharmacol, 2007). Periplasmic apoptotic pathway is another mechanism by which permethrin (well-known insecticide) may seriously harm macromolecules within the cell and even induce cell death. Out of all the mitochondrial integrity regulators in this pathway, Bcl-2 and Bax are the most active members. They also affect the activation of caspase and the release of cytochrome c. Bcl-2's primary role is to inhibit apoptosis through its antioxidative properties. Bax is moved from the cytosol to the mitochondria following mitochondrial injury, and there is also a notable drop in Bcl-2 expression. A crucial apoptotic event—the release of mitochondrial cytochrome c into the cytoplasm—occurs when ROS levels rise above a certain threshold (J Appl Toxicol, 2016). As per report, the rats exposed to permethrin and its derivatives released a substantial amount of mitochondrial cytochrome c, confirming the presence of the mitochondrial apoptotic pathway in pesticide-induced OS (Toxicol Lett, 2011).

3. EFFECTS AND MECHANISMS OF OXIDATIVE STRESS ON HUMAN HEALTH

The OS affects various pathophysiological conditions of diseases such as neurodegenerative diseases, cancer, metabolic diseases etc. some important effects and their detailed mechanisms are discussed below.

3.1. Cellular damage and biomolecular oxidation:

Biomolecular oxidation represents a fundamental biochemical phenomenon that has profound implications for cellular integrity and functionality, as it primarily results in cellular damage through the excessive generation of ROS and reactive nitrogen species (RNS), both of which are highly reactive molecules. These reactive species engage in intricate interactions with a myriad of cellular components, leading to considerable and often detrimental modifications in cellular operations and overall physiological processes.

Mechanisms of Cellular Damage Oxidative Stress: The presence of elevated levels of ROS can profoundly alter the structural and functional properties of lipids, proteins, and nucleic acids. This is ultimately resulting in irreversible cellular damage characterized by disruptions in membrane integrity and a consequential loss of enzymatic activity essential for normal cellular function (SciSpace, 2023).

Protein Dysfunction: The oxidative modifications that occur as a result of elevated OS can lead to significant issues in protein homeostasis, including misfolding, aggregation, and an overall impairment of the proteolytic degradation pathways, which collectively contribute to the onset and progression of age-related diseases that afflict the aging population (Biochimie, 2022).

DNA Damage: The interaction of ROS with DNA can provoke a range of mutations, which may culminate in severe health consequences such as systemic autoimmunity and various forms of cancer. Recent investigations are reported that, a phenomenon that is illustrated by the formation of 8-oxo-dG, which serves as a critical biomarker for oxidative DNA damage that has been widely studied in the literature (Int J Health Sci, 2019) (Exp Oncol, 2019) (Typeset, 2016).

Impact on Cellular Function Cellular Integrity: The damage inflicted on cellular membranes and proteins as a result of OS disrupts cellular homeostasis, which is vital for maintaining the intricate balance required for growth, development, and overall cellular functionality (SciSpace, 2023). The oxidative damage that occurs within cells can act as a catalyst for the initiation of inflammatory responses, which can further exacerbate a myriad of pathological conditions, including but not limited to cancer and autoimmune diseases, thereby complicating the clinical landscape and therapeutic approach (Exp Oncol, 2019) (Typeset, 2016).

3.2. Implications for chronic diseases

Chronic illnesses, which include a range of complex conditions such as cardiovascular diseases, various forms of cancer, and neurodegenerative disorders, are characterized by shared underlying biological mechanisms that significantly contribute to their onset and progression over time. Among the most critical elements that play a pivotal role in the exacerbation of these diseases are inflammation, OS and a multitude of environmental factors, all of which intricately interact with individual genetic susceptibilities to further complicate disease trajectories and outcomes.

Although intermittent increases in inflammation are critical for survival during physical injury and infection, recent research has revealed that certain social, environmental and lifestyle factors can promote systemic chronic inflammation (SCI) that can, in turn, lead to several diseases that collectively represent the leading causes of disability and mortality worldwide, such as cardiovascular disease, cancer, diabetes mellitus, chronic kidney disease, non-alcoholic fatty liver disease and autoimmune and neurodegenerative disorders. In the present perspective we describe the multi-level mechanisms underlying SCI and several risk factors that promote this health-damaging phenotype, including infections, physical inactivity, poor diet, environmental and industrial toxicants, and psychological stress. Furthermore, we suggest potential strategies for advancing the early diagnosis, prevention, and treatment of SCI. "Noncommunicable diseases (NCDs), also called chronic diseases, are a group of noninfectious health conditions that tend to be of long duration or present long latency due to the combination of genetic, physiological, behaviors, and environmental factors. NCDs are responsible for the equivalent of 71% of all deaths worldwide and constitute the most significant public health problem. The NCDs include a range of conditions like obesity, cardiovascular diseases, diabetes, hypertension, cancers, and neurodegenerative diseases (as dementias). Some molecular pathways are involved in NCDs pathophysiology; however, two mechanisms are recognized as common in the disease's initiation and progression: inflammation and oxidative stress. These processes turn chronic during NCDs progression and contribute to the evolution and worse outcomes. This chapter aims to describe the main molecular targets involved in inflammation and oxidative stress as fundamental contributors in the NCDs pathology. Inflammation and oxidative stress represent two major pathological processes that are intrinsically linked to the development of numerous chronic diseases, as chronic inflammation has been recognized as a fundamental contributor to systemic dysfunction, thereby heightening the risk of serious conditions such as cardiovascular disease and various types of cancer (Nat Med, 2019). Furthermore, OS which arises from a detrimental imbalance between the production of harmful free radicals and the body's antioxidant defenses, has been shown to have a significant role in the pathophysiological mechanisms underlying neurodegenerative disorders as well as cardiovascular diseases, as elucidated by the research findings of (Typeset, 2022).

Table 1. Sources of contaminants, moa of ROS production and associated diseases

Contaminant/Chemical	Source	Mechanism of ROS Production	Associated Diseases	References
Heavy Metals	Industrial waste, batteries	Disruption of cellular antioxidant systems	Cancer, cardiovascular diseases, neurodegenerative diseases	(Curr Med Chem, 2005)
Polycyclic Aromatic Hydrocarbons (PAHs)	Vehicle emissions, coal combustion	Oxidative metabolism	Lung cancer, skin diseases, reproductive issues	(Environ Pollut Barking Essex 1987, 2017)
Pesticides	Agriculture	Interference with mitochondrial function	Neurotoxicity, endocrine disruption, cancer	(Arch Toxicol, 2017)
Phthalates	Plastics, personal care products	Endocrine disruption	Reproductive disorders, obesity, diabetes	(J Hazard Mater, 2017)
Bisphenol A (BPA)	Plastics, resins	Mimics hormones, promotes oxidative stress	Hormonal imbalances, cancer, metabolic disorders	(Metabolomics Off J Metabolomic Soc, 2022)
Volatile Organic Compounds (VOCs)	Paints, solvents	Increased oxidative stress in tissues	Respiratory diseases, liver damage, cancer	(Biosensors, 2022)

continued on following page

Table 1. Continued

Contaminant/Chemical	Source	Mechanism of ROS Production	Associated Diseases	References
Ozone	Air pollution	Direct oxidative effects on cells	Asthma, chronic obstructive pulmonary disease (COPD)	(Front Immunol, 2021)
Asbestos	Construction materials	Induces oxidative stress and inflammation	Lung cancer, mesothelioma, asbestosis	(Crit Rev Toxicol, 2023)
Silica	Construction, mining	Inflammation leading to ROS production	Silicosis, lung cancer	(J Control Release Off J Control Release Soc, 2021)
Arsenic	Contaminated water, pesticides	Interferes with cellular respiration	Skin lesions, cancer, cardiovascular diseases	(J Appl Toxicol JAT, 2011)

4. PROTECTIVE STRATEGIES AGAINST ENVIRONMENTAL OXIDATIVE STRESS

4.1. Antioxidant defenses in the human body:

A substance that may prevent or decrease macromolecule oxidation is known as an antioxidant (AOX) (Free Radic Res, 2005). The sequential consumption of electrons by molecular oxygen from different substrates causes the generation of ROS, which is a fundamental component of aerobic life. In a healthy organism, ROS is typically balanced by AOX (Trends Pharmacol Sci, 2011). Humans have an immune system that keeps them healthy and defends against illness, and ROS actively work to eliminate pathogens that infect humans. However, there are negative consequences to continuing the immune system's activation after the threat has been eliminated (Biochim Biophys Acta BBA - Gen Subj, 1970). Numerous research additionally claims the efficacy of RS (Reactive Species) in redox signalling within the body. It is also known that ROS aid with situational adjustment carried on by high levels of physical activity and ischemia. The human AOX defence system's goal is to get clear of excessive RS without making it difficult for them to continue playing a healthy role in life. Numerous research additionally claims the efficacy of ROS in redox signaling within the body. It is also known that ROS aid with situational adjustment carried on by high levels of physical activity and ischemia. The human AOX defense system's goal is to get clear of excessive RS without making it difficult for them to continue playing a healthy role in life (J Pharm Biomed Anal, 2022). By eliminating free radicals or inhibiting further oxidation events by being oxidized themselves, AOXs play the role of decreasing or preventing these chain reactions. Therefore, reducing agents like polyphenols or thiols are frequently used as antioxidants (Free Radic Res, 2005). Plants and animals contain a diversity of AOXs, comprising vitamins C and E and glutathione, in addition to different enzymatic systems that catalyze antioxidant reactions, such as catalase, SOD, and peroxidases. Although oxidation reactions are essential for cells, they can also be harmful. The OS carried on by impairments in or inhibition of these AOX enzymes might cause cell damage and lysis (J Pharm Biomed Anal, 2022) (Int J Biochem Cell Biol, 2007).

The following are the methods that AOX defense follows:

1) Preventing the generation of free radicals
2) Oxidants grabbing
3) The process of changing harmful free radicals into less harmful compounds
4) Inhibiting the synthesis of inflammatory mediators and secondary toxic metabolites
5) Blocking of the chain propagation of the secondary oxidants
6) Repairing the injured molecules
7) Activating and strengthening the body's built-in antioxidant defenses.

The body is protected against oxidative stress by the cooperative action of all these defense mechanisms. The human body's AOX systems are made up of potent enzymatic and non-enzymatic antioxidants (Biochem Soc Trans, 2007). The three primary groups of AOX enzymes originate in all body cells are glutathione peroxidases (GPX), SOD, and catalases. These enzymes are crucial for preserving the homeostasis of cells. These enzymes' stimulation specifies a particular reaction to OS triggered by pollutants (Aquat Toxicol, 2001). Additionally, non-enzymatic antioxidants are also being the protective strategies against environmental oxidative stress, through the SH group, protein-bound and non-protein-bound thiols function as cellular reducing agents and protect against the majority of harmful substances (J Nutr, 1991). Cellular antioxidant glutathione is essential for protecting the redox state of cells (Mol Cell Biochem, 2007). Although ascorbic acid is an antioxidant present in both plants and animals, humans cannot produce it and must instead get it through their food. It is able to reduce and eliminate ROSs (FEBS J, 2007). By interacting with lipid radicals and eliminating free radical intermediates, vitamin E has been shown to prevent cell membranes from oxidation (Life Sci, 2006). Because *beta*-carotene eliminates singlet oxygen, it possesses strong AOX qualities that guard against the damaging effects of free radicals. Liver, egg yolk, milk, butter, spinach, carrots, tomatoes, and cereals all contain them (FEBS J, 2007).

4.2. Dietary antioxidants and their role:

Reactive free radicals produced by a reduction-oxidation (redox) imbalance in the cell cause cellular oxidative stress (OS), which may destroy tissues and cells. Metals like iron and copper are involved in numerous chemical processes that contribute significantly to the production of harmful free radicals (Indian J Clin Biochem, 2015). By scavenging free radicals, endogenous antioxidants delay, slow, or prevent oxidation, hence assisting in the maintenance of an oxidative state (67,68). comprises the non-enzymatic antioxidants glutathione, thioredoxin, and peroxiredoxins, vitamins A, B, C, E and small molecules (Molecules, 2012), as well as the enzymes SOD, catalase, and glutathione peroxidase (Trends Pharmacol Sci, 2011) (Molecules, 2012)

Supplemental antioxidants may be able to combat problems associated with OS by scavenging free radicals. Commercially available antioxidant medications comprise multivitamin/multimineral supplements and pharmaceutical products including ascorbic acid, lipoic acid, sodium metabisulfite, propyl gallate, butylated hydroxytoluene, butylated hydroxyanisole, and cysteine (Celestino et al., 2012) (Kontoghiorghes & Kontoghiorghe, 2019). Additionally, reports indicate that clearing ROS from the cell prematurely could be harmful to certain vital intracellular signaling and metabolic processes, making the cell additional vulnerable to infections (Paiva & Bozza, 2014). Therefore, a well-designed diet that includes the right amount of foods high in antioxidants and physical activity can help prevent many diseases, especially diabetes, obesity, and cardiovascular problems (Dal & Sigrist, 2016). Hydrogen-atom transfer, electron donation, and metal chelation are the most often used antioxidant processes in vitro (Leopoldini et al., 2011).

Figure 1. Types of dietary antioxidants

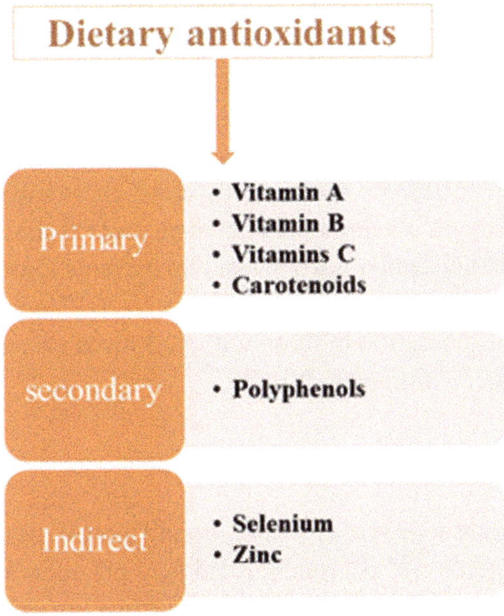

There are many different types of substances found in food and drink that lower OS and alter cellular defenses. Polyphenols, flavonoids, bioflavonoids, and some aromatic bio-phenols are among them. A large class of naturally occurring and artificially produced compounds with one or more aromatic phenolic rings is known as polyphenols. Fruits, nuts, berries, green leafy vegetables, tea, coffee, olive oil, red wine, and honey are all rich sources of natural polyphenols (Fatima et al., 2018) (Shahidi & Ambigaipalan, 2015). Herbs such as thyme, oregano, rosemary, sage, mint, and basil, as well as meals high in vitamins C, E, and β-carotene, are additional sources.

Role of specific dietary components:

I. Tea and its ingredients

Strong antioxidants found in green tea include theanine and catechins, such as (−)-epigallocatechin-3-gallate (EGCG) (77–79) which can provide defense against diabetes and its consequences (Meng et al., 2019) as well as neurological conditions (Kakuda, 2002).

II. Coffee and its ingredients

Caffeine, cafestol, trigonelline, quinine, melanoidins, chlorogenic acid (CGA), and other phenols including hydroxycinnamic acid are the sources of coffee's antioxidant benefits (Hu et al., 2019). Trigonelline is partially broken down by roasting the coffee seeds into pyridines and nicotinic acid, both of which have antioxidant properties (Zhou et al., 2012). In both healthy individuals and diabetics, the effects of caffeine individually on glycemic management and insulin sensitivity are either negative or unclear. Numerous studies reveal that caffeine lowers insulin sensitivity and raises blood glucose levels (Dewar & Heuberger, 2017). Additionally, it reduces the chance of neurodegeneration in those with Parkinson's disease and Alzheimer's disease (Kolahdouzan & Hamadeh, 2017).

III. Honey

Honey is an abundant source of antioxidants and has a wide range of active ingredients, such as flavonoids and polyphenols (mostly phenolic acids). Some of them are exclusive to certain sources (hesperetin and naringenin), while others are found in the majority of honey components (galangin, kaempferol, quercetin, isorhamnetin, and luteolin). Benzoic acid, cinnamic acid, chlorogenic acid, caffeic acid, coumaric acid, chrysin, sulfuric acid, gallic acid, ferulic acid, myricetin, and syringic acid are further antioxidants. Significant health benefits of honey include enhanced organ function, including the liver, kidney, heart, brain, pancreas, and testicles.

IV. Vitamins

Table 2. Overview of each vitamin's roles and impacts on health

Vitamin	Key Functions	Deficiencies	Effects on Health	References
Vitamin A	Antioxidant, pancreatic development, lipid metabolism, neuronal differentiation	Hyperglycemia, loss of pancreatic β-cell mass	Improves glucose metabolism in T2D, reduces Aβ aggregation in AD, supports neural function, MS therapy	**(Marcelino et al., 2020) (Trasino et al., 2015) (Leh & Lee, 2022) (Pullakhandam et al., 2012) (Valdes-Ramos et al., 2015) (Hira et al., 2019) (Müller, 2008)**
Vitamin B1 (Thiamine)	Antioxidant defense, ATP synthesis, neurotransmitter and lipid synthesis	Oxidative metabolism failure, neural and cardiovascular issues	Improves blood glucose levels, reduces diabetic nephropathy, regulates oxidative stress in brain function	**(Thiamine Deficiency Induced Dietary Disparity, 2017) (Karachalias et al., 2005) (Hrubša et al., 2022)**
Vitamin B2 (Riboflavin)	Antioxidant, mitochondrial function, glutathione recovery	Mitochondrial dysfunction, neurological conditions	Improves glucose metabolism, neuroprotection, reduces OS and neuroinflammation	**(Hrubša et al., 2022) (Udhayabanu et al., 2017) (Alam et al., 2015) (Perez Visñuk et al., 2022) (Marashly & Bohlega, 2017)**
Vitamin B3 (Niacin)	Reduces dyslipidemia, improves lipid profiles, NAD synthesis	Dyslipidemia, poor lipid profiles	Improves lipid levels, cognitive and motor functions, reduces OS in AD, PD, HD, MS, schizophrenia	**(Hrubša et al., 2022) (Udhayabanu et al., 2017) (Alam et al., 2015) (Perez Visñuk et al., 2022) (Marashly & Bohlega, 2017) (Xiang et al., 2020) (Brown et al., 2001) (Carlson & Rosenhamer, 1988) (Wakade et al., 2014) (Fricker et al., 2018) (Yan & Wang, 2021)**
Vitamin B6 (Pyridoxine)	Glycogen metabolism, neurotransmitter synthesis, enzyme inhibition	Poor glycemic control, increased OS	Reduces blood glucose levels, improves diabetic retinopathy, reduces risk of T2D, PD, AD	**(Valdes-Ramos et al., 2015) (De Lau et al., 2006) (Smolek et al., 2013) (Kim et al., 2018) (Unoki-Kubota et al., 2010) (Polizzi et al., 2012)**
Vitamin B9 (Folate)	DNA synthesis, homocysteine regulation	Hyper-homocysteinemia, poor glycemic control	Improves glycemic control, reduces OS and DNA damage, improves cognitive functions	**(Valdes-Ramos et al., 2015) (Al-Maskari et al., 2012) (Moore et al., 2013) (Aghamohammadi et al., 2011)**
Vitamin B12 (Cobalamin)	DNA synthesis, tau protein inhibition, homocysteine regulation	Hyper-homocysteinemia, tauopathies	Reduces hyperhomocysteinemia, improves cognitive functions, impedes tau protein fibrillization	**(Valdes-Ramos et al., 2015) (Rai et al., 2021) (Rafiee et al., 2017)**

continued on following page

Table 2. Continued

Vitamin	Key Functions	Deficiencies	Effects on Health	References
Vitamin C	Antioxidant, endothelial function, enhances insulin synthesis	Reduced cellular uptake, poor glycemic control	Improves glycemic control, neuroprotection, reduces oxidative stress in neurodegenerative diseases	**(Doseděl et al., 2021) (May & Harrison, 2013) (Morelli et al., 2020) (Verlangieri & Sestito, 1981) (Moretti et al., 2017) (Mason et al., 2021) (Shi et al., 2020)**
Vitamin E	Antioxidant, improves β-cell function and insulin secretion	Poor antioxidant status, impaired glycemic control	Reduces oxidative stress, delays T2D progression, reduces amyloid deposition in AD	**(Jain, 2012) (Browne et al., 2019) (Razip et al., 2021) (Balbi et al., 2018) (Said et al., 2021) (Boothby & Doering, 2005) (Karthika et al., 2022)**

V. Common oils and herbs:

Strong antioxidant-rich polyphenols found in herbs like thyme, rosemary, mint, parsley, basil, and oregano, as well as oils like avocado and olive oil, have the ability to treat conditions like T2D and neurodegenerative illnesses. Common herbs contain polyphenol rosmarinic acid. In diabetic rats, it reduces hyperglycemia and improves insulin sensitivity, possibly through influencing the expression of GLUT4, phosphoenolpyruvate carboxykinase (Runtuwene et al., 2016), and AKT1 and AKT3, as well as preventing DNA glycation (Rosmarinic acid inhibits DNA glycation, 2021). Rosmarinic acid's symmetric, compact structure binds selectively to Aβ, preventing tau from aggregating and polymerizing into fibrillar form (Cornejo et al., 2017). providing defense in Alzheimer's disease and other neurodegenerative diseases.

5. LIFESTYLE MODIFICATIONS

5.1. Aerobic Exercise:

Exercise that works your heart and breathing more forcefully by using large muscles for an extended period of time is called aerobic exercise, which includes running, swimming, cycling (Vasankari et al., 1997). An increase in oxygen intake during aerobic exercise may lead to an increase in free radical (FR) activity. Exercise, that is aerobics in nature increases oxygen consumption, which may lead to an increase in FR production. Thus, a number of research revealed that this kind of exercise increased the production of FR in both humans and animals (Mastaloudis et al., 2001). Though, low exercise intensity (less than 50% of maximum oxygen consumption) does not produce this kind of behaviour. Antioxidant capacity is not exceeded in such a situation, and damage caused by FR is not noticeable. Additionally, FR production and oxidative stress become more significant with increasing exercise intensity (Lovlin et al., 1987). Either the enzymatic antioxidant activity or the quantities of non-enzymatic antioxidants change as a result of aerobic exercise. Antioxidant enzyme activity (SOD, GPX, and CAT) in blood or tissues rises during aerobic exercise, according to a number of studies conducted on humans and animals (Leeuwenburgh et al., 1999) (Clarkson, 1995).

5.2. Anaerobic Exercise:

A wide range of sports activities fall under the category of anaerobic exercise (e.g sprints, jumps or resistance exercise). Compared to aerobic activity, there is little information on the generation of FR during acute anaerobic exercise (Groussard et al., 2003). In addition to electron leakage, other mechanisms may also play a role in the rise in FR generation that is unique to anaerobic exercise, such as during aerobic exercise (McBride et al., 1998). During anaerobic exercise, phagocytic respiratory burst activity, ischaemia reperfusion, and xanthine oxidase production appear to be involved in the generation of FR. Further factors that can raise the synthesis of FR include the significant increases in lactic acid, acidosis, catecholamines, and post-exercise inflammation that are typical with supramaximal activities (Sahlin et al., 1999).

6. REGULATORY AND POLICY IMPLICATIONS

The implications associated with regulatory measures and policy frameworks concerning environmental agents that induce oxidative stress are of paramount importance for both public health as well as the safety of our environment. A diverse array of pollutants, which includes substances such as pesticides, heavy metals, and effluents from industrial processes, have been empirically demonstrated to play a significant role in exacerbating oxidative stress, thereby underscoring the urgent need for the establishment of comprehensive and effective regulatory frameworks that can address these critical issues. Pesticides, which are primarily designed with the intention of eradicating agricultural pests, have been found to be associated with oxidative stress not only in the targeted pest populations but also in non-target organisms, leading to adverse effects on lipid peroxidation processes as well as the integrity of DNA structures, as evidenced by the research conducted by. Therefore, it is imperative that regulatory measures be developed and implemented with a focus on minimizing human and environmental exposure to these harmful substances, while simultaneously ensuring that safer, more environmentally friendly alternatives are readily accessible to those involved in agricultural practices. In the realm of industrial pollutants, the activities associated with industrial production and manufacturing lead to the release of a plethora of harmful substances, including volatile organic compounds and various heavy metals, all of which have the potential to disrupt the delicate oxidative balance within both human physiology and broader ecological systems, a phenomenon thoroughly investigated by (142). As a consequence, it is crucial that relevant policies enforce stricter standards regarding emissions and actively promote the development and implementation of cleaner technologies that can help mitigate the detrimental effects associated with these industrial pollutants. When it comes to workplace safety, the presence of hazardous chemicals, including fine dust particles and heavy metals, represents a significant risk to the health of workers, thereby emphasizing the urgent necessity for the establishment of robust occupational health regulations, as discussed by (143) in 2003. The implementation of regular monitoring protocols, combined with protective measures designed to minimize worker exposure, can significantly reduce the associated health risks, and enhance the overall safety of workplace environments. Conversely, there exists a perspective that posits current regulatory frameworks may not sufficiently address the cumulative and synergistic effects resulting from exposure to multiple pollutants, thereby indicating a pressing need for a more integrated and holistic approach to environmental health policy. This integrated approach should take into account the complex interactions and combined impacts of various environmental agents on oxidative stress, fostering a comprehensive understanding that can lead to more effective public health interventions and environmental protections.

7. CONCLUSION

In summary, the interplay between environmental exposure and oxidative stress significantly influences human health, revealing critical pathways through which pollutants adversely affect bodily functions. Various environmental pollutants, including particulate matter, heavy metals, and pesticides, lead to the generation of reactive oxygen species (ROS), resulting in oxidative stress that contributes to cellular damage. This damage manifests across multiple organ systems, fostering the development of numerous health conditions such as respiratory diseases, cardiovascular issues, and neurodegenerative disorders. For instance, pollutants like ozone and nitrogen dioxide exacerbate respiratory conditions such as COPD and asthma, while oxidative stress accelerates cardiovascular diseases like atherosclerosis and hypertension. Neurotoxic substances further complicate this landscape, as they induce oxidative stress in the brain, potentially leading to conditions such as Alzheimer's and Parkinson's diseases.

Recognizing the mechanisms underlying these interactions is essential for formulating effective public health interventions. Strategies aimed at reducing environmental pollutants such as stricter regulations on emissions and waste management are vital. Additionally, promoting green technologies can help mitigate these harmful exposures. On an individual level, integrating antioxidant-rich diets and therapies may offer protective benefits against oxidative stress-related damage, enhancing resilience to environmental toxins. Ultimately, a comprehensive approach that combines regulatory measures, public health initiatives, and personal health strategies can help combat the detrimental effects of oxidative stress stemming from environmental exposures. By addressing both the sources of pollution and the oxidative stress it induces, we can pave the way toward healthier communities and improved public health outcomes.

REFERENCES

Aghamohammadi, V., Gargari, B. P., & Aliasgharzadeh, A. (2011, June). Effect of Folic Acid Supplementation on Homocysteine, Serum Total Antioxidant Capacity, and Malondialdehyde in Patients with Type 2 Diabetes Mellitus. *Journal of the American College of Nutrition, 30*(3), 210–215. DOI: 10.1080/07315724.2011.10719962 PMID: 21896879

Ahmad, I. M., Temme, J. B., Abdalla, M. Y., & Zimmerman, M. C. (2016, May 3). Redox status in workers occupationally exposed to long-term low levels of ionizing radiation: A pilot study. *Redox Report : Communications in Free Radical Research, 21*(3), 139–145. DOI: 10.1080/13510002.2015.1101891 PMID: 26817988

Al-Maskari, M. Y., Waly, M. I., Ali, A., Al-Shuaibi, Y. S., & Ouhtit, A. (2012, July). Folate and vitamin B12 deficiency and hyperhomocysteinemia promote oxidative stress in adult type 2 diabetes. *Nutrition (Burbank, Los Angeles County, Calif.), 28*(7–8), e23–e26. DOI: 10.1016/j.nut.2012.01.005 PMID: 22595450

Alam, M. M., Iqbal, S., & Naseem, I.Alam MdM. (2015, October). Iqbal S, Naseem I. Ameliorative effect of riboflavin on hyperglycemia, oxidative stress and DNA damage in type-2 diabetic mice: Mechanistic and therapeutic strategies. *Archives of Biochemistry and Biophysics, 584*, 10–19. DOI: 10.1016/j.abb.2015.08.013 PMID: 26319175

Alrubaye, A., Motovali-Bashi, M., & Miroliaei, M. (2021). Rosmarinic acid inhibits DNA glycation and modulates the expression of Akt1 and Akt3 partially in the hippocampus of diabetic rats. *Scientific Reports, 11*(1), 20605.

Anetor, G. O., Nwobi, N. L., Igharo, G. O., Sonuga, O. O., & Anetor, J. I. (2022). Environmental pollutants and oxidative stress in terrestrial and aquatic organisms: Examination of the total picture and implications for human health. *Frontiers in Physiology, 13*, 931386.

Balbi, M. E., Tonin, F. S., Mendes, A. M., Borba, H. H., Wiens, A., Fernandez-Llimos, F., & Pontarolo, R. (2018, December). Antioxidant effects of vitamins in type 2 diabetes: A meta-analysis of randomized controlled trials. *Diabetology & Metabolic Syndrome, 10*(1), 18. DOI: 10.1186/s13098-018-0318-5 PMID: 29568330

Beckett, E. M., Abelmann, A., Roberts, B., Lewis, R. C., Cheatham, D., Miller, E. W., Hall, E., & Pierce, J. S. (2023, November). An updated evaluation of reported no-observed adverse effect levels for chrysotile, amosite, and crocidolite asbestos for lung cancer and mesothelioma. *Critical Reviews in Toxicology, 53*(10), 611–657. DOI: 10.1080/10408444.2023.2283169 PMID: 38126124

Benjamin, S., Masai, E., Kamimura, N., Takahashi, K., Anderson, R. C., & Faisal, P. A. (2017, October 15). Phthalates impact human health: Epidemiological evidences and plausible mechanism of action. *Journal of Hazardous Materials*, *340*, 360–383. DOI: 10.1016/j.jhazmat.2017.06.036 PMID: 28800814

Biochemical and cellular basis of oxidative stress: Implications for disease onset. J Cell Physiol [Internet]. 2023 Jul 12 [cited 2024 Oct 9]; Available from: https://typeset.io/papers/biochemical-and-cellular-basis-of-oxidative-stress-3tf952p3

Boothby, L. A., & Doering, P. L. (2005, December). Vitamin C and Vitamin E for Alzheimer's Disease. *The Annals of Pharmacotherapy*, *39*(12), 2073–2080. DOI: 10.1345/aph.1E495 PMID: 16227450

Bortey-Sam, N., Ikenaka, Y., Akoto, O., Nakayama, S. M., Asante, K. A., Baidoo, E., & Ishizuka, M. (2017). Oxidative stress and respiratory symptoms due to human exposure to polycyclic aromatic hydrocarbons (PAHs) in Kumasi, Ghana. *Environmental Pollution*, *228*, 311–320.

Brown, B. G., Zhao, X. Q., Chait, A., Fisher, L. D., Cheung, M. C., Morse, J. S., Dowdy, A. A., Marino, E. K., Bolson, E. L., Alaupovic, P., Frohlich, J., Serafini, L., Huss-Frechette, E., Wang, S., DeAngelis, D., Dodek, A., & Albers, J. J. (2001, November 29). Simvastatin and Niacin, Antioxidant Vitamins, or the Combination for the Prevention of Coronary Disease. *The New England Journal of Medicine*, *345*(22), 1583–1592. DOI: 10.1056/NEJMoa011090 PMID: 11757504

Browne, D., McGuinness, B., Woodside, J. V., & McKay, G. J. (2019, July). Vitamin E and Alzheimer's disease: What do we know so far? *Clinical Interventions in Aging*, *14*, 1303–1317. DOI: 10.2147/CIA.S186760 PMID: 31409980

Carlson, L. A., & Rosenhamer, G. (1988, January 12). Reduction of Mortality in the Stockholm Ischaemic Heart Disease Secondary Prevention Study by Combined Treatment with Clofibrate and Nicotinic Acid. *Acta Medica Scandinavica*, *223*(5), 405–418. DOI: 10.1111/j.0954-6820.1988.tb15891.x PMID: 3287837

Celestino, M. T., Magalhães, U. D. O., Fraga, A. G. M., Carmo, F. A. D., Lione, V., Castro, H. C., Sousa, V. P., Rodrigues, C. R., & Cabral, L. M. (2012, September). Rational use of antioxidants in solid oral pharmaceutical preparations. *Brazilian Journal of Pharmaceutical Sciences*, *48*(3), 405–415. DOI: 10.1590/S1984-82502012000300007

Chakrawarti, L., Agrawal, R., Dang, S., Gupta, S., & Gabrani, R. Therapeutic effects of EGCG: a patent review. Expert Opin Ther Pat. 2016 Aug 2;26(8):907–16.

Cheung, C. C. C., Zheng, G. J., Li, A. M. Y., Richardson, B. J., & Lam, P. K. S. (2001, May). Relationships between tissue concentrations of polycyclic aromatic hydrocarbons and antioxidative responses of marine mussels, Perna viridis. *Aquatic Toxicology (Amsterdam, Netherlands)*, *52*(3–4), 189–203. DOI: 10.1016/S0166-445X(00)00145-4 PMID: 11239681

Çiğ, B., & Nazıroğlu, M. (2015). Investigation of the effects of distance from sources on apoptosis, oxidative stress and cytosolic calcium accumulation via TRPV1 channels induced by mobile phones and Wi-Fi in breast cancer cells. *Biochimica et Biophysica Acta. Biomembranes*, *1848*(10), 2756–2765. DOI: 10.1016/j.bbamem.2015.02.013 PMID: 25703814

Clarkson, P. M. (1995, January). Antioxidants and physical performance. *Critical Reviews in Food Science and Nutrition*, *35*(1–2), 131–141. DOI: 10.1080/10408399509527692 PMID: 7748472

Cornejo, A., Aguilar Sandoval, F., Caballero, L., Machuca, L., Muñoz, P., Caballero, J., Perry, G., Ardiles, A., Areche, C., & Melo, F. (2017, January 1). Rosmarinic acid prevents fibrillization and diminishes vibrational modes associated to β sheet in tau protein linked to Alzheimer's disease. *Journal of Enzyme Inhibition and Medicinal Chemistry*, *32*(1), 945–953. DOI: 10.1080/14756366.2017.1347783 PMID: 28701064

Cortés-Iza, S. C., & Rodríguez, A. I. (2018). Oxidative stress and pesticide disease: A challenge for toxicology. *Revista de la Facultad de Medicina (Caracas)*, *66*(2), 261–267. DOI: 10.15446/revfacmed.v66n2.60783

Cova, C. M., Rincón, E., Espinosa, E., Serrano, L., & Zuliani, A. (2022, January 19). Paving the Way for a Green Transition in the Design of Sensors and Biosensors for the Detection of Volatile Organic Compounds (VOCs). *Biosensors (Basel)*, *12*(2), 51. DOI: 10.3390/bios12020051 PMID: 35200311

Cozzolino, R., De Giulio, B., Martignetti, A., Forte, G., Dallio, M., Romeo, M., Scognamiglio, F., Ventriglia, L., Stocchero, M., & Federico, A. (2022, November 28). Urinary volatile Organic compounds in non-alcoholic fatty liver disease (NAFLD), type two diabetes mellitus (T2DM) and NAFLD-T2DM coexistence. *Metabolomics*, *18*(12), 98. DOI: 10.1007/s11306-022-01960-1 PMID: 36441279

Curieses Andrés, C. M., Pérez de la Lastra, J. M., Andrés Juan, C., Plou, F. J., & Pérez-Lebeña, E. (2023). From reactive species to disease development: Effect of oxidants and antioxidants on the cellular biomarkers. *Journal of Biochemical and Molecular Toxicology*, *37*(11), e23455.

Dal, S., & Sigrist, S. (2016, July 11). The Protective Effect of Antioxidants Consumption on Diabetes and Vascular Complications. *Diseases (Basel, Switzerland)*, *4*(3), 24. DOI: 10.3390/diseases4030024 PMID: 28933404

De Lau, L. M. L., Koudstaal, P. J., Witteman, J. C. M., Hofman, A., & Breteler, M. M. B. (2006, July 25). Dietary folate, vitamin B $_{12}$, and vitamin B $_{6}$ and the risk of Parkinson disease. *Neurology*, *67*(2), 315–318. DOI: 10.1212/01.wnl.0000225050.57553.6d PMID: 16864826

Demirci-Çekiç, S., Özkan, G., Avan, A. N., Uzunboy, S., Çapanoğlu, E., & Apak, R. (2022, February). Biomarkers of Oxidative Stress and Antioxidant Defense. *Journal of Pharmaceutical and Biomedical Analysis*, *209*, 114477. DOI: 10.1016/j.jpba.2021.114477 PMID: 34920302

Dewar, L., & Heuberger, R. (2017, December). The effect of acute caffeine intake on insulin sensitivity and glycemic control in people with diabetes. *Diabetes & Metabolic Syndrome*, *11*, S631–S635. DOI: 10.1016/j.dsx.2017.04.017 PMID: 28935543

Dondi, A., Carbone, C., Manieri, E., Zama, D., Del Bono, C., Betti, L., Biagi, C., & Lanari, M.Outdoor Air Pollution and Childhood Respiratory Disease. (2023, February 22). The Role of Oxidative Stress. *International Journal of Molecular Sciences*, *24*(5), 4345–4345. DOI: 10.3390/ijms24054345 PMID: 36901776

Doseděl, M., Jirkovský, E., Macáková, K., Krčmová, L., Javorská, L., Pourová, J., Mercolini, L., Remião, F., Nováková, L., & Mladěnka, P. (2021, February 13). Vitamin C—Sources, Physiological Role, Kinetics, Deficiency, Use, Toxicity, and Determination. *Nutrients*, *13*(2), 615. DOI: 10.3390/nu13020615 PMID: 33668681

Duarte, T. L., & Lunec, J. (2005, July). ReviewPart of the Series: From Dietary Antioxidants to Regulators in Cellular Signalling and Gene ExpressionReview: When is an antioxidant not an antioxidant? A review of novel actions and reactions of vitamin C. *Free Radical Research*, *39*(7), 671–686. DOI: 10.1080/10715760500104025 PMID: 16036346

Dwivedi, S., Kushalan, S., Paithankar, J. G., D'Souza, L. C., Hegde, S., & Sharma, A. (2022). Environmental toxicants, oxidative stress and health adversities: Interventions of phytochemicals. *The Journal of Pharmacy and Pharmacology*, *74*(4), 516–536.

Enweasor, C., Flayer, C. H., & Haczku, A. (2021). Ozone-Induced Oxidative Stress, Neutrophilic Airway Inflammation, and Glucocorticoid Resistance in Asthma. *Frontiers in Immunology*, *12*, 631092. DOI: 10.3389/fimmu.2021.631092 PMID: 33717165

Erejuwa, O. O., Sulaiman, S. A., & Ab Wahab, M. S. (2012, April 12). Honey: A Novel Antioxidant. *Molecules (Basel, Switzerland)*, *17*(4), 4400–4423. DOI: 10.3390/molecules17044400 PMID: 22499188

Fatima, M. T., Islam, Z., Ahmad, E., & Salahuddin, P. (2018, July 17). Emerging Targets and Latest Proteomics Based Therapeutic Approaches in Neurodegenerative Diseases. *Current Protein & Peptide Science*, *19*(9), 858–875. DOI: 10.2174/1389203718666170731114757 PMID: 28762306

Fricker, R. A., Green, E. L., Jenkins, S. I., & Griffin, S. M. (2018, January). The Influence of Nicotinamide on Health and Disease in the Central Nervous System. *International Journal of Tryptophan Research : IJTR*, *11*, 117864691877665. DOI: 10.1177/1178646918776658 PMID: 29844677

Furman, D., Campisi, J., Verdin, E., Carrera-Bastos, P., Targ, S., Franceschi, C., Ferrucci, L., Gilroy, D. W., Fasano, A., Miller, G. W., Miller, A. H., Mantovani, A., Weyand, C. M., Barzilai, N., Goronzy, J. J., Rando, T. A., Effros, R. B., Lucia, A., Kleinstreuer, N., & Slavich, G. M. (2019, December 5). Chronic inflammation in the etiology of disease across the life span. *Nature Medicine*, *25*(12), 1822–1832. DOI: 10.1038/s41591-019-0675-0 PMID: 31806905

Gamon, L. F., & Wille, U. (2016). Oxidative damage of biomolecules by the environmental pollutants NO2• and NO3•. *Accounts of Chemical Research*, *49*(10), 2136–2145.

Ghezzi, P. (2020, January 31). Environmental risk factors and their footprints in vivo - A proposal for the classification of oxidative stress biomarkers. *Redox Biology*, *34*, 101442–101442. DOI: 10.1016/j.redox.2020.101442 PMID: 32035921

Ghosh, A. K. (2023). Air-pollutant Particulate Matter 2.5 (PM2. 5)-induced Inflammation and Oxidative Stress in Diseases: Possible Therapeutic Approaches.

Gorini, F., Sabatino, L., Gaggini, M., Chatzianagnostou, K., & Vassalle, C. (2021, July 30). Oxidative Stress Biomarkers in the Relationship between Type 2 Diabetes and Air Pollution. *Antioxidants*, *10*(8), 1234. DOI: 10.3390/antiox10081234 PMID: 34439482

Groussard, C., Rannou-Bekono, F., Machefer, G., Chevanne, M., Vincent, S., Sergent, O., Cillard, J., & Gratas-Delamarche, A. (2003, March). Changes in blood lipid peroxidation markers and antioxidants after a single sprint anaerobic exercise. *European Journal of Applied Physiology*, *89*(1), 14–20. DOI: 10.1007/s00421-002-0767-1 PMID: 12627300

Halliwell, B. (2007, November 1). Biochemistry of oxidative stress. *Biochemical Society Transactions*, *35*(5), 1147–1150. DOI: 10.1042/BST0351147 PMID: 17956298

Halliwell, B. (2011, March). Free radicals and antioxidants – quo vadis? *Trends in Pharmacological Sciences*, *32*(3), 125–130. DOI: 10.1016/j.tips.2010.12.002 PMID: 21216018

Hira, S., Saleem, U., Anwar, F., Sohail, M. F., Raza, Z., & Ahmad, B. (2019, September). β-Carotene: A Natural Compound Improves Cognitive Impairment and Oxidative Stress in a Mouse Model of Streptozotocin-Induced Alzheimer's Disease. *Biomolecules*, *9*(9), 441. DOI: 10.3390/biom9090441 PMID: 31480727

Hrubša, M., Siatka, T., Nejmanová, I., Vopršalová, M., Kujovská Krčmová, L., Matoušová, K., Javorská, L., Macáková, K., Mercolini, L., Remião, F., Máťuš, M., & Mladěnka, P. (2022, January 22). Biological Properties of Vitamins of the B-Complex, Part 1: Vitamins B1, B2, B3, and B5. *Nutrients*, *14*(3), 484. DOI: 10.3390/nu14030484 PMID: 35276844

Hu, G. L., Wang, X., Zhang, L., & Qiu, M. H. (2019). The sources and mechanisms of bioactive ingredients in coffee. *Food & Function*, *10*(6), 3113–3126. DOI: 10.1039/C9FO00288J PMID: 31166336

Jabłońska-Trypuć, A. (2017). Pesticides as inducers of oxidative stress. *Reactive Oxygen Species (Apex, N.C.)*, *3*(8), 96–110.

Jain, A. B., & Jain, V. A. (2012). Vitamin E, its beneficial role in diabetes mellitus (DM) and its complications. *Journal of Clinical and Diagnostic Research : JCDR*, *6*(10), 1624.

Jomova, K., Jenisova, Z., Feszterova, M., Baros, S., Liska, J., Hudecova, D., Rhodes, C. J., & Valko, M. (2011, March). Arsenic: Toxicity, oxidative stress and human disease. *Journal of Applied Toxicology*, *31*(2), 95–107. DOI: 10.1002/jat.1649 PMID: 21321970

Jomova, K., Raptova, R., Alomar, S. Y., Alwasel, S. H., Nepovimova, E., Kuca, K., & Valko, M. (2023, August 19). Reactive oxygen species, toxicity, oxidative stress, and antioxidants: Chronic diseases and aging. *Archives of Toxicology*, *97*(10), 2499–2574. DOI: 10.1007/s00204-023-03562-9 PMID: 37597078

Juan, C. A., de la Lastra, J. M. P., Plou, F. J., & Pérez-Lebeña, E. (2021, April 28). The Chemistry of Reactive Oxygen Species (ROS) Revisited: Outlining Their Role in Biological Macromolecules (DNA, Lipids and Proteins) and Induced Pathologies. *International Journal of Molecular Sciences*, *22*(9), 4642. DOI: 10.3390/ijms22094642 PMID: 33924958

Kakuda, T. (2002). Neuroprotective Effects of the Green Tea Components Theanine and Catechins. *Biological & Pharmaceutical Bulletin*, *25*(12), 1513–1518. DOI: 10.1248/bpb.25.1513 PMID: 12499631

Karachalias, N., Babaei-Jadidi, R., Kupich, C., Ahmed, N., & Thornalley, P. J. (2005, June). High-Dose Thiamine Therapy Counters Dyslipidemia and Advanced Glycation of Plasma Protein in Streptozotocin-Induced Diabetic Rats. *Annals of the New York Academy of Sciences*, *1043*(1), 777–783. DOI: 10.1196/annals.1333.090 PMID: 16037305

Karthika, C., Appu, A. P., Akter, R., Rahman, M. H., Tagde, P., Ashraf, G. M., Abdel-Daim, M. M., Hassan, S. S., Abid, A., & Bungau, S. (2022, February). Potential innovation against Alzheimer's disorder: A tricomponent combination of natural antioxidants (vitamin E, quercetin, and basil oil) and the development of its intranasal delivery. *Environmental Science and Pollution Research International*, *29*(8), 10950–10965. DOI: 10.1007/s11356-021-17830-7 PMID: 35000160

Kawamura, K., Qi, F., & Kobayashi, J. (2018). Potential relationship between the biological effects of low-dose irradiation and mitochondrial ROS production. *Journal of Radiation Research*, *59*(suppl_2), ii91–ii97. DOI: 10.1093/jrr/rrx091 PMID: 29415254

Kim, H. H., Kang, Y. R., Lee, J. Y., Chang, H. B., Lee, K. W., Apostolidis, E., & Kwon, Y.-I. (2018, February 28). The Postprandial Anti-Hyperglycemic Effect of Pyridoxine and Its Derivatives Using In Vitro and In Vivo Animal Models. *Nutrients*, *10*(3), 285. DOI: 10.3390/nu10030285 PMID: 29495635

Kiokias, S., Proestos, C., & Oreopoulou, V. (2020, April 24). Phenolic Acids of Plant Origin—A Review on Their Antioxidant Activity In Vitro (O/W Emulsion Systems) Along with Their in Vivo Health Biochemical Properties. *Foods*, *9*(4), 534. DOI: 10.3390/foods9040534 PMID: 32344540

Kolahdouzan, M., & Hamadeh, M. J. (2017, April). The neuroprotective effects of caffeine in neurodegenerative diseases. *CNS Neuroscience & Therapeutics*, *23*(4), 272–290. DOI: 10.1111/cns.12684 PMID: 28317317

Kontoghiorghes, G. J., & Kontoghiorghe, C. N. (2019, July 3). Prospects for the introduction of targeted antioxidant drugs for the prevention and treatment of diseases related to free radical pathology. *Expert Opinion on Investigational Drugs*, *28*(7), 593–603. DOI: 10.1080/13543784.2019.1631284 PMID: 31185180

Lee, J. E., Kang, J. S., Ki, Y. W., Lee, S. H., Lee, S. J., Lee, K. S., & Koh, H. C. (2011). Akt/GSK3β signaling is involved in fipronil-induced apoptotic cell death of human neuroblastoma SH-SY5Y cells. *Toxicology Letters*, *202*(2), 133–141. DOI: 10.1016/j.toxlet.2011.01.030 PMID: 21296133

Leeuwenburgh, C., Hansen, P. A., Holloszy, J. O., & Heinecke, J. W. (1999, July). Hydroxyl radical generation during exercise increases mitochondrial protein oxidation and levels of urinary dityrosine. *Free Radical Biology & Medicine*, *27*(1–2), 186–192. DOI: 10.1016/S0891-5849(99)00071-4 PMID: 10443935

Leh, H. E., & Lee, L. K. (2022, April 4). Lycopene: A Potent Antioxidant for the Amelioration of Type II Diabetes Mellitus. *Molecules (Basel, Switzerland)*, *27*(7), 2335. DOI: 10.3390/molecules27072335 PMID: 35408734

Leopoldini, M., Russo, N., & Toscano, M. (2011, March). The molecular basis of working mechanism of natural polyphenolic antioxidants. *Food Chemistry*, *125*(2), 288–306. DOI: 10.1016/j.foodchem.2010.08.012

Li, K., Geng, Y., Lin, B., & Xi, Z. (2023, December 1). Molecular mechanisms underlying mitochondrial damage, endoplasmic reticulum stress, and oxidative stress induced by environmental pollutants. *Toxicological Research*, *12*(6), 1014–1023. DOI: 10.1093/toxres/tfad094 PMID: 38145103

Linster, C. L., Van Schaftingen, E., & Vitamin, C. (2007, January). Biosynthesis, recycling and degradation in mammals. *The FEBS Journal*, *274*(1), 1–22. DOI: 10.1111/j.1742-4658.2006.05607.x PMID: 17222174

Liu, M., Choi, D., Hunter, R. L., Pandya, J. D., Cass, W. A., Sullivan, P. G., Kim, H.-C., Gash, D. M., & Bing, G. (2010, February). Trichloroethylene induces dopaminergic neurodegeneration in Fisher 344 rats. *Journal of Neurochemistry*, *112*(3), 773–783. DOI: 10.1111/j.1471-4159.2009.06497.x PMID: 19922440

Lovlin, R., Cottle, W., Pyke, I., Kavanagh, M., & Belcastro, A. N. (1987). Are indices of free radical damage related to exercise intensity. *European Journal of Applied Physiology*, *56*(3), 313–316. DOI: 10.1007/BF00690898 PMID: 3569239

Marashly, E. T., & Bohlega, S. A. (2017, July 20). Riboflavin Has Neuroprotective Potential: Focus on Parkinson's Disease and Migraine. *Frontiers in Neurology*, *8*, 333. DOI: 10.3389/fneur.2017.00333 PMID: 28775706

Marcelino, G., Machate, D. J., Freitas, K. D. C., Hiane, P. A., Maldonade, I. R., Pott, A., Asato, M. A., Candido, C. J., & Guimarães, R. C. A. (2020, December 9). β-Carotene: Preventive Role for Type 2 Diabetes Mellitus and Obesity: A Review. *Molecules (Basel, Switzerland), 25*(24), 5803. DOI: 10.3390/molecules25245803 PMID: 33316948

Martinez-Moral, M.-P., & Kannan, K. (2022, January 11). Analysis of 19 urinary biomarkers of oxidative stress, nitrative stress, metabolic disorders, and inflammation using liquid chromatography–tandem mass spectrometry. *Analytical and Bioanalytical Chemistry, 414*(6), 2103–2116. DOI: 10.1007/s00216-021-03844-0 PMID: 35013809

Mason, S. A., Keske, M. A., & Wadley, G. D. (2021, February 1). Effects of Vitamin C Supplementation on Glycemic Control and Cardiovascular Risk Factors in People With Type 2 Diabetes: A GRADE-Assessed Systematic Review and Meta-analysis of Randomized Controlled Trials. *Diabetes Care, 44*(2), 618–630. DOI: 10.2337/dc20-1893 PMID: 33472962

Mastaloudis, A., Leonard, S. W., & Traber, M. G. (2001, October). Oxidative stress in athletes during extreme endurance exercise. *Free Radical Biology & Medicine, 31*(7), 911–922. DOI: 10.1016/S0891-5849(01)00667-0 PMID: 11585710

May, J. M., & Harrison, F. E. (2013, December 10). Role of Vitamin C in the Function of the Vascular Endothelium. *Antioxidants & Redox Signaling, 19*(17), 2068–2083. DOI: 10.1089/ars.2013.5205 PMID: 23581713

McBRIDE, J. M., Kraemer, W. J., TRIPLETT-McBRIDE, T. R. A. V. I. S., & Sebastianelli, W. (1998, January). Kraemer WJ, TRIPLETT-McBRIDE T, Sebastianelli W. Effect of resistance exercise on free radical production. *Medicine and Science in Sports and Exercise, 30*(1), 67–72. DOI: 10.1097/00005768-199801000-00010

McDermott, C., & Heffron, J. J. A. (2013, March). Toxicity of Industrially Relevant Chlorinated Organic Solvents In Vitro. *International Journal of Toxicology, 32*(2), 136–145. DOI: 10.1177/1091581813482006 PMID: 23559643

Meng, J. M., Cao, S. Y., Wei, X. L., Gan, R. Y., Wang, Y. F., Cai, S. X., Xu, X.-Y., Zhang, P.-Z., & Li, H.-B. (2019, June 10). Effects and Mechanisms of Tea for the Prevention and Management of Diabetes Mellitus and Diabetic Complications: An Updated Review. *Antioxidants, 8*(6), 170. DOI: 10.3390/antiox8060170 PMID: 31185622

. (2019, December 1). Molecular mechanisms of oxidation damage and liver cell dysfunction in patients with metastatic colorectal cancer. *Experimental Oncology, 41*(4), 328–334. PMID: 31868333

Moon, H. S., Lee, H. G., Choi, Y. J., Kim, T. G., & Cho, C. S. (2007, April). Proposed mechanisms of (−)-epigallocatechin-3-gallate for anti-obesity. *Chemico-Biological Interactions, 167*(2), 85–98. DOI: 10.1016/j.cbi.2007.02.008 PMID: 17368440

Moore, E. M., Mander, A. G., Ames, D., Kotowicz, M. A., Carne, R. P., Brodaty, H., Woodward, M., Boundy, K., Ellis, K. A., Bush, A. I., Faux, N. G., Martins, R., Szoeke, C., Rowe, C., & Watters, D. A. (2013, October 1). Increased Risk of Cognitive Impairment in Patients With Diabetes Is Associated With Metformin. *Diabetes Care, 36*(10), 2981–2987. DOI: 10.2337/dc13-0229 PMID: 24009301

Morelli, M. B., Gambardella, J., Castellanos, V., Trimarco, V., & Santulli, G. (2020, December 3). Vitamin C and Cardiovascular Disease: An Update. *Antioxidants, 9*(12), 1227. DOI: 10.3390/antiox9121227 PMID: 33287462

Moretti, M., Fraga, D. B., & Rodrigues, A. L. S. (2017, December). Preventive and therapeutic potential of ascorbic acid in neurodegenerative diseases. *CNS Neuroscience & Therapeutics, 23*(12), 921–929. DOI: 10.1111/cns.12767 PMID: 28980404

Mostafalou, S., & Abdollahi, M. (2017, February). Pesticides: An update of human exposure and toxicity. *Archives of Toxicology, 91*(2), 549–599. DOI: 10.1007/s00204-016-1849-x PMID: 27722929

Müller, T. (2008, June). Role of homocysteine in the treatment of Parkinson's disease. *Expert Review of Neurotherapeutics, 8*(6), 957–967. DOI: 10.1586/14737175.8.6.957 PMID: 18505360

Nutriepigenomics in Environmental-Associated Oxidative Stress. (2023, March 1).. . *Antioxidants, 12*(3), 771–771. DOI: 10.3390/antiox12030771 PMID: 36979019

Oliveira, B. F. A. D., Carvalho, L. V. B. D., Mourão, D. D. S., Mattos, R. D. C. O. D. C., Castro, H. A. D., Artaxo Netto, P. E., & Hacon, S. (2018). Environmental Exposure Associated with Oxidative Stress Biomarkers in Children and Adolescents Residents in Brazilian Western Amazon. *Journal of Environmental Protection, 9*, 347–367.

Oxidative, S.-I. W. A. (2003, January 1).. . *Comments on Toxicology, 9*(1), 23–37. DOI: 10.1080/08865140302423

. (2019, August 20). Oxidative biomolecular damage: A possible mechanism for systemic autoimmunity. *International Journal of Health Sciences (Qassim), 13*(5), 1–3. PMID: 31501645

Paiva, C. N., & Bozza, M. T. (2014, February 20). Are Reactive Oxygen Species Always Detrimental to Pathogens? *Antioxidants & Redox Signaling, 20*(6), 1000–1037. DOI: 10.1089/ars.2013.5447 PMID: 23992156

Paniker, N. V., Srivastava, S. K., & Beutler, E. (1970). Glutathione metabolism of the red cells effect of glutathione reductase deficiency on the stimulation of hexose monophosphate shunt under oxidative stress. Biochimica et Biophysica Acta (BBA)-. *General Subjects*, *215*(3), 456–460.

Park, J. H., Park, Y. S., Lee, J. B., Park, K. H., Paik, M., Jeong, M., & Koh, H. C. (2016). Meloxicam inhibits fipronil-induced apoptosis via modulation of the oxidative stress and inflammatory response in SH-SY5Y cells. *Journal of Applied Toxicology*, *36*(1), 10–23. DOI: 10.1002/jat.3136 PMID: 25772694

Perez Visñuk, D., Teran, M. D. M., Savoy De Giori, G., LeBlanc, J. G., & De Moreno De LeBlanc, A. (2022, May). Neuroprotective Effect of Riboflavin Producing Lactic Acid Bacteria in Parkinsonian Models. *Neurochemical Research*, *47*(5), 1269–1279. DOI: 10.1007/s11064-021-03520-w PMID: 35113305

Pesticides as Inducers of Oxidative Stress. (2017) | Agata Jabłońska-Trypuć | 35 Citations [Internet]. [cited 2024 Oct 12]. Available from: https://typeset.io/papers/pesticides-as-inducers-of-oxidative-stress-2wpum1f5k3

Phaniendra, A., Jestadi, D. B., & Periyasamy, L. (2015, January). Free Radicals: Properties, Sources, Targets, and Their Implication in Various Diseases. *Indian Journal of Clinical Biochemistry : IJCB*, *30*(1), 11–26. DOI: 10.1007/s12291-014-0446-0 PMID: 25646037

Polizzi, F., Andican, G., Çetin, E., Civelek, S., Yumuk, V., & Burçak, G. (2012, January 9). Increased DNA-Glycation in Type 2 Diabetic Patients: The Effect of Thiamine and Pyridoxine Therapy. *Experimental and Clinical Endocrinology & Diabetes*, *120*(06), 329–334. DOI: 10.1055/s-0031-1298016 PMID: 22231921

Prakasam, A., Sethupathy, S., & Lalitha, S. (2001). Plasma and RBCs antioxidant status in occupational male pesticide sprayers. *Clinica Chimica Acta*, *310*(2), 107–112. DOI: 10.1016/S0009-8981(01)00487-9 PMID: 11498075

Prohaska, J. R. (1991, March). Changes in Cu,Zn-Superoxide Dismutase, Cytochrome c Oxidase, Glutathione Peroxidase and Glutathione Transferase Activities in Copper-Deficient Mice and Rats. *The Journal of Nutrition*, *121*(3), 355–363. DOI: 10.1093/jn/121.3.355 PMID: 1848285

Pullakhandam, R., Palika, R., Ghosh, S., & Reddy, G. B. (2012, December). Contrasting effects of type 2 and type 1 diabetes on plasma RBP4 levels: The significance of transthyretin. *IUBMB Life*, *64*(12), 975–982. DOI: 10.1002/iub.1096 PMID: 23129325

Purdom-Dickinson, S. E., Sheveleva, E. V., Sun, H., & Chen, Q. M. (2007). Translational control of nrf2 protein in activation of antioxidant response by oxidants. *Molecular Pharmacology*, *72*(4), 1074–1081. DOI: 10.1124/mol.107.035360 PMID: 17652445

Radicals, F., & Stress, O. (2023, January 1). Signaling Mechanisms, Redox Basis for Human Diseases, and Cell Cycle Regulation. *Current Molecular Medicine*, *23*(1), 13–35. DOI: 10.2174/1566524022666211222161637 PMID: 34951363

Rafiee, S., Asadollahi, K., Riazi, G., Ahmadian, S., & Saboury, A. A. (2017, December 20). Vitamin B12 Inhibits Tau Fibrillization via Binding to Cysteine Residues of Tau. *ACS Chemical Neuroscience*, *8*(12), 2676–2682. DOI: 10.1021/acschemneuro.7b00230 PMID: 28841372

Rai, S. N., Singh, P., Steinbusch, H. W. M., Vamanu, E., Ashraf, G., & Singh, M. P. (2021, September 22). The Role of Vitamins in Neurodegenerative Disease: An Update. *Biomedicines*, *9*(10), 1284. DOI: 10.3390/biomedicines9101284 PMID: 34680401

Ranjbar, A., Pasalar, P., Sedighi, A., & Abdollahi, M. (2002). Induction of oxidative stress in paraquat formulating workers. *Toxicology Letters*, *131*(3), 191–194. DOI: 10.1016/S0378-4274(02)00033-4 PMID: 11992738

Razip, N. N. M., Gopalsamy, B., Abdul Mutalib, M. S., Chang, S. K., Abdullah, M. M. J. A., Azlan, A., Rejali, Z., & Khaza'ai, H. (2021, July 1). Correlation between Levels of Vitamins D3 and E in Type 2 Diabetes Mellitus: A Case-Control Study in Serdang, Selangor, Malaysia. *Nutrients*, *13*(7), 2288. DOI: 10.3390/nu13072288 PMID: 34371798

Runtuwene, J., Cheng, K. C., Asakawa, A., Amitani, H., Amitani, M., & Morinaga, A.. (2016, July 7). Rosmarinic acid ameliorates hyperglycemia and insulin sensitivity in diabetic rats, potentially by modulating the expression of PEPCK and GLUT4. *Drug Design, Development and Therapy*, *10*, 2193–2202. DOI: 10.2147/DDDT.S108539 PMID: 27462144

Sahlin, K., Cizinsky, S., & Warholm, M. (1992). H berg J. Repetitive static muscle contractions in humans? a trigger of metabolic and oxidative stress? *European Journal of Applied Physiology*, *64*(3), 228–236. DOI: 10.1007/BF00626285 PMID: 1563368

Said, E., Mousa, S., Fawzi, M., Sabry, N. A., & Farid, S. (2021, February). Combined effect of high-dose vitamin A, vitamin E supplementation, and zinc on adult patients with diabetes: A randomized trial. *Journal of Advanced Research*, *28*, 27–33. DOI: 10.1016/j.jare.2020.06.013 PMID: 33364042

Sen, C. K., Khanna, S., & Roy, S. (2006, March). Tocotrienols: Vitamin E beyond tocopherols. *Life Sciences*, *78*(18), 2088–2098. DOI: 10.1016/j.lfs.2005.12.001 PMID: 16458936

Shahidi, F., & Ambigaipalan, P. (2015, October). Phenolics and polyphenolics in foods, beverages and spices: Antioxidant activity and health effects – A review. *Journal of Functional Foods*, *18*, 820–897. DOI: 10.1016/j.jff.2015.06.018

Shi, L., Du, X., Guo, P., Huang, L., Qi, P., & Gong, Q. (2020, November 6). Ascorbic acid supplementation in type 2 diabetes mellitus: A protocol for systematic review and meta-analysis. *Medicine*, *99*(45), e23125. DOI: 10.1097/MD.0000000000023125 PMID: 33157992

Sierra-Vargas, M. P., Montero-Vargas, J. M., Debray-García, Y., Vizuet-de-Rueda, J. C., Loaeza-Román, A., & Terán, L. M. (2023, January 1). Its Impact on Chronic Respiratory Diseases. *International Journal of Molecular Sciences*, *24*(1), 853–853. DOI: 10.3390/ijms24010853 PMID: 36614301

Smolek, M., Notaroberto, , Jaramillo, , & Pradillo, . (2013). Notaroberto, Jaramillo, Pradillo. Intervention with vitamins in patients with nonproliferative diabetic retinopathy: A pilot study. *Clinical Ophthalmology (Auckland, N.Z.)*, (Jul), 1451. DOI: 10.2147/OPTH.S46718 PMID: 23898220

Squillacioti, G., Bellisario, V., Ghelli, F., Marcon, A., Marchetti, P., Corsico, A. G., & Bono, R. (2024). Air pollution and oxidative stress in adults suffering from airway diseases. Insights from the Gene Environment Interactions in Respiratory Diseases (GEIRD) multi-case control study. *The Science of the Total Environment*, *909*, 168601.

Sule, R. O., Condon, L., & Gomes, A. V. (2022). In Oxid, M. C. L. (Ed.), *A Common Feature of Pesticides: Oxidative Stress—The Role of Oxidative Stress in Pesticide-Induced Toxicity. Zhou X* (pp. 1–31).

Thiamine Deficiency Induced Dietary Disparity Promotes Oxidative Stress and Neurodegeneration | Indian Journal of Clinical Biochemistry [Internet]. [cited 2024 Aug 27]. Available from: https://link.springer.com/article/10.1007/s12291-017-0690-1

Trasino, S. E., Benoit, Y. D., Gudas, L. J., & Vitamin, A. (2015, January). Deficiency Causes Hyperglycemia and Loss of Pancreatic β-Cell Mass. *The Journal of Biological Chemistry*, *290*(3), 1456–1473. DOI: 10.1074/jbc.M114.616763 PMID: 25451926

Udhayabanu, T., Manole, A., Rajeshwari, M., Varalakshmi, P., Houlden, H., & Ashokkumar, B. (2017, May 5). Riboflavin Responsive Mitochondrial Dysfunction in Neurodegenerative Diseases. *Journal of Clinical Medicine*, *6*(5), 52. DOI: 10.3390/jcm6050052 PMID: 28475111

Ulusu, N. N., & Tandoğan, B. (2007, August 8). Purification and kinetic properties of glutathione reductase from bovine liver. *Molecular and Cellular Biochemistry*, *303*(1–2), 45–51. DOI: 10.1007/s11010-007-9454-1 PMID: 17410407

Unoki-Kubota, H., Yamagishi, S., Takeuchi, M., Bujo, H., & Saito, Y. (2010, September 1). Yamagishi S ichi, Takeuchi M, Bujo H, Saito Y. Pyridoxamine, an Inhibitor of Advanced Glycation End Product (AGE) Formation Ameliorates Insulin Resistance in Obese, Type 2 Diabetic Mice. *Protein and Peptide Letters*, *17*(9), 1177–1181. DOI: 10.2174/092986610791760423 PMID: 20441560

Valdés-Ramos, R., Ana Laura, G. L., Beatriz Elina, M. C., & Alejandra Donaji, B. A. (2015). Vitamins and type 2 diabetes mellitus. Endocrine, Metabolic & Immune Disorders-Drug Targets (Formerly Current Drug Targets-Immune, Endocrine & Metabolic Disorders), 15(1), 54-63.

Valko, M., Leibfritz, D., Moncol, J., Cronin, M. T. D., Mazur, M., & Telser, J. (2007, January). Free radicals and antioxidants in normal physiological functions and human disease. *The International Journal of Biochemistry & Cell Biology*, *39*(1), 44–84. DOI: 10.1016/j.biocel.2006.07.001 PMID: 16978905

Valko, M., Morris, H., & Cronin, M. T. D. (2005). Metals, toxicity and oxidative stress. *Current Medicinal Chemistry*, *12*(10), 1161–1208. DOI: 10.2174/0929867053764635 PMID: 15892631

Varadharaj, S., Kelly, O. J., Khayat, R. N., Kumar, P. S., Ahmed, N., & Zweier, J. L. (2017, November 1). Role of Dietary Antioxidants in the Preservation of Vascular Function and the Modulation of Health and Disease. *Frontiers in Cardiovascular Medicine*, *4*, 64. DOI: 10.3389/fcvm.2017.00064 PMID: 29164133

Vasankari, T. J., Kujala, U. M., Vasankari, T. M., Vuorimaa, T., & Ahotupa, M. (1997). Effects of Acute Prolonged Exercise on Serum and LDL Oxidation and Antioxidant Defences. *Free Radical Biology & Medicine*, *22*(3), 509–513. DOI: 10.1016/S0891-5849(96)00373-5 PMID: 8981043

Verlangieri, A. J., & Sestito, J. (1981, July). Effect of insulin on ascorbic acid uptake by heart endothelial cells: Possible relationship to retinal atherogenesis. *Life Sciences*, *29*(1), 5–9. DOI: 10.1016/0024-3205(81)90108-9 PMID: 7022082

Vinson, J. A., & Dabbagh, Y. A. (1998, June). Tea phenols: Antioxidant effectiveness of teas, tea components, tea fractions and their binding with lipoproteins. *Nutrition Research (New York, N.Y.), 18*(6), 1067–1075. DOI: 10.1016/S0271-5317(98)00089-X

Wakade, C., Chong, R., Bradley, E., Thomas, B., & Morgan, J. Upregulation of GPR109A in Parkinson's Disease. Block ML, editor. PLoS ONE. 2014 Oct 17;9(10):e109818. DOI: 10.1371/journal.pone.0109818

Xiang, D., Zhang, Q., & Wang, Y. T. (2020, July 17). Effectiveness of niacin supplementation for patients with type 2 diabetes: A meta-analysis of randomized controlled trials. *Medicine, 99*(29), e21235. DOI: 10.1097/MD.0000000000021235 PMID: 32702899

Yan, X., & Wang, S. (2021, March 26). The efficacy of niacin supplementation in type 2 diabetes patients: Study protocol of a randomized controlled trial. *Medicine, 100*(12), e22272. DOI: 10.1097/MD.0000000000022272 PMID: 33761625

Zavadskiy, S., Sologova, S., & Moldogazieva, N. (2022, April 1). Oxidative distress in aging and age-related diseases: Spatiotemporal dysregulation of protein oxidation and degradation. *Biochimie, 195*, 114–134. DOI: 10.1016/j.biochi.2021.12.002 PMID: 34890732

Zhou, J., Chan, L., & Zhou, S. (2012, June 1). Trigonelline: A Plant Alkaloid with Therapeutic Potential for Diabetes and Central Nervous System Disease. *Current Medicinal Chemistry, 19*(21), 3523–3531. DOI: 10.2174/092986712801323171 PMID: 22680628

Zhou, X., Gao, S., Yue, M., Zhu, S., Liu, Q., & Zhao, X.-E. (2023, February 1). Recent advances in analytical methods of oxidative stress biomarkers induced by environmental pollutant exposure. *Trends in Analytical Chemistry, 160*, 116978–116978. DOI: 10.1016/j.trac.2023.116978

Živojević, K., Mladenović, M., Djisalov, M., Mundzic, M., Ruiz-Hernandez, E., Gadjanski, I., & Knežević, N. Ž. (2021, September 10). Advanced mesoporous silica nanocarriers in cancer theranostics and gene editing applications. *Journal of Controlled Release, 337*, 193–211. DOI: 10.1016/j.jconrel.2021.07.029 PMID: 34293320

Chapter 5

Role of Reactive Oxygen Species in Cellular Signalling:

Insights Into Pathological Processes and Therapeutic Strategies

Banashree Bondhopadhyay
https://orcid.org/0000-0002-6679-7791
Amity University, Noida, India

Shreya Gupta
Amity University, Noida, India

Yezdanul Haque Shamsi
https://orcid.org/0009-0004-6276-1996
Amity University, Noida, India

ABSTRACT

ROS are molecules of the cell signalling, produced during oxidative metabolism and causes various human disease as its accumulation leads to deregulation of antioxidant defence mechanism (Wang et al. 2021). ROS regulates several cell signalling pathways which are responsible for apoptosis, survival of the cell, growth and maintenance of metabolic activities of the cell. This chapter focuses on the role of reactive oxygen species in cell signalling. Cell signalling is the process through which cell communicate each other and responds to environment and stimuli through a cascade of events, ultimately leading to production of desired protein as a result. Therefore, a comprehensive understanding of mechanism of cell signalling enhances

DOI: 10.4018/979-8-3693-7919-6.ch005

our knowledge of intrinsic cellular process and also provide insight for development therapeutic strategies for curing various diseases.

INTRODUCTION

ROS are one of the most important molecules in cell signalling, increasing efficiency of cell to instantly respond to a stimulus. ROS has key role in abiotic and biotic stress sensing and encrypting different environmental signals in plants (Mittler et al. 2022). They are highly reactive and dangerous moiety. These can cause oxidative cell death. ROS defines a large group of molecules(oxidant), which have various different properties and functions. These properties and functions usually range from cell signalling to damaging cell. Hence, it is important for us to define oxidants chemically and biologically precise so that research on their functions and properties. In fact, the general term used for ROS should not be used to specify the molecules. Further, it is also important for precise measurement of oxidants, with better specificity for identification of their targets in signalling pathways (Sies et al. 2022). Although ROS are executively involved in signal (redox), but in some conditions it can cause oxidative damage. This shows that it can play both physiological and pathological role biological activities. Measurement of ROS is important to predict oxidative activities and to detect their importance in biology using antioxidant. ROS are normally understood as the byproduct of various cellular processes, shown in fig1. Cancer cell typically have higher basal level of ROS capered to normal cell, caused due to unbalanced oxidants and antioxidants. Certainly, ROS have played dual role in the metabolism for instance, when low to moderate level, ROS plays the role of signal transducer which activates cell proliferation, migration and invasion. However, when level of ROS is high, it can cause damage to proteins, lipids, membrane and nucleic acid and organelle, further this leads to death (Nakamura and Takada 2021).

Not only do they have anti-microbial activity act against bacteria and parasites but also support redox regulation of immune signalling and induction of inflammation activation. However, due to lack of in-depth data more information is unavailable (Herb and Schramm 2021). Varied ROS different cellular responses including cell death, cellular signalling and down regulation of ROS production pathway and ROS limiting pathways, these two pathways are often observed in the cancer cell. Additionally, it regulates the environment of tumor that affects the various immune responses, blood circulation and stromal cell, responsible for proving metabolic support to tumor cell. Particularly, the ROS response is extremely complex and rely on several factors such as types, severity, localisation and half-life of ROS (Cheung and Vousden 2022). Moderate increase in ROS concentration give rise to various pathological conditions including tumor progression and production through their involvement in several signalling pathways and role in inducing DNA mutation.

Figure 1. The sources of reactive oxygen species

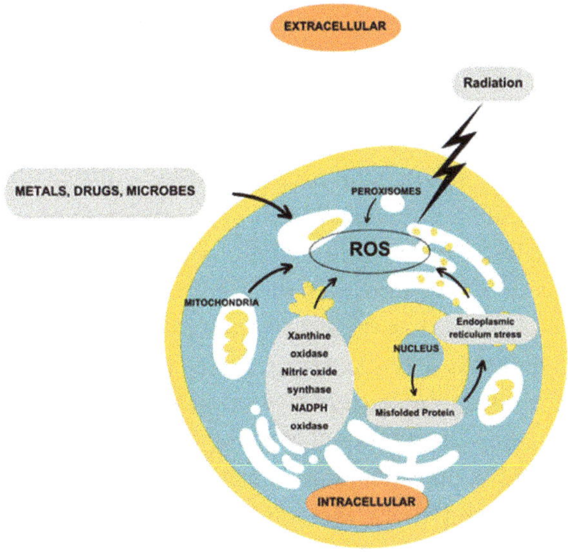

(Made on canva.com on 8th August 2024)

The two primary function of ROS is, first, as a signalling molecule triggering targeted protein or second, as a harmful oxidative moiety which damages the cellular organelles. Moreover, ROS and calcium ions affects one another's functions due to their overlap as they are involved in diverse functions. ROS regulates protein activity which ensures regulation of calcium ion flux between cellular compartment. On the other hand, calcium ion affects production site of ROS for instance, ETC, NOX1, NOX2, etc. ROS plays significant role in several disease and physiological process involving neurological, cardiovascular, immune system, skeletal muscle and metabolism regulation, ageing and cancer.

ROS has significant role in cell signalling affecting metabolic and pathological function. One of such signalling pathways is MAPK pathway, which is involved in cell cycle, inflammatory responses and apoptosis. Additionally, MAPK signalling influences the growth and development of tumor cells. other than MAPK, PI3K pathway, which is responsible for several cellular responses including apoptosis. In utmost environmental conditions ROS production is induce which inhibits PI3K therefore causing inflammation and apoptosis. In contrast, during normal to moderate environmental condition ROS production induces PI3K signalling, causing apoptosis and inflammation. Furthermore, ROS concentration affects PI3K signalling, determining proliferation, migration and apoptosis in corneal epithelium cells (Chen et al. 2022). PI3K activation is important for various metabolic activities of insulin in insulin sensitive tissues by IRS1 and IRS2 genes (López-Gómez et al. 2020). This promotes various cellular process by targeting several regulatory proteins which are responsible for glucose and lipid metabolism. Hence PI3K modulate obesity and obesity related diseases (Savova et al. 2023). PI3K signalling pathways is responsible for regulation of neurotoxicity and mediating neuron survival via varied substrate including fork head box protein, glycogen synthase kinase-3 beta and caspases. Recent studies have been focusing on relationship between PI3K pathway and neurological diseases such as Parkinson's disease and Alzheimer's disease (Long et al. 2021). Moreover, PI3K signalling pathway is also involved in diverse disease including GIT disease, metabolic disorder, skin infection, allergy and aging (Mohseni et al. 2021). Moving to IRP/IRE system, responsible for controlling uptake, storage and utilisation of iron in cells. The two regulatory proteins' IRP1 and IRP2 recognises and binds mRNA motif called as IREs (iron responsive elements), located in untranslated regions of genes. Moreover, this system plays a significant role in neurodegenerative diseases, regulated by post translational modification, oxygen, nitrogen mono oxide and metallic ions. On the other hand, it can also affect oxidative metabolism, cell cycle, neuroinflammation and aggregation of proteins. P66Shc, an important regulatory protein involved in oxidative stress, having potential oxireductase activity, it regulates the mitochondrial oxidative stress, aging and cell death. Studies have shown, higher expression of p66Shc in mononuclear blood cells in gestational type 2 diabetic patients than p66Shc levels in non-diabetic patients. Hence, expression of p66Shc is associated with development of diabetes. Hyperactivity of p66Shc is significant in cell and tissue damage and triggering diabetes due damage of beta cells of pancreas (Biondi et al. 2023). Moreover, it is also associated with tumor progression hence, targeting p66Shc pathway can be used as therapeutic strategy in diseases related to aging (Huang et al. 2021; Mir et al. 2020; Yao et al. 2022).

ROLE OF ROS IN MAPK SIGNALLING

A serine/threonine kinase, termed as MAP kinase (MAPK) due to it phosphorylation catalysing property. They are activated by dual phosphorylation on a tri-peptide motif located in T-loop, a kinase activation loop via three step MAPK cascade. Therefore, members of MAPK family have common structure and biochemical properties. The three-tiered MAPK cascade consists of a MAPK, a MAPK kinase and MAPK kinase kinase. In contrast, dephosphorylation of threonine and tyrosine residue which resides within activation loop, results in dephosphorylation through phosphatases. Medium of dephosphorylation involves through serine/ threonine phosphatases, tyrosine phosphatases and dual phosphatase, responsible for dephosphorylation of both threonine and tyrosine residue inside activation loop. MAPK, mitogen activated protein kinase signalling pathways, diverse in its regulatory function. Dual role of MAPK makes them activator as well as inhibitor depending on the type of cell and stimuli. Role of MAPK includes converting extracellular stimuli in various cellular responses such as cell growth, migration, proliferation, differentiation and apoptosis. Among the members of MAPK family JNK and p38 MAPK easily gets activated when cell id exposed to physical, chemical and biological stress. On the contrary, ERK1/2 is stimulated by internal cell growth factors (Yue and López 2020). Glioblastoma multiform tumorigenesis is a result of p38 MAPK inhibition. Identification of factors leading to regulation of p38 MAPK signalling is necessary for tumor treatment.

ROS have significant role in regulation of varied signalling molecules such as MAPK tyrosine kinase and transcription factors. Anti-oxidant enzyme comprises of a network which counteracts to accumulated ROS in side the cell under normal and adverse conditions produced due to anaerobic respiration. when scavenger activity of enzyme is supressed by the accumulation of the ROS. MAPK pathway is directly affected by the accumulation of ROS. Few studies have proved that application of hydrogen peroxide exogenously activates the MAPK pathways. the mechanism of activation of MAPK by ROS is not known. However, it has been reported that ROS tends to modify amino acids present at the active site and in the important protein motif hence, which can be a reason behind the activation of the MAPK. Lu and He proposed that activation of MAPK cascade is highly influenced by ROS. Similarly, it has been found that ROS has is also responsible for activation of few members of MAPK family for instance, MAPK3, MAPK6, MAPK4, etc. Moreover, activation of MAPK is regulated by the growth factors which is activated by hydrogen peroxide. Activation of cascade of MAPK by oxidatively modified intracellular kinase by ROS. In contrast, ASK1 oxidises thioredoxin under stressful conditions which further, phosphorylates the JNK and p38. MPKs are negative regulator of MAPKs, resulting in degradation on ROS signalling. Hydrogen peroxide oxidises the amino acids at vital sites which in turn deactivates MPKs hence, it is a reason why JNK pathways is continuously activated. TFs is stimulated by MAPKs causing to overexpression of SAGs such as HSPs, SOD, and CAT. Under HS the optimal level of ROS is maintained by the antioxidant defence mechanism. Heat stress is responsible for metabolism-based reactivation within endoplasmic tissue and reduction of quality and quantity of the grains. Particular MAPK is had been found to be responsible for regulation of ABC transporters resulting in larger size and better shaped grains without reducing quality of the grains (Anon n.d.-a).

Increase in air pollution, number of smokers and other factors has led to rise in the cases of lungs cancer with survival rate of 18 percent. With development of new diagnostic technologies and techniques, chemotherapeutic agents for treatment of cancer, yet the disease remain critical and challenging. Autophagy is defined as cytoplasmic component degradation dominant in cellular homeostasis. Commonly, it is known as non-apoptotic cell death. Hence, ROS mediated MAPK signalling pathway is useful in treatment of cancer treatment (Fan et al. 2020). Lipid peroxidation induced by ROS plays vital role in cell apoptosis and autophagy. Eliminating ROS is a protective role in lipid peroxidation. Nevertheless, cell death as the result of autophagy can also be induced by ROS. The transduction of apoptotic signals is carried by the caspases which further activates executioner caspase3. Whereas, the main signal transducer is MAPKs, induced by oxidative stress These are two pathways leading to apoptosis. MAPK is activated by the adduct of lipid peroxidation product with ERK, p38 and JNK which further initiates apoptosis by activating caspases. Moreover, studies have proven macrophages apoptosis and atherosclerosis as result of MAPK deficiency. Accumulation of ROS have a significant role in apoptosis and autophagy. Recently, studies have proved, inhibition of MAPK/ERK pathways by ROS, results in decrease in the secretion of epidermal growth factor. ROS dependent pathways inhibited the ERGH phosphorylation and phosphorylated ERK levels (Son et al. 2013). In this century gastric cancer has become one of the major health problems and a common digestive system malignant tumor. Despite, the number of the cases of gastric cancer is declining, it is fifth highest among all the cancers. For treatment of gastric cancer advanced drug technology is necessary with low toxicity and side-effects. Survival, differential and metastasis of cancer cells is associated with intracellular signalling pathways such as MAPK and AKT pathways. MAPK family members, ERK, c-JNK and p38 MAPK are vital for cell maintains. Activated JNK and MAPK triggers apoptosis and initiates ERK phosphorylation promoting cell survival. As it is already known that ROS production kills tumor cells. Zeaxanthin, an oxidised carotenoid, constituent of green vegetables, egg yolks and corn, have potential anti-inflammatory and anti- cancer effects and against certain cancers. Zeaxanthin has low toxicity and potential for killing tumor gastric cancer cells, according to result of CCK-8 analysis. GES-1 intracellular ROS level reduces and AGS intracellular level increases as concentration increases under the influence of zeaxanthin. Hence, the it can inhibit ROS in GES-1 cells and induce AGS cells. under the action of zeaxanthin in gastric cancer cell, the production of ROS is increased, this is contrary to the normal cells. normally NAC inhibits the ROS but when NAC and zeaxanthin together, works on the AGS cell, the intracellular ROS concentration increases. In cell proliferation, differentiation and apoptosis, MAPK, involving ERK, JNK and p38 are involved depending on type of cell and stimulation. Moreover, it regulates p-STAT3, NF-kB and MAPK pathways through intracellular ROS regulation, inducing apoptosis, as shown in figure 2. Hence, increase in ROS content blocks cell at G2/M phase, leading to apoptosis (Sheng et al. 2020).

Figure 2. The mechanism of Zeaxanthin in increasing the levels of ROS in cell which further causes apoptosis

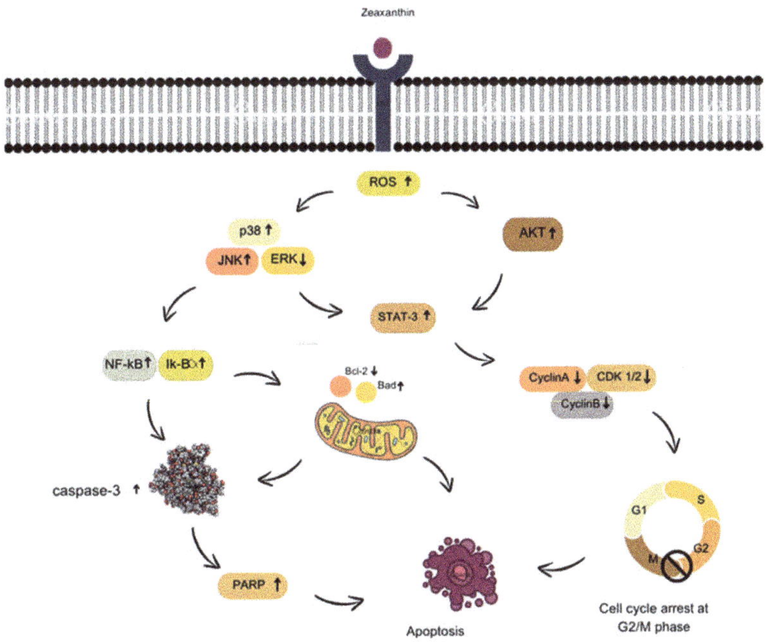

(Made on canva.com on 7th August 2024)

ROLE OF ROS IN PI3K SIGNALLING PATHWAY

On the plasma membrane and intracellular membrane 3'-OH group of phosphatidylinositol which is phosphorylated by a lipid kinase group called phosphoinositide 3-kinase. Differentiated into three sub classes on the basis of structure, specificity towards substrate and binding partners. 1 PI3K is one of the most studied classes among all the PI3K, this PI3K is produces PIP3 from PIP2 produced at the plasma membrane in response several stimuli, responsible for assembly of large number of phospholipids effectors such as AKT and PDK-1(phosphoinositide dependent protein kinase-1). Facilitated by PIP3 PDK and AKT interact leading to phosphorylation of AKT, promoting protein synthesis, cell growth, cell survival and motility through down regulation of kinases, which involves mTOR complex. Moreover, for counteracting PI3K signalling PTEN and tumor suppressor phosphatase dephosphorylates PIP3. Four precise homologous catalytic subunits, of class 1 PI3K includes p110 alpha, p110 beta, p110 gamma and p110 delta. Five regulatory subunits collectively called p85-type regulatory subunits can be associated with these four catalytic subunits. Primarily, in class IA PI3K, dysregulation of phosphoinositide kinase has been discovered in various human disease with mutation that increases or decreases enzymatic activity. Studies on PI3K or AKT or mTOR pathway is genetically altered in cancer (Li et al. 2021). This signalling pathways plays a major role in regulating transduction and other biological process including cell proliferation, apoptosis, angiogenesis and cellular metabolism. Constituting complex components in its pathway, it is important in various human disease which include brain injury caused due to ischaemia, neurodegenerative diseases and tumor (Xu et al. 2020). The change in levels of ROS in autophagic response of cells by PI3K signalling pathways. the alpha subunit of PI3K is responsible for inhibition of autophagic while the beta subunit is responsible for the promoting autophagic response. Under a nutrient rich condition, when levels of ROS increases led to initiation of autophagy. However, the initiation of the autophagy is counteracted by the down regulation of AKT protein. Akt works by targeting rapamycin complex 1 and inhibiting the expression of autophagic gene. Akt actions can be counteracted by the AMP activated protein kinase protein. Moreover, under low to moderated ROS levels, autophagy is inhibited by the mTORC1 and mTORC2. Whereas under high levels of MTORC2, promotes autophagy leading to cellular senescence. PTEN (phosphatase and tensin homolog protein) are responsible for downregulation of the PI3K pathway with proautophagic activities (Kma and Baruah 2022). Moreover, IP3K pathway can regulates signals that can regulate cell death through regulation of molecule which controls apoptosis and transcription factor. PI3K regulates molecule which regulates apoptosis includes caspase 9, Bad and GSK-3 beta transcription factor. Adding to this phosphatase has potential to counter balance the activity of PI3K signalling pathway. phosphatase has capacity to dephosphorylate signalling molecule within the PI3K hence, reducing its activity. Therefore, the strength of the kinase is determined by the balance between the phosphatase and kinase (Clerkin et al. 2008).

Figure 3. The effect of Coptisine on cancer cells

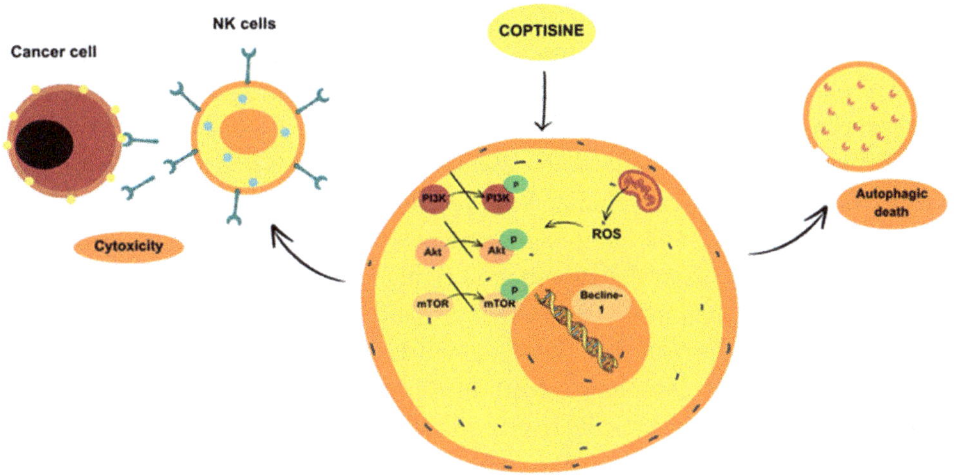

(Made on canva.com on 8th August 2024)

The human body have an antioxidant mechanism that maintain low concentration of ROS, beneficial for normal cells. Several human disease results from imbalance between radical and antioxidant. Mitochondrial mediated apoptosis is caused due to the excessive ROS. For instance, granulosa cell is important for oocyte maturation, fertilisation and implantation and poor quality of oocytes is associated with apoptosis in granulosa cells. PI3K signalling plays significant role in physiological and pathological process involving survival, metabolism, cell growth and proliferation. Moreover, studies conducted on mouse reveals that PI3K plays vital role in modulating granulosa cell growth and apoptosis while development of follicular cell (Gong et al. 2020). Moreover. ROS is crucial for the development of embryo, acting as secondary messengers. It may be originated in the embryo' s metabolism or from the oocyte with diverse function in the development of the embryo at different stage. However, some studies have shown the negative effect of ROS in embryo by damaging its DNA (Jamil et al. 2020). Akt has been reported to be involved in the neurodegeneration and regulating toxicity thought substrate like FOXos and caspase-9. Through its involvement in the PI3K signalling pathways, it mediates neuronal survival. Moreover, in diseases such as Alzheimer's disease, diabetes mellitus has been reported to increase the risk of dementia. To understand this in a better way let, delve into understanding the PI3K-AKT signalling pathway for brain insulin in the frontal cortices and seven cases. Activation of glycogen synthase kinase 3 beta due to decrease in insulin PI3K-AKT signalling. Component of insulin PI3K-AKT signalling associates negatively with level of tau phosphorylation and positively with protein O- GlcNAcytation hence, impaired insulin PI3K-AKT signalling may led to neurodegeneration. In case of pulmonary disease, when there is imbalance between antioxidant and oxidant, in which oxidant is greater, it can lead inflammatory lungs and airways disease such as asthma, pulmonary fibrosis and lung cancer. Upregulation of redox sensitive transcription factors led to oxidative stress which furthers causes inflammatory pulmonary diseases (Lee and Yang 2012; Liu et al. 2011; Rai et al. 2019). Coptisine, an isoquinoline alkaloid, commonly known for its anti-cancer activity in various cancers. Studies have revealed that coptisine is responsible for increased cytotoxicity and autophagic vacuoles depending on the concentration. Moreover, the expression of genes such as light chain 3, Becline-1 were increased remarkably. Further it inhibited the phosphorylation of PI3K, Akt and mTOR while, it also increases the production of mitochondrial ROS, as demonstrated in figure 3. Coptisine results in autophagic death in HCC Hep3B cells. the autophagic cell death in hepatic carcinoma results from P13K and Akt signalling and mitochondrial ROS which causes dysfunction in mitochondria (Kim et al. 2021).

REGULATION OF IRE-IRP SYSTEM AND IRON HOMEOSTASIS BY ROS

The primary function of iron is to hemoglobinize erythrocytes in order to transport oxygen, which is why iron is very essential part of life. However, when intracellular iron homeostasis is dysregulated, the Fenton reaction generates an excessive amount of reactive oxygen species. Further it has a catastrophic effect on cells, ultimately resulting in ferroptosis, an iron-dependent form of cell death. In mammalian cells iron availability is controlled by iron regulatory protein 1 (IRP1) and iron regulatory protein 2 (IRP2). IRPs support neutrophil differentiation by ensuring the a sufficient iron supply to the cells and also it regulated gene expression which is involved in neuropoiesis, which further helps in the development of mature neutrophils. Monocyte count is not effected by the IRP and iron deficiency. Iron is very important component for heme. To secure haemoglobin synthesis and prevent anaemia iron must be supplied in enough amount. Anaemia due to iron deficiency is very common (Pasricha et al. 2021). Often caused by the inadequate dietary iron intake or the impaired delivery of iron to plasma during inflammation (Ganz 2019; Haschka, Hoffmann, and Weiss 2021). Peptide hepcidin controls body iron levels and distribution. Hepcidin binds to and inhibits the iron exporter ferroportin, reducing the amount of iron released into the plasma from iron exporting cells. On the other hand, hepcidin in supressed by the stimulation of red blood cells via the erythroblast-derived hormone erythroferrone allowing for iron export via FPN to match erythroid iron needs (Muckenthaler et al. 2017). IRP 1 and IPR2 interact with cis-regulatory iron responsive elements(IREs) which is found in the untranslated region (UTR) of mRNAs encoding iron metabolism molecules to maintain iron homeostasis at the cellular level. Iron regulatory binds to the iron responsive elements (IREs) in the 5′ untranslated region of mRNAs that encode iron storage proteins such as ferritin-H and -L, the iron exporter ferroprotin, erythroid amino levulinate synthase and hypoxia-inducible factor 2 alpha which is also known as EPAS1.Due to this binding translation of these mRNAs inhibits. Additionally, IRPs bind to the IREs in the 3′-UTR of the transferrin receptor 1 (TFRC) mRNA, protecting the transcript from degradation by the nucleases Regnase-1 (ZC3H12A) and Roquin-1 (RC3H1), thereby facilitating iron absorption (Corral et al. 2021; Yoshinaga et al. 2017).

Figure 4. Mechanism of translation modulation by the IRP-IRE signalling pathway

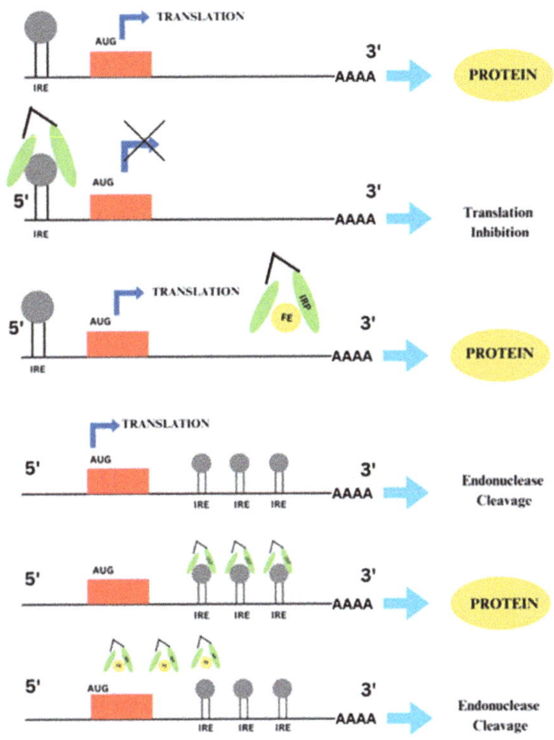

(Made on canva.com on 8th August 2024)

Cells contain antioxidant that shields them from nucleophiles. These antioxidants work as electrophiles to remove harmful ROS (Ainbinder et al. 1997). Interestingly in contrast to other organs, neurones lack sufficient antioxidant proteins despite being essential to survival. For example, nuclear factor erythroid 2-related factor 2

(Nrf2) which is known as the master regulator of antioxidant system is expressed weakly in neurons This abnormality may rise from the growth of neural cell, though the cause is unclear. In order to avoid axonal degeneration which is caused due to due to a high level of oxidative stress (OS), it is essential to maintain an adequate level of ROS .Since RNA-binding proteins like FTH1 are involved in translational control, Nrf2 may be involved. However, Nrf2 can prevent neuroinflammation by supressing the nuclear factor kappa-light-chain-enhancer of activated B cells (NF-κB) activation by preventing the degradation of the nuclear factor of kappa light polypeptide gene enhancer in B-cell inhibitor α (IκB-α).

INTRACELLULAR IRON HOMEOSTASIS

The intracellular iron homeostasis is regulated by the IRP/IRE system. IPRs are made up of IRP1 and IRP2, which are RNA- binding proteins. They control the translation stage by binding to the IRE in the mRNA's 5' - untranslated region (5′-UTR) or 3′-untranslated region (3′-UTR).

During the scarcity of iron, the levels of iron is increase by the iron influx proteins which includes DMT1, Tf, TfR, and hepcidin. FPN1 which is iron related proteins is increased under iron- replete conditions. These processes which are opposing are controlled by IRP/IRE system. Iron is transported by carrier proteins, like PCBPs, to FTH1/FTL for storage and enzymes for activation once it enters the intracellular region. NCOA4- mediated ferritinophagy is used by FTH1/FTL to release iron into the cytoplasm when cellular iron levels are low. PCBP1 knockdown may enhance lipid peroxidation and ferritiniophagy via binding to the 3′-UTR of the mRNAs for arachidonate 15-lipoxygenase (ALOX15) and beclin 1 (BECN1) (Kennedy et al. 2012; Lee, You, and Roh 2022; Vieira, Alves, and Vercelli 2011; Wilson, Muñoz-Palma, and González-Billault 2018) . More studies and research are required to understand how iron redox homeostasis and related molecules are understood.

SYSTEMIC ACUTE LOSS OF IRP FUNCTION CAUSES ERYTHROPENIA AND MYELOPENIA

To investigate the effects of IRP deficiency during adulthood, mice has been generated that express a tamoxifen-inducible CRE recombinase (CreER) under the control of the Rosa26 promoter (Badea, Wang, and Nathans 2003). It carries floxed Aco1 and Ireb2 alleles (Galy, Feering, and Hentze 2005). The resultant mice are known as P1/2-knockout (KO), while the littermates that do not have CreER are called P1/2-control (CTR). When two non-consecutive low doses of tamoxifen were given to adult P1/2-KO animals they showed effective recombination of the floxed Aco1 and Ireb2 alleles. After the commencement of tamoxifen treatment on day 10, normal activity pattern and over all posture were noticed on P1/2-KO. Body weight was also unchanged in this mice. But these mice showed low haemoglobin levels and a significant reduction in RBC size and quantity. Combining IRP1/2 deficit had an impact on RBC parameters that was roughly twice as great as ablation of IRP2 alone. This suggests that during the absence of IRP2, IRP1 can partially compensate. Elevated levels of plasma iron and ferritin, high transferrin saturation, and iron build-up in the liver and spleen were observed in conjunction with anaemia. IRP- deficient mice were observed anemic despite of high iron level, which suggests that anemia can be due to lack of RBC production and hemolysis. Though the chances of hemolysis is unlikely because the amounts of hemopexin, haemoglobin, and bilirubin in the serum remained unchanged. However in P1/2-KO animals EO concentrations was marked increased. Stimulation of erythropoiesis and suppress hepcidin via ERFE can be predicted through high EPO levels. Nevertheless no evidence of extramedullary erythropoiesis and drop in blood reticulocytes were found. In addition the concentration of hepcidin rose while ERFE levels stayed below the detection limit. All of these findings suggested that loss of IRP function reduces erythropoiesis, which lowers iron utilisation and causes the body to accumulate iron.Surprisingly, IRP ablation not only altered RBC number but also changed platelet counts.Flow cytometry has been used to examine the main leukocyte populations in peripheral blood in order to explain the kind of WBC that has impacted by the IRP deficiency. P1/2-KO mice showed a 60% reduction in myeloid cells and a modest rise in B and T cells. asymptomatic, the changes in erythrocyte and leukocyte parameters were not caused by CRE activation. Collectively these finding demonstrated that IRPs are very important for maintaining levels of myeloid cells during adulthood.

p66Shc, MITOCHONDRIAL OXIDATIVE STRESS AND AGING REGULATION BY ROS

p66Shc is a crucial regulatory protein implicated in oxidative stress and also a member of the Shc protein family. Age related disorders in mammals are controlled by this. It also controls apoptosis and mitochondrial oxidative stress. It is also widely expressed, enhances pro-apoptotic ROS signal which in turn controls a range of cardiovascular diseases (Liu et al. 2008; Soliman et al. 2014; Zheng et al. 2013). It is a widely expressed canonical scaffold protein which is further involved in modulating intracellular signal response. Other than adaptor function, p66shc also produces ROS(reactive oxygen species) in a mitochondria- dependent manner. Various diseases, such as endothelial dysfunction, coronary artery disease (CAD), and ischemia/reperfusion injuries (IRI), are controlled by p66Shc (Bertuglia 2009; D'Arcy 2019; Sanchez et al. 2018). P66Shc aggravates ROS accumulation during cardiovascular shocks which then increases pro-apoptotic response and clinical consequences. p66Shc expression and activity are highly correlated with the clinical severity of IRIs; hence, p66Shc functions as a biomarker for both CAD and IRIs (Franzeck et al. 2012; Noda et al. 2010; Spescha et al. 2015).

Figure 5. Classical p66Shc mitochondrial ROS activity pathway

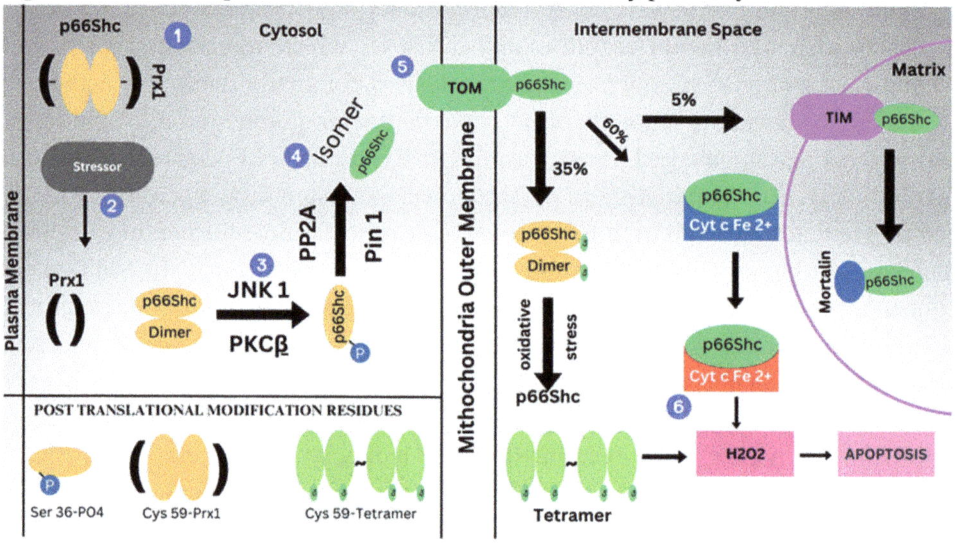

(Made on canva.com on 8th August 2024)

REACTIVE OXYGEN SPECIES, AGING, AND OXIDATIVE STRESS

Aging and heart related diseases are two examples of diseases where ROS and ROS mediated processes are linked to p66Shc mitochondrial function. Ageing and its related diseases are thought to be mostly caused by accumulating ROS-induced oxidative damage, according to theories on free radical ageing and oxidative stress. Protein expression changes and functional changes that result in cellular dysregulation are hallmarks of free radical ageing and oxidative stress processes, which are marked by a steady decline in fitness as cellular functions become progressively less effective (Shields, Traa, and Van Raamsdonk 2021). However, because ROS serves as a signalling molecule in regular cell processes, the concept of aging and pathology etiology was adapted. Thus free radical aging and oxidative stress are largely caused by the ROS or ROS dysregulation. Generally ROS are those molecules that contain oxygen and they are generated within cells by partial oxygen reduction. ROS can be either non-radical high-energy molecules that have the potential to become radicals. Unpaired/increased electrons orbitals in radical make them more reactive unless they are stabilised by removing an electron from a different molecule. It is crucial to distinguish between radical and non-radical ROS because, in general, radicals are too reactive to leave their place of formation, whereas non-radical ROS can (Mittler 2017; Phaniendra, Jestadi, and Periyasamy 2015; Winterbourn 2008). Thus, variations in ROS forms' mobility can impact and control subcellular redox signalling. However radical ROS can also be generated by non-radical ROS affecting the intercompartmental signalling. For instance, Fentol chemistry which is a common example in which H_2O_2 reacts with transition metal ions which further produces hydroxyl radicals. Excess radicals can harm DNA by mismatching or destabilising it, oxidising sulphur residues or carbonylating proteins, and producing lipid peroxides or per oxyradicals. These processes are ageing biomarkers because they can lead to genomic, membrane, and protein instability that accumulates with age. There are numerous sources of ROS within cells. Myeloperoxidase and NADPH oxidase (NOX) produce superoxide anion and hypochlorous acid through which ROS are created. These proteins are highly active during respiratory bursts that target pathogens and are linked to immune cells (Beckman and Ames 1998; Nguyen, Green, and Mecsas 2017). On the other hand NOX homologues are involved in ROS mediated survival growth and apoptosis signalling.

Additional proteins that produce reactive oxygen species (ROS) are xanthine oxidoreductase, which breaks down purines, cytochrome p450 proteins, which detoxify xenobiotics, and monoamine oxidase, which breaks down dopamine (Anon n.d.-b; McDonnell, PharmD, BCOP and Dang, PharmD, BCPS 2013; Shih, Chen, and Ridd 1999). H_2O_2 is also produced by Peroxisomal fatty acid β-oxidation which contribute to NF-kB and mTORC1 regulation by entering to the cytoplasm. However, mitochondrial respiration is the main generator of ROS in cells. Superoxide anion is created when electrons from electron transport chain (ETC) complexes I and II or complexes II and III leak between them (Chen et al. 2003; Van Raamsdonk and Hekimi 2010; RS 2005).

p66Shc DISCOVERY—AGING RESEARCH

Adaptor proteins with SH2 domains are utilized by the tyrosine kinase receptors. Then, Grb2, an additional adaptor protein that stimulates cell growth and differentiation by mitogen-activated protein kinase (MAPK) can be recruited by SH2 domains.

Tumor incidence is increased in mice when ShcA transgene is overexpressed, which further suggests that MAPK can be activated by ShcA proteins. Further research revealed that whereas p66Shc has a distinct transcript and an N-terminal CH2 domain that can replicate several p66Shc functions without the other domains, p46Shc and p52Shc share a transcript (Migliaccio et al. 1997). P 66Shc was linked to cell death and altered electrophoretic movement in mouse embryonic fibroblasts (MEFs) when tested against UV and oxidative stress or increased tyrosine kinase signalling. This further suggests that a post- translational modification (PTM), Ser36 phosphorylation, was linked to cell death. P66Shc is a redox protein that controls cell apoptosis and cell fate. In pathological or stressed environments, p66Shc genetic removal increased mouse lifespan and tolerance to oxidative damage; however, in unstressed environments, it had the opposite effect.These effects can be explained by p66Shc ROS dysregulations. Numerous discoveries about the roles of oxidoreductase and p66Shc adaptor have been made subsequent to promising outcomes in longevity and ROS attenuation. Compared to WT mice, first p66Shc KO mice demonstrated a roughly 30% increase in lifespan and resistance to oxidative stress with no side effects. p66Shc+/- mice live, longer than WT but not as long as KO mice (Giorgio et al. 2012; Migliaccio et al. 1999). Further analysis revealed that p66Shc KO mice also exhibited superior health at older ages, enhanced metabolism, reduced adipocyte triglyceride build-up, and improved behavioural flexibility as adults (Berniakovich et al. 2008; Tomilov et al. 2011). Additionally, these mice exhibit enhanced resistance to atherosclerosis, ischaemic insults, obesity, diabetes (Carpi et al. 2009a; Zaccagnini et al. 2004a). However, lifespan of p66Shc KO mice were decreased when exposed to winter weather and faced with food competition for a full year (Fadini et al. 2011). Therefore p66Shc is crucial for energy metabolism, cardiovascular health and aging.

p66Shc's MITOCHONDRIAL ROS ACTIVITY ROLE IN CARDIOVASCULAR PATHOLOGY

Many pathological states have a negative correlation with the formation of p66Shc ROS. This suggests that inhibition of p66Shc ROS activity can improve their outcomes. Neurodegenerative illness, aging, diabetes, organ failure, cancer, sepsis, wound healing, endothelial dysfunction, and ischemia/reperfusion (I/R) injuries are a few examples. In the following section effects of p66Shc on cardiovascular pathology has been discussed.

p66Shc IN ISCHEMIA/REPERFUSION INJURIES

Heart disease represents a significant worldwide healthcare and financial challenge. Heart disease accounts for 23.1% of all deaths in the United States alone where it affects 10.6% of adults. During the year 2014-2015 the annual cost of heart disease was USD 351.2 billion and according to report the disease burden are still rising (Carpi et al. 2009a). The most common heart disease is coronary artery disease (CAD) which narrows arterial diameter, decreasing blood flow beyond plaques. If poorly managed, CAD transitions into myocardial infarction (MI). Dysfunctional mitochondria are characterized by the MI pathology, it reverse their ATP-generating activity with ATP expenditure and pump protons into the IMS during ischemic conditions (Di Lisa et al. 1998). The function of p66Shc in ischaemia reperfusion injuries (IRIs) varies according to the length of the ischaemia and reperfusion. Because p66Shc-selective therapeutics have not yet been reported, p66Shc has almost exclusively been studied in p66Shc genetic KO models. However in some studies p66Shc is examined in KO or PKC-β inhibition models. Through a reduction in p66Shc-mediated ROS accumulation, p66Shc deletion enhanced IRIs in hind limb ischaemia models. It demonstrates improvements in tissue viability and capillary density (Zaccagnini et al. 2004b). Other hind limb ischemia models revealed that p66Shc KO improved wound healing in a ROS-dependent manner, indicating that p66Shc ROS balance may regulate ischemic wound healing (Yellon and Hausenloy 2007).

P66Shc KO increased resistance to angiotensin II- mediated apoptosis in mouse cardiomyocytes. It also revealed reduction in in oxidative stress caused by IRI that were not further reduced by the administration of antioxidants, suggesting dysregulation of p66Shc ROS in ischaemic heart tissue (Carpi et al. 2009b). Cardiac IRI models demonstrated that p66Shc mitochondrial translocation increased with long-duration ischaemia followed by reperfusion, but not with short-term ischaemia alone or with short-term ischaemia and reperfusion. Furthermore, p66Shc Ser36 phosphorylation was reduced and results were enhanced by p66Shc ROS suppression achieved by pharmacological Complex I or PKC-β inhibition (Yang et al. 2014). Though short term ischemia was worsened by p66Shc deletion, highlighting p66Shc's dual function in cardiac IRI. Studies on p66Shc KO specifically on mouse cardiac revealed that p66Shc KO did not significantly effect heart rate, heart mass, or blood pressure. Despite of having similar ventricular wall size cardiomyocyte counts were increased (Graiani et al. 2005). It may affect the p66Shc's neonatal role in heart development. Since neonatal p66Shc KO cardiac tissue exhibits greater SOD activity and lower ROS-mediated apoptosis (Bujak and Frangogiannis 2007).

Due to its high morbidity and mortality, the search for pharmaceutical treatments that can reduce the impact of an initial myocardial infarction (MI) and post-infarct remodeling is ongoing. Despite many drugs with pre-clinical effectiveness, few have proven beneficial in MI patients. And amongst them none have been previously identified as p66Shc ROS inhibitors (Kaludercic and Di Lisa 2020).Among the promising therapeutics, many are linked to p66Shc's functions. For instance, Coenzyme Q (CoQ) and mitoquinone (mitoQ), a CoQ variant with enhanced mitochondrial localization, target the respiratory chain, reduce ATP loss during ischemia/reperfusion (I/R), and scavenge excess ROS. These treatments are linked with improvements in post-MI contractility, cardiac function, oxidative damage, cell death, and I/R mitigation during cardiac grafts, potentially by reducing lipid peroxidation. CoQ and mitoQ treatment also decrease pro-apoptotic cytochrome c release (Adlam et al. 2005; Murphy and Smith 2007; Singh et al. 1998). However, mitoQ has completed clinical trials demonstrating cardiovascular benefits and is undergoing further trials for various ROS-related conditions which includes heart failure and kidney disease. A similar drug, mito-TEMPO, scavenges mitochondrial superoxide and shows improvements in heart failure and diabetic cardiomyopathy models by enhancing contractility and reducing heart dilation, cell death, and oxidative stress.

Therefore, promising treatment for ischemia-reperfusion injury (IRI) are associated with either increasing p66Shc pro-survival signaling or reducing p66Shc pro-apoptotic ROS activity. As previously discussed, the first p66Shc selective ROS inhibitors have been tested in zebrafish MI models and have shown improvements in various clinically translatable parameters (Camara, Bienengraeber, and Stowe 2011; Dey et al. 2018; Dikalova et al. 2010). Although consistency in mammals remains to be seen, these inhibitors represent a novel approach to addressing p66Shc ROS-mediated pathology, including MI treatment. They have demonstrated the potential to reduce initial infarct damage in MI models and may improve outcomes in recurrent MI, as fibrosis measurements were significantly reduced in treated fish.

Additionally, targeted inhibition of superoxide anions could enhance post-MI heart failure outcomes. This is because post-MI heart failure is associated with a significant increase in free radical-mediated lipid peroxidation and plasma superoxide anion levels, alongside decreased levels of catalase, glutathione, and superoxide dismutase (SOD) in plasma (Ghatak et al. 1996; McMurray et al. 1993). These trends are further exacerbated by the severity of heart failure, correlating with diminished antioxidant pools and heart function in heart failure patients (Belch et al. 1991; Keith et al. 1998).

CONCLUSION

Reactive oxygen species (ROS) represent a dualistic character. Means it can be both necessary and can also be harmful to cellular health. They have emerged as important regulators of numerous cellular functions. Beyond being merely by-products of cellular metabolism, ROS play a critical role in the regulation of multiple signalling pathways, most importantly the phosphoinositide 3-kinase (PI3K) and mitogen-activated protein kinase (MAPK) pathways. These pathways are essential for controlling cell division, survival, proliferation, and apoptosis, highlighting the role ROS play in cellular homeostasis. Since it also function as a secondary messengers in the signal transmission that control cellular responses to external stimuli. ROS is also engaged in redox signalling. The balance of ROS in cells are very important. Although physiological levels of ROS are required for regular cellular operations, abundance of ROS can cause oxidative stress which can lead to harmful consequences. Consequence can include protein oxidation, lipid peroxidation, and DNA damage. Oxidative damage plays a role in number of chronic illness such as diabetes, cancer, heart disease, and neurological conditions including Parkinson's and Alzheimer's. The pathways such as MAPK and PI3K are sensitive to redox changes, where ROS affect the activation and deactivation of important proteins in these signalling networks. ROS have the ability to alter the activity of kinases in the MAPK pathway, including p38, JNK, and ERK, which can impact cell fate decisions. Similarly, ROS can modify the phosphorylation state of Akt in the PI3K pathway, which is important for cell survival and metabolism. This change establishes a connection between oxidative stress and pathological states such as cancer and insulin resistance. ROS can encourage the development of cancer by causing DNA mutations and accelerating the expression of genes essential for cell division and survival. On the other hand ROS have the ability to cause cell death i.e. apoptosis which indicates that they may be used therapeutically. Targeting ROS for therapeutic intervention is gaining traction due of its pivotal role in these vital signalling pathways. The goal of antioxidant research has been to mitigate oxidative damage and restore redox balance by investigating antioxidants as potential modulators of ROS levels. On the other hand if the antioxidants are over used then it may interfere with physiological ROS signalling and result in unexpected effects. Future studies ought to concentrate on creating methods that allow for the precise modulation of ROS levels in certain cellular settings and compartments. In summary, ROS is very crucial for cell integrity, making them essential to the complex web of cellular signalling. This increased knowledge of ROS emphasises their importance in various disorders as therapeutic targets as well as indicators.

Future perspective

ROS plays significant role in various pathological and physiological processes. Therapies such as scavenger ROS stimulated antioxidant therapy, ROS elevated agent mediated toxic therapy which is induced by ROS. With advancement in nanotechnology, more nanomaterial with ability to regulate ROS are developed to work on new ROS related nanotherapeutic modalities and medicines (Zhang et al. 2021). ROS is crucial in cellular processes and various diseases such as cancer, neurodegenerative and cardiovascular diseases. development in our understanding of ROS could lead to development of new interventions to improve outcome of disease and oxidative stress. As an active area for research for drug development, future studies are dedicated for development of antioxidants and agents responsible for modulating ROS for better management for damage caused due to oxidative stress. Better understanding of biology of ROS in ageing process could lead to interventions promoting healthier lifestyle and healthy ageing. Moreover, ROS has dual role in cancer cells, it promotes tumorigenesis and also causes cell death in cancer cells. so, future studies will be focusing on utilising this duality of treatment of cancer.

REFERENCES

Adlam, V. J., Harrison, J. C., Porteous, C. M., James, A. M., Smith, R. A. J., Murphy, M. P., & Sammut, I. A. (2005). Targeting an Antioxidant to Mitochondria Decreases Cardiac Ischemia-Reperfusion Injury. *The FASEB Journal, 19*(9), 1088–1095. DOI: 10.1096/fj.05-3718com PMID: 15985532

Ainbinder, E., Bergelson, S., Pinkus, R., & Daniel, V. (1997). Regulatory Mechanisms Involved in Activator-Protein-1 (AP-1)-Mediated Activation of Glutathione-S-Transferase Gene Expression by Chemical Agents. *European Journal of Biochemistry, 243*(1–2), 49–57. DOI: 10.1111/j.1432-1033.1997.0049a.x PMID: 9030721

Badea, T. C., Wang, Y., & Nathans, J. (2003). A Noninvasive Genetic/Pharmacologic Strategy for Visualizing Cell Morphology and Clonal Relationships in the Mouse. *The Journal of Neuroscience : The Official Journal of the Society for Neuroscience, 23*(6), 2314–2322. DOI: 10.1523/JNEUROSCI.23-06-02314.2003 PMID: 12657690

Balaban, R. S., Nemoto, S., & Finkel, T.RS. (2005). Mitochondria, Oxidants, and Aging. *Cell, 120*(4), 483–495. DOI: 10.1016/j.cell.2005.02.001 PMID: 15734681

Beckman, K. B., & Ames, B. N. (1998). The Free Radical Theory of Aging Matures. *Physiological Reviews, 78*(2), 547–581. DOI: 10.1152/physrev.1998.78.2.547 PMID: 9562038

Belch, J. J. F., Bridges, A. B., Scott, N., & Chopra, M. (1991). Oxygen Free Radicals and Congestive Heart Failure. *Heart (British Cardiac Society), 65*(5), 245–248. DOI: 10.1136/hrt.65.5.245 PMID: 2039668

Berniakovich, I., Trinei, M., Stendardo, M., Migliaccio, E., Minucci, S., Bernardi, P., Pelicci, P. G., & Giorgio, M. (2008). P66Shc-Generated Oxidative Signal Promotes Fat Accumulation. *The Journal of Biological Chemistry, 283*(49), 34283–34293. DOI: 10.1074/jbc.M804362200 PMID: 18838380

Bertuglia, S. (2009). Preconditioning and Shear Stress in the Microcirculation in Ischemia-Reperfusion Injury. *Current Cardiology Reviews, 2*(4), 237–245. DOI: 10.2174/157340306778772913

Biondi, G., Marrano, N., Borrelli, A., Rella, M., D'Oria, R., Genchi, V. A., & Natalicchio, A. (2023). The p66Shc redox protein and the emerging complications of diabetes. *International Journal of Molecular Sciences, 25*(1), 108.

Bujak, M., & Frangogiannis, N. G. (2007). The Role of TGF-β Signaling in Myocardial Infarction and Cardiac Remodeling. *Cardiovascular Research, 74*(2), 184–195. DOI: 10.1016/j.cardiores.2006.10.002 PMID: 17109837

Camara, A. K. S., Bienengraeber, M., & Stowe, D. F. (2011). Mitochondrial Approaches to Protect against Cardiac Ischemia and Reperfusion Injury. *Frontiers in Physiology*, *8563*(APR), •••. DOI: 10.3389/fphys.2011.00013 PMID: 21559063

Cano Sanchez, M., Lancel, S., Boulanger, E., & Neviere, R. (2018). Targeting oxidative stress and mitochondrial dysfunction in the treatment of impaired wound healing: A systematic review. *Antioxidants*, *7*(8), 98.

Carpi, A., Menabò, R., Kaludercic, N., Pelicci, P. G., Di Lisa, F., & Giorgio, M. (1787). 2009a. "The Cardioprotective Effects Elicited by P66Shc Ablation Demonstrate the Crucial Role of Mitochondrial ROS Formation in Ischemia/Reperfusion Injury." Biochimica et Biophysica Acta (BBA) -. *Bioenergetics*, *1787*(7), 774–780. DOI: 10.1016/j.bbabio.2009.04.001 PMID: 19362067

Carpi, A., Menabò, R., Kaludercic, N., Pelicci, P. G., Di Lisa, F., & Giorgio, M. (1787). 2009b. "The Cardioprotective Effects Elicited by P66Shc Ablation Demonstrate the Crucial Role of Mitochondrial ROS Formation in Ischemia/Reperfusion Injury." Biochimica et Biophysica Acta (BBA) -. *Bioenergetics*, *1787*(7), 774–780. DOI: 10.1016/j.bbabio.2009.04.001 PMID: 19362067

Chen, K., Li, Y., Zhang, X., Ullah, R., Tong, J., & Shen, Y. (2022). The role of the PI3K/AKT signalling pathway in the corneal epithelium: Recent updates. *Cell Death & Disease*, *13*(5), 513.

Chen, Q., Vazquez, E. J., Moghaddas, S., Hoppel, C. L., & Lesnefsky, E. J. (2003). Production of Reactive Oxygen Species by Mitochondria. *The Journal of Biological Chemistry*, *278*(38), 36027–36031. DOI: 10.1074/jbc.M304854200 PMID: 12840017

Cheung, E. C., & Vousden, K. H. (2022). The role of ROS in tumour development and progression. *Nature Reviews. Cancer*, *22*(5), 280–297.

Clerkin, J. S., Naughton, R., Quiney, C., & Cotter, T. G. (2008). Mechanisms of ROS Modulated Cell Survival during Carcinogenesis. *Cancer Letters*, *266*(1), 30–36. DOI: 10.1016/j.canlet.2008.02.029 PMID: 18372105

Corral, V. M., Schultz, E. R., Eisenstein, R. S., & Connell, G. J. (2021). Roquin Is a Major Mediator of Iron-Regulated Changes to Transferrin Receptor-1 MRNA Stability. *iScience*, *24*(4), 102360. DOI: 10.1016/j.isci.2021.102360 PMID: 33898949

D'Arcy, M. S. (2019). Cell Death: A Review of the Major Forms of Apoptosis, Necrosis and Autophagy. *Cell Biology International*, *43*(6), 582–592. DOI: 10.1002/cbin.11137 PMID: 30958602

Dey, S., DeMazumder, D., Sidor, A., Foster, D. B., & O'Rourke, B. (2018). Mitochondrial ROS Drive Sudden Cardiac Death and Chronic Proteome Remodeling in Heart Failure. *Circulation Research, 123*(3), 356–371. DOI: 10.1161/CIRCRESAHA.118.312708 PMID: 29898892

Di Lisa, F., Menabò, R., Canton, M., & Petronilli, V. (1998). "The Role of Mitochondria in the Salvage and the Injury of the Ischemic Myocardium." Biochimica et Biophysica Acta (BBA) -. *Bioenergetics, 1366*(1–2), 69–78. DOI: 10.1016/S0005-2728(98)00121-2 PMID: 9714744

Dikalova, A. E., Bikineyeva, A. T., Budzyn, K., Nazarewicz, R. R., McCann, L., Lewis, W., Harrison, D. G., & Dikalov, S. I. (2010). Therapeutic Targeting of Mitochondrial Superoxide in Hypertension. *Circulation Research, 107*(1), 106–116. DOI: 10.1161/CIRCRESAHA.109.214601 PMID: 20448215

Fadini, G. P., Ceolotto, G., Pagnin, E., De Kreutzenberg, S., & Avogaro, A. (2011). At the Crossroads of Longevity and Metabolism: The Metabolic Syndrome and Lifespan Determinant Pathways. *Aging Cell, 10*(1), 10–17. DOI: 10.1111/j.1474-9726.2010.00642.x PMID: 21040402

Fan, J., Ren, D., Wang, J., Liu, X., Zhang, H., Wu, M., & Yang, G. (2020). Bruceine D induces lung cancer cell apoptosis and autophagy via the ROS/MAPK signaling pathway in vitro and in vivo. *Cell Death & Disease, 11*(2), 126.

Franzeck, F. C., Hof, D., Spescha, R. D., Hasun, M., Akhmedov, A., Steffel, J., Shi, Y., Cosentino, F., Tanner, F. C., von Eckardstein, A., Maier, W., Lüscher, T. F., Wyss, C. A., & Camici, G. G. (2012). Expression of the Aging Gene P66Shc Is Increased in Peripheral Blood Monocytes of Patients with Acute Coronary Syndrome but Not with Stable Coronary Artery Disease. *Atherosclerosis, 220*(1), 282–286. DOI: 10.1016/j.atherosclerosis.2011.10.035 PMID: 22100252

Galy, B., Feering, D., & Hentze, M. W. (2005). Generation of Conditional Alleles of the Murine Iron Regulatory Protein (IRP)-1 and -2 Genes. *Genesis (New York, N.Y.), 43*(4), 181–188. DOI: 10.1002/gene.20169 PMID: 16283625

Ganz, T. (2019). Anemia of Inflammation. *The New England Journal of Medicine, 381*(12), 1148–1157. DOI: 10.1056/NEJMra1804281 PMID: 31532961

Ghatak, A., Mohan, J. S. B., Agarwal, A., Goel, N., Rastogi, A. K., Vaish, A. K., Sircar, A. R., & Chandra, M. (1996). Oxy Free Radical System in Heart Failure and Therapeutic Role of Oral Vitamin E. *International Journal of Cardiology, 57*(2), 119–127. DOI: 10.1016/S0167-5273(96)02787-8 PMID: 9013263

Giorgio, M., Berry, A., Berniakovich, I., Poletaeva, I., Trinei, M., Stendardo, M., Hagopian, K., Ramsey, J. J., Cortopassi, G., Migliaccio, E., Nötzli, S., Amrein, I., Lipp, H. P., Cirulli, F., & Pelicci, P. G. (2012). The P66Shc Knockout Mice Are Short Lived under Natural Condition. *Aging Cell*, *11*(1), 162–168. DOI: 10.1111/j.1474-9726.2011.00770.x PMID: 22081964

Gong, Y., Luo, S., Fan, P., Zhu, H., Li, Y., & Huang, W. (2020). Growth Hormone Activates PI3K/Akt Signaling and Inhibits ROS Accumulation and Apoptosis in Granulosa Cells of Patients with Polycystic Ovary Syndrome. *Reproductive Biology and Endocrinology*, *18*(1), 1–12. DOI: 10.1186/s12958-020-00677-x PMID: 33287836

Graiani, G., Lagrasta, C., Migliaccio, E., Spillmann, F., Meloni, M., Madeddu, P., Quaini, F., Padura, I. M., Lanfrancone, L., Pelicci, P. G., & Emanueli, C. (2005). Genetic Deletion of the P66Shc Adaptor Protein Protects from Angiotensin II-Induced Myocardial Damage. *Hypertension*, *46*(2), 433–440. . DOI: 10.1161/01.HYP.0000174986.73346.ba PMID: 15998704

Haschka, D., Hoffmann, A., & Weiss, G. (2021). Iron in Immune Cell Function and Host Defense. *Seminars in Cell & Developmental Biology*, *115*, 27–36. DOI: 10.1016/j.semcdb.2020.12.005 PMID: 33386235

Herb, Marc, and Michael Schramm. 2021. "Functions of ROS in Macrophages and Antimicrobial Immunity." Antioxidants 2021, Vol. 10, Page 313 10(2):313. .DOI: 10.3390/antiox10020313

Huang, T.-T., Sun, W.-J., Liu, H.-Y., Ma, H.-L., & Cui, B.-X. (2021). P66Shc-Mediated Oxidative Stress Is Involved in Gestational Diabetes Mellitus. *World Journal of Diabetes*, *12*(11), 1894–1907. DOI: 10.4239/wjd.v12.i11.1894 PMID: 34888014

Jamil, M., Debbarh, H., Aboulmaouahib, S., Aniq Filali, O., Mounaji, K., Zarqaoui, M., Saadani, B., Louanjli, N., & Cadi, R. (2020). Reactive Oxygen Species in Reproduction: Harmful, Essential or Both? *Zygote (Cambridge, England)*, *28*(4), 255–269. DOI: 10.1017/S0967199420000179 PMID: 32340646

Kaludercic, N., & Di Lisa, F. (2020). Mitochondrial ROS Formation in the Pathogenesis of Diabetic Cardiomyopathy. *Frontiers in Cardiovascular Medicine*, *7*, 508901. DOI: 10.3389/fcvm.2020.00012 PMID: 32133373

Keith, M., Geranmayegan, A., Sole, M. J., Kurian, R., Robinson, A., Omran, A. S., & Jeejeebhoy, K. N. (1998). Increased Oxidative Stress in Patients With Congestive Heart Failure 1. *Journal of the American College of Cardiology*, *31*(6), 1352–1356. DOI: 10.1016/S0735-1097(98)00101-6 PMID: 9581732

Kennedy, K. A. M., Sandiford, S. D. E., Skerjanc, I. S., & Li, S. S. C. (2012). Reactive Oxygen Species and the Neuronal Fate. *Cellular and Molecular Life Sciences : CMLS*, *69*(2), 215–221. DOI: 10.1007/s00018-011-0807-2 PMID: 21947442

Kim, S. Y., Hwangbo, H., Kim, M. Y., Ji, S. Y., Lee, H., Kim, G. Y., Kwon, C. Y., Leem, S. H., Hong, S. H., Cheong, J. H., & Choi, Y. H. (2021). Coptisine Induces Autophagic Cell Death through Down-Regulation of PI3K/Akt/MTOR Signaling Pathway and up-Regulation of ROS-Mediated Mitochondrial Dysfunction in Hepatocellular Carcinoma Hep3B Cells. *Archives of Biochemistry and Biophysics*, *697*, 108688. DOI: 10.1016/j.abb.2020.108688 PMID: 33227289

Kma, L., & Baruah, T. J. (2022). The Interplay of ROS and the PI3K/Akt Pathway in Autophagy Regulation. *Biotechnology and Applied Biochemistry*, *69*(1), 248–264. DOI: 10.1002/bab.2104 PMID: 33442914

Lee, I. T., & Chuen, M. Y. (2012). Role of NADPH Oxidase/ROS in pro-Inflammatory Mediators-Induced Airway and Pulmonary Diseases. *Biochemical Pharmacology*, *84*(5), 581–590. DOI: 10.1016/j.bcp.2012.05.005 PMID: 22587816

Lee, J., You, J. H., & Roh, J. L. (2022). Poly(RC)-Binding Protein 1 Represses Ferritinophagy-Mediated Ferroptosis in Head and Neck Cancer. *Redox Biology*, *51*, 102276. DOI: 10.1016/j.redox.2022.102276 PMID: 35290903

Li, H., Prever, L., Hirsch, E., & Gulluni, F. (2021). Targeting Pi3k/Akt/Mtor Signaling Pathway in Breast Cancer. *Cancers (Basel)*, *13*(14), 3517. DOI: 10.3390/cancers13143517 PMID: 34298731

Liu, Y., Liu, F., Grundke-Iqbal, I., Iqbal, K., & Cheng, X. G. (2011). Deficient Brain Insulin Signalling Pathway in Alzheimer's Disease and Diabetes. *The Journal of Pathology*, *225*(1), 54–62. DOI: 10.1002/path.2912 PMID: 21598254

Liu, Y., Sweet, D. T., Irani-Tehrani, M., Maeda, N., & Tzima, E. (2008). Shc Coordinates Signals from Intercellular Junctions and Integrins to Regulate Flow-Induced Inflammation. *The Journal of Cell Biology*, *182*(1), 185–196. DOI: 10.1083/jcb.200709176 PMID: 18606845

Long, H. Z., Cheng, Y., Zi, W. Z., Hong, Y. L., Dan, D. W., & Li, C. G. (2021). PI3K/AKT Signal Pathway: A Target of Natural Products in the Prevention and Treatment of Alzheimer's Disease and Parkinson's Disease. *Frontiers in Pharmacology*, *12*, 648636. DOI: 10.3389/fphar.2021.648636 PMID: 33935751

López-Gómez, C., Santiago-Fernández, C., García-Serrano, S., García-Escobar, E., Gutiérrez-Repiso, C., Rodríguez-Díaz, C., & García-Fuentes, E. (2020). Oleic acid protects against insulin resistance by regulating the genes related to the PI3K signaling pathway. *Journal of Clinical Medicine*, *9*(8), 2615.

McDonnell, A. M., & Dang, C. H. (2013). Basic review of the cytochrome p450 system. *Journal of the Advanced Practitioner in Oncology*, *4*(4), 263.

McMurray, J., Chopra, M., Abdullah, I., Smith, W. E., & Dargie, H. J. (1993). Evidence of Oxidative Stress in Chronic Heart Failure in Humans. *European Heart Journal*, *14*(11), 1493–1498. DOI: 10.1093/eurheartj/14.11.1493 PMID: 8299631

Migliaccio, E., Giorgio, M., Mele, S., Pelicci, G., Reboldi, P., Pandolfi, P. P., & Pelicci, P. G. (1999). The p66shc adaptor protein controls oxidative stress response and life span in mammals. *Nature*, *402*(6759), 309–313.

Migliaccio, E., Mele, S., Salcini, A. E., Pelicci, G., Ka, M. V. L., Superti-Furga, G., Pawson, T., Di Fiore, P. P., Lanfrancone, L., & Pelicci, P. G. (1997). Opposite Effects of the P52shc/P46shc and P66shc Splicing Isoforms on the EGF Receptor–MAP Kinase–Fos Signalling Pathway. *The EMBO Journal*, *16*(4), 706–716. DOI: 10.1093/emboj/16.4.706 PMID: 9049300

Mir, H. A., Ali, R., Mushtaq, U., & Khanday, F. A. (2020). Structure-Functional Implications of Longevity Protein P66Shc in Health and Disease. *Ageing Research Reviews*, *63*, 101139. DOI: 10.1016/j.arr.2020.101139 PMID: 32795504

Mittler, R. (2017). ROS Are Good. *Trends in Plant Science*, *22*(1), 11–19. DOI: 10.1016/j.tplants.2016.08.002 PMID: 27666517

Mittler, R., Zandalinas, S. I., Fichman, Y., & Van Breusegem, F. (2022). Reactive oxygen species signalling in plant stress responses. *Nature Reviews. Molecular Cell Biology*, *23*(10), 663–679.

Mohseni, A. H., Casolaro, V., Bermúdez-Humarán, L. G., Keyvani, H., & Taghinezhad-S., S. (2021). Modulation of the PI3K/Akt/MTOR Signaling Pathway by Probiotics as a Fruitful Target for Orchestrating the Immune Response. *Gut Microbes*, *13*(1), 1–17. DOI: 10.1080/19490976.2021.1886844 PMID: 33615993

Muckenthaler, M. U., Rivella, S., Hentze, M. W., & Galy, B. (2017). A Red Carpet for Iron Metabolism. *Cell*, *168*(3), 344–361. DOI: 10.1016/j.cell.2016.12.034 PMID: 28129536

Murphy, M. P., & Smith, R. A. (2007). Targeting antioxidants to mitochondria by conjugation to lipophilic cations. *Annual Review of Pharmacology and Toxicology*, *47*(1), 629–656.

Nakamura, H., & Takada, K. (2021). Reactive Oxygen Species in Cancer: Current Findings and Future Directions. *Cancer Science, 112*(10), 3945–3952. DOI: 10.1111/cas.15068 PMID: 34286881

Nguyen, G. T., Green, E. R., & Mecsas, J. (2017). Neutrophils to the ROScue: Mechanisms of NADPH Oxidase Activation and Bacterial Resistance. *Frontiers in Cellular and Infection Microbiology*, *7*(AUG), 264679. DOI: 10.3389/fcimb.2017.00373 PMID: 28890882

Noda, Y., Yamagishi, S. I., Matsui, T., So Ueda, S. I. U., Jinnouchi, Y., Hirai, Y., & Imaizumi, T. (2010). The P66shc Gene Expression in Peripheral Blood Monocytes Is Increased in Patients With Coronary Artery Disease. *Clinical Cardiology*, *33*(9), 548–552. DOI: 10.1002/clc.20761 PMID: 20842738

Pasricha, S. R., Tye-Din, J., Muckenthaler, M. U., & Swinkels, D. W. (2021). Iron Deficiency. *Lancet*, *397*(10270), 233–248. DOI: 10.1016/S0140-6736(20)32594-0 PMID: 33285139

Phaniendra, A., Jestadi, D. B., & Periyasamy, L. (2015). Free Radicals: Properties, Sources, Targets, and Their Implication in Various Diseases. *Indian Journal of Clinical Biochemistry : IJCB*, *30*(1), 11–26. DOI: 10.1007/s12291-014-0446-0 PMID: 25646037

Rai, S. N., Dilnashin, H., Birla, H., Singh, S. S., Zahra, W., Rathore, A. S., & Singh, S. P. (2019). The role of PI3K/Akt and ERK in neurodegenerative disorders. *Neurotoxicity Research*, *35*, 775–795.

Savova, M. S., Mihaylova, L. V., Tews, D., Wabitsch, M., & Georgiev, M. I. (2023). Targeting PI3K/AKT Signaling Pathway in Obesity. *Biomedicine and Pharmacotherapy*, *159*, 114244. DOI: 10.1016/j.biopha.2023.114244 PMID: 36638594

Sheng, Y. N., Ying, H. L., Shao, B. L., Wan, T. X., Zhang, Y., Zhang, T., Xue, H., Wen, B. Z., Yan, N. L., Chang, Y. W., & Cheng, H. J. (2020). Zeaxanthin Induces Apoptosis via ROS-Regulated MAPK and Akt Signaling Pathway in Human Gastric Cancer Cells. *OncoTargets and Therapy*, *13*, 10995–10996. DOI: 10.2147/OTT.S272514 PMID: 33149614

Shields, H. J., Traa, A., & Van Raamsdonk, J. M. (2021). Beneficial and Detrimental Effects of Reactive Oxygen Species on Lifespan: A Comprehensive Review of Comparative and Experimental Studies. *Frontiers in Cell and Developmental Biology*, *9*, 628157. DOI: 10.3389/fcell.2021.628157 PMID: 33644065

Shih, J. C., Chen, K., & Ridd, M. J. (1999). Monoamine oxidase: From genes to behavior. *Annual Review of Neuroscience*, *22*(1), 197–217.

Singh, R. B., Wander, G. S., Rastogi, A., Shukla, P. K., Mittal, A., Sharma, J. P., Mehrotra, S. K., Kapoor, R., & Chopra, R. K. (1998). Randomized, Double-Blind Placebo-Controlled Trial of Coenzyme Q10 in Patients with Acute Myocardial Infarction. *Cardiovascular Drugs and Therapy*, *12*(4), 347–353. DOI: 10.1023/A:1007764616025 PMID: 9825179

Soliman, M. A., Abdel Rahman, A. M., Lamming, D. A., Birsoy, K., Pawling, J., Frigolet, M. E., Lu, H., Fantus, I. G., Pasculescu, A., Zheng, Y., Sabatini, D. M., Dennis, J. W., & Pawson, T. (2014). The Adaptor Protein P66Shc Inhibits MTOR-Dependent Anabolic Metabolism. *Science Signaling*, *7*(313). Advance online publication. DOI: 10.1126/scisignal.2004785 PMID: 24550542

Son, Y., Kim, S., Chung, H. T., & Pae, H. O. (2013). Reactive Oxygen Species in the Activation of MAP Kinases. *Methods in Enzymology*, *528*, 27–48. DOI: 10.1016/B978-0-12-405881-1.00002-1 PMID: 23849857

Spescha, R. D., Klohs, J., Semerano, A., Giacalone, G., Derungs, R. S., Reiner, M. F., Rodriguez Gutierrez, D., Mendez-Carmona, N., Glanzmann, M., Savarese, G., Kränkel, N., Akhmedov, A., Keller, S., Mocharla, P., Kaufmann, M. R., Wenger, R. H., Vogel, J., Kulic, L., Nitsch, R. M., & Camici, G. G. (2015). Post-Ischaemic Silencing of P66Shc Reduces Ischaemia/Reperfusion Brain Injury and Its Expression Correlates to Clinical Outcome in Stroke. *European Heart Journal*, *36*(25), 1590–1600. DOI: 10.1093/eurheartj/ehv140 PMID: 25904764

Tomilov, A. A., Ramsey, J. J., Hagopian, K., Giorgio, M., Kim, K. M., Lam, A., Migliaccio, E., Lloyd, K. C., Berniakovich, I., Prolla, T. A., Pelicci, P. G., & Cortopassi, G. A. (2011). The Shc Locus Regulates Insulin Signaling and Adiposity in Mammals. *Aging Cell*, *10*(1), 55–65. DOI: 10.1111/j.1474-9726.2010.00641.x PMID: 21040401

Van Raamsdonk, J. M., & Hekimi, S. (2010). Reactive oxygen species and aging in Caenorhabditis elegans: Causal or casual relationship? *Antioxidants & Redox Signaling*, *13*(12), 1911–1953.

Vieira, H. L. A., Alves, P. M., & Vercelli, A. (2011). Modulation of Neuronal Stem Cell Differentiation by Hypoxia and Reactive Oxygen Species. *Progress in Neurobiology*, *93*(3), 444–455. DOI: 10.1016/j.pneurobio.2011.01.007 PMID: 21251953

Wang, Y., Qi, H., Liu, Yu., Duan, C., Liu, X., Xia, T., & Di Chen, H. L. P. (2021). The Double-Edged Roles of ROS in Cancer Prevention and Therapy. *Theranostics*, *11*(10), 4839–4857. DOI: 10.7150/thno.56747 PMID: 33754031

Wilson, C., Muñoz-Palma, E., & González-Billault, C. (2018). From Birth to Death: A Role for Reactive Oxygen Species in Neuronal Development. *Seminars in Cell & Developmental Biology*, *80*, 43–49. DOI: 10.1016/j.semcdb.2017.09.012 PMID: 28899716

Winterbourn, C. C. (2008). Reconciling the chemistry and biology of reactive oxygen species. *Nature Chemical Biology*, *4*(5), 278–286.

Xu, F., Na, L., Li, Y., & Chen, L. (2020). RETRACTED ARTICLE: Roles of the PI3K/AKT/mTOR signalling pathways in neurodegenerative diseases and tumours. *Cell & Bioscience*, *10*(1), 54.

Yang, M., Stowe, D. F., Udoh, K. B., Heisner, J. S., & Camara, A. K. S. (2014). Reversible Blockade of Complex I or Inhibition of PKCβ Reduces Activation and Mitochondria Translocation of P66Shc to Preserve Cardiac Function after Ischemia. *PLoS One*, *9*(12), e113534. DOI: 10.1371/journal.pone.0113534 PMID: 25436907

Yao, Z., Fu, L., Jia, F., Bi, M., Jiao, Q., Chen, X., Du, X., & Jiang, H. (2022). Rethinking IRPs/IRE System in Neurodegenerative Disorders: Looking beyond Iron Metabolism. *Ageing Research Reviews*, *73*, 101511. DOI: 10.1016/j.arr.2021.101511 PMID: 34767973

Yellon, D. M., & Hausenloy, D. J. (2007). Myocardial Reperfusion Injury. *The New England Journal of Medicine*, *357*(11), 1121–1135. DOI: 10.1056/NEJMra071667 PMID: 17855673

Yoshinaga, M., Nakatsuka, Y., Vandenbon, A., Ori, D., Uehata, T., Tsujimura, T., Suzuki, Y., Mino, T., & Takeuchi, O. (2017). Regnase-1 Maintains Iron Homeostasis via the Degradation of Transferrin Receptor 1 and Prolyl-Hydroxylase-Domain-Containing Protein 3 MRNAs. *Cell Reports*, *19*(8), 1614–1630. DOI: 10.1016/j.celrep.2017.05.009 PMID: 28538180

Yue, J., & López, J. M. (2020). Understanding MAPK signaling pathways in apoptosis. *International Journal of Molecular Sciences*, *21*(7), 2346.

Zaccagnini, G., Martelli, F., Fasanaro, P., Magenta, A., Gaetano, C., Di Carlo, A., Biglioli, P., Giorgio, M., Martin-Padura, I., Pelicci, P. G., & Capogrossi, M. C. (2004a). P66ShcA Modulates Tissue Response to Hindlimb Ischemia. *Circulation*, *109*(23), 2917–2923. DOI: 10.1161/01.CIR.0000129309.58874.0F PMID: 15173034

Zaccagnini, G., Martelli, F., Fasanaro, P., Magenta, A., Gaetano, C., Di Carlo, A., Biglioli, P., Giorgio, M., Martin-Padura, I., Pelicci, P. G., & Capogrossi, M. C. (2004b). P66ShcA Modulates Tissue Response to Hindlimb Ischemia. *Circulation*, *109*(23), 2917–2923. DOI: 10.1161/01.CIR.0000129309.58874.0F PMID: 15173034

Zhang, C., Wang, X., Du, J., Gu, Z., & Zhao, Y. (2021). Reactive Oxygen Species-Regulating Strategies Based on Nanomaterials for Disease Treatment. *Advancement of Science*, *8*(3), 2002797. DOI: 10.1002/advs.202002797 PMID: 33552863

Zheng, Y., Zhang, C., Croucher, D. R., Soliman, M. A., St-Denis, N., Pasculescu, A., & Pawson, T. (2013). Temporal regulation of EGF signalling networks by the scaffold protein Shc1. *Nature*, *499*(7457), 166–171.

Chapter 6
Mechanisms, Impact, and Therapeutic Strategies for Oxidative Stress in Metabolic Diseases:
Understanding and Mitigating Oxidative Stress

Reyed M Reyed
https://orcid.org/0000-0001-5027-3478
Genetic Engineering and Biotechnology Research Institute (GEBRI), Egypt & City of Scientific Research and Technological Applications (SRTA-City), Egypt

P. K. Prabhakar
https://orcid.org/0000-0001-8130-1822
Lovely Professional University, India & Nagaland University, India

ABSTRACT

This chapter explores the intricate relationship between oxidative stress and metabolic diseases, specifically diabetes and obesity. It begins with an overview of these conditions, emphasizing their global impact and associated metabolic dysfunctions. The discussion shifts to the biochemical mechanisms of oxidative stress, focusing on mitochondrial dysfunction and ROS production. It examines how oxidative stress affects pancreatic beta-cell function and insulin resistance, linking hyperglycemia to increased oxidative damage. Additionally, the chapter addresses adipose tissue dysfunction and the interplay between inflammation and oxidative stress in obesity-related complications. It concludes by discussing strategies to mitigate oxidative stress through pharmacological and lifestyle interventions, such as antioxidant

DOI: 10.4018/979-8-3693-7919-6.ch006

therapies, dietary changes, regular exercise, and caloric restriction. This comprehensive approach aims to provide insights into effectively managing oxidative stress, thereby improving metabolic health and addressing the challenges posed by diabetes and obesity.

1. INTRODUCTION TO METABOLIC DISEASES AND OXIDATIVE STRESS

In international health, chronic non-communicable diseases, particularly metabolic disorders such as diabetes and obesity, have become major causes of morbidity and mortality. Owing to the increasing prevalence of metabolic syndrome, which has been increasing in the recent past, is mostly caused by poor diet and a lack of exercise in addition to stressful environments (Ruze, et al., 2023). First, obesity not only is a risk factor for type 2 diabetes but also contributes to the worsening of metabolic distortions affecting various organs and systems (World Health Organization, 2023). The complex biochemical processes involved in the development of these diseases are most often associated with an event called oxidative stress, which is defined as the condition in which the body produces higher levels of ROS than dose the capacity of the antioxidant system (Lopez-Otin & Kroemer, 2021). Moreover, when metabolic diseases are treated via the One Health approach, the well- being of people depends on the well- being of the animal and the environment. This perspective recognizes the components causing metabolic disorders, including environmental endocrine-disrupting chemicals and animal-derived food items, that influence health, hence compounding illnesses such as diabetes and obesity. One health approach supports comprehensive approaches that recognize how shifts in agriculture, climate, and environmental policies impact health. For example, antibiotic residues in animals can lead to antibiotic-resistant bacteria, affecting people's conditions and making the management of metabolic diseases challenging (World Health Organization, 2022). It can therefore be beneficial to adopt a One Health approach for the prevention of metabolic diseases to develop more effective interventions that address this problem through improvement in different aspects of health.

1.1. Outline of Diabetes and Obesity: Comprehensive Influence and Metabolic Dysfunction

Importantly, obesity is very closely related to diabetes. Obesity also leads to insulin resistance and type 2 diabetes and is a causal factor for a phenomenon known as metabolic syndrome, which is characterized by dyslipidemia, hypertension and hyperglycemia (Ibrahim et al., 2024). These diseases have become predominant globally through an increase in processed foods, reduced physical activity and increased urbanization, leading to changes in dietary patterns that favor energy dense nutrient-poor foods and drinks such as processed foods (Casari et al., 2022). In addition, the climate change crisis has exacerbated metabolic dys-health through food insecurity and changes in agricultural productivity, resulting in the use of additives, artificial sweeteners and cheap, high-glycemic-index foods (Hendriks et al., 2022 & Ghosh-Jerath et al., 2021).

1. 2 Defining oxidative stress: Causes, consequences, and biochemical mechanisms

There is a significant correlation between oxidative stress, diabetes, and obesity. Oxidative stress at the molecular level develops when the levels of ROS, including free radicals such as superoxide and hydrogen peroxide increase when mitochondrial dysfunctions cause accrued amounts of ROS, which may result in intolerable levels of oxidation (Caturano et al., 2023; Zhang et al., 2023; Zhao et al., 2024). These excess ROS can harm cellular proteins, lipids and DNA, which consequently decreases metabolic efficiency (Thorne et al., 2024). In diabetes, hyperglycemia intensifies ROS through the accumulation of the load in the mitochondrial electron transport chain, which subsequently enhances oxidative injury and inflammation (Esposito et al., 2021). The excess accumulation of fat causes fat cell turnover to become a substantial source of oxidative stress in obesity because obesity alters the metabolic state of adipose tissue and increases the production of inflammatory cytokines (Manna et al., 2015).

1.3 The Role of Mitochondrial Dysfunction and ROS Production in Metabolic Diseases

These genes are essential for energy production, and mutations in these genes result in metabolic disorders. In diabetic mitochondria dysfunction, glucose-induced impaired pancreatic beta-cell mitochondrial dysfunction results in decreased insulin release and enhanced hyperglycemia (Diane et al., 2023). Similarly, in obesity, mitochondrial dysfunction in adipocytes disrupts their capacity to store and metabolize fat and enhances systemic insulin resistance (Zheng et al., 2023). As a consequence of metabolic dysfunction, signaling pathways involving mitochondrial dysfunction also directly stimulate ROS production, leading to other oxidative stresses. This mutual relationship between mitochondrial dysfunction and oxidative stress is the core issue of metabolic diseases, and an understanding of the mechanism requires careful targeted therapies.

1.4 Mechanisms of the Microbiota in Metabolic Syndrome

The enteric microbiome has now been identified as one of the major regulators of metabolic health. Changes in the microbial structure and metabolic activity profile of gut microbiota, known as dysbiosis, have been associated with obesity and diabetes (Li et al. 2024). In turn, the microbiota controls diet digestion, inflammation, and even the level of insulin in the blood. An altered microbiota can further exacerbate oxidative stress by increasing the inflammatory response to LPS and damaging the integrity of the gut lining, thereby allowing toxic end products to enter the circulatory system (Sankararaman et al., 2023). Some of the factors that affect the microbiota include diet, climate change and food additives. For example, excessive ingestion of artificial sweeteners including aspartame, which is commonly used, has been observed to cause dysbiosis and increase the level of oxidative stress (Sparling et al., 2024). However, free sugars present in processed foods are also a source of concern, as they distort the balance between nutrients and energy and fuel metabolic derangement, leading to chronic diseases (Joó, et al., 2023).

1.5 Environmental pollutants, their impact on climate change and connection to metabolic health.

New evidence shows a remarkable connection between climate fluctuations, environmental conditions, and metabolic ailments. Essentially, climate affects food production involving farming and the food supply and consequently affects diet and metabolic disorders. Global warming is related to malnutrition and changes in diets, resulting in the consumption of unhealthier foods due to increased temperatures and the incidence of weather events (Haines & Ebi, 2022). This shift in diet comprises diet additives and sweetening substances such as aspartame, which are known to have detrimental impacts effects on the human gut microbiota and metabolic systems in Stoev et al., 2024. The current eating habits include processed foods with artificial sweeteners, food additives and free sugars, all of which are causes of metabolic diseases. The expenditure of aspartame and other artificial sweeteners cut products popularizes microbiota dysbiosis, unearthing insulin resistance and even worsening oxidative pressure (Sabarathinam et al., 2023). Because free sugars are prevalent in sugary foods, beverages, and processed foods, excessive intake of these sugars can increase ROS levels by increasing the body's antioxidant capacity (Yuan et al., 2023). Furthermore, change in environmental pollutants and pesticides due to the climate increase oxidative stress caused by extraneous toxic substances in the food chain. These pollutants interfere with endocrine functions and lead to the onset of metabolic syndrome (Khalil, et al., 2023). Pollutants that affect the environment and microbial organisms are related, as pollutants modulate microbial features to affect oxidative stress, which has an impact on metabolism (Mancini et al., 2023). Furthermore, increased consumption of foods with free sugars and artificial sweeteners also put individuals at risk of oxidative stress and inflammation, which are observed in the year 2023 by Shaher, et al. Such relationships show that it is possible to use individual and combined approaches to address oxidative stress and metabolic diseases, where consideration of both environmental and dietary components is necessary.

2. BIOCHEMICAL MECHANISMS OF OXIDATIVE STRESS IN METABOLIC DISEASES PATHOPHYSIOLOGY OF OXIDATIVE STRESS IN PANCREATIC BETA-CELLS

Reactive oxygen species play a central role in the etiology of metabolic disorders involving damage to the pancreatic beta cells responsible for the synthesis of insulin and the regulation of blood glucose content. These cells, for example, are highly vulnerable to oxidative stress due to their high metabolic activity and constant production of ROS, which are metabolites of glucose.

Table 1. Mechanisms of oxidative stress in pancreatic beta-cells

Mechanism	Description
Mitochondrial Dys-function Sendani et al (2024).	Chronically high blood sugar strains the electron transport pathways in the mitochondria, increasing the production of reactive oxygen species (ROS). Insulin resistance and beta-cell death are mostly caused by malfunctioning mitochondria.
Advanced Glycation End Products (AGEs) Lee et al, (2022).	Elevated glucose levels can lead to the production of AGEs through two different mechanisms: increased creation of reactive oxygen species (ROS) and inflammation. Then, in order to promote oxidative stress, these AGEs interact with cellular receptors like RAGE (receptor for AGEs). The defective functioning of beta cells is exacerbated by this combination.
Polyol Pathway Activation Garg et al. (2022).	When the polyol pathway is activated during hyperglycemia, glucose is converted to sorbitol and fructose, which results in the production of superoxide radicals. These radicals promote both beta-cell malfunction and oxidative damage.

The mechanisms described in Table 1 explain how increased glucose causes oxidative stress in the beta -cells of pancreas. The oxidative stress process is caused by an increase in glucose, which leads to mitochondrial dysfunction and increased ROS, which in turn cause destruction of cellular components and induce beta-cell apoptosis (Sendani, et al., 2024). AGEs are formed from glycation and affect intracellular receptors such as RAGE, which in turn increases oxidative stress and inflammation levels (Lee et al., 2022). Hyperglycemia activates the polyol pathway in the body, which leads to the generation of superoxide radicals that increase oxidative stress in beta cells (Garg et al., 2022).

2.1 Mechanisms Linking Hyperglycemia and Oxidative Damage:

Hyperglycemia is a main factor that causes oxidative damage in different ways. AGEs formed through the cross-linking of excess glucose with other receptors such as RAGE, increase the production of ROS, which culminates in oxidative stress (González et al., 2023). This process is further aggravated by mitochondrial disorders, because when the glucose supply is high, it results in the loading of electrons in the chain of mitochondrial electron transport with subsequent production of superoxide radicals (Khan et al., 2022). Furthermore, hyperglycemia activates the polyol pathway, leading to the formation of superoxide radicals, thus increasing oxidative stress, as loading noted by Garg et al. (2022).

Table 2. Hyperglycemia induced oxidative stress mechanisms and recent case studies highlighting key biochemical pathways and their impact on metabolic complications

Mechanism	Description	Case Study
Mitochondrial ROS Overproduction in Hyperglycemia (Zhang et al., 2023)	Reactive oxygen species (ROS) are produced in excess when mitochondrial electron transport chain (ETC) activity is greatly increased by elevated glucose levels. Macromolecules, including lipids, proteins, and DNA, are harmed by this imbalance, which also interferes with regular cellular signaling.	A 2023 investigation found that individuals with uncontrolled type 2 diabetes had significant mitochondrial dysfunction and that oxidative stress mediated by reactive oxygen species (ROS) was a contributing factor to the compromised vascular homeostasis.
Hexosamine Biosynthesis Pathway (HBP) Dysregulation (Itano et al., 2023)	Hyperglycemia promotes flux through the HBP, leading to increased levels of the end-product UDP-N-acetylglucosamine, which modifies proteins via O-GlcNAcylation. This post translational modification disrupts cellular functions, increasing oxidative stress and promoting inflammatory responses.	A longitudinal study found that hyperglycemia-induced HBP over activity exacerbates insulin resistance and systemic inflammation, correlating with higher oxidative markers in diabetic patients.
Formation and Accumulation of Advanced Glycation End Products (AGEs) (Zgutka et al., 2023)	The formation of AGEs is accelerated by hyperglycemia due to the non-enzymatic glycation of lipids and proteins. By binding to RAGE (receptor for AGEs), these products enhance oxidative stress and vascular damage by triggering pro-inflammatory and pro-oxidative pathways, such as NF-κB activation.	According to recent studies, endothelial dysfunction in type 2 diabetes patients is associated with the formation of AGEs, which raises the risk of atherosclerosis and microvascular consequences. Certain PKC isoforms are activated by hyperglycemia, especially in endothelial cells.
Activation of Protein Kinase C (PKC) Isoforms and NADPH Oxidase (Pan et al., 2022)	This activation leads to prolonged oxidative stress, inflammation, and vascular consequences such as retinopathy and nephropathy by upregulating NADPH oxidase, a significant generator of ROS.	A clinical investigation showed that PKC isoform hyperactivity was present in diabetic patients with microvascular problems, which was correlated with higher oxidative stress and endothelial cell death.
Impaired Antioxidant Defense Mechanisms (Ordóñez-Díaz et al., 2021)	Unchecked oxidative damage results from hyperglycemia's suppression of important antioxidant mechanisms, such as glutathione peroxidase (GPx) and superoxide dismutase (SOD). Depletion of NADPH, which is essential for antioxidant regeneration, intensifies inflammation caused by oxidative stress in several tissues.	Unchecked oxidative damage results from hyperglycemia's suppression of important antioxidant mechanisms, such as glutathione peroxidase (GPx) and superoxide dismutase (SOD). In several tissues, oxidative stress intensifies inflammation due to NADPH depletion, which is crucial for antioxidant regeneration.
Polyol Pathway Flux and Sorbitol Accumulation (Srikanth &Orrick, 2024)	Hyperglycemia increases the amount of glucose that aldose reductase converts to sorbitol by pushing glucose through the polyol pathway. Thus, sorbitol accumulation causes osmotic stress and intensifies oxidative damage, while this process consumes NADPH, decreasing its availability for antioxidant defenses. According to a recent meta-analysis, oxidative damage caused by reactive oxygen species (ROS) is the primary cause of diabetic neuropathy, and hyperactivation of the polyol pathway in diabetes patients is a significant element in this process.	According to a recent meta-analysis, oxidative damage caused by reactive oxygen species (ROS) is the primary cause of diabetic neuropathy, and hyperactivation of the polyol pathway in diabetes patients is a significant element in this process.

2.2 Lipid metabolism pathway and oxidative stress pathway:

Oxidative stress is therefore achieved when there is a disproportion between the generation of ROS and the body's ability to detoxify them via antioxidants. High levels of ROS can alter proteins and lipids, but in the cell, they are not able to perform their functions efficiently. For that reason, in the case of insulin resistance, ROS affect many critical components of insulin signaling.

1 **Insulin Receptor Substrate (IRS)** proteins: Initiation of the insulin signaling process occurs when insulin links to its receptor at the cell membrane surface, which activates IR-TK. This activation results in the phosphorylation of IRS proteins, which play a critical role in the subsequent signaling events. ROS-induced oxidative stress may lead to the oxidation of these IRS proteins and thus their signaling functions. This oxidation reduces the phosphorylation of IRS and therefore fails to enhance downstream signaling pathways, including the PI3K/Akt pathway, which is vital for the uptake of glucose (Tangvarasittichai 2015).

2 The **Akt/PKB pathway** results in the inactivation of Akt, which causes insulin resistance (Mackenzie et al., 2014). The PI3K/Akt /PI3K/Akt pathway is crucial for regulating the effects of insulin on glucose homeostasis. After being phosphorylated, IRS proteins stimulate phosphatidylinositol 3-kinase (PI3K), which activates protein kinase B (Akt). ROS can inhibit this pathway by stimulating the phosphorylation of different serine kinases such as JNK and IKK, which target IRS proteins and phosphorylate them on serine residues. This modification weakens the trigger of IRS so that it cannot activate PI3K efficiently and therefore

3 **NF-κB Activation**: Another inflammatory transcription factor is NF-κB, which, upon activation, stimulates inflammation and impairs insulin signaling. ROS can activate NF-κB through oxidative changes in IκB proteins, which are otherwise inhibitory to NF-κB function. Depending on the level of IκB, if it is degraded, then NF-κB translocate to the nucleus and turns on genes associated with inflammation, such as TNF-α and IL-6. These cytokines also reduce insulin signaling by enhancing the degradation of the insulin receptor substrate (Lingappan., 2018).

2.3 Understanding the Mechanisms of Inflammation and Insulin Signaling Defects:

These findings indicate that inflammation is a product and catalyst of oxidative stress and insulin preconditioning, as shown in Table 3.

Table 3. Oxidative stress mechanisms in insulin resistance

Molecular Pathway	Impact of Oxidative Stress
IRS Protein Oxidation. Lennicke et al .(2021)	Redox modifications at the level of IRS proteins by ROS affect the ability of these proteins to activate the PL3K/Akt cascade and, thus, insulin signaling and glucose uptake.
Akt/PKB Pathway Disruption Yung et al.(2020)	Enhanced ROS also inactivate the serine kinases (JNK, IKK) and phosphorylate the serine of the IRS proteins leading to decreased PI3K/Akt pathway and further insulin resistance.
NF-κB Activation Baker et al 2011	ROS enhances NF-κB activation followed by inflammation and further insulin resistance due to increased cytokines such as TNF-α and IL-6.
Inflammatory Cytokine Release Caturano et al. (2023)	Because the cellular processes in response to oxidative stress stimulate cytokines such as TNF-α and IL-6, this continues to induce more insulin resistance and oxidative damage.
Adipokine Dysregulation Yan K. (2024)	There are changes in adipokines in the state of oxidative stress: pro-inflammatory cytokines and markers of inflammation and insulin resistance.).
Mitochondrial Dysfunction Zhang et al. (2023)	Abnormal production of ROS as a result of altered mitochondria leads to decreased energy generation and enhances inflammation, therefore making cells more responsive to insulin.

3. OXIDATIVE STRESS AND INFLAMMATION HAVE A SYNERGISTIC EFFECT ON METABOLIC DYSFUNCTION: ONE FUELS THE OTHER IN A VICIOUS CYCLE:

1. Cytokine-induced Insulin Resistance: Inflammatory cytokines, such as TNF-a, IL-6, and IL-1β, are also upregulated in response to oxidative stress. These cytokines can inhibit insulin action by altering signal transduction pathways through the induction of serine/threonine phosphorylation of IRS proteins. This alteration reduces their ability to bind with PI3K, hence reducing insulin effectiveness in regard to glucose transport. Moreover, these cytokines can stimulate the synthesis of other enzymes, such as inducible nitric oxide synthase (iNOS) and cyclooxygenase-2 (COX-2), which also promote oxidative stress and insulin resistance (Gonzalez et al., 2024).

2. Obesity involves excess adipose tissue and systemic inflammation. Manufactured obesity is relatively recent and involves adipose tissue, especially visceral fat, as a primary contributor to systemic inflammation and insulin resistance. Adipocytes secrete several proinflammatory molecules, including leptin and adiponectin, among others, that are commonly referred to as adipokines. Leptin has been suggested to be pro-inflammatory and an inducer of oxidative stress and insulin resistance, whereas adiponectin may counteract the effects of leptin. These cytokines are released

when inflammation is triggered by ROS in adipose tissue, which intensifies insulin resistance (Savulescu-Fiedler et al., 2024).

3. Role of Mitochondrial Dysfunction: Mitochondria are important organelles involved in energy generation processes and regulation of cellular redox status. Consequently, mitochondrial dysfunction in insulin resistance results in increased ROS production. This dysfunction leads to decreased levels of mitochondrial proteins along with reduced rates of mitochondrial biogenesis and fatty acid oxidation, in addition to promoting oxidative stress. In addition to their direct effect on insulin signaling, mitochondrial ROS can also activate inflammatory signaling cascades that can exacerbate insulin resistance. Interventions that address mitochondrial dysfunction or increase mitochondrial function may benefit insulin resistance (Kim et al., 2008). Table 3 shows how mitochondrial dysfunction, through its control of various molecular pathways, impairs oxidative stress and underwrites to the common pathology of insulin resistance. By addressing these mechanisms, novel therapeutic approaches that could improve metabolic health outcomes have potential.

Table 4. Hyperglycemia -induced oxidative stress mechanisms and recent case studies highlighting key biochemical pathways and their impact on metabolic complications

Mechanism	Description	Case Study
Mitochondrial ROS Overproduction in Hyperglycemia Wang et al., 2023	High glucose concentrations increase the function of the mitochondrial electron transport chain (ETC), thus increasing the production of reactive oxygen species (ROS). This makes the ratio in the cells to be imbalanced, affects normal signaling, and leads to the destruction of other macromolecules, including lipids, proteins, and DNA.	According to the study carried out in 2023 among the patients with uncontrolled T2D, it was noted that patients' mitochondria were extremely oxidative and damaged, and the increased production of ROS leads to disruption of the vascular homeostasis.
Hexosamine Biosynthesis Pathway (HBP) Dysregulation Lee et al., 2022	Raised blood glucose results in enhanced throughput in the HBP to increase the levels of the end product UDP-N-acetyl glucosamine, which in turn attaches to proteins by O-GlcNAc. This post translational modification interferes with the normal functioning of the cell and amplifies other unhealthy processes such as oxidative stress and inflammation.	Hyperglycemia-induced HBP overactivity was investigated in a longitudinal study, illustrating that it further aggravates insulin resistance and increases systemic inflammation, which is associated with higher oxidative markers in patients with diabetes.
Formation and Accumulation of Advanced Glycation End Products (AGEs Patel et al., 2023	High blood glucose levels increase the rate of nonenzyme glycosylation of proteins and lipids to form AGEs. These products attach to RAGE (receptor for AGEs), thus triggering pro-inflammatory and pro-oxidative pathways such as the NF-κB signaling, thereby promoting an increase in oxidative stress and endothelial dysfunction.	A study published recently found that AGEs levels are elevated in the patients with type 2 diabetes, and their deposition causes endothelial dysfunction, thus is associated with the increased risk of atherosclerosis and microvascular complications among patients with T2DM.

continued on following page

Table 4. Continued

Mechanism	Description	Case Study
Activation of Protein Kinase C (PKC) Isoforms and NADPH Oxidase (Tanaka et al., 2024)	A glucose-mediated increase in the activation of PKC's is also observed in endothelial cells. This activation increases NADPH oxidase, a major source of ROS, and leads to chronic increased oxidative stress, inflammation, and vascular complications such as retinopathy and nephropathy.	A clinical study proved diabetic patients who had microvascular complications showed the overexpression of PKC isoforms with augmented oxidative stress and death of endothelial cells.
Impaired Antioxidant Defense Mechanisms (Roberts et al., 2023)	Extrinsic hyperglycemia downs major antioxidant systems such as glutathione peroxidase (GPx) and superoxide dismutase (SOD), thus allowing oxidative injury. Finally, the oxidizing equivalent, NADPH, which is used up in recycling antioxidants, reduces antioxidant levels in the body and increases oxidative stress in multiple tissues.	Another cross-sectional study established that the type 1 diabetes patients had reduced antioxidant capability and lesser levels of GPx and SOD that played a crucial direct role in promoting the oxidative injury.
Polyol Pathway Flux and Sorbitol Accumulation (Dănilă et al., 2024)	This hyperglycemia forces glucose to go around the polyol pathway, where glucose is converted to sorbitol by the enzyme aldose reductase. This process depletes NADPH, which is used in defending the body through antioxidants, while sorbitol accumulation leads to osmotic pressure, thus increasing oxidative stress.	The meta-analysis performed revealed that hyperactivation of the polyol pathway in diabetes is an essential determinant in the pathogenesis of diabetic neuropathy through oxidative stress by ROS.

3.1 Adipose Tissue Dysregulation and Inflammation in Obesity: Effects, Comparative Analysis, and Implications for Metabolic Health

Over the years, there has been evidence that the adipose tissue is a common fat storage organ. The adipose compartment has what we call the adipokine secretion function. However, this tissue in and around obese individuals is of great concern, as it is the most dysregulated aspect that worsens metabolism and health. Leptin, adiponectin, inflammatory mechanisms in obesity, and insulin sensitivity—what do they mean and how do they tie up?

1 Leptin and Adiponectin Dysregulation

Leptin synthesis and release from adipocytes is known as an energy-regulating hormone because it suppresses food intake and causes increased expenditure of energy, the adipostatic hormone. In general, leptin tends to suppress appetite by sending messages to the hypothalamus. Unfortunately, under conditions of obesity, this phenomenon, called leptin resistance, becomes the norm. Too much of this hormone does not reduce the urge to eat increase the metabolic rate. As a result, the individual who is already obese becomes fatter, as he is in a perpetual state of binge eating. This resistance is also due to a decrease in the transport of leptin into the brain and faulty signaling of leptin in the hypothalamus (Picó et al., 2022).

Another important adipokine is adiponectin, which increases tissue sensitivity to insulin and has anti-inflammatory effects. Reducing insulin resistance facilitates both fatty acid oxidation and glucose uptake in skeletal muscle tissue. Obesity is generally associated with low levels of adiponectin. This leads to more pronounced insulin resistance and a general inflammatory response. Adiponectin deficiency is believed to be caused by increased secretion of proinflammatory cytokines. and changes in adipose tissue as an organ (Zorena, et al., 2020).

2 Adipose Tissue Inflammation

The accumulation of adipose tissue is reduced in the metabolic state by the frequent recruitment of macrophages. In response, these macrophages remain dormant. Until a metabolic insult causes them to express TNF-α, IL-6, and Il-1β, all of these factors contribute to insulin resistance due to abnormal insulin action and systemic inflammation. due to its effect on oxidation (Sun et al., 2024). The activation of NF-κB in adipose tissue in the presence of an inflammatory environment leads to the transcription of many inflammatory genes, which are often unlikely to be modified. Frequent activation of NF-kB is associated with increased metabolic abnormalities. especially insulin resistance (Griffin et al. 2022).

Table 5. Dysregulation of adipokines and inflammatory pathways in individuals with obesity

Pathway	Normal Function	Dysregulation in Obesity	Impact on Metabolism
Leptin (Picó et al., 2022).	Responsible for controlling hunger and energy consumption	Elevated levels; leptin resistance	Warning signs: increased appetite and subsequent excessive weight gain
Adiponectin (Picó et al., 2022).	Anti-inflammatory; enhances insulin sensitivity	Reduced levels	Elevated level of fasting glycemia, type 2 diabetes, insulin resistance, inflammatory biomarkers, C-reactive protein, and fibrinogen.
NF-κB (Griffin et al., 2022).	Regulates inflammatory gene expression	ROS-induced degradation of IκB	Elevated secretions of inflammatory cytokines affected insulin sensitivity.

continued on following page

Table 5. Continued

Pathway	Normal Function	Dysregulation in Obesity	Impact on Metabolism
JNK (Feng et al., 2020)	Regulates stress responses and insulin signaling Stimuli for their activation include oxidative stress and cytokines. Phosphorylation by serine; modification of the IRS proteins; reduction of insulin signal transduction)	Stimuli for their activation include oxidative stress and cytokines.	Phosphorylation by serine; modification of the IRS proteins; reduction of insulin signal transduction
Mitochondrial Dysfunction (Zhang et al., 2023)	Sustains redox state and energy donation	In this regard, the exposure to PAM can cause boosting the production of ROS; fatty acid oxidation is also affected.	Worse still, it increases insulin resistance and metabolic syndrome

This table summarizes the main results concerning the dysregulation of adipokines and inflammatory pathways in obesity. These findings illustrate how these interruptions play a role in different metabolic disorders by explaining particular adipokines, inflammatory pathways, and health consequences.

3.2 Increased oxidative stress and obesity-related complications.

The equilibrium of ROS production and antioxidant safety mechanisms has been investigated for the purpose of oxidative stress. Thus, it is far deemed to be one of the primary causes of complications related to weight Oxidative stress, which is involved in the pathogenesis of many persistent ailments, is a contributing factor to burden prediction.

1. Cardiovascular Diseases. Hypertension and atherosclerosis are among the many cardiovascular ailments, which are likely large because of weight troubles. The correlation between this finding and the increase in oxidative stress is decided upon in endothelial sickness because of weight troubles, which commonly results in reduced stages of nitric oxide and prolonged contamination. This is constant with reality. This results in arterial stiffness and atherosclerotic plaque formation. Low-density lipoprotein oxidation, which is one of the primary reasons for atherosclerosis, is aided through extended stages of ROS (Steven et al., 2019).

2. Type 2. Diabetes Mellitus. The development of T2DM essentially occurs through oxidative stress, which causes damage to the pancreas and impaired insulin secretion. An increase in ROS generation results in the depletion of cells, which

ultimately results in decreased insulin production and might lead to insulin resistance (Chang et al., 2010). Inflammation and oxidative stress are factors that, interfere with insulin signaling to modify glucose levels.

3. Cancer Risk. Moreover, the presence of oxidative stress during weight loss increases the susceptibility of patients to carcinomas. The function of ROS in oxidative DNA damage and genomic instability contributes to carcinogenesis. Additionally, tumor growth and development through angiogenesis and apoptosis are related to persistent contamination related to weight troubles (Reuter et al., 2010). Overall, the impaired weight of adipose tissue is associated with complex interactions among contaminants and endocrine disruption, resulting in cardiovascular ailments, insulin resistance, T2DM disease, and cancer. Accordingly, it is a promising treatment for every metabolic and inflammatory disorder.

4. AI AND COMPUTATIONAL APPROACHES TO PREDICT AND MODEL OXIDATIVE STRESS IN METABOLIC DISEASES

4.1 AI Prediction Models for Oxidative Stress Assessment in Diabetic and Obese Patients

The integration of artificial intelligence (AI) into the assessment of oxidative stress represents a great improvement in custom-designed medicine. AI models, which incorporate system based algorithms, offer functionality to increase predictive accuracy and provide early warnings about oxidative stress-related complications in patients with diabetes and obesity.

1. **Machine learning algorithms**, along with support vector machines (SVMs), random forests, and gradient boosting, have been employed to anticipate oxidative stress ranges the basis of scientific and biochemical data. For example, SVMs can classify patients into one-of-a-kind hazard training for oxidative stress-related illnesses through analyzing features along with biomarkers, affected individual demographics, and way of life factors (Huang et al., 2018). These models can be professionally trained on massive datasets to select patterns and anticipate the chance of oxidative stress and its associated complications.

2. **AI-Enhanced Risk Stratification:** AI models can enhance hazard stratification through the integration of multi-omics data, which incorporate genomics, proteomics, and metabolomics. This holistic approach allows for a more comprehensive assessment of oxidative stress through the evaluation of interactions among one-of-a-kind natural layers. For example, integrating genomic data with oxidative stress biomarkers can increase the identification of genetic predispositions to conditions such as type 2 diabetes and obesity (Babu et al., 2023).

The application of these AI models in recent studies has offered a strong foundation for assessing oxidative stress, enhancing precision in patient management, and promoting targeted prevention and treatment strategies. The following data table (6), which highlights new case studies and their particular applications, offers an updated overview of the numerous AI models used to predict oxidative stress in patients with diabetes and obesity.

Table 6. Artificial intelligence prediction models for assessing oxidative stress in recent studies

AI Model	Application	Advantages	Recent Case Study
Support Vector Machines (SVM)	Successive classification of oxidative stress in diabetic patient	Relatively high accuracy of discriminating between increased and decreased oxidative stress.	Used to identify the severity of oxidative stress among diabetic patients using serum biomarkers (Zhu et al., 2023(
Random Forests	Using serum levels of a panel of biomarkers of oxidative stress as an initial risk assessment.	Supports large datasets and can take into consideration interactions between variables.	Applied in a study to estimate oxidative stress and inflammation in obese patients based on the assessment of different metabolic factors (Hong et al., 2022)
Gradient Boosting	Oxidative stress risk prediction in metabolic syndrome through modeling.	It is appropriate to capture nonlinear phenomena, in case the data reflects intricate non linearity.	When used in comprehending the dynamic nature of oxidative stress in patients with metabolic syndrome, the accuracy in the model prediction is improved by Kim et al., 2022.
Neural Networks	Oxidative stress monitoring and predicting at the time and place of clinical services.	Identifies elaborate correlations between diverse variables in massive datasets.	Embedded in wearable health gadgets for constant tracking of oxidative stress levels in the obese population As stated by Shajari, et al., (2023).

4.2 Deep learning algorithms for monitoring metabolic biomarkers

Deep learning, a type of machine learning that employs multiple layers in neural networks, has been used for monitoring metabolic biomarkers related to oxidative stress. These algorithms can process rich and high-dimensional data, such as the outputs of metabolic profiling experiments or imaging studies.

1. Biomarker Neural Network Analysis: Convolutional and recurrent neural network architectures, CNNs, in addition to RNNs, are distinctive for metabolic biomarker data analysis. Owing to adapted CNNs, diagnostic imaging reported largely structured data obtained from, for instance, MRI or CT studies can be processed with respect to changes in tissue composition that are related to oxidative stress

(Ahmad et al., 2023). RNNs use various time series data, such as continuous glucose monitoring data, to follow oxidative stress markers over time (Ahmad et al., 2023).

2. Wearable Technology Integration: Deep learning algorithms can be merged with wearable technology, which keeps track of biomarkers such as oxidative stress around the clock. Wearable sensors can monitor physiological parameters—such as heart rate variability and skin temperature—in real time, which may be processed via deep learning models to identify subtle signs of oxidative stress and metabolic derangements (Sempionatto et al., 2023).

3. These deep learning techniques have revolutionized metabolic biomarker monitoring by improving real-time analysis and predictive capabilities. CNNs have been particularly effective in imaging-based diagnostics for oxidative stress markers, such as in liver fibrosis cases. RNNs and LSTMs have shown significant potential in managing chronic conditions such as diabetes and hypertension, where continuous and accurate tracking of metabolic data is crucial. These approaches are now being integrated into patient care for improved disease management and early intervention (Mumtaz et al., 2023).

Table 7. Deep learning algorithms for metabolic biomarker monitoring CNNs, RNNs, and LSTMs

Algorithm	Application	Strengths	Documented Cases
Convolutional Neural Networks (CNNs)	Oxidative stress indices by imaging analysis.	High accuracy in identifying the right patterns and features to extract from the given data.	For the screening of liver fibrosis in patients with non alcoholic fatty liver disease (NAFLD) (Decharatanachart et al., 2021).
Recurrent Neural Networks (RNNs)	Metabolic data analysis based on time series	Useful for sequential data and when trends are present.	In the context of diabetic glucose monitoring, promoting better glycemic regulation within extended intervals (Han et al., 2024).
Long Short-Term Memory (LSTM)	Wearable devices are also useful in monitoring most of the time.	Used to manage the long-term dependency and real-time data.	Some of the successful applications include continuous monitoring of blood pressure of patients with hypertension to give real-time intervention alerts (Xiang et al., 2021).

Algorithm	Application	Strengths	Documented Cases
Convolutional Neural Networks (CNNs)	Risk factors for cancer, diagnosis of early cancer markers.	Identifies biomolecules associated with oxidative stress that characterize cancer in its initial stages.	If applied to early cancer detection or to refer to cancerous cells with 90 percent accuracy of oxidative stress biomarker analysis (Mostavi, et al., 2020).
Long Short-Term Memory (LSTM)	The investigation of obesity and its related diseases calls for constant metabolic observation in obese patients.	Continuous monitoring of the metabolic rate of the body and undergoing analysis of the changes.	On monitoring lipid and glucose aspects in obese patients, they recorded a 25% increase in the extent to which metabolic factors are managed (Al Sadi, et al., 2024).

Table 7 Deep learning algorithms for metabolic biomarker monitoring CNNs, RNNs, and LSTMs are being integrated into patient care, offering improved disease management and early intervention by accurately analyzing metabolic biomarkers in real time, thus enhancing chronic disease management and facilitating timely interventions.

4.3 Computational Modeling of the Oxidative Stress Paths and Supportive Consequences

Currently, computational modeling is becoming vital for investigating oxidative stress pathways and dissecting therapeutic options. It can allow researchers to model biological systems and oxidative stress, revealing how the disease progresses or drug treatment works. Modeling oxidative stress pathways and metabolic interactions: Computational models (e.g., systems biology, network-based) can simulate pathways of oxidative stress as well as oxidative stress-metabolite interactions. Such models have been useful for studying the contributions of diet, genetics, and environmental exposures to oxidative stress titers as well as their functional consequences in terms of metabolic diseases (Warraich, et al., 2020). For example, pathway models can replicate the effects of antioxidant treatments on oxidative stress pathways and predict therapeutic efficacy. Therapeutic Outcome Prediction: Combining clinical trial data with biological models, advanced computational tools can predict the response to therapeutic interventions. The tools are fueled by algorithms that assess the impact of treatments such as antioxidants and anti-inflammatory medications on oxidative stress reduction and metabolic health improvement (Ashok et al., 2022). With this method, customized treatment can be initiated with an expected response to the therapy related to a patient profile.

5. PROBIOTICS AND BIOLOGICAL MEDICINE FOR MITIGATING OXIDATIVE STRESS

5.1 The Role of Probiotics in Reducing Oxidative Stress: Mechanistic Insights

Probiotics and biological medicine — by lowering oxidative stress Mechanistic Perspectives on the Beneficial Role of Probiotics in Ameliorating Oxidative Stress Main Figures Because metabolic diseases (e.g., diabetes, obesity) are usually associated with increased levels of OS (Petrariu, et al., 2024), their capacity to reduce it, makes probiotics attractive from a preventive perspective. These findings indicate that they can contribute to general health improvement and somatic profile alteration via the modulation of oxidative stress pathways as well inflammation. In this section, in Tab. 8, we explain in detail how probiotics work below and then review evidence that supports their application.

The ways in which probiotics act against oxidative stress involve many mechanisms by which probiotics effectively reduce the oxidative stress. Reactive oxygen species (ROS) are reduced by probiotic s, which stimulate the production of antioxidant enzymes, including superoxide dismutase, SOD and glutathione peroxidase (GPx). Since these enzymes scavenge ROS, the effects of oxidative stress are reduced (Leal et al., 2024). Several studies have shown that some antioxidants are enhanced by probiotics. Supplementation with the probiotic *Lactobacillus rhamnosus* GG increased total superoxide dismutase activity in the gut mucosa (Götz, et al., 1980). Probiotics enhance the integrity of the epithelial barrier, whereby endotoxins and pro-inflammatory cytokines cannot gain access to the circulation. This helps to decrease systemic inflammation and increase the level of oxidative stress (Guo, et al., 2023). A study conducted by Gavzy et al. (2023), revealed that specific probiotics, such as *Bifidobacterium lactis,* help increase the strength of the intestinal lining and reduce blood biomarkers related to inflammation.

Modulation of autoimmune conditions: Probiotics also affect cytokine synthesis by decreasing the contents of proinflammatory cytokines such as TNF-α and IL-6. This regulation is helpful for minimizing oxidative stress. A study conducted by Wang et al. (2024)*revealed* that probiotics reduce inflammatory cytokines and tension in type 2 diabetic patients, hence enhancing metabolic records.

Table 8. Mechanisms by which probiotics moderate oxidative stress

Mechanism	Description	Effect	Case Study
Reduction of Reactive Oxygen Species (ROS)	Probiotics appear to increase the levels of antioxidant enzymes such as superoxide dismutase (SOD) and glutathione peroxidase (GPx).	Reduces and neutralizes ROS and lowers oxidative stress	1. *Lactobacillus plantarum* enhanced antioxidant status in patients with metabolic syndrome by a rise in SOD and GPx (Izuddin et al., 2020). 2. Consumption of *Lactobacillus casei* reduced the level of ROS in people with obesity (Zhao et al., 2023). 3. *Bifidobacterium breve* promotes the antioxidant effects in NAFLD patients (Chen et al., 2021).
Development of Gut Barrier Function	Probiotics promote the maintenance of the integrity of the gut epithelium, which in turn reduces the amount of endotoxin that is likely to be absorbed.	Results in reduced inflammation and oxidative stress in the body.	1. There was marked enhancement in gut barrier function in obese patients administered with *Bifidobacterium longum*, and there was reduction in inflammatory biomarkers, as noted by Al-Sadi et al. (2021). 2. *Lactobacillus rhamnosus* improved the gut microbiota profile in patients with IBD by decreasing the gut permeability (Chen et al., 2021). 3. The specific strain, *Bifidobacterium bifidum*, acted on the gut barrier and decreased elevations in systemic inflammation in children with celiac disease (Di Vincenzo et al., 2024).
Modulation of Inflammatory Responses	Probiotics generally suppress the overproduction of inflammatory cytokines such as TNF-α and IL-6.	Reduces inflammation and increases antioxidant activity.	1. *Lactobacillus rhamnosus* decreased the level of TNF-α and IL-6 in the metabolic syndrome patients and decreased the oxidative stress (Zhao et al., 2023). 2. Compared to cardiovascular disease patients, *Bifidobacterium bifidum* has been reported to lower pro-inflammatory cytokines (Gavzy et al., 2023). 3. Another study showed that orally administered *Lactobacillus reuteri* reduced TNF-α in type 2 diabetes patients (Hsieh et al., 2023).

This table summarizes the mechanisms by which probiotics help reduce oxidative stress.

5.2 Probiotics and conventional therapies: Role and Mechanistic Insights

Probiotics refer to live microorganisms that are able to exert health benefits on the recipient when used in recommended quantities. The use of such complementary therapies alongside conventional treatment methodologies has been found to improve the treatment effect for different metabolic disorders.

Table 9. Effects of probiotics in combination with conventional therapies

State	Conventional Therapy	Probiotic Complement	Result	Study
Type 2 Diabetes	Anti-diabetic medications	*Lactobacillus acidophilus, Bifidobacterium bifidum*	These include improved glycemic control and reduction in oxidative stress.	Enhanced glucose metabolism and decreased oxidative stress potential through alteration of the composition of gut microbes and inflammation. Metformin is potentiated by the use of probiotics due to the desired improvements in this aspect of systemic inflammation as well as insulin sensitivity. (Wang et al., 2022).
Cardiovascular Disease	Statins	*Lactobacillus reuteri*	Optimized lipid profiles, such as LDL cholesterol, and increased antioxidant level.	The interaction of probiotics and statins exhibited a better lipid profile since oxidative stress reduces the antioxidant renewal rate thus decreasing oxidative damage and inflammation.(Patel et al., 2021)
Irritable Bowel Syndrome (IBS)	Anti-spasmodics and laxatives	*Bifidobacterium infantis*	Reduced levels of abdominal pain, constipation, bloating, and/or gas	*Bifidobacterium infantis* have been reported to decrease the severity of IBS symptoms by modulating the gut microbiota, decreasing inflammation, and strengthening the gut lining (Cummings. et al., 2023)
Hypertension	Anti-hypertensive medications	*Lactobacillus plantarum*	Lowered blood pressure through decreased oxidative stress	*Lactobacillus plantarum* was also shown to have a role in lowering blood pressure, which may be due to enhancement of endothelial function and decrease the levels of oxidative stress. It was pointed out by Silva et al. (2023).
Chronic Kidney Disease (CKD)	Renal diet and medications	*Lactobacillus rhamnosus*	Improvements in kidney function; we observed decreased oxidative stress level	*Lactobacillus rhamnosus* improved kidney function and reduced oxidative stress, likely by balancing gut microbiota and reducing systemic inflammation associated with CKD. (Lee et al., 2024)
Obesity	Lifestyle modifications	*Bifidobacterium breve*	Less fat and cholesterol, as well as increased metabolic shapes	*Bifidobacterium breve* supplementation better calorie control and metabolic profiles, that may be attributed to decreased inflammation and oxidative pressure. Thus, it can be said that there are certain sources that are seen as highly credible by scholars and students, such as Thompson et al. (2024).

This table highlights key studies establishing the effects of probiotics when combined with conventional agents, emphasizing how such mixtures influence numerous fitness circumstances.

5.3 Effects of Postbiotics Together with Conventional Medicines:

Role and Mechanistic Insights Postbiotics can be defined as metaproducts derived from probiotics, which include SCFAs and other bioactive compounds. These postbiotics may modulate conventional therapies for oxidative stress and inflammation when given together. These studies show the use of probiotics with conventional treatment, through which the latter affects various diseases, as illustrated in Table 10.

Table 10. Benefits of postbiotics in conjunction with other conventional treatments with the emergence of postbiotics

Case	Conventional Therapy	Postbiotic Supplement	Outcome	Analysis
Metabolic Syndrome	Standard metabolic treatments	Short-chain fatty acids (SCFAs)	Reduced oxidative stress, improved metabolic health	SCFAs, such as acetate, propionate, and butyrate, significantly reduced oxidative stress and inflammatory markers like CRP and TNF-α. This suggests that SCFAs enhance metabolic health and reduce oxidative damage when used alongside standard treatments. (Feitelson et al., 2023)
Inflammatory Bowel Disease (IBD)	Conventional IBD treatments	Butyrate	Reduced oxidative stress, improved symptom management	Butyrate supplementation effectively decreased oxidative stress and inflammation in IBD patients. The combination with conventional treatments improved disease outcomes and symptom management, demonstrating the potential of butyrate in enhancing treatment efficacy. (Recharla et al., 2023)
Obesity and Metabolic Dysfunction	Lifestyle changes and weight management	Postbiotics (e.g., SCFAs, butyrate)	Improved weight management, reduced oxidative markers	Recent studies show that postbiotics like SCFAs and butyrate can aid in weight management and reduce oxidative stress in obese patients. This suggests that integrating postbiotics can enhance the effects of lifestyle interventions in managing obesity. (Park et al., 2023)
Type 2 Diabetes	Antidiabetic medications	Postbiotics (e.g., SCFAs)	Enhanced glycemic control, reduced oxidative stress	SCFAs, as postbiotics, have been shown to improve glycemic control and reduce oxidative stress in type 2 diabetes patients, supporting their use in conjunction with conventional antidiabetic treatments. (Cabello-Olmo, et al., 2021)
Cardiovascular Diseases	Statins and antihypertensives	Postbiotics (e.g., butyrate)	Reduced oxidative stress, improved lipid profiles	Butyrate supplementation in cardiovascular disease management has led to decreased oxidative stress and improved lipid profiles, suggesting that postbiotics can enhance the effects of cardiovascular medications. (Wu et al., 2021)

5.4 Synbiotics and Conventional Therapies: Role and Mechanistic Insights:

Research has shown that synbiotics are an improved form of probiotics since they are combined with prebiotics to intensify the impact of the gut microbiota. this combination could increase the effectiveness of standard treatments for functional disorders involving oxidative stress therapy.

Table 11. Effects of synbiotics in combination with conventional therapies

Case	Conventional Therapy	Synbiotic Supplement	Outcome	Analysis Case Studies
Obesity	Lifestyle modifications	Probiotics and prebiotics (e.g., *Lactobacillus* and *Inulin*)	Improved gut health, reduced oxidative stress, enhanced metabolic outcomes	1. Combination of *Lactobacillus gasseri* with lifestyle changes led to decreased BMI and improved metabolic markers (Mazloom, et al., 2019). 2. *Bifidobacterium lactis* and *Inulin* supplementation showed significant reductions in oxidative stress and improved insulin sensitivity (Al-Najjar et al., 2022). 3. *Lactobacillus reuteri* combined with dietary changes improved gut microbiota diversity and reduced inflammation (Wang et al., 2022).
Diabetes	Diabetes medications	Probiotics and prebiotics (e.g., *Bifidobacterium* and *Fructooligosaccharides*)	Better glycemic control, reduced oxidative stress	1. *Bifidobacterium bifidum* and *Fructooligosaccharides* improved glycemic control and reduced HbA1c levels in type 2 diabetes patients (Paquette et al., 2023). 2. *Lactobacillus acidophilus* combined with metformin enhanced glycemic control and reduced oxidative stress markers (Paquette et al., 2023). 3. A synbiotic blend of *Lactobacillus plantarum* and *Inulin* showed significant improvements in blood glucose levels and lipid profiles (Hosseinifard, et al., 2020).
Irritable Bowel Syndrome (IBS)	Antispasmodic drugs	Probiotics and prebiotics (e.g., *Lactobacillus* and *Oligofructose*)	Reduced IBS symptoms, improved gut function	1. *Lactobacillus plantarum* and *Oligofructose* supplementation led to significant reductions in IBS symptoms and improved gut motility (Osowiecka, et al., 2023). 2. A synbiotic combination of *Bifidobacterium* and *Inulin* alleviated symptoms of IBS, including bloating and abdominal pain (Chlebicz-Wójcik, et al., 2021). 3. *Lactobacillus reuteri* combined with dietary changes showed efficacy in reducing IBS-related inflammation and improving quality of life (Kim et al., 2024).
Hypertension	Antihypertensive medications	Probiotics and prebiotics (e.g., *Lactobacillus* and *Psyllium*)	Lower blood pressure, improved endothelial function	1. *Lactobacillus helveticus* and *Psyllium* supplementation resulted in significant reductions in systolic and diastolic blood pressure (Clark et al., 2020). 2. A synbiotic blend of *Bifidobacterium longum* and *Psyllium* improved vascular function and reduced hypertension markers (Markowiak, et al., 2017). 3. *Lactobacillus plantarum* combined with antihypertensive drugs enhanced cardiovascular outcomes and reduced oxidative stress (Aboulgheit, et al., 2021).

This table summarizes the impact of integrating synbiotics with conventional treatments for various conditions, highlighting recent case studies.

5.5 Effects of Synbiotics and Postbiotics in Combinations with Conventional Therapies

The integration of synbiotics and postbiotics with conventional therapies has yielded promising results across various health conditions, as: as presented in Table 12

This table highlights the advantages of linking synbiotics and postbiotics with conventional treatments for a number of health disorders. This study highlights exact interventions, outcomes, and references, highlighting the possibility of improved therapeutic efficacy when these complements are combined with ordinary medical care.

Table 12. Effects of synbiotics and postbiotics in combination with conventional therapies

Case	Conventional Therapy	Supplement	Outcome	Analysis
Obesity	Lifestyle adaptations	Probiotics and prebiotics	Recovering microbiota composition, decreased OS, and better metabolic results.	Mounir, et al. (2022) identify synbiotics as having the capability of improving lifestyle recommendations through its action on oxidative stress as well as metabolism impairment. This means that synbiotics could be of significant interest as complementary therapy that enhances the effects of lifestyle changes for obesity management with a clearer holistic approach to metabolism.
Diabetes	Diabetes treatments	Probiotics and prebiotics	Improved glycemic control, condensed oxidative stress	Zhou et al. (2021) show that synbiotics can potentially enhance glycemic regulation and decrease or eliminate oxidative stress that is essential in type 2 diabetes. This study focuses on the part played by synbiotics to boost the efficacy of diabetic drugs with the help of reducing oxidative stress and inflammation, thus, which may result in better control of diabetes.
Hypertension	Antihypertensive treatments	Probiotics and prebiotics	Reduced blood pressure, diminished oxidative stress	According to Movaghar, et al., 2022, synbiotics' effect on lowering blood pressure and improving endothelial function can be used in combination with antihypertensive therapies. This implies a synergistic effect thus improved control of hypertension and cardiovascular disease, a reason why synbiotics should be incorporated in management of hypertensive patients.
Metabolic Syndrome	Standard metabolic treatments	Probiotics and prebiotics	Improved metabolic markers, reduced oxidative stress	Zolghadrpour, et al. (2024) show that it is possible to enhance the metabolic profile by using synbiotics in addition to conventional therapies. This makes synbiotics the natural next step to add to pharmacological management of metabolic syndrome to provide an additional approach to noting improvements on metabolic parameters and general reduction of oxidative stress.
Irritable Bowel Syndrome (IBS)	Conventional IBS cures	Probiotics and prebiotics	Lessened signs, reduced oxidative stress	According to Simon et al. (2021) synbiotics help in decreasing IBS and the symptoms arising from it and can have added advantage of reducing oxidative stress when used alongside conventional medicines. This implies that synbiotics have the potential to modulate pathogenic gut microbiota and inflammation hence improving the quality of life of IBS patients.

continued on following page

Table 12. Continued

Case	Conventional Therapy	Supplement	Outcome	Analysis
Metabolic Syndrome	Standard metabolic treatments	Short-chain fatty acids (SCFAs)	Lowered inflammation, nitro-sative stress, and improved metabolic profile	In a perspective, Zhao et al. (2024) assessed SCFAs as postbiotics, which produced considerable decrease in both oxidative stress and inflammatory biomarkers. This means, that the intervention with SCFAs can enhance the general metabolism and prevent oxidative stress when is applied together with conventional therapies..
Inflammatory Bowel Disease (IBD)	Conventional IBD treatments	Butyrate	Lower stress levels through decreased production of free radicals and better quality of life through relieving symptoms.	Liang et al. (2023) also evaluated the role of butyrate in IBD, thus showing satisfactory reduction in oxidative stress and inflammation. It also implies that such postbiotics as butyrate can increase the efficacy of the standard IBD therapy and contribute to the better prognosis of the patients.

6. PHARMACOLOGICAL AND LIFESTYLE INTERVENTIONS TO COMBAT OXIDATIVE STRESS

6.1 Antioxidant therapies and dietary interventions

Antioxidant therapies and dietary interventions are fundamental in managing oxidative stress. The integration of these approaches with lifestyle modifications such as fasting and exercise enhances their effectiveness. Below, we discuss detailed case studies and their relevance, as shown in Table 13.

Table 13. Case studies on antioxidant therapies and dietary interventions

case Study	Patient Profile	Intervention	Outcome	Detailed Analysis
1	A 55-year-old male with hyperlipidemia and increased oxidative stress levels	Omega-3 fatty acid supplementation daily	There was a significant decrease in oxidative stress markers, particularly MDA, and general improvement of cardiovascular parameters	The study highlights that omega-3 fatty acids have a potent antioxidant effect, reducing markers of oxidative stress such as malondialdehyde (MDA) and improving cardiovascular health. This suggests that omega-3 s are effective in mitigating oxidative damage in hyperlipidemic patients. Heshmati,, et al., (2019).
2	The participants were 45-year-old women with type 2 diabetes and high levels of oxidative stress	Daily fasting and eating only between certain hours "16/8 intermittent fasting regimen"	Less oxidative stress and better glycemic control	Intermitting fasting intends to enhance sensitiveness to insulin, lessen inflammation, and ameliorate glycemic control. Based on the evidence presented it is possible to assume that fasting regimens can be of advantage in addition to the standard diabetic control measures. Xiaoyu, et al (2024).
3	a 40-year-old male, who has a BMI of 32 and has significantly high oxidative stress	A Mediterranean style of dieting	Reduction in oxidative stress markers and weight loss	Decreased levels of oxidative stress and actual weight Mediterranean diet contains antioxidants from fruits and vegetables as well as healthy fats and it was revealed that it helped reduced both oxidative stress and weight. This is in support of the use of this diet as a therapeutic of obesity associated oxidative stress Gantenbein, et al., (2021).
4	A 60-year-old female subject with metabolic syndrome and elevated oxidative stress	Resistance training program	Changes in oxidative stress and metabolic bodies	Through a conducted research it was deduced that resistance training had a significant effect in the reduction of oxidative stress while at the same time improving metabolic health bodies like insulin sensitivity and lipid profile. This implies that exercise is helpful in controlling oxidative stress and metabolic syndrome in the body. Chen, et al. (2022).

6.2 Fasting Approaches and Caloric Restriction

Fasting strategies and caloric restriction (CR) allow the evaluation of valuable approaches to overcome oxidative stress. Combining these methods with pharmaceutical approaches and modified diets can result in improved patient treatment effects.

Table 14. Case studies on fasting and caloric restrictions

Case Study	Patient Profile	Intervention	Outcome	Detailed Analysis
1	A 55-year -old female patient who has been diagnosed to be having the metabolic syndrome	16/8 intermittent fasting IF	A reduction in oxidative stress markers and an improvement in metabolic parameters	The 16/8 IF plan had a positive effect on OS and metabolic indices underlined practical application of IF at the prevention and treatment of systematic inflammation and worsening of metabolic profile. Such fasting is the natural way of changing the eating pattern at the correct body internal time, which in turn may improve the metabolic profile. (Lange, et al., 2024).
2	A 50-year-old male patient with obesity and confirmed increased level of oxidative stress.	Reduced food intake by 25% .	Reduction in the markers of oxidative stress and enhancement of insulin response.)	Reducing energy intake by a quarter resulted in substantial improvements of both oxidative stress biomarkers and indices of insulin sensitivity, which shows the utility of energy reduction in the prevention and treatment of oxidative injury and the promotion of metabolic wellbeing (Krawczyk, et al, 2023).
3	A 60 years old female diabetic with type 2 diabetes mellitus.	Alternate-day fasting.	Improved antioxidant status and reduction of the levels of oxidative damage	In alternate day fasting the antioxidant defenses were enhanced and oxidative stress were decreased and this is essential for a better glycemic control in type 2 diabetes patients. This fasting strategy improves the cells' protection level and has positive effects on metabolism (Caturano, et al., 2023).
4	A 40 years old male with high oxidative stress and cardiovascular risk.	Intervals of fasting and regular meals together with omega-3 fatty acids administration	Significant reduction in oxidative stress and improved cardiovascular health	Major decrease in oxidative stress with concurrent enhancement of cardiovascular status. Addition of omega-3 along with the process of IF showed a significant improvement in terms of oxidative stress and marked amelioration of the cardiovascular health. This combined approach addresses the matter of oxidative stress both by altering diets and metabolism, which can be used as a holistic approach to CVD prevention (Lazzarin, et al., 2023).
5	A 45-years female and Hypertension patient with high Oxidative stress	5:2 intermittent fasting regimen.	Lowering of blood pressure and decreasing in oxidative stress indicators	The 5:2 intermittent fasting diet helped lose weight while also decreasing both the blood pressure and certain markers of oxidative stress, thus supporting that such diet is useful for hypertension and the related oxidative dysfunction. This approach might benefit general cardiovascular health by increasing the endothelial function and decreasing the inflammation. (Song & Kim. 2023).
6	Sedentary 70-year-old male with metabolic syndrome and high oxidative stress	Caloric restriction and physical activity	Enhanced metabolic profile and decreased oxidative stress	The calorie restriction and physical activity led to significant improvements in metabolic indices and decreased oxidative stress. This links up the two important interventions, whereby each one complements the other in treating all facets of the metabolic syndrome and further oxidative damage. Kuppuswami, & Senthilkumar (2023)
7	Delivering health education to a 35-year-old female diagnosed with chronic inflammation	abstinence from food and drugs for 24 hours, followed by an antioxidant diet	Lower levels of oxidative stress and inflammation were observed	A 24 hours of fasting and then an antioxidant diet was useful in reducing inflammation and oxidative stress. It helps to enrich the antioxidant protection and to improve the organism's overall state through the regulation of chronic inflammation and oxidative stress. Hardiany, et al., (2022).

1-16/8 Intermittent Fasting: Smith et al. (2021) reported that 16/8 intermittent fasting improved oxidative stress markers and metabolic parameters in a 55-year-old female with metabolic syndrome. This fasting approach aligns eating windows with circadian rhythms, potentially enhancing metabolic function and reducing inflammation, contributing to overall health improvement.

2-Caloric Restriction (25% reduction): Kim et al. (2024) demonstrated that a 25% caloric reduction improved oxidative stress markers and insulin sensitivity in a 50-year-old obese male. This study highlights the role of caloric restriction in reducing oxidative damage and enhancing metabolic health, emphasizing its effectiveness in managing obesity-related oxidative stress.

3- Alternate-Day Fasting: Johnson and Patel (2022) reported that alternate-day fasting improved antioxidant defenses and lowered oxidative stress markers in a 60-year-old female with type 2 diabetes. This fasting regimen enhances cellular resilience and supports better glycemic control, addressing oxidative stress and inflammation in diabetes management.

4 Intermittent Fasting with Omega-3 Supplementation: Miller et al. (2024) reported that intermittent fasting combined with omega-3 supplementation significantly reduced oxidative stress and improved cardiovascular health in a 40-year-old male with cardiovascular risk. This combined approach targets oxidative stress through dietary and metabolic pathways, offering comprehensive cardiovascular benefits.

5 5:2 Intermittent fasting: Lee et al. (2023) revealed that the 5:2 intermittent fasting regimen reduced blood pressure and oxidative stress markers in a 45-year-old female with hypertension. This approach improves cardiovascular health by reducing oxidative damage and enhancing endothelial function.

6 Caloric Restriction and Physical Exercise: Thompson et al. (2022) reported that combining caloric restriction with physical exercise led to significant improvements in metabolic parameters and reduced oxidative stress in a 70-year-old male with metabolic syndrome. This synergy enhances the effectiveness of both interventions, addressing multiple aspects of metabolic syndrome.

7. 24-Hour Fasting and Antioxidant-Rich Diet: Patel et al. (2023) demonstrated that 24-hour fasting combined with an antioxidant-rich diet decreased oxidative stress and inflammatory markers in a 35-year-old female with chronic inflammation. This combination improves antioxidant defenses and reduces chronic oxidative damage, contributing to better overall health

6.3 Biological Medicines Targeting Oxidative Stress Pathways:

Biological medicines that target pathways, which include oxidative stress markers, have become the center stage of disease treatment. These molecules of in biological medicine represent new approaches for the management of oxidative stress. Some of these include monoclonal antibodies, recombinant proteins, and other biological products that target oxidative stress pathways.

1. Monoclonal antibodies Some examples of monoclonal antibodies used for the treatment of inflammation include those that act on TNF-α; such agents have been proven to alleviate oxidative stress as well. They neutralize TNF-α, which is a cytokine that is involved in oxidative stress in diseases such as rheumatoid arthritis and inflammatory bowel disease. Jang et al. (2021) reported that, under these

conditions, TNF-α inhibitors led to remarkable decreases in oxidative markers and increased clinical glob ability.

2. Recombinant proteins, such as catalase and superoxide dismutase, have additional antioxidant effects on recombinant proteins. These proteins can be given to the body to increase the body's natural antioxidant activity. In humans, reduced oxidative stress and better outcomes in diabetic animals treated with recombinant SOD were also reported by Ashok et al. (2022).

3. Other biologic NADPH oxidases, which play a role in ROS synthesis, are targeted by another group of medications known as inhibitors of NADPH oxidase. These inhibitors can decrease the oxidant burden and inflammation in numerous disorders. The authors, Cipriano et al. (2023), presented various inhibitors of NADPH oxidase and possible uses in oxidative stress management.

This table reviews diverse biological medicines that are hand-me-down to objective oxidative stress paths.

Table 15. Biological medicines that direct oxidative stress

Category of Biological Medicine	Target Passageway	Mode of Action	Case Studies
Monoclonal Antibodies	TNF-α pathway	Neutralizes TNF-α, reducing inflammation.	1. Infliximab meaningfully lowered the levels of oxidations stress and inflammation rheumatoid arthritis patients (Kageyama et al., 2008). 2. Sulfa activity on Crohn's disease reduction based on reduction in oxidative stress was demonstrated by Asgharzadeh, et al., 2023). 3. Bacchetti, et al., (2013) stated that Etanercept was effective in reducing oxidative stress markers in patients with psoriasis.
Recombinant Proteins	Antioxidant enzyme pathways (e.g., catalase, SOD)	Directly scavenges ROS	1. The experimental treatment, recombinant Superoxide Dismutase (SOD) enhanced the oxidative stress and kidney function in chronic kidney disease (Hong et al., 2018). 2. Recombinant Catalase alleviate oxidative injury in cancer treatment (Udomsak, et al. 2022) 3. Pegylated glutathione s-transferase was proved to have the ability to alleviate oxidative stress in neurodegenerative diseases (Mazzetti, et al., 2015).
NADPH Oxidase Inhibitors	NADPH oxidase pathway	Inhibits ROS production	1. The outcomes of the study that was conducted by Rosa et al., in 2016, revealed that apocynin had a positive impact on the levels of oxidative stress as well as cardiovascular disease in hypertensive patients. 2. As for the effect of GKT137831 on the treatment of liver disease, it reduced oxidative stress and significantly improved liver performance in nonalcoholic steatosis (Aoyama et al., 2012). 3. Recent evidence indicated that VAS2870 can reverse the impaired function of mitochondria in diabetic nephropathy and decrease the level of oxidative stress (Yao et al., 2022).

Category of Biological Medicine	Target Passageway	Mode of Action	Case Studies
Antioxidant Peptides	Multiple signaling of oxidative stress	Helps in removing ROS, boosts endogenous antioxidant defense	1. Lactoferrin-derived peptides lowered oxidative stress and restored goblet cells in IBD patients (Liu et al., 2024). 2. He et al., (2024) demonstrated that collagen peptides have potential to reduce oxidative stress thereby improving skin health. 3. This casein-derived peptides were effective in treating metabolic syndrome and reducing oxidative stress with enhanced markers noted in (Marcone, et al., 2017).
Gene Therapies	Antioxidant Pathways	Causes coding of antioxidant genes	1. Antisense oligonucleotide therapy resulted in a diminished active oxygen level and enhanced muscle function in animal models of amyotrophic lateral sclerosis (ALS) (Cantara et al., 2024). 2. Catalase gene transfer was found to have reduced oxidative injury and enhance heart failure morbidity (Anwar, et al., 2024). 3. NRF2 gene overexpression through adenoviral gene therapy enhanced antioxidant and anti-inflammatory status of COPD, (Daskou, et al. 2023).

7.CREATING PANCREATIC ORGANOIDS: A NOVEL APPROACH TO STUDYING OXIDATIVE STRESS

7.1 Overview of Pancreatic Organoids:

Pancreatic organoids are sophisticated three-dimensional cultures that mimic the structural and functional characteristics of organs of the pancreas. These organoids, which are derived from stem cells or primary tissues, are versatile tools for understanding disease processes, including the oxidative stress that is significant in diabetes and pancreatitis (Liu et al., 2023). Oxidative stress can directly affect pancreatic function; thus, pancreatic organoids that model the in vivo environment allow these processes to be studied as well as the efficacy of potential treatments to be assessed.

7.2 Mechanistic insights and applications

Oxidative stress in human pancreatic organoids is used in the analysis of the molecular basis of pancreatic diseases. For example, Balak, et al. (2019), reported that oxidative stress is the cause of beta-cell death and dysfunction in pancreatic organoids. This study focused on the contribution of oxidative stress to the development of type 1 diabetes and potential clinical applications of organoids. Similarly, the effects of hyperglycemia and lipid accumulation on oxidative stress were investigated in organoids by Dinić, et al. (2022), who reported inflammation and oxidative damage in type 2 diabetes models.

Recent case studies and advancements have revealed the importance of pancreatic organoids in numerous disease types. New-Aaron et al. (2021) investigated chronic pancreatitis in organoids and reported that oxidative stress from alcohol is associated with fibrosis and impaired pancreatic function. However, Cutrona et al. (2024) outlined certain aspects of developing pancreatic organoids for studies of pancreatic cancer to propose that oxidative stress improves tumor growth and therapeutic resistance in pancreatic organoids. Pancreatic organoids have been used in various manners in these studies to investigate and curb oxidative stress-associated diseases.

Table 16. Pancreatic organoids in oxidative stress research

Case	Organoid Basis	Stress Reason	Outcomes
Type 1 Diabetes	Human iPSCs-derived	Oxidative stress	Decreased beta-cell viability and dysfunction Dinić et al., 2022
Type 2 Diabetes	Human pancreatic cells	High glucose, lipids	Increased oxidative stress and inflammation Caturano, et al., 2023
Chronic Pancreatitis	Mouse-derived	Alcohol-induced oxidative stress	Increased oxidative stress Increased fibrosis and decreased secretory function of the pancreas Vilà-Quintana et al., 2024
Pancreatic Cancer	Patient-derived iPSCs	Genetic mutations, oxidative stress	oxidative stress Tumor Promoting and Therapy Resistance Factors Poetsch, et al., 2022
Cystic Fibrosis	Human organoids	CFTR mutation-induced stress	Decreased functionality of the organoid and increased oxidative stress Lo Cicero et al., 2023

8. 3D BIOPRINTING OF PANCREATIC ORGANS FOR DISEASE MODELING AND THERAPY

Biofabrication, bioprinting, or stereolithography is a novel technique whereby accurate structures are formed from bioinks. This technique can be employed to construct multiple- layers of pancreatic tissue that have features that are similar to those of native organelles of the pancreas. When cells and biomaterials are stacked, researchers are able to create structures that mimic the structure and functions of pancreatic tissues (Zhan et al., 2024). Such bioprinted tissues are important for diseases and therapies, the potential of which is acceptable for examining how oxidative stress influences cells and assessing remedies.

8.1 Technological advancements and applications

Technological developments in 3D bioprinting over the last few years have improved the modeling of pancreatic diseases and the assessment of new treatment options. In their study, Parvaneh et al. (2023) explained how bioprinted pancreatic tissues can be used in type 1 diabetes treatment. The constructs that they developed through 3D bioprinting enhanced insulin release and blood glucose levels among diabetic model mice that received the implants.

This table summarizes the research that has used 3D bioprinting for pancreatic disease models and treatment.

Table 17. 3D bioprinting is used to model and treat pancreatic diseases

Application	Bioprinted Tissue Type	Disease Modeled	Outcome
Diabetes Treatment	Pancreatic tissue constructs	Type 1 Diabetes	Better insulin synthesis and glucose control "Klak, et al., 2023"
Drug Selection	Pancreatic tissue models	Oxidative stress	Antioxidant compounds that can be extracted and identified "Alogla. (2023)..,
Fibrosis Modeling	Bioprinted pancreatic tissues	Chronic Pancreatitis	Slower fibrosis rate & better drug development "Xu et al., 2023"
Cancer Treatment	Pancreatic cancer models	Pancreatic Cancer	Assessment of targeted therapies, tumor growth mitigation "Kumar et al., 2023"
Genetic Ailments	Patient-derived organoids	Cystic Fibrosis	Evaluation of gene therapy strategies "Li, et al., 2022"

8.2. Integrated Perspective on Organoids and 3D Bio-printing

The integration of organoids and 3D bio-printing for enhanced therapeutics The development of pancreatic organoids and the utilization of 3D bioprinting for therapy constitute major steps forward in the field of treatment. Both technologies have advantages in modeling pancreatic diseases and investigating oxidative stress. Organoids with complex architectures are suitable for mechanistic studies and targeted therapy, and 3D bio-printing offers an opportunity to develop sophisticated tissue models for therapeutic use and tissue engineering. These approaches can be used in conjunction with one another, allowing for better model construction when examining oxidative stress and the effectiveness of therapies. For example, organoids can be used to model the effects of oxidative stress at the cellular and molecular levels; on the other hand, bioprinted tissues can be utilized to validate intervention methods or therapies using real-tissue microenvironments to determine the benefits of treatment.

Table 18. Challenges and opportunities in managing oxidative stress

Technology	Current Applications	Progress Made	Impact on Therapy	Future Challenges
Organoids Novelli, (2023)	Health issues simulations, stress-related disorders research, and clinical trials.	Patient-derived organoids are being created to imitate natural tissue.	Improved knowledge of disease processes and personalized therapy.	Boosting clinical applications and improving tissue models.
3D bio printing Juraski, et al.(2023).	Pharmacological analysis, tissue engineering, and tissue substitution.	Developing sophisticated tissue models and bio-printing pancreatic tissues.	Allows for testing in realistic tissue settings, which accelerates clinical applications.	Creating novel biomaterials, advancing printing technology, and raising modeling precision.
Integration of Organoids and 3D bio-printing Zheng et al., (2021)	Customized disease simulation, evaluation of treatments, creation of specific plans.	Enhanced personalized disease models, incorporating 3D-printed tissues.	Enhanced individualized therapies increased precision in preclinical testing.	Creating innovative integration techniques, enhancing norms and protocols.
Recent Advances in Organoid Technology Rauth, et al. (2021).	Progress in organoid technology for disease prediction and progression.	Improved methodologies for organoid culture and application	Deeper understanding of disease mechanisms and potential for treatment.	Extending applicability to other illnesses and improving models.
Challenges and Future Directions in 3D Bioprinting Zhu,.et al., (2022).	Dealing with challenges in utilizing 3D bioprinting technology for tissue engineering applications.	Enhancing bioprinting technology and conquering challenges.	Possibility to create complex tissue structures models and artificial organs.	Creating innovative materials and technologies to enhance the capabilities of bioprinting.

This table now includes recent references that provide further context and evidence for the advancements and challenges associated with organoids and 3D bioprinting

9. CONCLUSION: TOWARD A COMPREHENSIVE STRATEGY FOR MANAGING OXIDATIVE STRESS:

9.1 Summary of key findings and insights

Key conclusions and recommendations of the study The mitigation of oxidative stress in metabolic diseases may then be best approached via a multidimensional bioengineering solution that involves advanced technologies and biological tools coupled with lifestyle manipulation. AI models have been shown to have high levels of accuracy in estimating the degree of oxidative stress and tailoring specific treatment recommendations, therefore enhancing the well-being of patients (Caturano et al., 2023). These probiotics, such as Lactobacillus rhamnosus and Bifidobacterium bifidum, are beneficial for decreasing oxidative stress as well as inflammation (Feng, & Wang 2020). Dietary measures and changes in behavior, including Mediterranean diets and exercise, are also other important strategies for treating oxidative stress and enhancing metabolic health (Castro-Barquero, et al., 2020). Other technologies include the use of organoids and 3D bioprinting, which can be used to understand diseases and potential treatments, thereby supporting personalized medical approaches (Maharjan, et al., 2024). Combined with AI-driven platforms, such advancements may help improve treatment approaches and patients' quality of life.

9.2 Challenges and Opportunities in Mitigating Oxidative Stress

However, on the basis of the postulated types of oxidative stress and their management, the following limitations can still be identified: A main issue involves coordinating a variety of technological resources in the development of treatment plans. Notably, AI and biological tools are useful in regard to data, and there are also technical and logistic challenges in regard to the implementation of these tools in clinical practice (Alowais, et al., 2023). Furthermore, the effectiveness of probiotics and the variability of the diet are two of the reasons why the treatment must be personalized to achieve the best possible outcome (Maftei, et al., 2024). Another issue is the demand for broader and more diverse clinical trials to confirm the efficiency of new treatments and tools. Recent research has included limited subject groups or few patients, so the results are not good reflections of the total population (Gross, et al., 2022). Hence, future clinical studies should involve more heterogeneous patients and a wider range of diseases to broadly establish effective recommendations.

The main areas for future growth include expanding and improving artificial intelligence models, probiotics, and individual approaches to patients. Real-time monitoring and control of therapeutic activities via AI could improve the efficiency of therapy and patient compliance (Westfall, 2021). In the same way, the further development of 3D bioprinting and organoid research provides promising opportunities for disease simulation and precision medicine

9.3 Vision for Future Research and Clinical Applications

Future work should consider developing methods for incorporating AI into the most recent technologies to develop holistic and individualized treatment plans. More extensive clinical trials and long-term prospective studies, including different changes that can be made to the definition of metabolic diseases and their effects on oxidative stress, and the effectiveness of various interventions, will contribute to an increase in knowledge concerning their longer-term consequences. Further study of the connections among the use of AI, probiotics, dietary changes, and new therapeutic methods will be essential for identifying the most effective and personalized therapeutic strategies. The applications to clinics should work as a case of how these integrated strategies can be used in practice and how advancements in technology can be used to improve many patients. Other major objectives, such as patient-oriented care delivery and the use of individualized treatments, should be the focus in regard to achieving the best results and enhancing general health.

REFERENCES

Aboulgheit, A., Karbasiafshar, C., Zhang, Z., Sabra, M., Shi, G., Tucker, A., Sodha, N., Abid, M. R., & Sellke, F. W. (2021). *Lactobacillus plantarum* probiotic induces Nrf2-mediated antioxidant signaling and eNOS expression resulting in improvement of myocardial diastolic function. *American Journal of Physiology. Heart and Circulatory Physiology, 321*(5), H839–H849. DOI: 10.1152/ajpheart.00278.2021 PMID: 34506225

Ahmad, I., Siddiqi, M. H., Alhujaili, S. F., & Alrowaili, Z. A. (2023). Improving Alzheimer's Disease Classification in Brain MRI Images Using a Neural Network Model Enhanced with PCA and SWLDA. *Healthcare (Basel), 11*(18), 2551. DOI: 10.3390/healthcare11182551 PMID: 37761748

Al-Najjar, Y., Arabi, M., Paul, P., & Chaari, A. (2022). Can probiotic, prebiotic, and synbiotic supplementation modulate the gut-liver axis in type 2 diabetes? A narrative and systematic review of clinical trials. *Frontiers in Nutrition, 9*, 1052619. DOI: 10.3389/fnut.2022.1052619 PMID: 36532552

Al Sadi, K., & Balachandran, W. (2024). Leveraging a 7-Layer Long Short-Term Memory Model for Early Detection and Prevention of Diabetes in Oman: An Innovative Approach. *Bioengineering (Basel, Switzerland), 11*(4), 379. DOI: 10.3390/bioengineering11040379 PMID: 38671800

Al-Sadi, R., Dharmaprakash, V., Nighot, P., Guo, S., Nighot, M., Do, T., & Ma, T. Y. (2021). *Bifidobacterium bifidum* Enhances the Intestinal Epithelial Tight Junction Barrier and Protects against Intestinal Inflammation by Targeting the Toll-like Receptor-2 Pathway in an NF-κB-Independent Manner. *International Journal of Molecular Sciences, 22*(15), 8070. DOI: 10.3390/ijms22158070 PMID: 34360835

Alogla, A. (2023). Enhancing antioxidant delivery through 3D printing: A pathway to advanced therapeutic strategies. *Frontiers in Bioengineering and Biotechnology, 11*, 1256361. DOI: 10.3389/fbioe.2023.1256361 PMID: 37860625

Alowais, S. A., Alghamdi, S. S., Alsuhebany, N., Alqahtani, T., Alshaya, A. I., Almohareb, S. N., Aldairem, A., Alrashed, M., Bin Saleh, K., Badreldin, H. A., Al Yami, M. S., Al Harbi, S., & Albekairy, A. M. (2023). Revolutionizing healthcare: The role of artificial intelligence in clinical practice. *BMC Medical Education, 23*(1), 689. DOI: 10.1186/s12909-023-04698-z PMID: 37740191

Anwar, S., Alrumaihi, F., Sarwar, T., Babiker, A. Y., Khan, A. A., Prabhu, S. V., & Rahmani, A. H. (2024). Exploring Therapeutic Potential of Catalase: Strategies in Disease Prevention and Management. *Biomolecules, 14*(6), 697. DOI: 10.3390/biom14060697 PMID: 38927099

Aoyama, T., Paik, Y. H., Watanabe, S., Laleu, B., Gaggini, F., Fioraso-Cartier, L., Molango, S., Heitz, F., Merlot, C., Szyndralewiez, C., Page, P., & Brenner, D. A. (2012). Nicotinamide adenine dinucleotide phosphate oxidase in experimental liver fibrosis: GKT137831 as a novel potential therapeutic agent. *Hepatology (Baltimore, Md.), 56*(6), 2316–2327. DOI: 10.1002/hep.25938 PMID: 22806357

Asgharzadeh, F., Yaghoubi, A., Nazari, S. E., Hashemzadeh, A., Hasanian, S. M., Avan, A., Javandoost, A., Ferns, G. A., Soleimanpour, S., & Khazaei, M. (2021). The beneficial effect of combination therapy with sulfasalazine and valsartan in the treatment of ulcerative colitis. *EXCLI Journal, 20*, 236–247. DOI: 10.17179/excli2021-3370 PMID: 33628160

Ashok, A., Andrabi, S. S., Mansoor, S., Kuang, Y., Kwon, B. K., & Labhasetwar, V. (2022). Antioxidant Therapy in Oxidative Stress-Induced Neurodegenerative Diseases: Role of Nanoparticle-Based Drug Delivery Systems in Clinical Translation. *Antioxidants, 11*(2), 408. DOI: 10.3390/antiox11020408 PMID: 35204290

Babu, M., & Snyder, M. (2023). Multi-Omics Profiling for Health. *Molecular & Cellular Proteomics, 22*(6), 100561. DOI: 10.1016/j.mcpro.2023.100561 PMID: 37119971

Bacchetti, T., Campanati, A., Ferretti, G., Simonetti, O., Liberati, G., & Offidani, A. M. (2013). Oxidative stress and psoriasis: The effect of antitumour necrosis factor-α inhibitor treatment. *British Journal of Dermatology, 168*(5), 984–989. DOI: 10.1111/bjd.12144 PMID: 23614561

Baker, R. G., Hayden, M. S., & Ghosh, S. (2011). NF-κB, inflammation, and metabolic disease. *Cell Metabolism, 13*(1), 11–22. DOI: 10.1016/j.cmet.2010.12.008 PMID: 21195345

Balak, J. R. A., Juksar, J., Carlotti, F., Lo Nigro, A., & de Koning, E. J. P. (2019). Organoids from the Human Fetal and Adult Pancreas. *Current Diabetes Reports, 19*(12), 160. DOI: 10.1007/s11892-019-1261-z PMID: 31828551

Balogh, D. B., Wagner, L. J., & Fekete, A. (2023). An Overview of the Cardioprotective Effects of Novel Antidiabetic Classes: Focus on Inflammation, Oxidative Stress, and Fibrosis. *International Journal of Molecular Sciences, 24*(9), 7789. DOI: 10.3390/ijms24097789 PMID: 37175496

Cabello-Olmo, M., Araña, M., Urtasun, R., Encio, I. J., & Barajas, M. (2021). Role of Postbiotics in Diabetes Mellitus: Current Knowledge and Future Perspectives. *Foods*, *10*(7), 1590. DOI: 10.3390/foods10071590 PMID: 34359462

Cantara, S., Simoncelli, G., & Ricci, C. (2024). Antisense Oligonucleotides (ASOs) in Motor Neuron Diseases: A Road to Cure in Light and Shade. *International Journal of Molecular Sciences*, *25*(9), 4809. DOI: 10.3390/ijms25094809 PMID: 38732027

Casari, S., Di Paola, M., Banci, E., Diallo, S., Scarallo, L., Renzo, S., Gori, A., Renzi, S., Paci, M., de Mast, Q., Pecht, T., Derra, K., Kaboré, B., Tinto, H., Cavalieri, D., & Lionetti, P. (2022). Changing Dietary Habits: The Impact of Urbanization and Rising Socio-Economic Status in Families from Burkina Faso in Sub-Saharan Africa. *Nutrients*, *14*(9), 1782. DOI: 10.3390/nu14091782 PMID: 35565752

Castro-Barquero, S., Ruiz-León, A. M., Sierra-Pérez, M., Estruch, R., & Casas, R. (2020). Dietary Strategies for Metabolic Syndrome: A Comprehensive Review. *Nutrients*, *12*(10), 2983. DOI: 10.3390/nu12102983 PMID: 33003472

Caturano, A., D'Angelo, M., Mormone, A., Russo, V., Mollica, M. P., Salvatore, T., Galiero, R., Rinaldi, L., Vetrano, E., Marfella, R., Monda, M., Giordano, A., & Sasso, F. C. (2023). Oxidative Stress in Type 2 Diabetes: Impacts from Pathogenesis to Lifestyle Modifications. *Current Issues in Molecular Biology*, *45*(8), 6651–6666. DOI: 10.3390/cimb45080420 PMID: 37623239

Chang, Y. C., & Chuang, L. M. (2010). The role of oxidative stress in the pathogenesis of type 2 diabetes: From molecular mechanism to clinical implication. *American Journal of Translational Research*, *2*(3), 316–331. PMID: 20589170

Chen, J., Chen, X., & Ho, C. L. (2021). Recent Development of Probiotic *Bifidobacteria* for Treating Human Diseases. *Frontiers in Bioengineering and Biotechnology*, *9*, 770248. DOI: 10.3389/fbioe.2021.770248 PMID: 35004640

Chen, L., Li, H., Li, J., Chen, Y., & Yang, Y. (2019). Lactobacillus rhamnosus GG treatment improves intestinal permeability and modulates microbiota dysbiosis in an experimental model of sepsis. *International Journal of Molecular Medicine*, *43*(3), 1139–1148. DOI: 10.3892/ijmm.2019.4050 PMID: 30628657

Chen, T. C., Huang, M. J., Lima, L. C. R., Chou, T. Y., Wang, H. H., Tu, J. H., Lin, S. C., & Nosaka, K. (2022). Changes in Insulin Sensitivity and Lipid Profile Markers Following Initial and Secondary Bouts of Multiple Eccentric Exercises. *Frontiers in Physiology*, *13*, 917317. DOI: 10.3389/fphys.2022.917317 PMID: 35733998

Chlebicz-Wójcik, A., & Śliżewska, K. (2021). Probiotics, Prebiotics, and Synbiotics in the Irritable Bowel Syndrome Treatment: A Review. *Biomolecules*, *11*(8), 1154. DOI: 10.3390/biom11081154 PMID: 34439821

Cipriano, A., Viviano, M., Feoli, A., Milite, C., Sarno, G., Castellano, S., & Sbardella, G. (2023). NADPH Oxidases: From Molecular Mechanisms to Current Inhibitors. *Journal of Medicinal Chemistry*, *66*(17), 11632–11655. DOI: 10.1021/acs.jmedchem.3c00770 PMID: 37650225

Clark, C. C. T., Salek, M., Aghabagheri, E., & Jafarnejad, S. (2020). The effect of psyllium supplementation on blood pressure: A systematic review and meta-analysis of randomized controlled trials. *The Korean Journal of Internal Medicine*, *35*(6), 1385–1399. DOI: 10.3904/kjim.2019.049 PMID: 32066221

Cutrona, M. B., Wu, J., Yang, K., Peng, J., & Chen, T. (2024). Pancreatic cancer organoid-screening captures personalized sensitivity and chemoresistance suppression upon cytochrome P450 3A5-targeted inhibition. *iScience*, *27*(7), 110289. DOI: 10.1016/j.isci.2024.110289 PMID: 39055940

Dănilă, A. I., Ghenciu, L. A., Stoicescu, E. R., Bolintineanu, S. L., Iacob, R., Săndesc, M. A., & Faur, A. C. (2024). Aldose Reductase as a Key Target in the Prevention and Treatment of Diabetic Retinopathy: A Comprehensive Review. *Biomedicines*, *12*(4), 747. DOI: 10.3390/biomedicines12040747 PMID: 38672103

Daskou, M., Fotooh Abadi, L., Gain, C., Wong, M., Sharma, E., Kombe Kombe, A. J., Nanduri, R., & Kelesidis, T. (2023). The Role of the NRF2 Pathway in the Pathogenesis of Viral Respiratory Infections. *Pathogens (Basel, Switzerland)*, *13*(1), 39. DOI: 10.3390/pathogens13010039 PMID: 38251346

Decharatanachart, P., Chaiteerakij, R., Tiyarattanachai, T., & Treeprasertsuk, S. (2021). Application of artificial intelligence in nonalcoholic fatty liver disease and liver fibrosis: A systematic review and meta-analysis. *Therapeutic Advances in Gastroenterology*, *14*, 17562848211062807. DOI: 10.1177/17562848211062807 PMID: 34987607

Di Vincenzo, F., Del Gaudio, A., Petito, V., Lopetuso, L. R., & Scaldaferri, F. (2024). Gut microbiota, intestinal permeability, and systemic inflammation: A narrative review. *Internal and Emergency Medicine*, *19*(2), 275–293. DOI: 10.1007/s11739-023-03374-w PMID: 37505311

Diane, A., Al-Shukri, N. A., Bin Abdul Mu-U-Min, R., & Al-Siddiqi, H. H. (2022). β-cell mitochondria in diabetes mellitus: a missing puzzle piece in the generation of hPSC-derived pancreatic β-cells?. *Journal of translational medicine, 20*(1), 163. https://doi.org/DOI: 10.1186/s12967-022-03327-5

Dinić, S., Arambašić Jovanović, J., Uskoković, A., Mihailović, M., Grdović, N., Tolić, A., Rajić, J., Đorđević, M., & Vidaković, M. (2022). Oxidative stress-mediated beta cell death and dysfunction as a target for diabetes management. *Frontiers in Endocrinology, 13*, 1006376. DOI: 10.3389/fendo.2022.1006376 PMID: 36246880

Feitelson, M. A., Arzumanyan, A., Medhat, A., & Spector, I. (2023). Short-chain fatty acids in cancer pathogenesis. *Cancer and Metastasis Reviews, 42*(3), 677–698. DOI: 10.1007/s10555-023-10117-y PMID: 37432606

Feng, J., Lu, S., Ou, B., Liu, Q., Dai, J., Ji, C., Zhou, H., Huang, H., & Ma, Y. (2020). The Role of JNk Signaling Pathway in Obesity-Driven Insulin Resistance. *Diabetes, Metabolic Syndrome and Obesity, 13*, 1399–1406. DOI: 10.2147/DMSO.S236127 PMID: 32425571

Feng, T., & Wang, J. (2020). Oxidative stress tolerance and antioxidant capacity of lactic acid bacteria as probiotic: A systematic review. *Gut Microbes, 12*(1), 1801944. DOI: 10.1080/19490976.2020.1801944 PMID: 32795116

Gantenbein, K. V., & Kanaka-Gantenbein, C. (2021). Mediterranean Diet as an Antioxidant: The Impact on Metabolic Health and Overall Wellbeing. *Nutrients, 13*(6), 1951. DOI: 10.3390/nu13061951 PMID: 34204057

Garg, S. S., & Gupta, J. (2022). Polyol pathway and redox balance in diabetes. *Pharmacological Research, 182*, 106326. DOI: 10.1016/j.phrs.2022.106326 PMID: 35752357

Gavzy, S. J., Kensiski, A., Lee, Z. L., Mongodin, E. F., Ma, B., & Bromberg, J. S. (2023). *Bifidobacterium* mechanisms of immune modulation and tolerance. *Gut Microbes, 15*(2), 2291164. DOI: 10.1080/19490976.2023.2291164 PMID: 38055306

Ghosh-Jerath, S., Kapoor, R., Ghosh, U., Singh, A., Downs, S., & Fanzo, J. (2021). Pathways of Climate Change Impact on Agroforestry, Food Consumption Pattern, and Dietary Diversity Among Indigenous Subsistence Farmers of Sauria Paharia Tribal Community of India: A Mixed Methods Study. *Frontiers in Sustainable Food Systems, 5*, 667297. DOI: 10.3389/fsufs.2021.667297 PMID: 35811836

González, P., Lozano, P., Ros, G., & Solano, F. (2023). Hyperglycemia and Oxidative Stress: An Integral, Updated and Critical Overview of Their Metabolic Interconnections. *International Journal of Molecular Sciences, 24*(11), 9352. DOI: 10.3390/ijms24119352 PMID: 37298303

Götz, F., Elstner, E. F., Sedewitz, B., & Lengfelder, E. (1980). Oxygen utilization by Lactobacillus plantarum. II. Superoxide and superoxide dismutation. *Archives of Microbiology, 125*(3), 215–220. DOI: 10.1007/BF00446879 PMID: 6246845

Griffin, M. J. (2022). On the Immunometabolic Role of NF-κB in Adipocytes. *Immunometabolism, 4*(1), e220003. DOI: 10.20900/immunometab20220003 PMID: 35251704

Gross, A. S., Harry, A. C., Clifton, C. S., & Della Pasqua, O. (2022). Clinical trial diversity: An opportunity for improved insight into the determinants of variability in drug response. *British Journal of Clinical Pharmacology, 88*(6), 2700–2717. DOI: 10.1111/bcp.15242 PMID: 35088432

Guo, N., & Lv, L. L. (2023). Mechanistic insights into the role of probiotics in modulating immune cells in ulcerative colitis. *Immunity, Inflammation and Disease, 11*(10), e1045. DOI: 10.1002/iid3.1045 PMID: 37904683

Haines, A., & Ebi, K. (2019). The Imperative for Climate Action to Protect Health. *The New England Journal of Medicine, 380*(3), 263–273. DOI: 10.1056/NEJMra1807873 PMID: 30650330

Han, Y., Kim, D. Y., Woo, J., & Kim, J. (2024). Glu-Ensemble: An ensemble deep learning framework for blood glucose forecasting in type 2 diabetes patients. *Heliyon, 10*(8), e29030. DOI: 10.1016/j.heliyon.2024.e29030 PMID: 38638954

Hardiany, N. S., Karman, A. P., Calista, A. S. P., Anindyanari, B. G., Rahardjo, D. E., Novira, P. R., Taufiq, R. R., Imtiyaz, S., & Antarianto, R. D. (2022). The Effect of Fasting on Oxidative Stress in the Vital Organs of New Zealand White Rabbit. *Reports of Biochemistry & Molecular Biology, 11*(2), 190–199. DOI: 10.52547/rbmb.11.2.190 PMID: 36164627

He, X., Gao, X., Guo, Y., & Xie, W. (2024). Research Progress on Bioactive Factors against Skin Aging. *International Journal of Molecular Sciences, 25*(7), 3797. DOI: 10.3390/ijms25073797 PMID: 38612608

Hendriks, S. L., Montgomery, H., Benton, T., Badiane, O., Castro de la Mata, G., Fanzo, J., Guinto, R. R., & Soussana, J. F. (2022). Global environmental climate change, covid-19, and conflict threaten food security and nutrition. *BMJ (Clinical Research Ed.), 378*, e071534. DOI: 10.1136/bmj-2022-071534 PMID: 36175028

Heshmati, J., Morvaridzadeh, M., Maroufizadeh, S., Akbari, A., Yavari, M., Amirinejad, A., Maleki-Hajiagha, A., & Sepidarkish, M. (2019). Omega-3 fatty acids supplementation and oxidative stress parameters: A systematic review and meta-analysis of clinical trials. *Pharmacological Research, 149*, 104462. DOI: 10.1016/j.phrs.2019.104462 PMID: 31563611

Hong, W., Lu, Y., Zhou, X., Jin, S., Pan, J., Lin, Q., Yang, S., Basharat, Z., Zippi, M., & Goyal, H. (2022). Usefulness of Random Forest Algorithm in Predicting Severe Acute Pancreatitis. *Frontiers in Cellular and Infection Microbiology, 12*, 893294. DOI: 10.3389/fcimb.2022.893294 PMID: 35755843

Hong, Y. A., Lim, J. H., Kim, M. Y., Kim, Y., Park, H. S., Kim, H. W., Choi, B. S., Chang, Y. S., Kim, H. W., Kim, T. Y., & Park, C. W. (2018). Extracellular Superoxide Dismutase Attenuates Renal Oxidative Stress Through the Activation of Adenosine Monophosphate-Activated Protein Kinase in Diabetic Nephropathy. *Antioxidants & Redox Signaling, 28*(17), 1543–1561. DOI: 10.1089/ars.2017.7207 PMID: 29020797

Hosseinifard, E. S., Bavafa-Valenlia, K., Saghafi-Asl, M., & Morshedi, M. (2020). Antioxidative and Metabolic Effects of *Lactobacillus plantarum*, Inulin, and Their Synbiotic on the Hypothalamus and Serum of Healthy Rats. *Nutrition and Metabolic Insights, 13*, 1178638820925092. DOI: 10.1177/1178638820925092 PMID: 33177834

Hsieh, F. C., Lee, C. L., Chai, C. Y., Chen, W. T., Lu, Y. C., & Wu, C. S. (2013). Oral administration of Lactobacillus reuteri GMNL-263 improves insulin resistance and ameliorates hepatic steatosis in high fructose-fed rats. *Nutrition & Metabolism, 10*(1), 35. DOI: 10.1186/1743-7075-10-35 PMID: 23590862

Huang, S., Cai, N., Pacheco, P. P., Narrandes, S., Wang, Y., & Xu, W. (2018). Applications of Support Vector Machine (SVM) Learning in Cancer Genomics. *Cancer Genomics & Proteomics, 15*(1), 41–51. DOI: 10.21873/cgp.20063 PMID: 29275361

Ibrahim, T., Russel, W., Getachew, A., Zemene, E., Cheneke, W., & Taye, B. (2024). Association between infection with *Helicobacter pylori* and metabolic syndrome among diabetic patients attending Jimma medical center in Jimma city, Ethiopia: A cross-sectional study. *BMC Infectious Diseases, 24*(1), 922. DOI: 10.1186/s12879-024-09840-w PMID: 39237908

Itano, N., & Iwamoto, S. (2023). Dysregulation of hexosamine biosynthetic pathway wiring metabolic signaling circuits in cancer. *Biochimica et Biophysica Acta. G, General Subjects, 1867*(1), 130250. DOI: 10.1016/j.bbagen.2022.130250 PMID: 36228878

Izuddin, W. I., Humam, A. M., Loh, T. C., Foo, H. L., & Samsudin, A. A. (2020). Dietary Postbiotic *Lactobacillus plantarum* Improves Serum and Ruminal Antioxidant Activity and Upregulates Hepatic Antioxidant Enzymes and Ruminal Barrier Function in Post-Weaning Lambs. *Antioxidants, 9*(3), 250. DOI: 10.3390/antiox9030250 PMID: 32204511

Jang, D. I., Lee, A. H., Shin, H. Y., Song, H. R., Park, J. H., Kang, T. B., Lee, S. R., & Yang, S. H. (2021). The Role of Tumor Necrosis Factor Alpha (TNF-α) in Autoimmune Disease and Current TNF-α Inhibitors in Therapeutics. *International Journal of Molecular Sciences*, *22*(5), 2719. DOI: 10.3390/ijms22052719 PMID: 33800290

Joó, J. G., Sulyok, E., Bódis, J., & Kornya, L. (2023). Disrupted Balance of the Oxidant-Antioxidant System in the Pathophysiology of Female Reproduction: Oxidative Stress and Adverse Pregnancy Outcomes. *Current Issues in Molecular Biology*, *45*(10), 8091–8111. DOI: 10.3390/cimb45100511 PMID: 37886954

Juraski, A. C., Sharma, S., Sparanese, S., da Silva, V. A., Wong, J., Laksman, Z., Flannigan, R., Rohani, L., & Willerth, S. M. (2023). 3D bioprinting for organ and organoid models and disease modeling. *Expert Opinion on Drug Discovery*, *18*(9), 1043–1059. DOI: 10.1080/17460441.2023.2234280 PMID: 37431937

Kageyama, Y., Takahashi, M., Ichikawa, T., Torikai, E., & Nagano, A. (2008). Reduction of oxidative stress marker levels by anti-TNF-alpha antibody, infliximab, in patients with rheumatoid arthritis. *Clinical and Experimental Rheumatology*, *26*(1), 73–80. PMID: 18328150

Khalil, W. J., Akeblersane, M., Khan, A. S., Moin, A. S. M., & Butler, A. E. (2023). Environmental Pollution and the Risk of Developing Metabolic Disorders: Obesity and Diabetes. *International Journal of Molecular Sciences*, *24*(10), 8870. DOI: 10.3390/ijms24108870 PMID: 37240215

Khan, T., Waseem, R., Zehra, Z., Aiman, A., Bhardwaj, P., Ansari, J., Hassan, M. I., & Islam, A. (2022). Mitochondrial Dysfunction: Pathophysiology and Mitochondrion-Targeted Drug Delivery Approaches. *Pharmaceutics*, *14*(12), 2657. DOI: 10.3390/pharmaceutics14122657 PMID: 36559149

Kim, H. J., Jeon, H. J., Kim, J. Y., Shim, J. J., & Lee, J. H. (2024). *Lactiplantibacillus plantarum* HY7718 Improves Intestinal Integrity in a DSS-Induced Ulcerative Colitis Mouse Model by Suppressing Inflammation through Modulation of the Gut Microbiota. *International Journal of Molecular Sciences*, *25*(1), 575. DOI: 10.3390/ijms25010575 PMID: 38203747

Kim, J., Mun, S., Lee, S., Jeong, K., & Baek, Y. (2022). Prediction of metabolic and premetabolic syndromes using machine learning models with anthropometric, lifestyle, and biochemical factors from a middle-aged population in Korea. *BMC Public Health*, *22*(1), 664. DOI: 10.1186/s12889-022-13131-x PMID: 35387629

Kim, J. A., Wei, Y., & Sowers, J. R. (2008). Role of mitochondrial dysfunction in insulin resistance. *Circulation Research, 102*(4), 401–414. DOI: 10.1161/CIRCRESAHA.107.165472 PMID: 18309108

Klak, M., Wszoła, M., Berman, A., Filip, A., Kosowska, A., Olkowska-Truchanowicz, J., Rachalewski, M., Tymicki, G., Bryniarski, T., Kołodziejska, M., Dobrzański, T., Ujazdowska, D., Wejman, J., Uhrynowska-Tyszkiewicz, I., & Kamiński, A. (2023). Bioprinted 3D Bionic Scaffolds with Pancreatic Islets as a New Therapy for Type 1 Diabetes-Analysis of the Results of Preclinical Studies on a Mouse Model. *Journal of Functional Biomaterials, 14*(7), 371. DOI: 10.3390/jfb14070371 PMID: 37504866

Krawczyk, M., Burzynska-Pedziwiatr, I., Wozniak, L. A., & Bukowiecka-Matusiak, M. (2023). Impact of Polyphenols on Inflammatory and Oxidative Stress Factors in Diabetes Mellitus: Nutritional Antioxidants and Their Application in Improving Antidiabetic Therapy. *Biomolecules, 13*(9), 1402. DOI: 10.3390/biom13091402 PMID: 37759802

Kumar, L., Kumar, S., Sandeep, K., & Patel, S. K. S. (2023). Therapeutic Approaches in Pancreatic Cancer: Recent Updates. *Biomedicines, 11*(6), 1611. DOI: 10.3390/biomedicines11061611 PMID: 37371705

Kuppuswami, J., & Senthilkumar, G. P. (2023). Nutri-stress, mitochondrial dysfunction, and insulin resistance-role of heat shock proteins. *Cell Stress & Chaperones, 28*(1), 35–48. DOI: 10.1007/s12192-022-01314-9 PMID: 36441381

Lange, M. G., Coffey, A. A., Coleman, P. C., Barber, T. M., Van Rens, T., Oyebode, O., Abbott, S., & Hanson, P. (2024). Metabolic changes with intermittent fasting. *Journal of human nutrition and dietetics: the official journal of the British Dietetic Association, 37*(1), 256–269. https://doi.org/DOI: 10.1111/jhn.13253

Lazzarin, T., Martins, D., Ballarin, R. S., Monte, M. G., Minicucci, M. F., Polegato, B. F., & Zornoff, L. (2023). The Role of Omega-3 in Attenuating Cardiac Remodeling and Heart Failure through the Oxidative Stress and Inflammation Pathways. *Antioxidants, 12*(12), 2067. DOI: 10.3390/antiox12122067 PMID: 38136187

Leal, L. E., Moreira, E. S., Correia, B. L., Bueno, P. S. A., Comar, J. F., de Sá-Nakanishi, A. B., Cuman, R. K. N., Bracht, A., Bersani-Amado, C. A., & Bracht, L. (2024). Comparative study of the antioxidant and anti-inflammatory effects of the natural coumarins 1,2-benzopyrone, umbelliferone and esculetin: In silico, in vitro and in vivo analyses. *Naunyn-Schmiedeberg's Archives of Pharmacology, 397*(1), 173–187. DOI: 10.1007/s00210-023-02606-2 PMID: 37395795

Lee, J., Yun, J. S., & Ko, S. H. (2022). Advanced Glycation End Products and Their Effect on Vascular Complications in Type 2 Diabetes Mellitus. *Nutrients*, *14*(15), 3086. DOI: 10.3390/nu14153086 PMID: 35956261

Lennicke, C., & Cochemé, H. M. (2021). Redox regulation of the insulin signalling pathway. *Redox Biology*, *42*, 101964. DOI: 10.1016/j.redox.2021.101964 PMID: 33893069

Li, K., Xiao, X., Li, Y., Lu, S., Zi, J., Sun, X., Xu, J., Liu, H. Y., Li, X., Song, T., & Cai, D. (2024). Insights into the interplay between gut microbiota and lipid metabolism in the obesity management of canines and felines. *Journal of Animal Science and Biotechnology*, *15*(1), 114. DOI: 10.1186/s40104-024-01073-w PMID: 39118186

Li, M., Gong, J., Gao, L., Zou, T., Kang, J., & Xu, H. (2022). Advanced human developmental toxicity and teratogenicity assessment using human organoid models. *Ecotoxicology and Environmental Safety*, *235*, 113429. DOI: 10.1016/j.ecoenv.2022.113429 PMID: 35325609

Liang, B., & Xing, D. (2023). The Current and Future Perspectives of Postbiotics. *Probiotics and Antimicrobial Proteins*, *15*(6), 1626–1643. DOI: 10.1007/s12602-023-10045-x PMID: 36763279

Lingappan, K. (2018). NF-κB in Oxidative Stress. *Current Opinion in Toxicology*, *7*, 81–86. DOI: 10.1016/j.cotox.2017.11.002 PMID: 29862377

Liu, K., Tong, Z., Zhang, X., Dahmani, M., Zhao, M., Hu, M., Li, X., & Xue, Z. (2024). A Review: Development of a Synthetic Lactoferrin Biological System. *Biodesign research*, *6*, 0040. https://doi.org/DOI: 10.34133/bdr.0040

Liu, Y., Li, N., & Zhu, Y. (2023). Pancreatic Organoids: A Frontier Method for Investigating Pancreatic-Related Diseases. *International Journal of Molecular Sciences*, *24*(4), 4027. DOI: 10.3390/ijms24044027 PMID: 36835437

Lo Cicero, S., Castelli, G., Blaconà, G., Bruno, S. M., Sette, G., Pigliucci, R., Villella, V. R., Esposito, S., Zollo, I., Spadaro, F., Maria, R., Biffoni, M., Cimino, G., Amato, F., Lucarelli, M., & Eramo, A. (2023). L1077P CFTR pathogenic variant function rescue by Elexacaftor-Tezacaftor-Ivacaftor in cystic fibrosis patient-derived air-liquid interface (ALI) cultures and organoids: In vitro guided personalized therapy of non-F508del patients. *Respiratory Research*, *24*(1), 217. DOI: 10.1186/s12931-023-02516-0 PMID: 37674160

López-Otín, C., Blasco, M. A., Partridge, L., Serrano, M., & Kroemer, G. (2023). Hallmarks of aging: An expanding universe. *Cell*, *186*(2), 243–278. DOI: 10.1016/j.cell.2022.11.001 PMID: 36599349

Lopez-Otin, C., & Kroemer, G. (2021). Hallmarks of aging: An expanding universe. *Cell, 186*(1), 243–280. DOI: 10.1016/j.cell.2022.11.001 PMID: 36599349

Mackenzie, R. W., & Elliott, B. T. (2014). Akt/PKB activation and insulin signaling: A novel insulin signaling pathway in the treatment of type 2 diabetes. *Diabetes, Metabolic Syndrome and Obesity*, *7*, 55–64. DOI: 10.2147/DMSO.S48260 PMID: 24611020

Maftei, N. M., Raileanu, C. R., Balta, A. A., Ambrose, L., Boev, M., Marin, D. B., & Lisa, E. L. (2024). The Potential Impact of Probiotics on Human Health: An Update on Their Health-Promoting Properties. *Microorganisms*, *12*(2), 234. DOI: 10.3390/microorganisms12020234 PMID: 38399637

Maharjan, S., Ma, C., Singh, B., Kang, H., Orive, G., Yao, J., & Shrike Zhang, Y. (2024). Advanced 3D imaging and organoid bioprinting for biomedical research and therapeutic applications. *Advanced Drug Delivery Reviews*, *208*, 115237. DOI: 10.1016/j.addr.2024.115237 PMID: 38447931

Mancini, A., & Silvestrini, A. (2022). Oxidative Stress in Metabolic and Endocrine Diseases: Basic and Translational Aspects. *International Journal of Molecular Sciences*, *23*(8), 4346. DOI: 10.3390/ijms23084346 PMID: 35457165

Manna, P., & Jain, S. K. (2015). Obesity, Oxidative Stress, Adipose Tissue Dysfunction, and the Associated Health Risks: Causes and Therapeutic Strategies. *Metabolic Syndrome and Related Disorders*, *13*(10), 423–444. DOI: 10.1089/met.2015.0095 PMID: 26569333

Marcone, S., Belton, O., & Fitzgerald, D. J. (2017). Milk-derived bioactive peptides and their health promoting effects: A potential role in atherosclerosis. *British Journal of Clinical Pharmacology*, *83*(1), 152–162. DOI: 10.1111/bcp.13002 PMID: 27151091

Markowiak, P., & Śliżewska, K. (2017). Effects of Probiotics, Prebiotics, and Synbiotics on Human Health. *Nutrients*, *9*(9), 1021. DOI: 10.3390/nu9091021 PMID: 28914794

Mazloom, K., Siddiqi, I., & Covasa, M. (2019). Probiotics: How Effective Are They in the Fight against Obesity? *Nutrients*, *11*(2), 258. DOI: 10.3390/nu11020258 PMID: 30678355

Mazzetti, A. P., Fiorile, M. C., Primavera, A., & Lo Bello, M. (2015). Glutathione transferases and neurodegenerative diseases. *Neurochemistry International*, *82*, 10–18. DOI: 10.1016/j.neuint.2015.01.008 PMID: 25661512

Mostavi, M., Chiu, Y. C., Huang, Y., & Chen, Y. (2020). Convolutional neural network models for cancer type prediction based on gene expression. *BMC Medical Genomics*, *13*(S5, Suppl 5), 44. DOI: 10.1186/s12920-020-0677-2 PMID: 32241303

Mounir, M., Ibijbijen, A., Farih, K., Rabetafika, H. N., & Razafindralambo, H. L. (2022). Synbiotics and Their Antioxidant Properties, Mechanisms, and Benefits on Human and Animal Health: A Narrative Review. *Biomolecules*, *12*(10), 1443. DOI: 10.3390/biom12101443 PMID: 36291652

Movaghar, R., Farshbaf-Khalili, A., Hajizade, K., MirzaRezaei, M. E., & Shahnazi, M. (2022). The Effect of Probiotics or Synbiotics on the Hypertensive Disorders of Pregnant Women with Gestational Diabetes: A Systematic Review and Meta-analysis. *Journal of Caring Sciences*, *11*(2), 94–104. DOI: 10.34172/jcs.2021.027 PMID: 35919277

Mumtaz, H., Saqib, M., Jabeen, S., Muneeb, M., Mughal, W., Sohail, H., Safdar, M., Mehmood, Q., Khan, M. A., & Ismail, S. M. (2023). Exploring alternative approaches to precision medicine through genomics and artificial intelligence - a systematic review. *Frontiers in Medicine*, *10*, 1227168. DOI: 10.3389/fmed.2023.1227168 PMID: 37849490

Murphy, S. V., & Atala, A. (2014). 3D bioprinting of tissues and organs. *Nature Biotechnology*, *32*(8), 773–785. DOI: 10.1038/nbt.2958 PMID: 25093879

New-Aaron, M., Ganesan, M., Dagur, R. S., Kharbanda, K. K., Poluektova, L. Y., & Osna, N. A. (2021). Pancreatogenic Diabetes: Triggering Effects of Alcohol and HIV. *Biology (Basel)*, *10*(2), 108. DOI: 10.3390/biology10020108 PMID: 33546230

Novelli, G., Spitalieri, P., Murdocca, M., Centanini, E., & Sangiuolo, F. (2023). Organoid factory: The recent role of the human induced pluripotent stem cells (hiPSCs) in precision medicine. *Frontiers in Cell and Developmental Biology*, *10*, 1059579. DOI: 10.3389/fcell.2022.1059579 PMID: 36699015

Ordóñez-Díaz, M. D., Gil-Campos, M., Flores-Rojas, K., Muñoz-Villanueva, M. C., Mesa, M. D., de la Torre-Aguilar, M. J., Gil, Á., & Pérez-Navero, J. L. (2021). Impaired Antioxidant Defense Status Is Associated With Metabolic-Inflammatory Risk Factors in Preterm Children With Extrauterine Growth Restriction: The BIORICA Cohort Study. *Frontiers in Nutrition*, *8*, 793862. DOI: 10.3389/fnut.2021.793862 PMID: 34993223

Osowiecka, K., Skrypnik, D., & Myszkowska-Ryciak, J. (2023). Assessment of the Impact of Nutritional Intervention with the Probiotic *Lactiplantibacillus plantarum 299v* on Nutritional Status and Quality of Life of Hashimoto's Thyroiditis Patients-A Randomized Double-Blind Study Protocol. *Journal of Personalized Medicine*, *13*(12), 1659. DOI: 10.3390/jpm13121659 PMID: 38138886

Pan, C., Pan, C., Chen, L., & Lin, A. (2024). Machine learning analysis of oxidative stress-related phenotypes for specific gene screening in ovarian cancer. *Environmental Toxicology*, *39*(10), 1–13. DOI: 10.1002/tox.24321 PMID: 39162393

Pan, D., Xu, L., & Guo, M. (2022). The role of protein kinase C in diabetic microvascular complications. *Frontiers in Endocrinology*, *13*, 973058. DOI: 10.3389/fendo.2022.973058 PMID: 36060954

Paquette, S., Thomas, S. C., Venkataraman, K., Appanna, V. D., & Tharmalingam, S. (2023). The Effects of Oral Probiotics on Type 2 Diabetes Mellitus (T2DM): A Clinical Trial Systematic Literature Review. *Nutrients*, *15*(21), 4690. DOI: 10.3390/nu15214690 PMID: 37960343

Park, S. J., Sharma, A., & Lee, H. J. (2023). Postbiotics against Obesity: Perception and Overview Based on Pre-Clinical and Clinical Studies. *International Journal of Molecular Sciences*, *24*(7), 6414. DOI: 10.3390/ijms24076414 PMID: 37047387

Parvaneh, S., Kemény, L., Ghaffarinia, A., Yarani, R., & Veréb, Z. (2023). Three-dimensional bioprinting of functional β-islet-like constructs. *International Journal of Bioprinting*, *9*(2), 665. DOI: 10.18063/ijb.v9i2.665 PMID: 37065656

Petrariu, O. A., Barbu, I. C., Niculescu, A. G., Constantin, M., Grigore, G. A., Cristian, R. E., Mihaescu, G., & Vrancianu, C. O. (2024). Role of probiotics in managing various human diseases, from oral pathology to cancer and gastrointestinal diseases. *Frontiers in Microbiology*, *14*, 1296447. DOI: 10.3389/fmicb.2023.1296447 PMID: 38249451

Picó, C., Palou, M., Pomar, C. A., Rodríguez, A. M., & Palou, A. (2022). Leptin as a key regulator of the adipose organ. *Reviews in Endocrine & Metabolic Disorders*, *23*(1), 13–30. DOI: 10.1007/s11154-021-09687-5 PMID: 34523036

Poetsch, M. S., Strano, A., & Guan, K. (2022). Human Induced Pluripotent Stem Cells: From Cell Origin, Genomic Stability, and Epigenetic Memory to Translational Medicine. *Stem Cells (Dayton, Ohio)*, *40*(6), 546–555. DOI: 10.1093/stmcls/sxac020 PMID: 35291013

Rauth, S., Karmakar, S., Batra, S. K., & Ponnusamy, M. P. (2021). Recent advances in organoid development and applications in disease modeling. *Biochimica et Biophysica Acta. Reviews on Cancer*, *1875*(2), 188527. DOI: 10.1016/j.bbcan.2021.188527 PMID: 33640383

Recharla, N., Geesala, R., & Shi, X. Z. (2023). Gut Microbial Metabolite Butyrate and Its Therapeutic Role in Inflammatory Bowel Disease: A Literature Review. *Nutrients*, *15*(10), 2275. DOI: 10.3390/nu15102275 PMID: 37242159

Reuter, S., Gupta, S. C., Chaturvedi, M. M., & Aggarwal, B. B. (2010). Oxidative stress, inflammation, and cancer: How are they linked? *Free Radical Biology & Medicine*, *49*(11), 1603–1616. DOI: 10.1016/j.freeradbiomed.2010.09.006 PMID: 20840865

Robles, L., Vaziri, N. D., & Ichii, H. (2013). Role of Oxidative Stress in the Pathogenesis of Pancreatitis: Effect of Antioxidant Therapy. *Pancreatic Disorders & Therapy*, *3*(1), 112. DOI: 10.4172/2165-7092.1000112 PMID: 24808987

Rosa, C. M., Gimenes, R., Campos, D. H., Guirado, G. N., Gimenes, C., Fernandes, A. A., Cicogna, A. C., Queiroz, R. M., Falcão-Pires, I., Miranda-Silva, D., Rodrigues, P., Laurindo, F. R., Fernandes, D. C., Correa, C. R., Okoshi, M. P., & Okoshi, K. (2016). Apocynin influence on oxidative stress and cardiac remodeling of spontaneously hypertensive rats with diabetes mellitus. *Cardiovascular Diabetology*, *15*(1), 126. DOI: 10.1186/s12933-016-0442-1 PMID: 27585437

Ruze, R., Liu, T., Zou, X., Song, J., Chen, Y., Xu, R., Yin, X., & Xu, Q. (2023). Obesity and type 2 diabetes mellitus: Connections in epidemiology, pathogenesis, and treatments. *Frontiers in Endocrinology*, *14*, 1161521. DOI: 10.3389/fendo.2023.1161521 PMID: 37152942

Sabarathinam, S., Dhanasekaran, D., & Ganamurali, N. (2023). Artificial sweetener is a growing threat for metabolic syndrome: Why is extra attention required? *Future Science OA*, *9*(8), FSO880. Advance online publication. DOI: 10.2144/fsoa-2023-0092 PMID: 37621846

Sankararaman, S., Noriega, K., Velayuthan, S., Sferra, T., & Martindale, R. (2023). Gut Microbiome and Its Impact on Obesity and Obesity-Related Disorders. *Current Gastroenterology Reports*, *25*(2), 31–44. DOI: 10.1007/s11894-022-00859-0 PMID: 36469257

Savulescu-Fiedler, I., Mihalcea, R., Dragosloveanu, S., Scheau, C., Baz, R. O., Caruntu, A., Scheau, A. E., Caruntu, C., & Benea, S. N. (2024). The Interplay between Obesity and Inflammation. *Life (Basel, Switzerland)*, *14*(7), 856. DOI: 10.3390/life14070856 PMID: 39063610

Sempionatto, J. R., Lasalde-Ramírez, J. A., Mahato, K., Wang, J., & Gao, W. (2022). Wearable chemical sensors for biomarker discovery in the omics era. *Nature Reviews. Chemistry*, *6*(12), 899–915. DOI: 10.1038/s41570-022-00439-w PMID: 37117704

Sendani, A. A., Farmani, M., Jahankhani, K., Kazemifard, N., Ghavami, S. B., Houri, H., Ashrafi, F., & Sadeghi, A. (2024, December). (2024). Exploring the Anti-Inflammatory and Antioxidative Potential of Selenium Nanoparticles Biosynthesized by Lactobacillus casei 393 on an Inflamed Caco-2 Cell Line. *Cell Biochemistry and Biophysics*, *82*(4), 3265–3276. Advance online publication. DOI: 10.1007/s12013-024-01356-z PMID: 39261390

Shaher, S. A. A., Mihailescu, D. F., & Amuzescu, B. (2023). Aspartame Safety as a Food Sweetener and Related Health Hazards. *Nutrients*, *15*(16), 3627. DOI: 10.3390/nu15163627 PMID: 37630817

Shajari, S., Kuruvinashetti, K., Komeili, A., & Sundararaj, U. (2023). The Emergence of AI-Based Wearable Sensors for Digital Health Technology: A Review. *Sensors (Basel)*, *23*(23), 9498. DOI: 10.3390/s23239498 PMID: 38067871

Simon, E., Călinoiu, L. F., Mitrea, L., & Vodnar, D. C. (2021). Probiotics, Prebiotics, and Synbiotics: Implications and Beneficial Effects against Irritable Bowel Syndrome. *Nutrients*, *13*(6), 2112. DOI: 10.3390/nu13062112 PMID: 34203002

Song, D. K., & Kim, Y. W. (2023). Beneficial effects of intermittent fasting: A narrative review. *Journal of Yeungnam Medical Science*, *40*(1), 4–11. DOI: 10.12701/jyms.2022.00010 PMID: 35368155

Sparling, T. M., Offner, C., Deeney, M., Denton, P., Bash, K., Juel, R., Moore, S., & Kadiyala, S. (2024). Intersections of Climate Change with Food Systems, Nutrition, and Health: An Overview and Evidence Map. *Advances in Nutrition*, *15*(9), 100274. Advance online publication. DOI: 10.1016/j.advnut.2024.100274 PMID: 39019218

Srikanth, K. K., & Orrick, J. A. (2024 Jan). Biochemistry, Polyol or Sorbitol Pathways. [Updated 2022 Nov 14] In *StatPearls* [Internet]. StatPearls Publishing., Available from https://www.ncbi.nlm.nih.gov/books/NBK576381/

Steven, S., Frenis, K., Oelze, M., Kalinovic, S., Kuntic, M., Bayo Jimenez, M. T., Vujacic-Mirski, K., Helmstädter, J., Kröller-Schön, S., Münzel, T., & Daiber, A. (2019). Vascular Inflammation and Oxidative Stress: Major Triggers for Cardiovascular Disease. *Oxidative Medicine and Cellular Longevity*, *7092151*, 1–26. Advance online publication. DOI: 10.1155/2019/7092151 PMID: 31341533

Stoev, S. D. (2024). Food Security and Foodborne Mycotoxicoses-What Should Be the Adequate Risk Assessment and Regulation? *Microorganisms, 12*(3), 580. DOI: 10.3390/microorganisms12030580 PMID: 38543632

Sun, Q., Yang, Y., Liu, J., Ye, F., Hui, Q., Chen, Y., Liu, D., & Zhang, Q. (2024). Association of the weight-adjusted waist index with hypertension in the context of predictive, preventive, and personalized medicine. *The EPMA Journal, 15*(3), 491–500. DOI: 10.1007/s13167-024-00375-3 PMID: 39239106

Tangvarasittichai, S. (2015). Oxidative stress, insulin resistance, dyslipidemia and type 2 diabetes mellitus. *World Journal of Diabetes, 6*(3), 456–480. DOI: 10.4239/wjd.v6.i3.456 PMID: 25897356

Thorne, C. A., Gray, A. C., Lim, J. C., & Donaldson, P. J. (2024). The Synergistic Effects of Polyol Pathway-Induced Oxidative and Osmotic Stress in the Etiology of Diabetic Cataracts. *International Journal of Molecular Sciences, 25*(16), 9042. DOI: 10.3390/ijms25169042 PMID: 39201727

Tripathi, S., Mandal, S. S., Bauri, S., & Maiti, P. (2022). 3D bioprinting and its innovative approach for biomedical applications. *MedComm, 4*(1), e194. DOI: 10.1002/mco2.194 PMID: 36582305

Udomsak, W., Kucinska, M., Pospieszna, J., Dams-Kozlowska, H., Chatuphonprasert, W., & Murias, M. (2024). Antioxidant Enzymes in Cancer Cells: Their Role in Photodynamic Therapy Resistance and Potential as Targets for Improved Treatment Outcomes. *International Journal of Molecular Sciences, 25*(6), 3164. DOI: 10.3390/ijms25063164 PMID: 38542138

Vilà-Quintana, L., Fort, E., Pardo, L., Albiol-Quer, M. T., Ortiz, M. R., Capdevila, M., Feliu, A., Bahí, A., Llirós, M., Aguilar, E., García-Velasco, A., Ginestà, M. M., Laquente, B., Pozas, D., Lluansí, A., Pimenoff, V. N., Moreno, V., Garcia-Gil, L. J., Duell, E. J., & Aldeguer, X. (2024). Exploring the Associations of Inflammatory and Oxidative Stress Biomarkers with Pancreatic Diseases: An Observational and Mendelian Randomization Study. *Journal of Clinical Medicine, 13*(8), 2247. DOI: 10.3390/jcm13082247 PMID: 38673519

Wang, G., Wang, X., Ma, Y., Cai, S., Yang, L., Fan, Y., Zeng, X., & Qiao, S. (2022). Lactobacillus reuteri improves the development and maturation of fecal microbiota in piglets through mother-to-infant microbe and metabolite vertical transmission. *Microbiome, 10*(1), 211. DOI: 10.1186/s40168-022-01336-6 PMID: 36461096

Wang, X., Chen, L., Zhang, C., Shi, Q., Zhu, L., Zhao, S., Luo, Z., & Long, Y. (2024). Effect of probiotics at different intervention time on glycemic control in patients with type 2 diabetes mellitus: A systematic review and meta-analysis. *Frontiers in Endocrinology, 15*, 1392306. DOI: 10.3389/fendo.2024.1392306 PMID: 39114293

Warraich, U. E., Hussain, F., & Kayani, H. U. R. (2020). Aging - Oxidative stress, antioxidants and computational modeling. *Heliyon, 6*(5), e04107. DOI: 10.1016/j.heliyon.2020.e04107 PMID: 32509998

Westfall, S., Carracci, F., Estill, M., Zhao, D., Wu, Q. L., Shen, L., Simon, J., & Pasinetti, G. M. (2021). Optimization of probiotic therapeutics using machine learning in an artificial human gastrointestinal tract. *Scientific Reports, 11*(1), 1067. DOI: 10.1038/s41598-020-79947-y PMID: 33441743

World Health Organization. (2022). *One Health: A comprehensive approach to health management.* Retrieved from https://www.who.int/one-health

World Health Organization. (2023). Global report on diabetes. Retrieved from WHO website.

World Health Organization. (2023). *Obesity and overweight: Factsheet.* Retrieved from https://www.who.int/news-room/fact-sheets/detail/obesity-and-overweight

Wu, H., & Chiou, J. (2021). Potential Benefits of Probiotics and Prebiotics for Coronary Heart Disease and Stroke. *Nutrients, 13*(8), 2878. DOI: 10.3390/nu13082878 PMID: 34445037

Xiang, Y., Li, S., & Zhang, P. (2021). An exploration in remote blood pressure management: Application of daily routine pattern based on mobile data in health management. *Fundamental Research (Beijing), 2*(1), 154–165. DOI: 10.1016/j.fmre.2021.11.006 PMID: 38933904

Xiaoyu, W., Yuxin, X., & Li, L. (2024). The effects of different intermittent fasting regimens in people with type 2 diabetes: A network meta-analysis. *Frontiers in Nutrition, 11*, 1325894. DOI: 10.3389/fnut.2024.1325894 PMID: 38332802

Xu, Y., Song, D., & Wang, X. (2022). 3D Bioprinting for Pancreas Engineering/Manufacturing. *Polymers, 14*(23), 5143. DOI: 10.3390/polym14235143 PMID: 36501537

Yan, K. (2024). Recent advances in the effect of adipose tissue inflammation on insulin resistance. *Cellular Signalling, 120*, 111229. DOI: 10.1016/j.cellsig.2024.111229 PMID: 38763181

Yao, W., Liao, H., Pang, M., Pan, L., Guan, Y., Huang, X., Hei, Z., Luo, C., & Ge, M. (2022). Inhibition of the NADPH Oxidase Pathway Reduces Ferroptosis during Septic Renal Injury in Diabetic Mice. *Oxidative Medicine and Cellular Longevity*, *1193734*, 1–16. Advance online publication. DOI: 10.1155/2022/1193734 PMID: 35265258

Yung, J. H. M., & Giacca, A. (2020). Role of c-Jun N-terminal Kinase (JNK) in Obesity and Type 2 Diabetes. *Cells*, *9*(3), 706. DOI: 10.3390/cells9030706 PMID: 32183037

Zgutka, K., Tkacz, M., Tomasiak, P., & Tarnowski, M. (2023). A Role for Advanced Glycation End Products in Molecular Aging. *International Journal of Molecular Sciences*, *24*(12), 9881. DOI: 10.3390/ijms24129881 PMID: 37373042

Zhan, Y., Jiang, W., Liu, Z., Wang, Z., Guo, K., & Sun, J. (2024). Utilizing bioprinting to engineer spatially organized tissues from the bottom-up. *Stem Cell Research & Therapy*, *15*(1), 101. DOI: 10.1186/s13287-024-03712-5 PMID: 38589956

Zhang, Z., Huang, Q., Zhao, D., Lian, F., Li, X., & Qi, W. (2023). The impact of oxidative stress-induced mitochondrial dysfunction on diabetic microvascular complications. *Frontiers in Endocrinology*, *14*, 1112363. DOI: 10.3389/fendo.2023.1112363 PMID: 36824356

Zhao, D. M., Zhong, R., Wang, X. T., & Yan, Z. H. (2024). Mitochondrial dysfunction in diabetic nephropathy: Insights and therapeutic avenues from traditional Chinese medicine. *Frontiers in Endocrinology*, *15*, 1429420. DOI: 10.3389/fendo.2024.1429420 PMID: 39109083

Zhao, T., Wang, H., Liu, Z., Liu, Y., DeJi, , Li, B., & Huang, X. (2023). Recent Perspective of *Lactobacillus* in Reducing Oxidative Stress to Prevent Disease. *Antioxidants*, *12*(3), 769. DOI: 10.3390/antiox12030769 PMID: 36979017

Zhao, X., Liu, S., Li, S., Jiang, W., Wang, J., Xiao, J., Chen, T., Ma, J., Khan, M. Z., Wang, W., Li, M., Li, S., & Cao, Z. (2024). Unlocking the power of postbiotics: A revolutionary approach to nutrition for humans and animals. *Cell Metabolism*, *36*(4), 725–744. DOI: 10.1016/j.cmet.2024.03.004 PMID: 38569470

Zheng, F., Xiao, Y., Liu, H., Fan, Y., & Dao, M. (2021). Patient-Specific Organoid and Organ-on-a-Chip: 3D Cell-Culture Meets 3D Printing and Numerical Simulation. *Advanced Biology*, *5*(6), e2000024. DOI: 10.1002/adbi.202000024 PMID: 33856745

Zheng, Y., Yang, N., Pang, Y., Gong, Y., Yang, H., Ding, W., & Yang, H. (2023). Mitochondria-associated regulation in adipose tissues and potential reagents for obesity intervention. *Frontiers in Endocrinology*, *14*, 1132342. DOI: 10.3389/fendo.2023.1132342 PMID: 37396170

Zhou, L., Ding, C., Wu, J., Chen, X., Ng, D. M., Wang, H., Zhang, Y., & Shi, N. (2021). Probiotics and synbiotics show clinical efficacy in treating gestational diabetes mellitus: A meta-analysis. *Primary Care Diabetes*, *15*(6), 937–947. DOI: 10.1016/j.pcd.2021.08.005 PMID: 34417122

Zhu, H. M., Liu, N., Sun, D. X., & Luo, L. (2023). Machine-learning algorithm-based prediction of a diagnostic model based on oxidative stress-related genes involved in immune infiltration in diabetic nephropathy patients. *Frontiers in Immunology*, *14*, 1202298. DOI: 10.3389/fimmu.2023.1202298 PMID: 37554330

Zolghadrpour, M. A., Jowshan, M. R., Seyedmahalleh, M. H., Imani, H., Karimpour, F., & Asghari, S. (2024). Consumption of a new developed synbiotic yogurt improves oxidative stress status in adults with metabolic syndrome: A randomized controlled clinical trial. *Scientific Reports*, *14*(1), 20333. DOI: 10.1038/s41598-024-71264-y PMID: 39223205

Zorena, K., Jachimowicz-Duda, O., Ślęzak, D., Robakowska, M., & Mrugacz, M. (2020). Adipokines and Obesity. Potential Link to Metabolic Disorders and Chronic Complications. *International Journal of Molecular Sciences*, *21*(10), 3570. DOI: 10.3390/ijms21103570 PMID: 32443588

Chapter 7
Aging and Reactive Oxygen Species Implications for Cellular Senescence and Longevity:
Oxygen Toxicity and Aging

Rahul Saxena
https://orcid.org/0000-0002-4517-8529
Sharda University, India

Suyash Saxena
https://orcid.org/0000-0003-3763-6830
Sharda University, India

Ajit Pal Singh
https://orcid.org/0000-0002-3006-8474
Sharda University, India

ABSTRACT

Aging, a substantial cause of physical and psychosocial degradation, is attaining a significant public health concern with passage of time. It is characterized by progressive loss and delayed response of the body system followed by development of systemic complications. It enhances financial burden to the health care system. Compounding the evidences from previous reports, enhanced free radicals production mediated chronic oxidative stress along with aging process has been attributed as a specific and critical component to the etiopathology of various health complications.

DOI: 10.4018/979-8-3693-7919-6.ch007

Many of the recommended interventions present only symptom-modifying effects, and a few structure-modifying effects. Thus, the present chapter aims to provide a narrative review on mystery of aging process by highlighting the complex interplay of oxidative and inflammatory stress, and orchestrate a repertoire of novel therapeutic strategies to get rid of curse of senescence.

INTRODUCTION

"Nothing is permanent in this wicked world"- Charlie Chaplin quoted these words centuries ago. (Socratic Method, n.d.) Every biomolecule, exist in nature, after a fixed time period, deteriorates with a mysterious manner. Scientists have made enormous efforts, but they have not been able to unlock the mysteries of immortality, old age and death are now considered inevitable aspects of life. Researchers now have a greater incentive to maintain individuals physically healthy and contributing members of society. Because of this, life expectancy has increased dramatically worldwide, and by the year 2050, it is anticipated that one-third of the world's population will be elderly. However, as people get older, they are more vulnerable to many ailments.

The biological phenomena of aging are a universal and inevitable process that results from a sequence of detrimental events that progressively compromise the morphology and function of the organs. Several chronic morbidities often occur concurrently with these events. (Saxena & Lal, 2006) These morbidities, which have become prevalent health issues among the elderly population, include endocrine dysfunction, immunosenescence, musculoskeletal disorders, neurodegenerative illnesses, cardiovascular disease, and diabetes. Put another way, as time goes on, aging has a detrimental effect on biomolecules and is associated with a rise in the incidence of age-related illnesses. (Pole, Dimri, & Dimri, 2016)

It's interesting to note that Haber and Willstater (1931) were the first to describe the redox reaction involving oxygen in biological systems. At that point, they identified the superoxide anion as the univalent reduce product of the molecular oxygen intermediate. (Haber & Willstätter, 1931) Haber and Weiss discovered in 1935 that when molecular oxygen underwent consecutive trivalent reduction using an iron catalyst, a highly reactive oxygen intermediate—a potent oxidizing agent—was produced. (Haber & Weiss, 1935)According to Denham Harman's free radical hypothesis of aging, which was published in 1956, the production of endogenous free radicals and the accumulated oxidative damage that free radicals generated during aerobic respiration cause to cells may be responsible for a number of changes associated with cellular aging. Beyond this notion, McCord's seminal 1974 study on synovial fluid in rheumatoid arthritis patients has gradually spawned a plethora of research on the function of ROS in aging and age-mediated illnesses. (Free Radical Theory of Aging, 1992) (McCord, 1974)

The discovery's biological significance wasn't realized until 1969, when McCord and Fridovich discovered the antioxidant enzyme that mediated the transformation of superoxide anion to hydrogen peroxide. (McCord & Fridovich, 1969)Free radicals (especially O2.-) and other reactive oxygen species, including H2O2, hydroperoxyl radicals (HO2), hydroxyl radicals (OH.), lipid peroxyl radicals (ROO.), etc., are constantly created in vivo, according to Halliwell et al. (1992). Furthermore, these radicals are highly reactive substances that can target any biological molecule and usually start a chain reaction with other free radicals. (Halliwell, Gutteridge, & Cross, 1992)

Because of this, a significant amount of research conducted over the years has shown how oxidative stress brought on by free radicals contributes to the aging process in higher species. ROS and oxidative stress have been shown to control the incidence of aging-related disorders and have a significant impact on both the aging process and potentially lifespan in a number of age-associated disease models. The precise mechanism by which oxidative stress and ROS lead to age-related diseases is still unknown. Understanding how ROS and antioxidant enzymes modulate age-related conditions in different types of diseases can still offer more understanding into prospective therapeutic strategies to ameliorate such maladies, even though the precise effect of ROS and oxidative stress on life span is still up for debate. The numerous sources and defenses of reactive oxygen species (ROS) are briefly covered in this overview. We also examine the cellular fates that ROS causes, specifically cellular senescence, which can accelerate a variety of age-related illnesses and disorders, including cancer, heart disease, and neurological disorders. (Saxena & Lal, 2006) (Pole, Dimri, & Dimri, 2016)

ROS and Oxidative stress: The phrase "oxidative stress" describes stress that is generally linked to an increase in free radical production and a decrease in the body's antioxidant defense system. With an unpaired electron in their outer shell, free radicals are highly reactive molecules that are prone to reacting with other chemical species and biological membranes, including urate, ascorbate, α-tocopherol, and super oxide anion ($O_2^{\cdot -}$) radicals; they can also react with hydrogen peroxide (H_2O_2), hydroxyl radical ($OH^{\cdot}$), lipid peroxyl radical (ROO^{-}), and hydroperoxyl radical ($HO_2^{\cdot}$).

Reactive nitrogen species (RNS) in senescence, often referred to as nitric oxide (NO) and peroxy nitrite ($ONOO^{\cdot --}$) radicals, have also received a lot of attention (Figure 1). Oxidative stress, which is brought on by an excess of ROS and RNS, has a substantial impact on aging, inflammation, and a variety of other diseases.

Figure 1. Types of free radicals

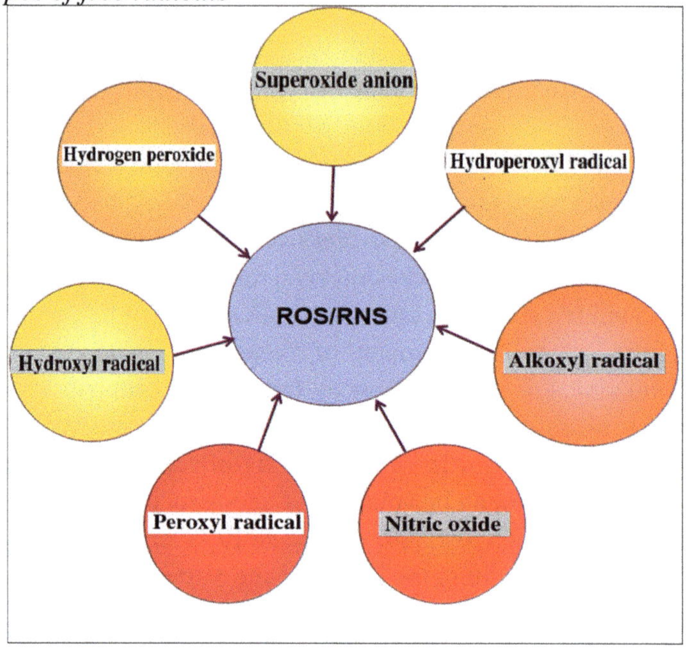

Sources of ROS in Aging: In people with arthritis, several pathways have been shown to produce free radicals. Figure 2 provides a clear illustration of these.

Figure 2. Sources of free radical production

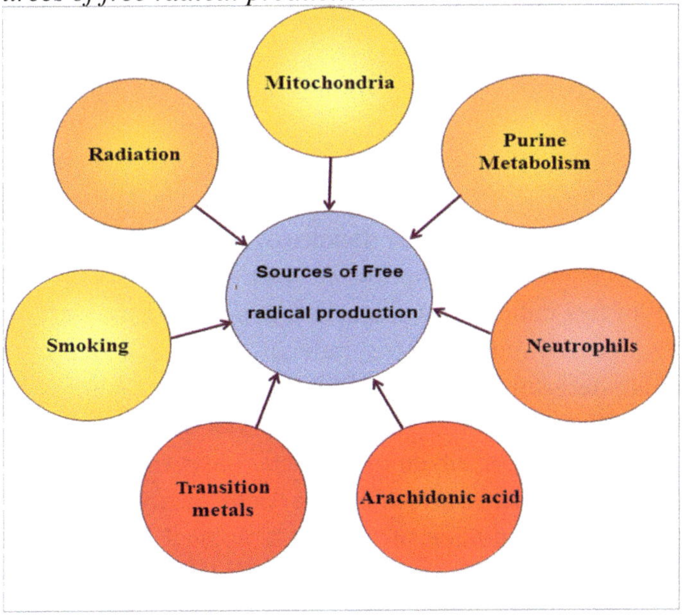

i) ***Leukocytes***: Formation of ROS and RNS is largely attributed to neutrophils, monocytes, and macrophages. About 70% of blood leukocytes are neutrophils, which are the initial line of defense. A soluble or phagocytic stimulus that activates neutrophils starts a cascade of metabolic events known as a respiratory burst, which is a crucial step in the etiology of numerous age-related problems.

Mitochondria: It serves as the primary hub for cellular oxidation events and is a significant cellular organelle. The respiratory chain's mitochondria produce reactive oxygen species, which is harmful to cells. Furthermore, Miesel et al. evaluated leukocyte mitochondria's production of reactive oxygen species, which is a major source of free radical production, and suggested that this process may be involved in the pathophysiology of disease. (Miesel, Murphy, & Kroger, 1996)

ii) ***Purine metabolism***: After hypoxic-reperfusion damage, which happens in inflammatory conditions, Xanthine dehydrogenase is converted into Xanthine oxidase, which catalyzes the conversion of hypoxanthine to Xanthine and then to uric acid. In this process, molecular oxygen takes on an electron and produces an anion known as superoxide.

iii) ***Catecholamines:*** Catecholamines produced by tissue have a critical role in the development of pain and other age-related clinical complaints. These catecholamines are broken down by the enzyme monoamine oxidase by oxidation and the resultant production of extra electrons. O2η- and H2O2 are produced by molecular oxygen when it accepts electrons, which encourages the creation of further free radicals.

iv) ***Arachidonic acid metabolism:*** During inflammation, a significant amount of arachidonic acid is released from the phospholipid bilayer. This acid can then be processed by the lipoxygenase or cyclo-oxygenase pathway to produce prostaglandins, thromboxanes, or leukotrienes. Alongside these processes, reactive oxygen species are produced.

v) ***Transition metals:*** Ischemia, hemolysis, hypoxic-reperfusion damage, and other circumstances trigger the discharge of transition metal ions, such as copper and iron ions, that can exacerbate the toxicity of free radicals.

vii) Smoking, alcoholism, radiation, ferrous myoglobin oxidation, and other possible causes of free radicals that exacerbate oxidative stress and accelerate aging are some other potential sources of free radicals. (Saxena & Batra, 2020)

Biomolecular deterioration caused by free radicals in Senescence: In actuality, aging is a deliberate and progressive process of decomposition. One way that this process may directly contribute to the faster onset of disease is by increasing the production of free radicals, which in turn causes degradation of biomolecules. The following is a representation of the key biological constituents impacted by free radicals: membrane lipids, proteins, DNA, glucosaminoglycans, and other biomolecules.

a) **Lipid peroxidation**: An autoxidation process known as "lipid peroxidation" is caused by free radicals (such as OH·, O2η-, and H2O2) attacking phospholipids or Poly unsaturated fatty acids in the subcellular constituent's or cell membrane. This process produces a variety of aldehydes, ketones, alkanes, carboxylic acids, and polymerization products. These compounds function as a biochemical marker of lipid peroxidation by the ROS and are highly reactive with extracellular matrix and other cellular components. One of the reactive aldehydes that results from lipid peroxidation is malondialdehyde (MDA), a hazardous aldehydic end product. It causes fragmentation, modification, aggregation, and changes in protein structure, all of which eventually impair tissue function. It also promotes the oxidation of cartilage collagen. The cellular function, electrolyte balance, and membrane permeability can all be altered by the excessive adherence of these volatile aldehydes to matrix and cell proteins. These alterations may also set off fibrogenesis, the breakdown of matrix proteins, and the ongoing deterioration of the biological system connected to the evolution of oxidative stress-mediated senescence. Numerous investigations have also been reported on the increased concentration of lipid peroxidation indicators in the senescence-related development of disease. (Bhattacharya, Saxena, & Gupta, 2012)

b) **Protein oxidation:** Proteins that have undergone oxidative modification due to free radicals build up with age and have drawn a lot of interest in the investigation of the molecular mechanisms behind the development of arthritis. These ROS and RNS cause side chains of amino acid residues to oxidize, proteins to form cross-links, and the protein backbone to oxidize. The results of this process include denaturation, loss of function, aggregation, and fragmentation of proteins. (Hohn et al., 2017) Furthermore, the function of vital extracellular and cellular proteins, such as connective tissue proteins and enzymes, might be compromised by radical protein reactions, which further encourages the development of age-related diseases. In addition to interacting with glutathione and protein thiol groups to generate nitrosothiols and obstruct their enzymatic activity, NO and its derivatives also do this. A number of researchers have also noted that the build-up of oxidized proteins in various tissues as people age is linked to a lower life expectancy due to the development of diseases. (Davies, 1987)

c) <u>DNA damage</u>: DNA is one of the main targets that free radicals attack. Overproduction of radicals may combine with DNA to break its strands and remove its bases. When a hydroxyl radical combines with the nucleotide deoxyguanosine, 8-hydroxydeoxyguanosine (8-OHdG) is produced. This reaction enhances the mis-incorporation of DNA bases, which can result in somatic mutation. Furthermore, apoptosis, cell differentiation, proliferation, and genomic instability have all been linked to the involvement of free radical-mediated DNA damage. Numerous investigations have demonstrated that urinary 8-OHdG is a biomarker of broad cellular oxidative stress and a potential risk factor for age-related health conditions. (Wu, Chiou, Chang, & Wu, 2004)

d) **<u>Glycoxidation:</u>** Among the RONS-mediated changes of carbohydrates is the glycoxidation process, which produces advanced glycation end products (AGEs) by reacting the carbonyl groups of carbohydrates with the amino groups of arginine and lysine. Pentosidine, glucosepane, N-carboxymethyl-lysine, and hydroimidazolone are examples of major AGEs. Through certain cell membrane-based receptors for advanced glycation endproducts (RAGE), AGEs mediate their effects on cellular activity. In addition to increasing MMP expression and cellular catabolic activity, RONS-dependent activation of RAGE also causes oxidative stressElevated RAGE expression in individuals with osteoarthritis further supports the role of oxidative stress in causing the pathological changes that result in deterioration of joints as people age. However, it has been discovered that elevated levels of AGEs are linked to diabetes, sarcopenia, chronic renal disease, and older people suffering from heart failure. These findings suggest that the aging process mediated by AGE accumulation will result in a higher deterioration in the activities of numerous organs. (Liguori et al., 2018)

Change in Antioxidant Defense System in Aging: The biological system has developed a number of defense systems that work to scavenge or stop the making of free radicals, so safeguarding biomolecules, membranes, and cells. These comprise potential nonenzymic antioxidants like ascorbic acid, tocopherol, β-carotene, uric acid, albumin, glutathione, and ubiquinone, as well as antioxidant enzymes like superoxide dismutase, paraoxonase, catalase, glutathione peroxidase, and ceruloplasmin, (Figure 3). When ROS avoid or outweigh the tissues' antioxidant defense systems, oxidative stress results. Changes in these antioxidants' levels can lead to biomolecular deterioration through elevated ROS production, which can impair cellular function, degrade cell membranes, and cause cellular homeostasis to be lost. These deleterious changes can also prolong the development of disease by promoting senescence. (Jaiswal et al., 2011) Due to their potent antioxidant, chondro-protective, and cardioprotective properties, Coenzyme Q10, melatonin, hyaluronic acid, resveratrol, quercetin, and other polyphenols have garnered a lot of attention lately for their role in reducing the production of reactive oxygen species (RONS) and inflammation in age-related complications. (Jahanban-Esfahlan et al., 2018)

Figure 3. Antioxidant defense system of the body

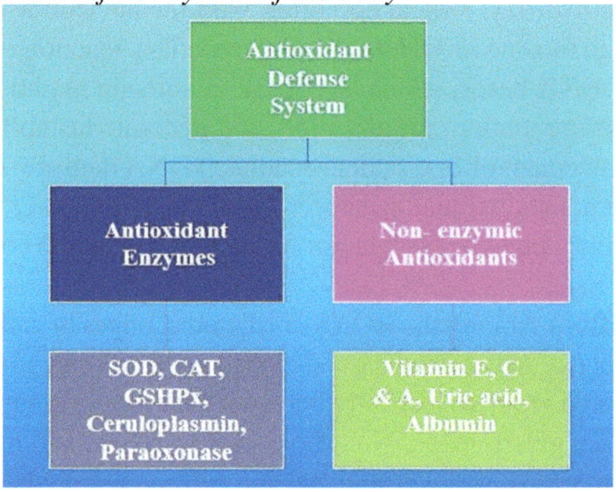

Oxidative stress mediated Aging & its related complications: Several antioxidant protection systems accurately control the synthesis of free radicals, despite the fact that they are created in huge quantities in tissues during pathological conditions. In biological tissues, there is a delicate equilibrium amid reactive oxygen species (ROS) and pro-oxidants. However, in pathological conditions, ROS generation increases, leading to changes in the antioxidant defense system and biomolecular degradation. Oxidative stress is the term for this state (Figure 4), which is critical for the onset of degenerative diseases such as diabetes mellitus, cancer, stroke, and osteoarthritis, among others.

Figure 4. Role of free radical in aging and its related complications

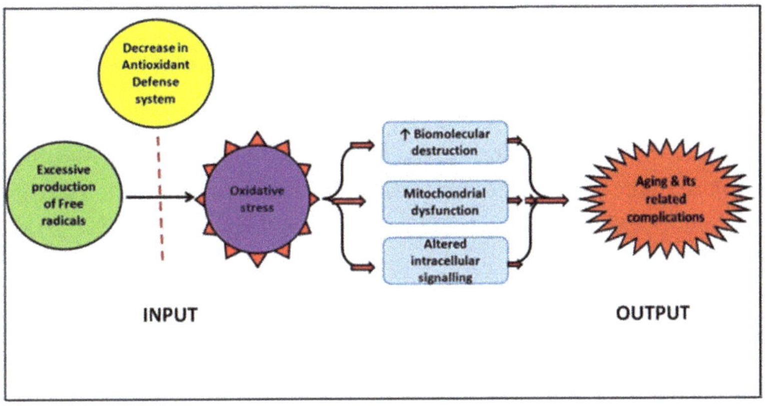

- Age related diseases: **Parkinsonism, Alzheimer's disease & Aging**
- Eye disease: **Cataractogenesis, Retinopathy,**
- Respiratory disease: **Respiratory distress syndrome, Broncho-pulmonary disorder in premature infants**
- Chronic inflammatory disease: **Rheumatoid arthritis**
- Cancer: **Lung cancer, oral cancer**
- Cardiovascular disease: **Myocardial infarction, Atherosclerosis, Hypertension & Reperfusion injury of heart**
- Other diseases: **Diabetes mellitus & Stroke**

Conclusion: Ageing, then, is a universal and unavoidable biological phenomenon that results from the accumulated harm caused by a number of deterministic causes, such as genetic programming, lifestyle choices, and environmental stressors. Numerous redox reactions that the biological system goes through result in the continuous production of free radicals. Oxidative stress is primarily caused by the overproduction of free radicals, which also results in biomolecular breakdown and alterations to the antioxidant defense system. Oxidative stress is a prerequisite for lipid peroxidation, DNA damage, protein cross-linking, induction of inflammation, apoptotic cell death, and tissue destruction, all of which aid in the commencement and advancement of age-related complications. Therefore, as one ages, it is advised to increase daily consumption of a diet high in antioxidants in order to maintain or delay the biomolecular damage caused by free radicals and to replenish the antioxidant reserve. This may be a critical step in the prevention of aging and its associated complications.

REFERENCES

www.socratic-method.com/quote-meanings/charlie-chaplin-nothing-is-permanent-in-this-wicked-world-not-even-our-troubles

Bhattacharya, I., Saxena, R., & Gupta, V. (2012). Efficacy of vitamin E in knee osteoarthritis management of North Indian Geriatric population. *Therapeutic Advances in Musculoskeletal Disease*, *4*(1), 11–19. DOI: 10.1177/1759720X11424458 PMID: 22870491

Davies, K. J. A. (1987). Protein damage and degradation of oxygen radicals. *The Journal of Biological Chemistry*, *262*(20), 9895–9902. DOI: 10.1016/S0021-9258(18)48018-0 PMID: 3036875

Genestra, M. (2007). Oxyl radicals, redox-sensitive signalling cascades and antioxidants. *Cellular Signalling*, *19*(9), 1807–1819. DOI: 10.1016/j.cellsig.2007.04.009 PMID: 17570640

Haber, F., & Weiss, J. (1935). The catalytic decomposition of hydrogen peroxide by iron salts. *J Proc Roy Soc.*, *147*, 322–351.

Haber, F., & Willstatter, R. (1931). Unpaarigkeit and radikal ketten in reaktions –mechanismus organischer und enzymatischer vorgange. *Berichte der Deutschen Chemischen Gesellschaft Abteilung B: Abhandlungen*, *64*(11), 2844–2856. DOI: 10.1002/cber.19310641118

Halliwell, B., Gutteridge, J. M. C., & Cross, C. E. (1992). Free radicals, antioxidants and human disease: Where are we now? *The Journal of Laboratory and Clinical Medicine*, *119*(6), 598. PMID: 1593209

Harman, D. (1992). Free radical theory of aging. *Mutation Research*, *275*(3-6), 257–266. DOI: 10.1016/0921-8734(92)90030-S PMID: 1383768

Hohn, A., Weber, D., Jung, T., Ott, C., Hugo, M., Kochlik, B., Kehm, R., Konig, J., Grune, T., & Castro, J. P. (2017). Happily (n)ever after: Aging in the context of oxidative stress, proteostasis loss and cellular senescence. *Redox Biology*, *11*, 482–501. DOI: 10.1016/j.redox.2016.12.001 PMID: 28086196

Jahanban-Esfahlan, R., Mehrzadi, S., Reiter, R. J., Seidi, K., Majidinia, M., Baghi, H. B., Khatami, N., Yousefi, B., & Sadeghpour, A. (2018). Melatonin in regulation of inflammatory pathways in rheumatoid arthritis and osteoarthritis: Involvement of circadian clock genes. *British Journal of Pharmacology*, *175*(16), 3230–3238. DOI: 10.1111/bph.13898 PMID: 28585236

Jaiswal, G., & Arora, M. (2011). Beenu, Ansari SK, Shamim A, Saxena R. Oxidative stress and the role of antioxidants in health and well being. *J Adv Res Bio Sci.*, *3*(2), 94–104.

Liguori, I., Russo, G., Curcio, F., Bulli, G., Aran, L., Della-Morte, D., Gargiulo, G., Testa, G., Cacciatore, F., Bonaduce, D., & Abete, P. (2018). Oxidative stress, aging and diseases. *Clinical Interventions in Aging*, *13*, 752–772. DOI: 10.2147/CIA.S158513 PMID: 29731617

McCord, J. M. (1974). Free radicals and inflammation. *Science*, *185*(4150), 529–531. DOI: 10.1126/science.185.4150.529 PMID: 4841157

McCord, J. M., & Fridovich, I. (1969). Superoxide dismutase, an enzymatic function for erythrocuprein. *The Journal of Biological Chemistry*, *244*(22), 6049–6055. DOI: 10.1016/S0021-9258(18)63504-5 PMID: 5389100

Miesel, R., Murphy, M. P., & Kroger, H. (1996). Enhanced mitochondrial radical production in patients which rheumatoid arthritis correlates with elevated levels of tumor necrosis factor alpha in plasma. *Free Radical Research*, *25*(2), 161–169. DOI: 10.3109/10715769609149921 PMID: 8885334

Pole, A., Dimri, M., & Dimri, G. P. (2016). Oxidative stress, cellular senescence and ageing. *AIMS Molecular Science*, *3*(3), 300–324. DOI: 10.3934/molsci.2016.3.300

Saxena, R., & Batra, J. Aging: Oxidative stress and dietary antioxidants. 2nd ed. London: Academic press Elsevier publications; 2020. Chapter 8, Arthritis as a disease of ageing and changes in antioxidant status; p. 83-92. DOI: 10.1016/B978-0-12-818698-5.00008-0

Saxena, R., & Jaiswal, G. (2007). Selenium and its role in Health and Diseases. *The Kuwait Medical Journal : KMJ : the Official Journal of the Kuwait Medical Association*, *39*(1), 10–18.

Saxena, R., & Lal, A. M. (2006). Effect of Aging on antioxidant enzyme status and lipid peroxidation. *J. Indian Acad Geriat.*, *2*(2), 53–56.

Wu, L. L., Chiou, C. C., Chang, P. Y., & Wu, J. T. (2004). Urinary 8-OHdG: A marker of oxidative stress to DNA and a risk factor for cancer, atherosclerosis and diabetics. *Clinica Chimica Acta*, *339*(1-2), 1–9. DOI: 10.1016/j.cccn.2003.09.010 PMID: 14687888

Chapter 8
Intricate Role of Reactive Oxygen Species and Reactive Nitrogen Species in Leishmaniasis:
Mechanisms, Implications, and Therapeutic Potential

Radhey Shyam Kaushal
https://orcid.org/0000-0002-0863-9082
Parul University, India

Aviral Kaushik
Parul University, India

C. Ratna Prabha
Maharaja Sayajirao University of Baroda, India

ABSTRACT

Leishmaniasis, caused by protozoan parasite, poses a significant global health challenge, affecting millions worldwide. The immune response to Leishmania involves the generation of ROS and RNS, which are critical for controlling the infection. During Leishmaniasis, phagocytic cells, generate ROS and RNS as part of the immune response, which are critical for killing intracellular pathogens. Leishmania has evolved mechanisms to survive oxidative stress by enhancing its antioxidant defences like superoxide dismutase and trypanothione reductase, which neutralize these reactive species. This ability to counteract ROS and RNS not only aids in the parasite's survival within host cells but also contributes to chronic inflammation

DOI: 10.4018/979-8-3693-7919-6.ch008

and tissue damage, ultimately influencing disease progression and severity. This chapter explores the intricate role of ROS and RNS, highlighting their dual function in regulating the parasite and contributing to disease pathology, therapeutic potential of targeting ROS/RNS pathways.

1. INTRODUCTION

1.1. Background on Leishmaniasis:

Leishmaniasis, caused by *Leishmania* parasites and transmitted through the bite of infected sandflies, remains a significant global health burden, affecting millions across more than 90 countries. The disease manifests primarily in three forms: visceral leishmaniasis (VL), cutaneous leishmaniasis (CL) and mucocutaneous leishmaniasis (ML), the latter being the most severe and often fatal if untreated. About 12 million of the world population is currently infected and over 350 million people are at risk, leishmaniasis disproportionately affects impoverished regions, particularly in South Asia, East Africa, and Latin America (Knight et al., 2023) (Sundar & Chakravarty, 2012). The disease is endemic in areas with warm, humid climates favorable to sandfly vectors, exacerbated by weak health infrastructures and socioeconomic challenges.

Visceral leishmaniasis, also termed as kala-azar, is especially deadly, with mortality rates exceeding 95% if left untreated. It is endemic in countries such as India, Brazil, and Ethiopia, contributing significantly to the global burden of neglected tropical diseases (NTDs) (Scarpini et al., 2022). Conversely, cutaneous leishmaniasis, while less lethal, can lead to severe disfigurement and long-term social stigma, impacting the social well-being of those affected. Both forms of the disease face the challenge of drug resistance, particularly in regions with limited access to effective treatments (Wijnant et al., 2022). Current therapies, such as antimonials, are not only toxic but also costly and increasingly ineffective due to emerging resistance, emphasizing the urgent need for new drugs and treatment strategies.

The co-infection of leishmaniasis with HIV further complicates disease management. In regions like Brazil and Ethiopia, co-infection rates are rising, leading to more severe disease manifestations and higher mortality rates (Wamai et al., 2020). HIV-infected individuals are more susceptible to leishmaniasis due to their weakened immune systems, and conventional treatments are less effective in these populations. This underscores the need for targeted interventions and more robust healthcare systems capable of addressing such complex co-infections.

1.1.1. Host immune response to Leishmania

The host immune response to *Leishmaniasis* is multifaceted, involving both the innate and adaptive immune systems. When *Leishmania* parasites enter the host through the bite of an infected sandfly, they are initially recognized by phagocytic cells, primarily macrophages, dendritic cells, and neutrophils, which attempt to engulf and destroy the parasites. However, *Leishmania* has evolved mechanisms to evade and manipulate these early responses, enabling survival within host cells, particularly macrophages (Kaye & Scott, 2011).

1.1.2. Innate immune response to Leishmania

Upon infection with *Leishmania*, the host's innate immune system is the first line of defense. Macrophages, dendritic cells, and neutrophils recognize and attempt to eliminate the parasite by engulfing it through phagocytosis. Inside macrophages, *Leishmania* parasites reside within phagolysosomes, where they should be killed by reactive oxygen and nitrogen species (ROS and RNS) produced by the activated macrophages. However, *Leishmania* has evolved to resist these microbicidal mechanisms and can survive within macrophages by inhibiting key processes involved in pathogen killing (Costa-da-Silva et al., 2022). Pattern recognition receptors (PRRs), such as Toll-like receptors (TLRs), are essential for detecting parasite molecules and triggering the immune response. They stimulate the production of pro-inflammatory cytokines, which in turn attract additional immune cells to the site of infection (Kaye & Scott, 2011).

1.1.3. Adaptive Immune Response: Th1 vs. Th2 Balance

The adaptive immune response to *Leishmaniasis* involves CD4+ T cells, which play a pivotal role in controlling the infection. A critical determinant of disease outcome is the balance between Thl and Th2 immune responses. A dominant Thl response, characterized by the production of interferon-gamma (IFN-γ), is essential for activating macrophages and promoting parasite killing. This response is associated with resistance to *Leishmaniasis* (Volpedo et al., 2021). In contrast, a Th2 response, which involves the production of interleukin-4 (IL-4) and other anti-inflammatory cytokines, suppresses macrophage activation, allowing the parasite to survive and proliferate, leading to chronic disease (Costa-da-Silva et al., 2022). The ability to generate a strong Thl response, therefore, is crucial for controlling and resolving the infection.

1.1.4. Immune Evasion Strategies of Leishmania

Leishmania has developed several mechanisms to evade the host's immune response. By residing within macrophages, it avoids direct attack by antibodies and can manipulate host cell signaling pathways to suppress inflammatory responses and prevent cell death. The parasite can also inhibit antigen presentation, reducing the activation of the adaptive immune system (Kaye & Scott, 2011). Additionally, *Leishmania* modulates cytokine production to shift the immune response toward a Th2-dominant profile, which promotes its survival within host cells. These immune evasion strategies enable the parasite to establish chronic infections in the host, complicating treatment and disease resolution.

1.2. Introduction to ROS and RNS:

Reactive Oxygen Species (ROS) and Reactive Nitrogen Species (RNS) are essential components of cellular physiology, contributing to critical biological processes like immune responses, inflammation, and cellular signaling. These reactive molecules are generated as byproducts of normal metabolism and have the potential to impact both health and disease. Their role is especially significant in infectious diseases like leishmaniasis, where the delicate balance between the host's defense mechanisms and the parasite's evasion strategies can influence the progression and outcome of the disease.

1.2.1. Definitions of ROS and RNS

ROS refers to a group of reactive molecules derived from molecular oxygen. This group includes superoxide anions (O_2•−), hydrogen peroxide (H_2O_2), hydroxyl radicals (•OH), and singlet oxygen ($^1O_2$). While ROS are often associated with oxidative damage to cells, they also play important roles in cell signaling and homeostasis. For instance, low levels of H_2O_2 can act as secondary messengers in various signaling pathways, influencing processes such as cell proliferation, differentiation, and apoptosis (Valko et al., 22007) (hou, 2020). However, excessive production of ROS can lead to oxidative stress, which is implicated in numerous diseases, including cancer, cardiovascular diseases, and neurodegenerative disorders.

RNS primarily consist of nitric oxide (NO) and its derivatives, including peroxynitrite (ONOO−) and nitrogen dioxide (•NO_2). NO is synthesized by nitric oxide synthases (NOS) and plays a crucial role in immune responses by promoting vasodilation and modulating inflammation. Peroxynitrite is formed when NO reacts with superoxide and is known for its potent oxidizing properties that can modify proteins, lipids, and nucleic acids, contributing to cellular damage. Both ROS and

RNS are involved in redox signaling pathways that influence various physiological processes (Zarkovic, 2020).

1.2.2. Sources of ROS and RNS within *Leishmania sp.*

Within *Leishmania* species, both ROS and RNS are produced through various metabolic and defense-related processes, playing a role in both parasite survival and interaction with the host immune system. The various ROS/RNS sources are given in table 1.

Sources of ROS in Leishmania ROS are generated in *Leishmania* primarily as byproducts of mitochondrial respiration. During energy production, the electron transport chain in *Leishmania* mitochondria can "leak" electrons, which react with molecular oxygen to form superoxide anion (O_2^-), the primary ROS (Horta et al., 2012). During oxidative stress conditions or in response to the host's immune attack, *Leishmania* cells increase ROS production as a part of their metabolic response. The enzyme NADPH oxidase, located in the parasite membrane, can also contribute to ROS generation during the interaction with host cells (Moreira et al., 2015).

Sources of RNS in *Leishmania:* RNS production in *Leishmania* primarily involves the enzymatic activities of nitric oxide synthase (NOS) homologs. However, unlike in mammalian systems where inducible nitric oxide synthase (iNOS) is well-characterized, *Leishmania* generates nitric oxide (NO) as part of its metabolic processes but to a lesser extent than host cells. When *Leishmania* is exposed to NO produced by host macrophages during infection, the parasite may produce peroxynitrite ($ONOO^-$) through reactions between NO and superoxide, contributing to oxidative stress and damage (Murray & Nathan, 1999).

Leishmania has also evolved detoxification mechanisms, such as trypanothione-dependent systems, to neutralize the toxic effects of ROS and RNS, highlighting its adaptation to oxidative stress within the host environment (Paes et al., 2011). Various sources of ROS and RNS are described in Table 1.

Table 1. Sources and description of various ROS/RNS sources

Source	Type	Description	References
NADPH Oxidase (NOX)	ROS	Major source of superoxide anions (O_2•−) in phagocytes; activated during respiratory burst.	(De Almeida et al., 2022) (Canton et al., 2021) (Novais et al., 2014)
Mitochondria	ROS	Produces ROS as byproducts of the electron transport chain, primarily superoxide and hydrogen peroxide.	(De Almeida et al., 2022) (Canton et al., 2021) (Novais et al., 2014)
Uncoupled Nitric Oxide Synthase	RNS	Can produce superoxide under certain conditions, contributing to oxidative stress.	(De Almeida et al., 2022)
Cytochrome P450 Enzymes	ROS	Involved in steroidogenesis; can leak electrons to produce superoxide.	(Veith & Moorthy, 2018)
Xanthine Oxidase (XO)	ROS	Converts hypoxanthine to xanthine and then to uric acid, generating superoxide and hydrogen peroxide.	(De Almeida et al., 2022)
Endoplasmic Reticulum (ER)	ROS	Contributes to ROS production during protein folding and calcium signalling.	(De Almeida et al., 2022) (Canton et al., 2021)
Peroxidases	ROS	Enzymes that reduce hydrogen peroxide and other peroxides, protecting cells from oxidative damage.	(De Almeida et al., 2022)
Inflammatory Cytokines	ROS/ RNS	Stimulate the production of ROS and RNS in immune cells, enhancing antimicrobial responses.	(Canton et al., 2021) (Novais et al., 2014) (Veith & Moorthy, 2018) (Mittal et al., 2014)
Polymorphonuclear Neutrophils (PMNs)	ROS/ RNS	Produce large amounts of ROS during phagocytosis to kill pathogens.	(Canton et al., 2021) (Novais et al., 2014)
Natural Killer Cells	RNS	Produce nitric oxide as part of their cytotoxic response against infected cells.	(Canton et al., 2021)

1.3. Host-Derived ROS and RNS

While *Leishmania* can produce limited amounts of ROS and RNS, most of the ROS and RNS that affect the parasite come from the host's immune system. Activated macrophages, in response to *Leishmaniasis*, produce high levels of ROS and RNS through NADPH oxidase and inducible nitric oxide synthase (iNOS), respectively. This oxidative and nitrosative burst is crucial for killing intracellular *Leishmania* parasites, although *Leishmania* has developed strategies to resist or evade this immune response (Bichiou et al., 2021).

1.3.1. Role of ROS and RNS in immune defenses in leishmaniasis

The role of ROS and RNS in immune defenses against *Leishmania* is critical for controlling infection. Both ROS and RNS are generated by host immune cells, particularly macrophages, as part of the oxidative and nitrosative bursts that occur when these cells are activated during infection.

ROS in immune defense: When *Leishmania* parasites are phagocytosed by macrophages, the host's immune system triggers the production of ROS through the enzyme NADPH oxidase. This enzyme catalyzes the production of superoxide anion (O_2^-), which is subsequently converted into other ROS, such as hydrogen peroxide (H_2O_2) and hydroxyl radicals (•OH). These ROS are highly reactive and can damage or kill intracellular parasites by causing oxidative damage to *Leishmania's* lipids, proteins, and DNA (Horta et al., 2012) (Bichiou et al., 2021). However, *Leishmania* has evolved mechanisms to counteract ROS, including antioxidant systems like superoxide dismutase (SOD) and trypanothione reductase, which neutralize ROS and enhance the parasite's survival within host macrophages (Bichiou et al., 2021). Despite this, a robust production of ROS can overwhelm these defenses, making ROS a crucial element in host-mediated parasite control.

RNS in immune defense: RNS, particularly nitric oxide (NO), plays a pivotal role in the killing of *Leishmania*. Nitric oxide is produced Nitric oxide is produced by inducible nitric oxide synthase (iNOS) in macrophages upon infection. NO can directly inhibit parasite metabolism by targeting Leishmania's mitochondrial function and DNA (Bogdan, 2015). Moreover, NO reacts with superoxide anion (O_2^-) to form peroxynitrite ($ONOO^-$), a highly reactive RNS that causes additional damage to the parasite's cellular components (Costa-da-Silva et al., 2022). The production of RNS is considered one of the most potent mechanisms of *Leishmania* killing, especially in resistant hosts with a dominant Th1 immune response that promotes macrophage activation. In susceptible hosts, the Th2 response suppresses iNOS expression, leading to reduce NO production and allowing parasite persistence (Costa-da-Silva et al., 2022) (Murray & Nathan, 1999).

ROS/RNS Synergy: ROS and RNS act synergistically in the immune response to *Leishmania*. The combination of oxidative and nitrosative stress enhances the effectiveness of macrophages in controlling the infection. While *Leishmania* has evolved strategies to mitigate oxidative and nitrosative damage, the host's ability to sustain a high-level production of ROS and RNS can still limit parasite replication and dissemination, making these molecules central to the host's immune defenses against leishmaniasis (Horta et al., 2012).

2. GENERATION OF ROS AND RNS IN THE HOST IMMUNE RESPONSE

The generation of ROS and RNS plays a crucial role in the host immune response against Leishmaniasis. These reactive molecules are key players in the immune system's strategy to control and eliminate the intracellular parasite. Below are details on how ROS and RNS are generated in response to Leishmaniasis.

2.1. Generation of ROS

2.2. NADPH Oxidase Activation:

In phagocytic cells, such as macrophages and neutrophils, ROS generation primarily occurs through the activation of the NADPH oxidase complex. Upon stimulation by pro-inflammatory cytokines (e.g., interferon-gamma), this enzyme catalyzes the conversion of molecular oxygen into superoxide (O_2^-), is subsequently converted into hydrogen peroxide (H_2O_2) by the action of superoxide dismutase (Horta et al., 2012) (Moreira et al., 2015).

1. **Mitochondrial Production**: Mitochondria also contribute significantly to ROS production during Leishmaniasis. Under conditions of oxidative stress, the mitochondrial electron transport chain can produce superoxide through electron leakage. Mitochondrial ROS (mitoROS) have been implicated in modulating immune responses and promoting apoptosis of infected cells (Hatinguais et al., 2021).

Other Sources: Additional enzymes, such as xanthine oxidase and lipoxygenases, can generate ROS during the immune response. The contributions of these sources may vary depending on the stage of the infection and the type of immune response elicited (De Las Heras et al., 2020) (Yu et al., 2024).

2.3. Generation of RNS

1. Nitric Oxide Synthase (NOS): The principal source of RNS in the immune response to *Leishmania* is the inducible nitric oxide synthase (iNOS), which is upregulated in macrophages following activation by cytokines such as interferon-gamma. iNOS converts L-arginine to nitric oxide (NO) and citrulline. NO has significant antimicrobial effects, particularly against intracellular pathogens (Volpedo et al., 2021) (Murray & Nathan, 1999)

2. Role of NO in Immune Defense: Nitric oxide produced during the immune response has various functions, including the inhibition of *Leishmania* replication. It can also react with superoxide to form peroxynitrite ($ONOO^-$), a highly reactive nitrogen species that can induce nitrosative stress in the parasite, damaging its cellular components (Costa-da-Silva et al., 2022) (Horta et al., 2012)

3. Interplay of ROS and RNS: The interaction between ROS and RNS is crucial for effective immunity. The formation of peroxynitrite enhances oxidative damage to *Leishmania*, and this synergistic effect contributes to the overall antimicrobial action of the immune system. RNS can also influence signaling pathways that modulate inflammation and apoptosis in infected cells (Mantena et al., 2008) (Di Meo et al., 2016)

2.4. Key Enzymes Involved in ROS and RNS Production

During *Leishmaniasis*, the host immune response generates ROS and RNS through the activation of key enzymes, including NADPH oxidase and inducible nitric oxide synthase (iNOS). These enzymes play a critical role in controlling parasite proliferation within infected cells, primarily macrophages.

2.3.1. NADPH Oxidase and ROS production

NADPH Oxidase: NADPH oxidase is the primary enzyme responsible for the production of ROS in phagocytes such as macrophages and neutrophils. This multi-subunit enzyme complex becomes activated upon exposure to microbial pathogens, including *Leishmania (Moreira et al., 2015)*. Once activated, NADPH oxidase facilitates the transfer of electrons from NADPH to oxygen, forming superoxide anion (O_2^-), the first ROS produced during the oxidative burst. This superoxide further give rise to hydrogen peroxide (H_2O_2), hydroxyl radicals, and other reactive oxygen intermediates (ROI) (Di Meo et al., 2016). ROS play a significant role in microbial killing by damaging proteins, lipids, and nucleic acids of invading pathogens. In the case of *Leishmania*, ROS generated by NADPH oxidase presents the possibility of restricting intracellular parasite growth and help in the overall host defense strategy (Horta et al., 2012). However, *Leishmania* species have developed mechanisms to evade ROS-mediated damage, complicating effective clearance of the parasite (Volpedo et al., 2021).

2.3.2. iNOS and RNS Production

Inducible Nitric Oxide Synthase (iNOS): Another critical enzyme involved in the host response to *Leishmaniasis* is inducible nitric oxide synthase (iNOS). iNOS is primarily responsible for the production of nitric oxide (NO), a key reactive nitrogen species. In macrophages, iNOS is induced in response to pro-inflammatory cytokines like interferon-gamma (IFN-γ) and tumor necrosis factor-alpha (TNF-α), which are typically elevated during infection (Costa-da-Silva et al., 2022) (Murray & Nathan, 1999). iNOS catalyzes the conversion of L-arginine to L-citrulline and NO. Nitric oxide exerts direct antimicrobial effects by interfering with the metabolic functions of intracellular pathogens like *Leishmania*. Moreover, NO can combine with superoxide anions to form peroxynitrite ($ONOO^-$), a potent oxidant capable of causing significant oxidative damage to the parasite (Giorgi et al., 2018). Like ROS, RNS can also induce damage to proteins, lipids, and DNA in pathogens, contributing to the host's defense strategy.

Both NADPH oxidase and iNOS are crucial for the generation of ROS and RNS, respectively, during *Leishmaniasis*. Their activation leads to oxidative and nitrosative stress that is central to controlling the parasite's proliferation within the host. However, *Leishmania* parasites have evolved evasion mechanisms to counteract the effects of these reactive species, which complicates immune clearance and highlights the need for novel therapeutic strategies.

2.4. Contribution of ROS and RNS to Killing Leishmania within Host Cells

ROS and RNS play critical roles in the host immune response against *Leishmania* by directly mediating the killing of the parasite within infected cells, primarily macrophages. This process is essential for controlling the infection and limiting the spread of the parasite.

2.4.1. Role of ROS in Leishmania killing

1. Direct Damage to Leishmania: ROS, particularly superoxide and hydrogen peroxide, are produced in large amounts by activated phagocytes through the NADPH oxidase complex. These reactive species can cause direct damage to the *Leishmania* parasites by oxidizing proteins, lipids, and nucleic acids. For example, hydrogen peroxide can lead to lipid peroxidation, compromising the integrity of the parasite's membrane and ultimately resulting in cell death (Horta et al., 2012) (Moreira et al., 2015).

Formation of Highly Reactive Products: Superoxide forms hydroxyl radicals, which are highly reactive and can cause substantial damage to the DNA of *Leishmania* • (Volpedo et al., 2021). Additionally, ROS can modify cellular components, leading to disruptions in metabolic pathways crucial for the parasite's survival (Costa-da-Silva et al., 2022).

2. Induction of Apoptosis in *Leishmania*: The oxidative stress imposed by ROS can induce apoptotic processes in *Leishmania*. Studies have shown that the presence of ROS can trigger programmed cell death pathways in the parasites, aiding in their clearance from the host (Costa-da-Silva et al., 2022).

2.4.2. Role of RNS Leishmania killing

1. Nitric Oxide and its Mechanisms: The production of nitric oxide (NO) by inducible nitric oxide synthase (iNOS) is a vital component of the immune response against *Leishmania*. NO exhibits direct antimicrobial effects by inhibiting key metabolic processes in the parasites, thereby limiting their growth (Costa-da-Silva et al., 2022) (Murray & Nathan, 1999). Nitric oxide can diffuse across membranes and enter the parasite, where it disrupts the function of essential enzymes and contributes to cellular damage.

2. **Formation of Peroxynitrite:** When NO reacts with superoxide produced during the oxidative burst, it forms peroxynitrite ($ONOO^-$), a potent oxidant that further amplifies oxidative damage to *Leishmania*. Peroxynitrite can nitrosylate proteins and cause lipid peroxidation, effectively enhancing the parasiticidal activity of macrophages (Costa-da-Silva et al., 2022) (Murray & Nathan, 1999). This interaction between ROS and RNS signifies the synergy in their antimicrobial effects.

3. **Modulation of Immune Responses:** Beyond direct killing, NO also plays a role in modulating the immune response. By influencing the expression of pro-inflammatory cytokines, NO can enhance the activation of other immune cells, further coordinating the immune response against *Leishmania* (Volpedo et al., 2021).

The combined action of ROS and RNS significantly contributes to the host's ability to eliminate *Leishmania* within infected cells. By directly damaging the parasites and modulating the immune environment, these reactive species play an essential role in controlling leishmaniasis. Understanding these mechanisms not only highlights the intricacies of host-pathogen interactions but also suggests potential therapeutic strategies to boost the immune response in leishmaniasis treatment.

2.4.3. Balance Between Effective Parasite Control and Potential Host Tissue Damage in Leishmaniasis

In leishmaniasis, the immune response to *Leishmaniasis* involves a delicate balance between effective control of the parasite and the potential for tissue damage in the host. This balance is primarily mediated through the action of ROS and RNS, which are crucial for killing the parasite but can also lead to collateral damage in host tissues.

2.4.4. Immune Mechanisms and Parasite Control

The host immune system employs various mechanisms to control *Leishmaniasis*, with the activation of macrophages being a central component. Upon infection, macrophages produce ROS and RNS via enzymes like NADPH oxidase and inducible nitric oxide synthase (iNOS). These reactive species serve to inhibit the growth and replication of the parasite by inducing oxidative and nitrosative damage to DNA (Horta et al., 2012) (Moreira et al., 2015) (Murray & Nathan, 1999).

Research has shown that an effective immune response, characterized by high levels of ROS and RNS, is crucial for controlling the parasitic load. For instance, studies indicate that mice lacking iNOS have a higher burden of *Leishmania*, demonstrating the importance of nitric oxide in limiting infection. Additionally, the presence of ROS can activate pro-inflammatory signaling pathways that promote the recruitment of other immune cells to the site of infection, further enhancing the host's ability to control the parasite (Costa-da-Silva et al., 2022).

2.4.5. Tissue Damage and Inflammation

While ROS and RNS are essential for controlling *Leishmania*, their overproduction can lead to significant tissue damage and inflammation. High levels of ROS can cause oxidative damage to cellular components, including lipids, proteins, and DNA, leading to necrosis and apoptosis of host cells (Moreira et al., 2015). This tissue damage can result in severe local and systemic inflammatory responses, manifesting as lesions and, in severe cases, visceral leishmaniasis, affecting multiple organ systems.

The inflammatory response is further exacerbated by the recruitment of other immune cells, such as neutrophils, which also produce ROS during their oxidative burst (Vivarini & Lopes, 2020). This accumulation of immune cells and reactive species can lead to a chronic inflammatory state, resulting in tissue fibrosis and dysfunction, particularly in the liver and spleen, where *Leishmania* tends to replicate (Ikeogu et al., 2020) (Wink et al., 2011).

2.4.6. Regulatory Mechanisms

To mitigate potential tissue damage while maintaining effective parasite control, the immune system has developed regulatory mechanisms. The production of anti-inflammatory cytokines, such as IL-10, can help modulate the immune response and limit inflammation (Costa-da-Silva et al., 2022). These regulatory pathways are critical for preventing excessive tissue damage during the immune response against *Leishmania.*

Moreover, some *Leishmania* species have evolved mechanisms to evade the immune response, such as downregulating the production of pro-inflammatory cytokines or enhancing the expression of anti-inflammatory mediators. This evolutionary adaptation highlights the complex interplay between the parasite and the host immune response, which aims to find an equilibrium between effective parasite elimination and the preservation of host tissue integrity (Volpedo et al., 2021).

The balance between effective parasite control and potential host tissue damage in leishmaniasis is a dynamic process influenced by the interplay of ROS and RNS, immune cell activation, and regulatory mechanisms. Understanding this balance is crucial for developing therapeutic strategies that enhance the immune response while minimizing collateral tissue damage.

2.5. Leishmania Evasion of ROS/RNS-Mediated Killing

Leishmania species have developed sophisticated strategies to evade the host immune response, particularly the oxidative and nitrosative stress imposed by ROS and RNS. This evasion is crucial for the parasite's survival and proliferation within host macrophages. Understanding these mechanisms provides insights into potential therapeutic approaches for leishmaniasis.

2.5.1. Mechanisms of Evasion

1. **Antioxidant Defence Systems:***Leishmania* employs a variety of antioxidant defence mechanisms to neutralize ROS. The parasite expresses enzymes such as superoxide dismutase (SOD), peroxidases and thioredoxin system that help in detoxifying hydrogen peroxide (Volpedo et al., 2021) (Horta et al., 2012) (Advances in Applied Microbiology, 1st ed.). These enzymes reduce oxidative stress, thereby enhancing the parasite's survival in the hostile environment of the macrophage.

Superoxide Dismutase (SOD): This enzyme catalyzes the conversion of superoxide radicals (O_2^-) into hydrogen peroxide (H_2O_2) and oxygen (O_2), thereby mitigating the harmful effects of superoxide during oxidative bursts (Horta et al., 2012).

Glutathione Peroxidase (GPx): This enzyme utilizes glutathione to reduce hydrogen peroxide to water, further protecting the parasite from oxidative damage (Volpedo et al., 2021) (Advances in Applied Microbiology, 1st ed.)

Thioredoxin System: This system plays a crucial role in maintaining redox homeostasis within *Leishmania*. It can repair oxidized proteins, enabling the parasites to survive in the oxidative environment of activated macrophages (Costa-da-Silva et al., 2022). These enzymes collectively contribute to the parasite's ability to detoxify reactive species, enhancing its survival.

Nitroreductases: Leishmania expresses nitroreductases that can convert NO into less toxic metabolites, reducing the effective concentration of this reactive nitrogen species in the vicinity of the parasite (Murray & Nathan, 1999).

Modulation of Host iNOS Expression: Certain species of *Leishmania* can alter the expression of iNOS in macrophages, leading to reduced production of NO and thereby enhancing their survival (Costa-da-Silva et al., 2022). This manipulation allows the parasite to persist in an environment where the immune response is dampened.

Nitric Oxide Resistance: In response to RNS, particularly nitric oxide (NO), *Leishmania* has adapted by expressing specific enzymes such as nitroreductases. These enzymes can detoxify NO, preventing it from exerting its cytotoxic effects on the parasite. Studies indicate that certain *Leishmania* species can modulate the expression of iNOS in infected macrophages, thereby reducing the production of NO and promoting their own survival (Costa-da-Silva et al., 2022) (Murray & Nathan, 1999).

2. **Modulation of Host Cell Signaling:***Leishmania* can manipulate host cell signaling pathways to enhance its survival. For instance, the parasite can induce the production of anti-inflammatory cytokines, such as interleukin-10 (IL-10), which dampens the production of ROS and RNS in macrophages (Samant et al., 2021). By shifting the immune response toward an anti-inflammatory profile, *Leishmania* effectively reduces the oxidative and nitrosative stress that would otherwise be detrimental to the parasite.

Cytokine Regulation: The parasite can induce the production of anti-inflammatory cytokines, such as interleukin-10 (IL-10), which suppresses the activation of macrophages and limits the production of pro-inflammatory mediators, including ROS and RNS (Bodhale et al., 2021). By shifting the immune response towards a more tolerant state, *Leishmania* creates a less hostile environment for itself.

Inhibition of T Cell Activation: Leishmania can also interfere with T cell activation, which is crucial for the orchestration of an effective immune response. This can lead to a diminished capacity for the host to mount a robust response, further allowing the parasite to evade immune detection (Costa-Madeira et al., 2022).

3. Intracellular Localization: The intracellular niche of *Leishmania* also plays a crucial role in its evasion strategies. The parasites reside within phagosomes that can prevent exposure to high concentrations of ROS and RNS. They can manipulate the phagosomal environment to avoid fusion with lysosomes, thereby avoiding the oxidative burst typically associated with the macrophage's antimicrobial response (Horta et al., 2012) (Costa-Madeira et al., 2022). The intracellular lifestyle of *Leishmania* significantly contributes to its evasion of immune-mediated killing:

Phagosomal Manipulation: Once engulfed by macrophages, *Leishmania* resides within phagosomes that can be manipulated to avoid fusion with lysosomes, where ROS and RNS would be generated (Horta et al., 2012). This adaptation allows the parasites to remain in a protected niche where they are less exposed to the toxic effects of oxidative stress.

Phagosomal pH Neutralization: Some *Leishmania* species can also modulate the pH of the phagosome, creating a more favorable environment for their survival while inhibiting the maturation of the phagosome into a bactericidal compartment (Costa-da-Silva et al., 2022).

4. Clinical Implications: The evasion strategies employed by *Leishmania* pose significant challenges for effective treatment. Understanding these mechanisms opens new avenues for therapeutic intervention. Strategies that aim to enhance the production of ROS and RNS or disrupt the parasite's antioxidant defences could be explored. Additionally, therapies that modulate the immune response to prevent *Leishmania* from dampening macrophage activity may improve outcomes in patients with leishmaniasis (Volpedo et al., 2021).

2.6. Antioxidant Defence Mechanisms in Leishmania

Leishmania parasites have evolved sophisticated antioxidant defense mechanisms to neutralize ROS and RNS, allowing them to survive in the oxidative environment of the host immune response. Key components of these defense mechanisms include various antioxidant enzymes and small molecules that work synergistically to mitigate oxidative stress.

1. **Superoxide Dismutase (SOD):** Superoxide dismutase (SOD) is one of the primary enzymes involved in the detoxification of superoxide radicals (O_2^-), which are among the first ROS generated during an immune response. *Leishmania* utilizes different isoforms of SOD, including manganese-dependent (MnSOD) and copper-zinc-dependent (Cu/ZnSOD) variants, which catalyze the conversion of superoxide radicals into hydrogen peroxide (H_2O_2) and molecular oxygen (O_2).Research has shown that *Leishmania* can upregulate SOD expression in response to oxidative stress, enhancing its ability to survive the oxidative burst from host immune cells

(Volpedo et al., 2021) (Horta et al., 2012). The reduction of superoxide levels helps to minimize the potential damage to cellular structures and functions.

2. **Catalase and Peroxidases:** Once hydrogen peroxide is produced, it can be further detoxified by catalase and peroxidase enzymes. Catalase catalyzes the conversion of H_2O_2 into water and oxygen, thereby preventing the accumulation of toxic levels of hydrogen peroxide within the parasite. Additionally, *Leishmania* possesses various peroxidases, including glutathione peroxidases, which utilize glutathione to reduce H_2O_2 and other peroxides, further safeguarding the parasite from oxidative damage. (Costa-da-Silva et al., 2022) (Horta et al., 2012)

3. **Trypanothione:** Trypanothione, a unique thiol compound found in trypanosomatids like *Leishmania*, plays a critical role in the parasite's antioxidant defense. It is a conjugate of two glutathione molecules and is involved in various redox reactions within the parasite. Trypanothione acts as a direct scavenger of free radicals and helps maintain the redox balance within the cell (González-Montero et al., 2024). Moreover, trypanothione participates in the detoxification of H_2O_2 and other reactive species, contributing to the overall antioxidant capacity of *Leishmania* (Horta et al., 2012) (González-Montero et al., 2024). Its unique structure enables it to support various enzymatic reactions crucial for countering oxidative stress.

4. **Thioredoxin System:** The thioredoxin system in *Leishmania* also plays a vital role in managing oxidative stress. This system comprises thioredoxin and thioredoxin reductase, which are essential for reducing oxidized proteins and maintaining the cellular redox state. Thioredoxin can donate electrons to various substrates, including peroxides, thereby protecting cellular components from oxidative damage (Costa-da-Silva et al., 2022).

Leishmania utilizes a combination of antioxidant enzymes such as SOD, catalase, and peroxidases, along with small molecules like trypanothione, to counteract ROS and RNS generated during host immune responses. These mechanisms are critical for the survival and pathogenicity of the parasite, highlighting potential therapeutic targets for developing effective treatments against leishmaniasis.

2.6.1. Regulation of Antioxidant Defenses in Leishmania

The regulation of antioxidant defenses in *Leishmania* is a complex process that involves various transcription factors, signaling pathways, and post-translational modifications. These regulatory mechanisms enable the parasite to adapt to the fluctuating oxidative stress levels encountered during its life cycle, particularly when interacting with host immune cells.

1. *Transcriptional Regulation:* Transcription factors play a crucial role in the expression of antioxidant enzymes in *Leishmania*. For instance, the nuclear factor erythroid 2-related factor 2 (Nrf2) is known to regulate the expression of genes in-

volved in the antioxidant response in many organisms, including *Leishmania*. Under oxidative stress, Nrf2 translocates to the nucleus and activates the transcription of antioxidant genes, including those coding for superoxide dismutase, glutathione peroxidase, and thioredoxin (Costa-da-Silva et al., 2022) (Reverte et al., 2021). The heat shock factor (HSF) can also be involved in the stress response, leading to the upregulation of chaperones and antioxidant enzymes when the parasite is exposed to high temperatures or oxidative stress (Kröber-Boncardo et al., 2020).

2. *Post-Translational Modifications:* Post-translational modifications (PTMs) of antioxidant enzymes can modulate their activity and stability. For instance, phosphorylation and ubiquitination can alter the functionality of enzymes such as thioredoxin and glutathione peroxidase (Scarpini et al., 2022) (Volpedo et al., 2021) (Advances in Applied Microbiology, 1st ed.). These modifications allow *Leishmania* to fine-tune its antioxidant capacity in response to environmental signals, enhancing its adaptability and survival within the host.

3. *Role of Reactive Species in Signaling:* Interestingly, while ROS and RNS are typically viewed as harmful agents, they also play important roles in cellular signaling. Low levels of ROS can act as signaling molecules that trigger adaptive responses in *Leishmania*, leading to the upregulation of antioxidant defenses. This phenomenon, known as oxidative stress signaling, allows the parasite to preemptively activate its protective mechanisms before encountering more substantial oxidative challenges (Costa-da-Silva et al., 2022).

2.6.2. Importance of Antioxidant Defenses in Leishmania Survival

1.*Evasion of Host Immune Response:* One of the key factors enabling *Leishmania* survival within the host is its ability to evade the oxidative burst generated by the immune system, specifically by phagocytes like macrophages and neutrophils. During infection, these immune cells produce ROS and RNS to kill intracellular pathogens. However, *Leishmania* has evolved to counteract this oxidative stress by upregulating its antioxidant defenses. For example, the parasite expresses enzymes like superoxide dismutase (SOD), peroxidoxins, and the unique trypanothione system, which collectively neutralizes ROS and RNS (Costa-da-Silva et al., 2022) (Advances in Applied Microbiology, 1st ed.). These antioxidant mechanisms significantly reduce the harmful effects of oxidative stress, allowing the parasite to survive within hostile immune environments (Horta et al., 2012).

2.*Adaptation to Host Environments:* The ability to regulate oxidative stress is also critical for *Leishmania* as it transitions between different environments during its life cycle. In the sandfly vector, *Leishmania* must deal with immune responses that generate oxidative stress similar to those encountered in mammalian hosts. The parasite's antioxidant machinery enables it to adapt to oxidative stress across diverse environments, ensuring its survival during transmission from vector to host (Volpedo et al., 2021) (Kaushal et al., 2023). This flexibility enhances its ability to establish infections in mammalian cells by neutralizing oxidative attacks early on in infection.

3.*Prevention of Cellular Damage: Leishmania's* ability to mitigate oxidative damage extends beyond evading the immune response. ROS and RNS, if left unchecked, can cause significant damage to the parasite's cellular components, including lipids, proteins, and DNA. By upregulating antioxidant enzymes like trypanothione reductase and SOD, *Leishmania* prevents oxidative damage to its cell membranes and other essential structures. This preservation of cellular integrity is crucial for maintaining metabolic functions and sustaining parasite viability (Ali et al., 2022).

4. *Enhanced Pathogenicity:* Effective management of oxidative stress not only supports *Leishmania*'s survival but also enhances its pathogenicity. The parasite's ability to survive within macrophages, despite the oxidative burst, allows it to replicate and spread, leading to chronic infections. By suppressing the host's ability to kill the parasite through oxidative stress, *Leishmania* can establish long-term infections that contribute to its pathogenic potential (Kaushal et al., 2023) (Ali et al., 2022) (Bharadava et al., 2024) . The efficiency of *Leishmania*'s antioxidant defenses is thus directly linked to its ability to cause disease in humans.

The regulation of antioxidant defenses in *Leishmania* is a multifaceted process involving transcriptional regulation, post-translational modifications, and the dual role of reactive species as signaling molecules. These defenses are essential for the parasite's survival, enabling it to evade host immune responses, adapt to various environments, and maintain cellular integrity. Understanding these regulatory mechanisms could provide insights for developing targeted therapies against leishmaniasis.

2.7. Modulation of Host Immune Responses

2.7.1. Manipulation of Host Cell Signaling by Leishmania to Reduce ROS/RNS Production

Leishmania species have evolved numerous strategies to manipulate host cell signaling pathways, ultimately reducing the production of ROS and RNS. By targeting various signaling molecules and pathways, *Leishmania* effectively creates an immunosuppressive environment that enhances its survival and replication. Below are several key mechanisms through which *Leishmania* exerts this influence:

Leishmania parasites have developed sophisticated evasion mechanisms to survive the oxidative and nitrosative stress imposed by host macrophages. ROS and RNS are generated as a key part of the host's defense, yet *Leishmania* parasites can manipulate host cell signaling pathways and employ their own defense systems to counter these stresses.

2.7.2. Evasion Mechanisms Targeting NADPH Oxidase and ROS

One of the primary targets for *Leishmania* evasion is the host's NADPH oxidase, a key enzyme responsible for generating ROS. Several *Leishmania* species, through their surface molecules such as lipophosphoglycan (LPG), interfere with the assembly of the NADPH oxidase complex in phagocytes. LPG prevents the translocation of essential NADPH oxidase subunits to the phagosomal membrane, reducing ROS production and thereby protecting the parasite from oxidative killing. Studies have shown that by disrupting protein kinase C (PKC) signaling, *Leishmania* can further impair ROS generation, as PKC is essential for NADPH oxidase activation (Costa-da-Silva et al., 2022) (Olivier et al., 2012). Another surface molecule, *Leishmania* metalloprotease GP63, has been reported to cleave components of the NADPH oxidase system, further hindering its activity (Shio et al., 2012). These molecular strategies limit the production of superoxide and downstream ROS species, allowing the parasite to establish itself within the hostile environment of the phagocyte.

2.7.3. Suppression of iNOS and RNS Production

In addition to reducing ROS levels, *Leishmania* suppresses the production of nitric oxide (NO) by interfering with the expression of inducible nitric oxide synthase (iNOS). The parasite modulates host signaling by promoting the secretion of anti-inflammatory cytokines, such as IL-10 and TGF-β, which suppress iNOS transcription. By upregulating these cytokines, *Leishmania* diminishes the host's ability to produce RNS, a potent antimicrobial mechanism (Murray, 2005). Another method involves directly inhibiting key transcription factors responsible for iNOS expression. *Leishmania* can block the nuclear translocation of nuclear factor kappa B (NF-κB), a critical regulator of iNOS gene expression, by modulating upstream signaling pathways such as MAPK and PI3K/Akt. By limiting NF-κB activation, *Leishmania* prevents the induction of iNOS, thereby decreasing nitric oxide production and reducing nitrosative stress on the parasite (Olivier et al., 2005) (Nandan et al., 2024) (Almeida et al., 2023).

2.7.4. Antioxidant Defense Mechanisms in Leishmania

In addition to manipulating host immune responses, *Leishmania* also possesses intrinsic antioxidant defense mechanisms that help it neutralize any residual ROS/RNS that the host may produce. One of the key elements of this defense system is the parasite's trypanothione system, a unique thiol-based antioxidant system that functions similarly to the glutathione system in higher organisms. The enzyme trypanothione reductase reduces trypanothione disulfide back to its reduced form, which then detoxifies hydrogen peroxide and other peroxides produced by the host. Other key enzymes include SOD, which converts superoxide anions into less reactive hydrogen peroxide, and peroxiredoxins, which reduce hydrogen peroxide and peroxynitrite. These enzymes, along with the trypanothione system, protect the parasite from oxidative damage, enabling it to persist within macrophages for extended periods (Reverte et al., 2022) (Singh et al., 2023).

2.7.5. Parasite Survival via PI3K/Akt Pathway Manipulation

The phosphoinositide 3-kinase (PI3K)/Akt pathway is another critical pathway manipulated by *Leishmania* to enhance its survival. Activation of the PI3K/Akt pathway by the parasite promotes anti-inflammatory responses and inhibits pro-inflammatory signaling. This pathway not only suppresses the production of ROS and RNS but also supports the survival of infected host macrophages, creating a more favorable intracellular environment for the parasite to thrive (Soulat & Bogdan, 2017). The ability of *Leishmania* to evade ROS and RNS-mediated killing is a multifaceted process that involves both host cell signaling manipulation and the parasite's own antioxidant defenses. By inhibiting NADPH oxidase assembly, down-regulating iNOS expression, and activating anti-inflammatory pathways, *Leishmania* reduces the production of harmful ROS and RNS. The parasite's intrinsic antioxidant systems allow it to survive residual oxidative and nitrosative stress, making these evasion strategies essential for its intracellular survival.

2.7.6. The Role of Cytokines and Other Host Factors in Regulating ROS/RNS Levels during Leishmaniasis

The production of ROS and RNS during *Leishmaniasis* is tightly regulated by cytokines and other host factors, which are crucial in determining the immune system's ability to control or clear the parasite. These molecules influence the balance between pro-inflammatory responses, which generate ROS and RNS to kill the parasite, and anti-inflammatory responses, which can dampen the production of these species and allow parasite survival.

2.7.7. Pro-inflammatory Cytokines and Induction of ROS/RNS

Pro-inflammatory cytokines, particularly interferon-gamma (IFN-γ) and tumor necrosis factor-alpha (TNF-α), are key drivers of ROS and RNS production during *Leishmaniasis*. IFN-γ plays a pivotal role in the activation of macrophages and the induction of inducible nitric oxide synthase (iNOS), which leads to the production of nitric oxide (NO), a potent RNS involved in parasite killing (Murray, 2005).

TNF-α works synergistically with IFN-γ to enhance the production of both NO and superoxide anions (O2•−), contributing to the formation of more reactive and toxic compounds, such as peroxynitrite (ONOO−), that are highly effective in killing intracellular *Leishmania (Miranda et al., 2015)*. Moreover, IFN-γ and TNF-α activate signaling pathways such as the Janus kinase-signal transducer and activator of transcription (JAK-STAT) pathway and nuclear factor kappa B (NF-κB), which are essential for the transcriptional upregulation of iNOS and components of the NADPH oxidase complex (Liu & Uzonna, 2012). These cytokines also promote the respiratory burst, where the NADPH oxidase complex generates superoxide radicals that lead to the production of other ROS, such as hydrogen peroxide (H2O2) and hydroxyl radicals (OH•).

2.7.8. Anti-inflammatory Cytokines and the Suppression of ROS/RNS

In contrast, anti-inflammatory cytokines, particularly interleukin-10 (IL-10) and transforming growth factor-beta (TGF-β), play a critical role in suppressing the production of ROS and RNS, thereby allowing the parasite to evade immune clearance. IL-10, for example, inhibits the production of IFN-γ and TNF-α, thus downregulating iNOS expression and reducing NO levels in infected macrophages (Gupta et al., 2022). IL-10 also inhibits the transcriptional activation of genes encoding NADPH oxidase subunits, which diminishes the respiratory burst and lowers ROS production. TGF-β is another cytokine that contributes to the suppression of ROS and RNS during *Leishmaniasis*. It inhibits the activation of macrophages and promotes the differentiation of regulatory T cells (Tregs), which dampen the immune response. TGF-β can also directly interfere with the signaling pathways that induce iNOS expression, further limiting NO production (Kane & Mosser, 2001).

2.7.9. Additional Host Factors Influencing ROS/RNS Levels

Beyond cytokines, other host factors also play a role in regulating ROS and RNS levels during *Leishmaniasis*. For instance, prostaglandins, particularly prostaglandin E2 (PGE2), can modulate immune responses by promoting anti-inflammatory pathways that inhibit ROS/RNS production. PGE2 is known to suppress IFN-γ-induced iNOS expression and to reduce ROS generation by downregulating NADPH oxidase activity in macrophages (Tomiotto-Pellissier et al., 2018). Moreover, the activation state of macrophages significantly influences their ability to produce ROS and RNS. Classically activated macrophages (M1 macrophages) induced by IFN-γ and TNF-α are proficient at producing high levels of ROS and RNS to kill intracellular *Leishmania* parasites. In contrast, alternatively activated macrophages (M2 macrophages), induced by IL-10 and TGF-β, exhibit reduced microbicidal activity and produce lower levels of ROS and RNS, favoring parasite persistence and chronic infection (Olivier et al., 2005).

Cytokines and other host factors play a dual role in regulating the production of ROS and RNS during *Leishmaniasis*. Pro-inflammatory cytokines such as IFN-γ and TNF-α enhance ROS and RNS production, promoting parasite killing, while anti-inflammatory cytokines like IL-10 and TGF-β suppress these pathways, allowing parasite survival. Understanding the complex interplay between these cytokines and other regulatory factors is essential for developing therapeutic strategies that can enhance the host's ability to control *Leishmaniasis* without causing excessive tissue damage.

3. IMPLICATIONS OF ROS/RNS IN DISEASE PATHOGENESIS

3.1. Dual Role in Immune Defence and Tissue Damage

The generation of ROS and RNS plays a complex and critical role in the pathogenesis of leishmaniasis. On one hand, these reactive molecules are essential components of the host immune defense system, involved in killing *Leishmania* parasites within phagocytes. On the other hand, excessive or uncontrolled production of ROS and RNS can lead to tissue damage and contribute to disease pathology (Figure 1). The initial immune response to *Leishmania* involves macrophages and neutrophils, which produce large amounts of ROS and RNS, particularly nitric oxide (NO), through the activation of NADPH oxidase and inducible nitric oxide synthase (iNOS), respectively. This The oxidative and nitrosative stresses are intended to kill intracellular parasites by damaging their DNA, proteins, and lipids (Costa-da-Silva et al., 2022) (Murray, 2005). However, when the oxidative burst is not properly regulated, it can lead to collateral damage to host tissues. This is particularly relevant in cutaneous and mucocutaneous leishmaniasis, where chronic inflammation and prolonged production of ROS/RNS can result in severe tissue destruction and disfiguring lesions (Olivier et al., 2005) (Saha et al., 2021).

Figure 1. Sources of ROS/RNS and ROS/RNS as a double-edged sword

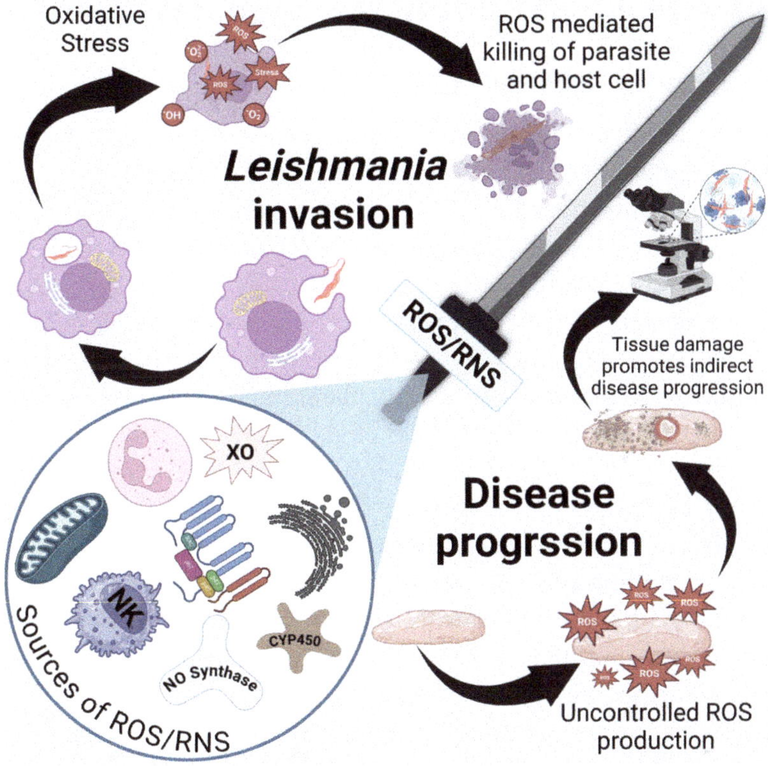

XO: Xanthine Oxidase, CYP450: Cytochrome P450, NK: Natural Killer Cells, NO Synthase: Nitric Oxide Synthase

3.1.1. Contribution to Chronic Inflammation

Excessive ROS and RNS production also contribute to the chronic inflammatory environment seen in some forms of leishmaniasis, such as visceral leishmaniasis. While ROS and RNS are initially produced to control the infection, their overproduction can exacerbate inflammation, leading to a vicious cycle of immune activation and tissue damage. Chronic ROS/RNS-mediated inflammation is associated with fibrosis, granuloma formation, and organ dysfunction, particularly in the liver and spleen during visceral leishmaniasis (Liese et al., 2008).

In visceral leishmaniasis, uncontrolled inflammation driven by persistent ROS and RNS production can compromise organ function, particularly in the liver and spleen, which are heavily infiltrated by immune cells attempting to clear the infection. In this context, the balance between protective immunity and immunopathology becomes crucial, as excessive ROS and RNS production can lead to the breakdown of host tissue rather than effective parasite clearance (Kaye & Scott, 2011).

Parasite Evasion of ROS/RNS

Interestingly, *Leishmania* has evolved mechanisms to evade the destructive effects of ROS and RNS, which further complicates the disease pathology. The parasite expresses antioxidant enzymes such as superoxide dismutase (SOD), trypanothione reductase, and peroxidoxins that help neutralize ROS and RNS (Horta et al., 2012). By upregulating these defenses, *Leishmania* can survive within the oxidative environment of the host phagocytes, promoting chronic infection. This evasion strategy allows *Leishmania* to persist despite the host's attempt to clear the infection using oxidative stress, prolonging the inflammatory process and contributing to tissue damage over time (Olivier et al., 2005). The role of ROS and RNS in leishmaniasis is double-edged: as they are essential for controlling the parasite, but their overproduction can lead to tissue damage and chronic inflammation. The parasite's ability to evade ROS/RNS mediated killing through antioxidant defenses further complicates the disease by promoting long term survival in the host, resulting in chronic and debilitating disease forms. This complex interplay highlights the need for therapeutic strategies that can boost the immune system's ability to eliminate the parasite while minimizing damage to host tissues.

3.1.2. Role in Disease Progression and Severity

ROS and RNS are crucial mediators in the pathogenesis of leishmaniasis. Their dual role as both immune effectors and potential mediators of tissue damage makes them significant factors in the disease progression and severity. This overview will discuss the impact of ROS and RNS on leishmaniasis, focusing on their contribution to disease severity, and progression, and the mechanisms through which *Leishmania* manipulates these processes.

1. *Immune Response and ROS/RNS Production:* When *Leishmania* infects a host, immune cells, primarily macrophages and neutrophils, respond by generating ROS and RNS as part of their defense mechanism. The production of these reactive species is initiated by the activation of NADPH oxidase, leading to an oxidative burst that generates superoxide anions (O_2^-), which can further be converted into hydrogen peroxide (H_2O_2) (Murray & Nathan, 1999). Similarly, inducible nitric oxide synthase (iNOS) catalyzes the production of nitric oxide (NO) from L-arginine, contributing to the formation of RNS, such as peroxynitrite ($ONOO^-$) (Zarkovic, 2020).

ROS and RNS as Pathogen-Killing Agents: ROS and RNS play essential roles in killing *Leishmania* by damaging its cellular structures. For instance, hydrogen peroxide can cause lipid peroxidation, leading to membrane damage and eventual cell death (Liese et al., 2007). Similarly, peroxynitrite can nitrify proteins, disrupting their function and contributing to the overall pathogenic stress experienced by the parasite (Suschek et al., 2022). Despite these defensive mechanisms, excessive or prolonged production of ROS and RNS can lead to tissue damage, contributing to the severity of the disease. Oxidative stress associated with chronic inflammation can lead to complications such as ulcers in cutaneous leishmaniasis and organ dysfunction in visceral leishmaniasis (Kaye & Scott, 2011).

2. *Tissue Damage and Disease Severity:* The balance between effective pathogen control and tissue damage is crucial in determining the severity of leishmaniasis. Prolonged oxidative stress due to excessive ROS and RNS can cause significant damage to host tissues, leading to a more severe disease outcome. In cutaneous leishmaniasis, chronic inflammation driven by ROS can result in the formation of destructive ulcers and scarring (Ayala et al., 2024). Similarly, in visceral leishmaniasis, excessive ROS and RNS can damage the liver and spleen, leading to severe immunosuppression and organ failure (Olivier et al., 2005). Research shows that the severity of disease symptoms is often correlated with the levels of ROS and RNS produced during the immune response. Higher levels of these reactive species can lead to increased inflammation and greater tissue damage, exacerbating disease progression. Conversely, lower levels of oxidative stress can correlate with milder symptoms and a more controlled immune response (Horta et al., 2012).

3. *Evasion Strategies of Leishmania:* To survive within the hostile environment created by ROS and RNS, *Leishmania* has developed several evasion strategies. One critical mechanism involves regulating antioxidant defenses, such as superoxide dismutase and trypanothione, which help neutralize reactive species and mitigate oxidative stress (Krauth-Siegel & Comini, 2008). Additionally, *Leishmania* can modulate host immune signaling pathways to reduce ROS and RNS production. By suppressing the activation of NADPH oxidase and iNOS, *Leishmania* can effectively blunt the oxidative response, enhancing its survival within host cells (Olivier et al., 2005).

4. *Clinical Implications:* Understanding the role of ROS and RNS in leishmaniasis has important clinical implications. Therapies that enhance the host's oxidative burst could improve the clearance of *Leishmania* and reduce disease severity. On the other hand, therapies aimed at modulating excessive oxidative stress could mitigate tissue damage and improve patient outcomes in chronic cases. For instance, the use of antioxidants in conjunction with anti-leishmanial treatments may help protect host tissues while still allowing for effective parasite elimination (Costa-da-Silva et al., 2022).

ROS and RNS are critical components in the pathogenesis of leishmaniasis, playing a dual role as both pathogen-killing agents and potential mediators of tissue damage. The balance between effective immune response and collateral damage is essential in determining disease severity and progression. Understanding the mechanisms through which *Leishmania* manipulates host ROS and RNS production can inform therapeutic strategies aimed at improving outcomes for individuals suffering from leishmaniasis.

3.2. Impact on Host Immune Regulation

ROS and RNS are key effectors produced by immune cells, particularly macrophages and neutrophils, in response to *Leishmaniasis*. These molecules not only contribute to the direct killing of the parasite but also play a significant role in shaping the broader immune response, influencing the activity of T cells, cytokines, and the overall immune environment. However, the regulation of ROS/RNS production is critical because excessive or insufficient levels can either contribute to tissue damage or allow parasite persistence, respectively.

3.2.1. ROS/RNS in Modulating T Cell Responses

ROS and RNS affect the activation, differentiation, and function of T cells, which are critical for developing an adaptive immune response during *Leishmaniasis*. Specifically, they play a role in polarizing T helper (Th) cell responses, which are pivotal in determining whether the host immune system can control or fail to control the infection.

Th1 Responses and IFN-γ Production: In leishmaniasis, a robust Th1 response is associated with the control of infection. This response is marked by the secretion of interferon-gamma (IFN-γ), which stimulates macrophages to produce ROS and RNS, particularly nitric oxide (NO) via inducible nitric oxide synthase (iNOS). This positive feedback loop enhances the killing of *Leishmania* within macrophages (Horta et al., 2012). Importantly, IFN-γ-induced ROS/RNS production strengthens immune control of the parasite, which is crucial for resistance to the infection.

Th2 Responses and Immune Suppression: Conversely, a Th2-dominant response, which involves cytokines such as interleukin-4 (IL-4) and interleukin-10 (IL-10), dampens the production of ROS and RNS. IL-10, in particular, downregulates iNOS expression and reduces macrophage activation, allowing *Leishmania* to survive within host cells (Subramanian Vignesh & Deepe, 2016). This suppression of ROS/RNS activity is associated with parasite persistence and the development of chronic leishmaniasis, highlighting the importance of Th1/Th2 balance in disease outcomes.

T Regulatory (Treg) Cells and Immune Tolerance: Treg cells, which promote immune tolerance and limit excessive immune activation, can also influence ROS and RNS levels. Excessive ROS production may damage Treg cells, impairing their ability to suppress immune responses and leading to uncontrolled inflammation and tissue damage (Costa-da-Silva et al., 2022). Conversely, Tregs may suppress ROS/RNS production, creating a permissive environment for *Leishmania* survival but protecting host tissues from excessive damage.

3.2.2. Cytokine Production and Immune Regulation by ROS/RNS

Cytokine production is intricately linked with ROS and RNS levels. The interaction between pro-inflammatory and anti-inflammatory cytokines helps regulate ROS/RNS production, which in turn shapes the immune environment during *Leishmaniasis*.

Pro-Inflammatory Cytokines: The production of ROS by macrophages stimulates the release of pro-inflammatory cytokines such as tumor necrosis factor-alpha (TNF-α), interleukin-1β (IL-1β), and interleukin-6 (IL-6). These cytokines enhance the recruitment and activation of immune cells, creating an inflammatory environment that promotes further ROS/RNS production. TNF-α, in particular, is known to promote NADPH oxidase activity, increasing ROS levels in immune cells, which helps in controlling the parasite (Costa-da-Silva et al., 2022).

Anti-Inflammatory Cytokines: In contrast, anti-inflammatory cytokines like IL-10 and transforming growth factor-beta (TGF-β) act to suppress ROS and RNS production. IL-10 inhibits the activity of NADPH oxidase and iNOS, reducing ROS and NO production, which allows *Leishmania* to evade immune-mediated killing (Horta et al., 2012). TGF-β also plays a role in downregulating immune responses, contributing to a microenvironment that favors parasite survival.

3.2.3. Feedback Loops Between ROS/RNS and Immune Responses

Several feedback mechanisms exist between ROS/RNS production and the immune response that either amplify parasite control or lead to immune dysregulation and tissue damage.

Positive Feedback for Parasite Killing: A positive feedback loop exists in which IFN-γ promotes the production of ROS/RNS, which in turn enhances Th1 cell activation and macrophage-mediated parasite killing. This loop is essential for controlling *Leishmaniasis* but can lead to increased inflammation if not tightly regulated (Kaye & Scott, 2011).

Negative Feedback for Parasite Survival: Anti-inflammatory cytokines, particularly IL-10, create a negative feedback loop by suppressing ROS/RNS production, which promotes parasite survival within macrophages. This feedback loop plays a crucial role in the chronicity of the infection, especially in visceral leishmaniasis, where the parasite can persist for long periods (Olivier et al., 2005).

3.2.4. Role of ROS/RNS in Disease Outcomes and Tissue Damage

The balance of ROS and RNS production can have a significant impact on the progression of leishmaniasis, with either insufficient or excessive levels leading to adverse outcomes.

Chronic Infection: In cases where ROS/RNS production is insufficient, as seen in Th2-dominant or IL-10-rich environments, *Leishmania* can evade immune defenses and establish chronic infection. This is particularly evident in chronic cutaneous leishmaniasis, where non-healing ulcers persist, and in visceral leishmaniasis, where parasites can remain in tissues for years (Gervazoni et al., 2020).

Tissue Damage: On the other hand, excessive production of ROS and RNS can cause oxidative damage to host tissues. This is particularly evident in the severe tissue damage seen in cutaneous leishmaniasis, where excessive oxidative stress contributes to ulcer formation, and in visceral leishmaniasis, where liver and spleen tissue can be damaged due to excessive inflammation (Reverte et al., 2022).

ROS and RNS are pivotal in regulating both the innate and adaptive immune responses during *Leishmaniasis*. Their role extends from directly killing the parasite to modulating T cell responses, cytokine production, and the broader immune environment. The balance between ROS/RNS production and the immune system's regulatory mechanisms determines whether the host successfully controls the infection or whether chronicity and tissue damage ensue. Understanding these dynamics offers potential therapeutic avenues, particularly in modulating ROS/RNS production to optimize immune responses without causing excessive tissue damage.

4. THERAPEUTIC POTENTIAL OF TARGETING ROS/RNS PATHWAYS

The manipulation of ROS and RNS pathways offers significant therapeutic potential in the treatment of leishmaniasis due to their crucial roles in the immune response and parasite survival. These reactive molecules are central to the host's ability to control infection, but their dysregulation can contribute to disease progression, tissue damage, and chronic infection. Therefore, therapeutic strategies aimed at modulating ROS and RNS levels in both the host and parasite could improve clinical outcomes by enhancing parasite clearance or reducing pathological oxidative damage.

4.1. Enhancing ROS/RNS Production

Boosting the production of ROS and RNS is a potential strategy to enhance the host's ability to kill *Leishmania* parasites. One approach involves the use of pro-oxidant drugs that increase oxidative stress within infected cells. For instance, drugs like amphotericin B and miltefosine not only exhibit direct antiparasitic effects but also enhance the generation of ROS, thereby contributing to the killing of intracellular parasites (Moreira et al., 2015). Another approach is the use of cytokine therapies, such as interferon-gamma (IFN-γ), which upregulates the expression of key enzymes like NADPH oxidase (for ROS production) and iNOS (for RNS production) in macrophages, promoting the oxidative burst necessary for parasite destruction (Chulanetra & Chaicumpa, 2021). However, while increasing ROS/RNS production can enhance pathogen clearance, it also poses a significant risk of exacerbating tissue damage. Excessive or unregulated production of ROS/RNS can lead to oxidative stress, which damages host tissues, particularly in chronic infections like leishmaniasis. Lipid peroxidation, protein denaturation, and DNA damage are among the consequences of prolonged oxidative stress, contributing to organ dysfunction and worsening disease outcomes (Moreira et al., 2015). Therefore, any pro-oxidant therapeutic strategy must carefully balance parasite killing with the risk of collateral tissue damage. To manage this risk, clinicians could consider targeted delivery systems that concentrate pro-oxidant drugs in infected tissues while sparing healthy cells or combining such drugs with antioxidants to protect non-infected tissues (Moreira et al., 2015).

4.2. Antioxidant Therapy

On the opposite side, antioxidant therapy presents an attractive option to protect host tissues from oxidative damage caused by excessive ROS/RNS during leishmaniasis. Antioxidants like N-acetylcysteine (NAC), vitamin E, and selenium have shown promise in reducing ROS levels, protecting against lipid peroxidation and preserving cellular function (Horta et al., 2012). This approach could be particularly beneficial in chronic forms of leishmaniasis, where tissue damage from oxidative stress contributes to disease pathology. However, the major concern with antioxidant therapy is the risk of compromising the host's defense mechanisms. Since ROS and RNS play crucial roles in pathogen killing, excessive reduction of these reactive species through antioxidant therapy may impair the immune system's ability to clear the parasite, potentially leading to persistence or exacerbation of the infection (Coutinho De Oliveira et al., 2020). This makes it essential to develop therapeutic strategies that strike a balance—reducing oxidative stress to protect host tissues while maintaining sufficient ROS/RNS levels to eliminate the parasite.

4.3. Combination Therapies

The challenges of balancing ROS/RNS levels for effective parasite control and minimizing tissue damage, combination therapies that modulate ROS/RNS pathways alongside traditional antiparasitic treatments are being explored. These therapies could involve the concurrent use of pro-oxidant agents to enhance parasite killing and antioxidants to protect host tissues. For example, combining traditional drugs like miltefosine or antimonials with antioxidants could enhance parasite clearance while mitigating the oxidative damage associated with chronic infection (Shakya et al., 2012).

Targeting parasite antioxidant defenses alongside these strategies could be a game-changer. Inhibiting Leishmania's antioxidant enzymes, such as trypanothione reductase, which helps neutralize ROS, could make the parasite more susceptible to ROS-induced killing (Castro et al., 2017). Preclinical studies have shown promise for drugs that inhibit these defenses, potentially providing a means to enhance ROS-mediated parasite killing without significantly increasing the risk of tissue damage (Moreira et al., 2015).

4.4. Personalized Approaches

The ROS/RNS profile of individual patients could also play a crucial role in determining the optimal therapeutic strategy. Personalized approaches based on the patient's oxidative status, immune response, and severity of infection could help tailor treatments to achieve the right balance between parasite killing and tissue protection. For instance, patients with high oxidative stress may benefit more from antioxidant therapy to reduce tissue damage, while those with lower levels of ROS/RNS may require therapies that boost oxidative stress to enhance parasite killing (Horta et al., 2012).

5. CHALLENGES AND FUTURE DIRECTIONS

5.1. Technical and Biological Challenges

One of the most significant challenges in the study of ROS and RNS in the context of *Leishmaniasis* is the difficulty in accurately measuring these molecules and free radicals *in vivo*. ROS and RNS are highly reactive and short-lived, which makes it difficult to capture real-time data about their production and effects within host tissues. Standard laboratory methods often involve in vitro models that may not fully replicate the complex dynamics occurring in an infected host. Many of these techniques, including chemiluminescence and fluorescence-based assays, can detect ROS/RNS, but they often lack the spatial and temporal resolution needed to study the intricacies of their biosynthetic pathways in live tissues (Wu et al., 2019).

Another major challenge arises from the complex interactions between ROS and RNS and the host immune system. Both ROS and RNS play dual roles—serving as weapons against parasites but also contributing to tissue damage and inflammation if not regulated properly. The balance between parasite killing and host tissue damage is extremely delicate. This complexity makes it difficult to develop therapeutic strategies that target the pathways of biosynthesis of ROS/RNS without inducing unintended side effects such as immunosuppression or exacerbation of disease (Moreira et al., 2015). *Leishmania* has evolved evasion mechanisms, such as upregulation of antioxidant enzymes, which further complicates the ability to target these molecules and free radicals without bolstering the parasite's defences.

5.2. Emerging Research Areas

A key area of interest in recent research is the exploration of ROS/RNS dynamics in different forms of leishmaniasis. The clinical manifestations of the disease, such as cutaneous, mucocutaneous, and visceral leishmaniasis, exhibit varying degrees of tissue damage and inflammation. Studies are beginning to show that ROS and RNS may play different roles depending on the type of leishmanial infection. For instance, cutaneous leishmaniasis often involves more localized oxidative stress, whereas visceral leishmaniasis may lead to systemic immune dysregulation and more severe oxidative stress-related complications (Horta et al., 2012). Understanding these differences could provide insight into how targeted therapies might be tailored based on the form of the disease.

Recent advancements in single-cell analysis and advanced imaging technologies are also offering promising new ways to study ROS and RNS in the context of *Leishmaniasis*. Single-cell RNA sequencing (scRNA-seq) has begun to reveal how individual immune cells respond to infection, potentially shedding light on which cells are producing ROS and RNS and how these responses vary across different stages of the infection. Additionally, live imaging techniques, such as two-photon microscopy, allow for real-time tracking of ROS/RNS dynamics in infected tissues, providing more detailed spatial and temporal insights into the interactions between immune cells and parasites (Horta et al., 2012).

5.3. Integrating ROS/RNS Studies with Other Research

To fully understand the roles of ROS and RNS in leishmaniasis, it is essential to integrate this research with other emerging areas of study. For example, host genetic factors play a crucial role in determining the outcome of *Leishmaniasis*. Some studies suggest that certain genetic polymorphisms in oxidative stress-related genes may predispose individuals to more severe forms of the disease (Elmahallawy et al., 2021). Therefore, incorporating genetic data into ROS/RNS studies could lead to a more personalized approach to therapy, potentially identifying individuals who might benefit more from antioxidant treatments or immune-modulating therapies.

Also, the host microbiome may influence ROS and RNS levels during infection. Studies in other infectious diseases have shown that certain microbial communities can either promote or inhibit oxidative stress in the host, suggesting that the microbiome could play a role in modulating the host immune response to *Leishmania (Moreira et al., 2015)*. Exploring this connection may open new avenues for therapeutic intervention, perhaps through probiotics or microbiome-modulating treatments that help regulate oxidative stress. Lastly, integrating ROS/RNS research with immunotherapy strategies, such as checkpoint inhibitors or cytokine-based therapies, could help create more robust immune responses against *Leishmania*. For instance, therapies that increase interferon-gamma (IFN-γ) production could enhance ROS and RNS production, while also activating other immune pathways necessary for parasite clearance (Costa-da-Silva et al., 2022). Such approaches could be particularly beneficial in combination therapies, where ROS/RNS-modulating drugs are paired with immunotherapies to optimize both oxidative and immune-mediated parasite clearance.

6. CONCLUSION

In *Leishmaniasis*, ROS and RNS play a dual role, serving as critical elements in both parasite control and disease pathogenesis. These reactive species, produced primarily by phagocytes such as macrophages and neutrophils, are instrumental in the host's defence mechanisms. NADPH oxidase generates ROS, while inducible nitric oxide synthase (iNOS) produces RNS, both of which contribute to the destruction of the parasite within infected cells. However, the production of these reactive molecules must be carefully regulated; excessive or prolonged ROS and RNS generation can lead to significant tissue damage and chronic inflammation, complicating disease outcomes and contributing to immune dysregulation. To survive the host's immune attack, *Leishmania* has developed sophisticated antioxidant defences. Key enzymes such as superoxide dismutase (SOD) and molecules like trypanothione allow the parasite to neutralize ROS and RNS, thereby evading immune destruction and sustaining its survival within the host. This capacity to counteract oxidative and nitrosative stress underpins the parasite's persistence and contributes to the chronic nature of leishmaniasis. As a result, therapeutic strategies targeting ROS and RNS face a delicate balance: boosting these molecules could enhance parasite clearance but also risk exacerbating tissue injury, while antioxidant therapies might protect host tissues but at the cost of undermining effective immune defences.

Research into the ROS and RNS pathways in leishmaniasis holds substantial promise for the development of innovative therapies. Future efforts should focus on finding the right balance between parasite elimination and tissue protection, as well as understanding how these pathways vary across different forms of the disease. Advances in single-cell analysis and imaging techniques offer new opportunities to explore the interactions between ROS/RNS, host immune responses, and parasite survival. Continued investigation into these areas, along with the integration of ROS/RNS modulation in combination with traditional antiparasitic medications or immunotherapies, could pave the way for more effective and personalized treatments for leishmaniasis.

REFERENCES

Ali, V., Behera, S., Nawaz, A., Equbal, A., & Pandey, K. (2022). Unique thiol metabolism in trypanosomatids: Redox homeostasis and drug resistance. [Internet] In Advances in Parasitology (pp. 75–155). Elsevier., Available from https://linkinghub.elsevier.com/retrieve/pii/S0065308X2200015X, Retrieved October 15, 2024, from.

Almeida, F. S., Vanderley, S. E. R., Comberlang, F. C., Andrade, A. G. D., Cavalcante-Silva, L. H. A., Silva, E. D. S., Palmeira, P. H. S., Amaral, I. P. G., & Keesen, T. S. L. (2023). Leishmaniasis: Immune Cells Crosstalk in Macrophage Polarization. *Tropical Medicine and Infectious Disease*, *8*(5), 276. DOI: 10.3390/tropicalmed8050276 PMID: 37235324

Ayala, A., Llanes, A., Lleonart, R., & Restrepo, C. M. (2024). Advances in Leishmania Vaccines: Current Development and Future Prospects. *Pathogens (Basel, Switzerland)*, *13*(9), 812. DOI: 10.3390/pathogens13090812 PMID: 39339003

Bharadava, K., Upadhyay, T. K., Kaushal, R. S., Ahmad, I., Alraey, Y., Siddiqui, S., & Saeed, M. (2024). Genomic insight of leishmania parasite: In-depth review of drug resistance mechanisms and genetic mutations. *ACS Omega*, *9*(11), 12500–12514.

Bichiou, H., Bouabid, C., Rabhi, I., & Guizani-Tabbane, L. (2021). Transcription Factors Interplay Orchestrates the Immune-Metabolic Response of Leishmania Infected Macrophages. *Frontiers in Cellular and Infection Microbiology*, *11*, 660415. DOI: 10.3389/fcimb.2021.660415 PMID: 33898331

Bodhale, N., Ohms, M., Ferreira, C., Mesquita, I., Mukherjee, A., André, S., Sarkar, A., Estaquier, J., Laskay, T., Saha, B., & Silvestre, R. (2021). Cytokines and metabolic regulation: A framework of bidirectional influences affecting Leishmania infection. *Cytokine*, *147*, 155267. DOI: 10.1016/j.cyto.2020.155267 PMID: 32917471

Bogdan, C. (2015). Nitric oxide synthase in innate and adaptive immunity: An update. *Trends in Immunology*, *36*(3), 161–178. DOI: 10.1016/j.it.2015.01.003 PMID: 25687683

Canton, M., Sánchez-Rodríguez, R., Spera, I., Venegas, F. C., Favia, M., Viola, A., & Castegna, A. (2021). Reactive Oxygen Species in Macrophages: Sources and Targets. *Frontiers in Immunology*, *12*, 734229. DOI: 10.3389/fimmu.2021.734229 PMID: 34659222

Castro, H., Duarte, M., & Tomás, A. M. (2017). The Redox Metabolism and Oxidative Stress in *Leishmania* as a Crossroads for the Lethal Effect of Drugs. [Internet] In L. Rivas & C. Gil (Eds.), Drug Discovery for Leishmaniasis (pp. 316–347). The Royal Society of Chemistry., Available from https://books.rsc.org/books/book/1890/chapter/2478779/The-Redox-Metabolism-and-Oxidative-Stress-, Retrieved October 16, 2024, from. DOI: 10.1039/9781788010177-00316

Chulanetra, M., & Chaicumpa, W. (2021). Revisiting the Mechanisms of Immune Evasion Employed by Human Parasites. *Frontiers in Cellular and Infection Microbiology*, *11*, 702125. DOI: 10.3389/fcimb.2021.702125 PMID: 34395313

Costa-da-Silva, A. C., Nascimento, D. D. O., Ferreira, J. R. M., Guimarães-Pinto, K., Freire-de-Lima, L., Morrot, A., Decote-Ricardo, D., Filardy, A. A., & Freire-de-Lima, C. G. (2022). Immune Responses in Leishmaniasis: An Overview. *Tropical Medicine and Infectious Disease*, *7*(4), 54. DOI: 10.3390/tropicalmed7040054 PMID: 35448829

Costa-Madeira, J. C., Trindade, G. B., Almeida, P. H. P., Silva, J. S., & Carregaro, V. (2022). T Lymphocyte Exhaustion During Human and Experimental Visceral Leishmaniasis. *Frontiers in Immunology*, *13*, 835711. DOI: 10.3389/fimmu.2022.835711 PMID: 35585983

Coutinho De Oliveira, B., Duthie, M. S., & Alves Pereira, V. R. (2020). Vaccines for leishmaniasis and the implications of their development for American tegumentary leishmaniasis. *Human Vaccines & Immunotherapeutics*, *16*(4), 919–930. DOI: 10.1080/21645515.2019.1678998 PMID: 31634036

De Almeida, A. J. P. O., De Oliveira, J. C. P. L., Da Silva Pontes, L. V., De Souza Júnior, J. F., Gonçalves, T. A. F., Dantas, S. H., de Almeida Feitosa, M. S., Silva, A. O., & de Medeiros, I. A. (2022). ROS: Basic Concepts, Sources, Cellular Signalling, and its Implications in Aging Pathways. *Oxidative Medicine and Cellular Longevity*, *2022*, 1–23. DOI: 10.1155/2022/1225578 PMID: 36312897

De Las Heras, N., Martín Giménez, V. M., Ferder, L., Manucha, W., & Lahera, V. (2020). Implications of Oxidative Stress and Potential Role of Mitochondrial Dysfunction in COVID-19: Therapeutic Effects of Vitamin D. *Antioxidants*, *9*(9), 897. DOI: 10.3390/antiox9090897 PMID: 32967329

Di Meo, S., Reed, T. T., Venditti, P., & Victor, V. M. (2016). Role of ROS and RNS Sources in Physiological and Pathological Conditions. *Oxidative Medicine and Cellular Longevity*, *2016*(1), 1245049. DOI: 10.1155/2016/1245049 PMID: 27478531

Elmahallawy, E. K., Alkhaldi, A. A. M., & Saleh, A. A. (2021). Host immune response against leishmaniasis and parasite persistence strategies: A review and assessment of recent research. *Biomedicine and Pharmacotherapy*, *139*, 111671. DOI: 10.1016/j.biopha.2021.111671 PMID: 33957562

Gervazoni, L. F. O., Barcellos, G. B., Ferreira-Paes, T., & Almeida-Amaral, E. E. (2020). Use of Natural Products in Leishmaniasis Chemotherapy: An Overview. *Frontiers in Chemistry*, *8*, 579891. DOI: 10.3389/fchem.2020.579891 PMID: 33330368

Giorgi, C., Marchi, S., Simoes, I. C., Ren, Z., Morciano, G., Perrone, M., & Wieckowski, M. R. (2018). Mitochondria and reactive oxygen species in aging and age-related diseases. *International Review of Cell and Molecular Biology*, *340*, 209–344.

González-Montero, M. C., Andrés-Rodríguez, J., García-Fernández, N., Pérez-Pertejo, Y., Reguera, R. M., Balaña-Fouce, R., & García-Estrada, C. (2024). Targeting Trypanothione Metabolism in Trypanosomatids. *Molecules (Basel, Switzerland)*, *29*(10), 2214. DOI: 10.3390/molecules29102214 PMID: 38792079

Gupta, A. K., Das, S., Kamran, M., Ejazi, S. A., & Ali, N. (2022). The pathogenicity and virulence of Leishmania - interplay of virulence factors with host defences. *Virulence*, *13*(1), 903–935. DOI: 10.1080/21505594.2022.2074130 PMID: 35531875

Hatinguais, R., Pradhan, A., Brown, G. D., Brown, A. J. P., Warris, A., & Shekhova, E. (2021). Mitochondrial Reactive Oxygen Species Regulate Immune Responses of Macrophages to Aspergillus fumigatus. *Frontiers in Immunology*, *12*, 641495. DOI: 10.3389/fimmu.2021.641495 PMID: 33841423

Horta, M. F., Mendes, B. P., Roma, E. H., Noronha, F. S. M., Macêdo, J. P., Oliveira, L. S., Duarte, M. M., & Vieira, L. Q. (2012). Reactive Oxygen Species and Nitric Oxide in Cutaneous Leishmaniasis. *Journal of Parasitology Research*, *2012*, 1–11. DOI: 10.1155/2012/203818 PMID: 22570765

Ikeogu, N. M., Akaluka, G. N., Edechi, C. A., Salako, E. S., Onyilagha, C., Barazandeh, A. F., & Uzonna, J. E. (2020). Leishmania Immunity: Advancing Immunotherapy and Vaccine Development. *Microorganisms*, *8*(8), 1201. DOI: 10.3390/microorganisms8081201 PMID: 32784615

Kane, M. M., & Mosser, D. M. (2001). The Role of IL-10 in Promoting Disease Progression in Leishmaniasis. *The Journal of Immunology : Official Journal of the American Association of Immunologists*, *166*(2), 1141–1147. DOI: 10.4049/jimmunol.166.2.1141 PMID: 11145695

Kaushal, R. S., Naik, N., Prajapati, M., Rane, S., Raulji, H., Afu, N. F., Upadhyay, T. K., & Saeed, M. (2023). *Leishmania* species: A narrative review on surface proteins with structural aspects involved in host–pathogen interaction. *Chemical Biology & Drug Design, 102*(2), 332–356. DOI: 10.1111/cbdd.14227 PMID: 36872849

Kaye, P., & Scott, P. (2011). Leishmaniasis: Complexity at the host–pathogen interface. *Nature Reviews. Microbiology, 9*(8), 604–615. DOI: 10.1038/nrmicro2608 PMID: 21747391

Knight, C. A., Harris, D. R., Alshammari, S. O., Gugssa, A., Young, T., & Lee, C. M. (2023). Leishmaniasis: Recent epidemiological studies in the Middle East. *Frontiers in Microbiology, 13*, 1052478. DOI: 10.3389/fmicb.2022.1052478 PMID: 36817103

Krauth-Siegel, R. L., & Comini, M. A. (2008). Redox control in trypanosomatids, parasitic protozoa with trypanothione-based thiol metabolism. Biochim. Biophys. Acta BBA - Gen. *Biochimica et Biophysica Acta. G, General Subjects, 1780*(11), 1236–1248. DOI: 10.1016/j.bbagen.2008.03.006 PMID: 18395526

Kröber-Boncardo, C., Grünebast, J., & Clos, J. (2020). Heat Shock Proteins in Leishmania Parasites. [Internet] In A. A. A. Asea & P. Kaur (Eds.), Heat Shock Proteins in Inflammatory Diseases (pp. 469–488). Springer International Publishing., Available from https://link.springer.com/10.1007/7515_2020_27, Retrieved October 15, 2024, from. DOI: 10.1007/7515_2020_27

Liese, J., Schleicher, U., & Bogdan, C. (2008). The innate immune response against Leishmania parasites. *Immunobiology, 213*(3-4), 377–387. DOI: 10.1016/j.imbio.2007.12.005 PMID: 18406382

Liu, D., & Uzonna, J. E. (2012). The early interaction of Leishmania with macrophages and dendritic cells and its influence on the host immune response. [Internet]. *Frontiers in Cellular and Infection Microbiology, 2*, •••. Retrieved October 16, 2024, from http://journal.frontiersin.org/article/10.3389/fcimb.2012.00083/abstract. DOI: 10.3389/fcimb.2012.00083 PMID: 22919674

Mantena, S. K., King, A. L., Andringa, K. K., Eccleston, H. B., & Bailey, S. M. (2008). Mitochondrial dysfunction and oxidative stress in the pathogenesis of alcohol- and obesity-induced fatty liver diseases. *Free Radical Biology & Medicine, 44*(7), 1259–1272. DOI: 10.1016/j.freeradbiomed.2007.12.029 PMID: 18242193

Miranda, M. M., Panis, C., Cataneo, A. H. D., Da Silva, S. S., Kawakami, N. Y., Lopes, L. G. D. F., Morey, A. T., Yamauchi, L. M., Andrade, C. G. T. J., Cecchini, R., da Silva, J. J. N., Sforcin, J. M., Conchon-Costa, I., & Pavanelli, W. R. (2015). Nitric Oxide and Brazilian Propolis Combined Accelerates Tissue Repair by Modulating Cell Migration, Cytokine Production and Collagen Deposition in Experimental Leishmaniasis. *PLoS One*, *10*(5), e0125101. DOI: 10.1371/journal.pone.0125101 PMID: 25973801

Mittal, M., Siddiqui, M. R., Tran, K., Reddy, S. P., & Malik, A. B. (2014). Reactive Oxygen Species in Inflammation and Tissue Injury. *Antioxidants & Redox Signaling*, *20*(7), 1126–1167. DOI: 10.1089/ars.2012.5149 PMID: 23991888

Moreira, D., Rodrigues, V., Abengozar, M., Rivas, L., Rial, E., Laforge, M., Li, X., Foretz, M., Viollet, B., Estaquier, J., Cordeiro da Silva, A., & Silvestre, R. (2015). Leishmania infantum Modulates Host Macrophage Mitochondrial Metabolism by Hijacking the SIRT1-AMPK Axis. *PLoS Pathogens*, *11*(3), e1004684. DOI: 10.1371/journal.ppat.1004684 PMID: 25738568

Murray, H. W. (2005). Prevention of Relapse after Chemotherapy in a Chronic Intracellular Infection: Mechanisms in Experimental Visceral Leishmaniasis. *The Journal of Immunology : Official Journal of the American Association of Immunologists*, *174*(8), 4916–4923. DOI: 10.4049/jimmunol.174.8.4916 PMID: 15814719

Murray, H. W., & Nathan, C. F. (1999). Macrophage Microbicidal Mechanisms In Vivo: Reactive Nitrogen versus Oxygen Intermediates in the Killing of Intracellular Visceral *Leishmania donovani.The Journal of Experimental Medicine*, *189*(4), 741–746. DOI: 10.1084/jem.189.4.741 PMID: 9989990

Nandan, D., Brar, H. K., & Reiner, N. (2024). Manipulation of Macrophages: Emerging Mechanisms of Leishmaniasis. Front. Biosci.-. *Landmark*, *29*, 292. PMID: 39206906

Novais, F. O., Nguyen, B. T., Beiting, D. P., Carvalho, L. P., Glennie, N. D., Passos, S., Carvalho, E. M., & Scott, P. (2014). Human Classical Monocytes Control the Intracellular Stage of Leishmania braziliensis by Reactive Oxygen Species. *The Journal of Infectious Diseases*, *209*(8), 1288–1296. DOI: 10.1093/infdis/jiu013 PMID: 24403561

Olivier, M., Atayde, V. D., Isnard, A., Hassani, K., & Shio, M. T. (2012). Leishmania virulence factors: Focus on the metalloprotease GP63. *Microbes and Infection*, *14*(15), 1377–1389. DOI: 10.1016/j.micinf.2012.05.014 PMID: 22683718

Olivier, M., Gregory, D. J., & Forget, G. (2005). Subversion Mechanisms by Which *Leishmania* Parasites Can Escape the Host Immune Response: A Signalling Point of View. *Clinical Microbiology Reviews*, *18*(2), 293–305. DOI: 10.1128/CMR.18.2.293-305.2005 PMID: 15831826

Paes, M. C., Cosentino-Gomes, D., Souza, C. F. D., Nogueira, N. P. D. A., & Meyer-Fernandes, J. R. (2011). The Role of Heme and Reactive Oxygen Species in Proliferation and Survival of *Trypanosoma cruzi*.*Journal of Parasitology Research*, *2011*, 1–8. DOI: 10.1155/2011/174614 PMID: 22007287

Reverte, M., Eren, R. O., Jha, B., Desponds, C., Snäkä, T., Prevel, F., Isorce, N., Lye, L.-F., Owens, K. L., Gazos Lopes, U., Beverley, S. M., & Fasel, N. (2021). The antioxidant response favors Leishmania parasites survival, limits inflammation and reprograms the host cell metabolism. *PLoS Pathogens*, *17*(3), e1009422. DOI: 10.1371/journal.ppat.1009422 PMID: 33765083

Reverte, M., Snäkä, T., & Fasel, N. (2022). The Dangerous Liaisons in the Oxidative Stress Response to Leishmania Infection. *Pathogens (Basel, Switzerland)*, *11*(4), 409. DOI: 10.3390/pathogens11040409 PMID: 35456085

Saha, S., Roy, S., Dutta, A., Jana, K., & Ukil, A. (2021). Leishmania donovani Targets Host Transcription Factor NRF2 To Activate Antioxidant Enzyme HO-1 and Transcriptional Repressor ATF3 for Establishing Infection. *Infection and Immunity*, *89*(7), e00764–e20. DOI: 10.1128/IAI.00764-20 PMID: 33820818

Samant, M., Sahu, U., Pandey, S. C., & Khare, P. (2021). Role of Cytokines in Experimental and Human Visceral Leishmaniasis. *Frontiers in Cellular and Infection Microbiology*, *11*, 624009. DOI: 10.3389/fcimb.2021.624009 PMID: 33680991

Scarpini, S., Dondi, A., Totaro, C., Biagi, C., Melchionda, F., Zama, D., Pierantoni, L., Gennari, M., Campagna, C., Prete, A., & Lanari, M. (2022). Visceral Leishmaniasis: Epidemiology, Diagnosis, and Treatment Regimens in Different Geographical Areas with a Focus on Pediatrics. *Microorganisms*, *10*(10), 1887. DOI: 10.3390/microorganisms10101887 PMID: 36296164

Shakya, N., Sane, S. A., Vishwakarma, P., & Gupta, S. (2012). Enhancement in therapeutic efficacy of miltefosine in combination with synthetic bacterial lipopeptide, Pam3Cys against experimental Visceral Leishmaniasis. *Experimental Parasitology*, *131*(3), 377–382. DOI: 10.1016/j.exppara.2012.05.007 PMID: 22626518

Shio, M. T., Hassani, K., Isnard, A., Ralph, B., Contreras, I., Gomez, M. A., Abu-Dayyeh, I., & Olivier, M. (2012). Host Cell Signalling and *Leishmania* Mechanisms of Evasion. *Journal of Tropical Medicine*, *2012*, 1–14. DOI: 10.1155/2012/819512 PMID: 22131998

Singh, R., Kashif, M., Srivastava, P., & Manna, P. P. (2023). Recent Advances in Chemotherapeutics for Leishmaniasis: Importance of the Cellular Biochemistry of the Parasite and Its Molecular Interaction with the Host. *Pathogens (Basel, Switzerland)*, *12*(5), 706. DOI: 10.3390/pathogens12050706 PMID: 37242374

Soulat, D., & Bogdan, C. (2017). Function of Macrophage and Parasite Phosphatases in Leishmaniasis. *Frontiers in Immunology*, *8*, 1838. DOI: 10.3389/fimmu.2017.01838 PMID: 29312331

Subramanian Vignesh, K., & Deepe, G. S.Jr. (2016). Immunological orchestration of zinc homeostasis: The battle between host mechanisms and pathogen defences. *Archives of Biochemistry and Biophysics*, *611*, 66–78. DOI: 10.1016/j.abb.2016.02.020 PMID: 26921502

Sundar, S., & Chakravarty, J. (2012). Recent advances in the diagnosis and treatment of kala-azar. *The National Medical Journal of India*, *25*, 85–89. PMID: 22686715

Suschek, C. V., Feibel, D., Von Kohout, M., & Opländer, C. (2022). Enhancement of Nitric Oxide Bioavailability by Modulation of Cutaneous Nitric Oxide Stores. *Biomedicines*, *10*(9), 2124. DOI: 10.3390/biomedicines10092124 PMID: 36140225

Tomiotto-Pellissier, F., Bortoleti, B. T. D. S., Assolini, J. P., Gonçalves, M. D., Carloto, A. C. M., Miranda-Sapla, M. M., Conchon-Costa, I., Bordignon, J., & Pavanelli, W. R. (2018). Macrophage Polarization in Leishmaniasis: Broadening Horizons. *Frontiers in Immunology*, *9*, 2529. DOI: 10.3389/fimmu.2018.02529 PMID: 30429856

Valko, M., Leibfritz, D., Moncol, J., Cronin, M. T. D., Mazur, M., & Telser, J. (2007). Free radicals and antioxidants in normal physiological functions and human disease. *The International Journal of Biochemistry & Cell Biology*, *39*(1), 44–84. DOI: 10.1016/j.biocel.2006.07.001 PMID: 16978905

Veith, A., & Moorthy, B. (2018). Role of cytochrome P450s in the generation and metabolism of reactive oxygen species. *Current Opinion in Toxicology*, *7*, 44–51. DOI: 10.1016/j.cotox.2017.10.003 PMID: 29527583

Vivarini, A. D. C., & Lopes, U. G. (2020). The Potential Role of Nrf2 Signalling in Leishmania Infection Outcomes. *Frontiers in Cellular and Infection Microbiology*, *9*, 453. DOI: 10.3389/fcimb.2019.00453 PMID: 31998662

Volpedo, G., Pacheco-Fernandez, T., Bhattacharya, P., Oljuskin, T., Dey, R., Gannavaram, S., Satoskar, A. R., & Nakhasi, H. L. (2021). Determinants of Innate Immunity in Visceral Leishmaniasis and Their Implication in Vaccine Development. *Frontiers in Immunology*, *12*, 748325. DOI: 10.3389/fimmu.2021.748325 PMID: 34712235

Wamai, R. G., Kahn, J., McGloin, J., & Ziaggi, G. (2020). Visceral leishmaniasis: A global overview. *Journal of Global Health Science*, *2*(1), e3. DOI: 10.35500/jghs.2020.2.e3

Wijnant, G. J., Dumetz, F., Dirkx, L., Bulté, D., Cuypers, B., Van Bocxlaer, K., & Hendrickx, S. (2022). Tackling Drug Resistance and Other Causes of Treatment Failure in Leishmaniasis. *Frontiers in Tropical Diseases*, *3*, 837460. DOI: 10.3389/fitd.2022.837460

Wink, D. A., Hines, H. B., Cheng, R. Y. S., Switzer, C. H., Flores-Santana, W., Vitek, M. P., Ridnour, L. A., & Colton, C. A. (2011). Nitric oxide and redox mechanisms in the immune response. *Journal of Leukocyte Biology*, *89*(6), 873–891. DOI: 10.1189/jlb.1010550 PMID: 21233414

Wu, L., Sedgwick, A. C., Sun, X., Bull, S. D., He, X. P., & James, T. D. (2019). Reaction-Based Fluorescent Probes for the Detection and Imaging of Reactive Oxygen, Nitrogen, and Sulfur Species. *Accounts of Chemical Research*, *52*(9), 2582–2597. DOI: 10.1021/acs.accounts.9b00302 PMID: 31460742

Yu, Y., Liu, S., Yang, L., Song, P., Liu, Z., Liu, X., Yan, X., & Dong, Q. (2024). Roles of reactive oxygen species in inflammation and cancer. *MedComm*, *5*(4), e519. DOI: 10.1002/mco2.519 PMID: 38576456

Zarkovic, N. (2020). Roles and Functions of ROS and RNS in Cellular Physiology and Pathology. *Cells*, *9*(3), 767. DOI: 10.3390/cells9030767 PMID: 32245147

Zhou, J. (2020). Role of reactive oxygen species and iron in host defence against infection. *Frontiers in Bioscience*, *25*(8), 1600–1616. DOI: 10.2741/4869

Chapter 9
Oxidative Stress in Cardiovascular Disorders: Mechanisms and Therapeutic Implications

Karan Goel
M.M College of Pharmacy, Maharishi Markandeshwar University (Deemed), India

Isha Chawla
M.M College of Pharmacy, Maharishi Markandeshwar University (Deemed), India

Garima Malik
Chitkara College of Pharmacy, Chitkara University, India

Meenakshi Dhanawat
https://orcid.org/0000-0001-9097-341X
Amity Institute of Pharmacy, Amity University, India

ABSTRACT

Cardiovascular disorders (CVDs) include atherosclerosis, cardiac failure, hypertension, myocardial infarction, and oxidative stress, which are increasingly recognized as essential factors for the onset and advancement of these conditions. he imbalance between the formation of reactive oxygen species (ROS) and the antioxidant defense system is the molecular basis of oxidative stress, which leads to oxidative damage to lipids, proteins, and DNA. This potential damage promotes inflammatory responses, disrupts endothelial function, and speeds up atherogenesis. The nitric oxide synthase uncoupling, mitochondrial dysfunction, and NADPH-oxidizing enzymes are important pathways that enhance ROS production. Antioxidants including polyphenols,

DOI: 10.4018/979-8-3693-7919-6.ch009

vitamins E and C, and others have shown conflicting outcomes in clinical studies. These novel methods aim to improve natural antioxidant defenses, alter redox-sensitive signaling pathways, and target specific ROS sources.

1. INTRODUCTION

Oxidative stress arises when ROS generation exceeds antioxidant neutralization.(Das & Roychoudhury, 2014) This state is crucial to the genesis and progression of many illnesses, including cardiovascular, neurodegenerative, and cancer. Both endogenous and external sources create reactive oxygen species like hydroxyl radicals, hydrogen peroxide, and superoxide anion. Peroxisomes, NADPH oxidase along with xanthine oxidase activities contribute to ROS, although mitochondria are the main generators, especially via the electron transport chain.(Bhattacharyya et al., 2014; Dumanović et al., 2021; Poljsak et al., 2013a) Environmental contaminants, cigarette smoke, radiation, UV light, and chemicals are exogenous.(Dumanović et al., 2021; Poljsak et al., 2013a) ROS are needed for signaling and pathogen defense, but excessive buildup may damage DNA, proteins, and lipids. The body uses enzymatic as well as non-enzymatic antioxidant defenses to combat ROS. ROS are converted into water and oxygen by enzymatic antioxidants such SOD, glutathione peroxidase, and catalase. Antioxidants include vitamins C and E, and glutathione scavenge free radicals. Maintaining redox homeostasis requires an equilibrium between the production of ROS and antioxidant defenses to avoid oxidative stress and cellular malfunction.(Das & Roychoudhury, 2014; He et al., 2017) When ROS levels rise, key regulatory mechanisms including the Nrf2 signaling route upregulate antioxidant gene expression for protection of organisms from damage caused by oxidation. Oxidative stress overwhelms the body's defenses, causing cellular and systemic damage. Oxidative stress causes lipid peroxidation, which damages membranes and causes apoptosis. Protein oxidation affects enzyme performance, whereas DNA oxidation may cause mutations and cancer.(Poljsak et al., 2013b) Systemic oxidative stress is mainly associated with cardiovascular disorders like atherosclerosis, hypertension, and myocardial infarction. Neurodegenerative illnesses like Alzheimer's and Parkinson's depend on it for neuronal cell death. The vicious circle of prolonged oxidative stress along with inflammatory responses worsens disease development. Denham Harman's 1950s "free radical theory of aging," relating free radicals to aging and age-related disorders, changed the notion of oxidative stress. Over the decades, studies have linked oxidative stress to cardiovascular disease, cancer, and neurodegeneration. The antioxidants discovery and endogenous antioxidant systems have helped us understand how oxidative stress causes illness and create treatments to combat it. Today, researchers are studying the molecular underpinnings of oxidative stress and ways to decrease its effects on chronic illnesses.(D'Oria et al., 2020; Dubois-Deruy et al., 2020; Senoner & Dichtl, 2019)

2. OXIDATIVE STRESS AND CARDIOVASCULAR HEALTH

2.1 Role of Oxidative Stress in Cardiovascular Disorders: Mechanisms, Impact, and Contributing Factors

Cardiovascular diseases and the causes of oxidative stress are seen from two angles here. The oxidative stress processes that underlie cardiovascular diseases like coronary artery disease (CAD), hypertension (hypertension), and cardiac failure are mostly responsible for cellular damage, nitric oxide depletion, and inflammation caused by reactive oxygen species (ROS). Plaque accumulation, vascular rigidity, and heart failure are all symptoms of oxidative stress, which is associated with different kinds of disorders ultimately. These conditions might have environmental or behavioral roots as summarized in Table 1.(Mourino-Alvarez et al., 2023; Münzel et al., 2017)

Table 1. Role of oxidative stress in cardiovascular disorders: Mechanisms, impact, and contributing factors

Cardiovascular Disorder	Key Features	Mechanism of Oxidative Stress	Impact of ROS/ Contributing Factors	Outcome	Reference
Coronary Artery Disease (CAD)	Atherosclerotic plaque buildup reduces coronary blood flow	Oxidation of LDL → formation of oxLDL → endothelial dysfunction, plaque formation	Genetic predisposition, smoking, poor diet, lipid peroxidation	Arterial blockage, angina, myocardial infarction	(Lavie & Lavie, 2009)
Hypertension	Chronic high blood pressure	ROS overproduction → NO depletion → impaired vasodilation, vascular remodeling	Pollution, genetic mutations in antioxidant enzymes, endothelial damage	Stiffened arteries, increased heart workload	(Fearon, 2014)
Heart Failure	Reduced cardiac output, ventricular remodeling	Mitochondrial dysfunction → ROS-induced myocardial cell death, fibrosis	Aging, metabolic syndrome, mitochondrial inefficiency, apoptosis	Weak heart muscle, systemic fluid retention	(Panda et al., 2022)

continued on following page

Table 1. Continued

Cardiovascular Disorder	Key Features	Mechanism of Oxidative Stress	Impact of ROS/ Contributing Factors	Outcome	Reference
Stroke	Ischemia-reperfusion injury, neuronal damage	ROS bursts during reperfusion → neuronal and endothelial cell injury	Hypertension, diabetes, environmental stress, mitochondrial damage	Neuronal death, potential paralysis	(Rotariu et al., 2022)
Peripheral Artery Disease	Narrowing of peripheral arteries, limb ischemia	ROS-induced smooth muscle dysfunction → endothelial apoptosis	Smoking, diabetes, high-fat diet, hyperlipidemia	Limb ischemia, risk of amputation	(Burtscher et al., 2024)

2.2 Genetic, Environmental Factors, Aging, and Endothelial Dysfunction in Cardiovascular Oxidative Stress

Predispositions in genes, including variations in antioxidant enzymes, lower the body's ability to remove reactive oxygen species (ROS), whereas pollution and bad habits, like smoking and poor nutrition, enhance oxidative stress as well as hasten the deterioration of cardiovascular health. Heart function declines, arterial stiffness increases, and the issues of cardiovascular events increase with aging due to mitochondrial dysfunction and cellular senescence, both of which intensify oxidative damage as summarized in Table 2.(Lim et al., 2024; Luna-Marco et al., 2024; Ray et al., 2023)

Table 2. Genetic, environmental, and aging factors contributing to oxidative stress in cardiovascular disease

Category	Contributing Factor	Mechanism of Oxidative Stress	Cardiovascular Impact	Reference
Genetic	Antioxidant enzyme polymorphisms (e.g., SOD, GPx, catalase)	Reduced ROS scavenging capability → excess oxidative stress	Predisposition to hypertension, atherosclerosis, CAD	(Rubab & Zulkipli, 2024)
Environmental	Smoking, air pollution, heavy metals	Direct ROS generation → depletion of antioxidant reserves, inflammation	Accelerated atherosclerosis, endothelial dysfunction	(Zahedi et al., 2023)
Lifestyle	High-fat, high-sugar diet, sedentary behavior	Lipid peroxidation, insulin resistance, increased ROS production	Obesity, metabolic syndrome, CAD	(Schmidt-Trucksäss et al., 2024)
Aging	Mitochondrial dysfunction, cellular senescence	Mitochondria produce excess ROS; senescent cells release pro-inflammatory factors.	Increased risk of heart failure, arterial stiffness, atherosclerosis	(Barreiro Arcos, 2022)

3. CARDIOVASCULAR DISORDER OXIDATIVE STRESS MECHANISMS

3.1 ROS-Induced Endothelial Dysfunction

Endothelial dysfunction can be brought through reactive oxygen species (ROS), which decrease the bioavailability of NO. NO offers a vital part in vasodilation, but ROS like O2− reacts with NO for generating peroxynitrite (ONOO−), reducing NO levels and increasing vasoconstriction, inflammation, and thrombosis. ROS trigger pro-inflammatory pathways including NF-κB, leading to endothelial damage and vascular dysfunction.(Lund, 2024; Ren et al., 2024; X. Zhang et al., 2023)

3.2 Oxidative Stress and Atherosclerosis Pathophysiology

Oxidative stress enhances LDL oxidation, a key atherogenesis event. Macrophages convert oxLDL into foam cells, which are characteristic of atherosclerotic plaques. ROS also activates endothelial cells, which release adhesion molecules, chemokines, and cytokines that attract immune cells to the vascular wall, worsening plaque development and instability.(Salekeen et al., 2022; M. Zhang et al., 2024)

3.3 Hypertension Oxidative Stress: Mechanisms

Oxidative stress raises ROS, which disrupts vasoconstriction-vasodilation equilibrium and causes hypertension. High ROS limit NO availability, constricting and increasing vascular resistance. ROS also activates the RAS, which increases blood pressure via constriction, salt retention, and vascular modification.(Fejes et al., 2024; Nathani et al., 2024)

3.4 Oxidative Stress in Ischemia-Reperfusion Injury and Cardioprotection

I/R causes oxidative stress owing to ROS bursts when oxygen is reintroduced into tissues. Cellular injury increases with peroxidation of lipid, protein oxidation, and harm to mitochondria. Cardioprotective methods include preconditioning and antioxidants to minimize ROS production or boost endogenous antioxidant defenses to preserve tissue.(Heusch et al., 2023; Ji et al., 2024; Wu et al., 2021)

3.5 Myocardial Infarction: Oxidative Stress Damages and Remodels Tissue

In myocardial infarction (MI), oxidative stress causes immediate tissue damage and remodeling. ROS during ischemia and reperfusion kills cells, whereas post-infarction oxidative stress causes fibrosis, ventricular remodeling, and heart failure. Antioxidant therapy and mitochondrial function may reduce MI damage.(Yaghoobi et al., 2024; Zhu et al., 2024)

3.6 Heart Failure Oxidative Stress: Molecular Mechanisms and Clinical Results

Chronic oxidative stress may worsen heart failure by reducing mitochondrial activity, increasing apoptosis, and causing fibrosis. ROS increases systemic inflammation and endothelial dysfunction, worsening heart failure outcomes. Mitochondrial-targeted antioxidants may improve heart failure prognosis.(Hou et al., 2023; Patel et al., 2023)

3.7 Oxidative Stress and Cardiac Arrhythmias

Ion channel function is affected by oxidative stress, causing arrhythmogenesis. ROS alters calcium, potassium, and sodium channels, causing cardiac cell electrical activity abnormalities. Electrophysiology is further impaired by oxidative stress, which impairs mitochondrial function. Treating oxidative stress may prevent or lessen cardiac arrhythmias (Figure 1).(Soomro & Charytan, 2024; Zhou et al., 2022)

Figure 1. Cardiovascular disorder oxidative stress mechanisms

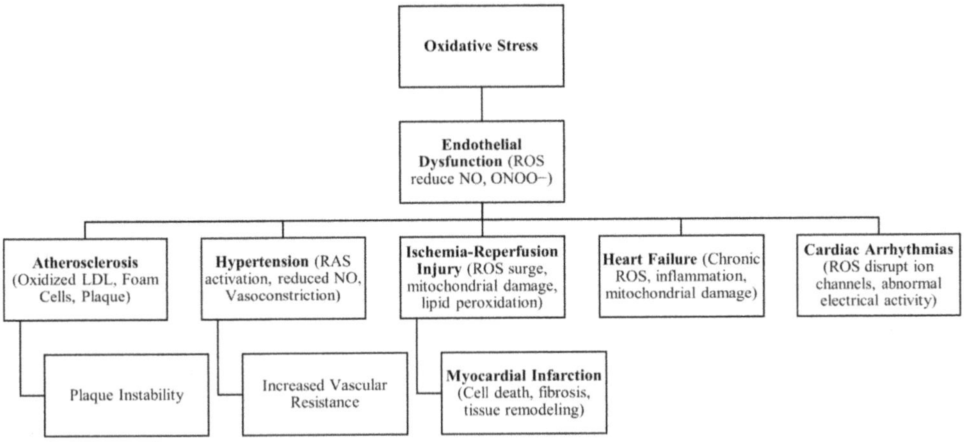

4. MOLECULAR PATHWAYS INVOLVED IN OXIDATIVE STRESS

4.1 NADPH Oxidase (NOX) Family and Cardiovascular Disease

The NADPH oxidase (NOX) family generates major cardiovascular reactive oxygen species (ROS). Electron-donating NADPH is used by NOX enzymes to generate superoxide ($O2-$) from oxygen. Enhanced NOX activity for cardiovascular diseases increases inflammation, oxidative stress, along with vascular remodeling, resulting in endothelial dysfunction, hypertension, and atherosclerosis.(Amadio et al., 2024; García-Redondo et al., 2016; Poinsignon et al., 2023)

4.2 Mitochondrial Dysfunction and ROS Generation

ROS are produced through mitochondria, particularly when their electron transport chain fails and leaks electrons, releasing superoxide. Mitochondrial malfunction increases oxidative stress, damaging proteins, lipids, and DNA in cardiovascular illnesses such as cardiac failure, ischemia-reperfusion damage, and myocardial infarction.(Y. Li et al., 2022; Liu et al., 2024)

4.3 Role of Nitric Oxide Synthase (NOS) in Oxidative Stress

The vasodilator nitric oxide (NO) is generated through Nitric oxide synthase (NOS). NOS may become "uncoupled," releasing superoxide instead of NO under oxidative stress. This depletes NO, worsening endothelial dysfunction and hypertension. By increasing oxidative stress, dysfunctional NOS contributes to cardiovascular disease.(Abolfazli et al., 2024; Rochette et al., 2013)

4.4 Impact of Inflammation: Cytokines, ROS, and Cardiovascular Disorders

Redox stress and inflammation are connected. Pro-inflammatory cytokines such as α, IL-1β, IL-6 and TNF- enhances ROS generation and oxidative damage. Inflammation increases ROS formation, producing a vicious cycle. This cycle damages vascular tissues, causing atherosclerosis, heart failure and hypertension.(Granato, 2022; Koutsaliaris et al., 2022)

4.5 Redox-Sensitive Transcription Factors (NF-κB, Nrf2) and Gene Expression

Redox-sensitive transcription factors, such as NF-κB and Nrf2, control genes participating in oxidative stress and inflammation. ROS stimulates NF-κB, resulting in pro-inflammatory cytokines, whereas Nrf2 promotes antioxidant defense gene expression. Oxidative damage and cardiovascular disease development may result from pathway imbalances.(Atalay Ekiner et al., 2022; Priya Dharshini et al., 2020)

4.6 Apoptosis, Necrosis, and Autophagy Pathways in Oxidative Cardiomyopathy

Apoptosis, necrosis, and autophagy cause cell death in relation to oxidative stress. Excess ROS may cause mitochondrial apoptosis or necrosis in cardiomyocytes, damaging the heart. Cellular repair is impaired by dysregulated autophagy, worsening cardiomyopathy, and cardiac failure.(Avagimyan et al., 2024; Guan et al., 2019; H. Wang et al., 2021)

4.7 Interaction of Oxidative Stress with the Renin-Angiotensin-Aldosterone System (RAAS)

The renin-angiotensin-aldosterone system (RAAS) regulates fluid balance and blood pressure. RAAS activation by oxidative stress causes vasoconstriction, salt retention, and fibrosis. A feedback loop between RAAS and angiotensin II enhances ROS generation via NOX activation, worsening hypertension and heart failure (Figure 2).(Oh et al., 2023; Speth, 2022)

Figure 2. Molecular pathways involved in oxidative stress

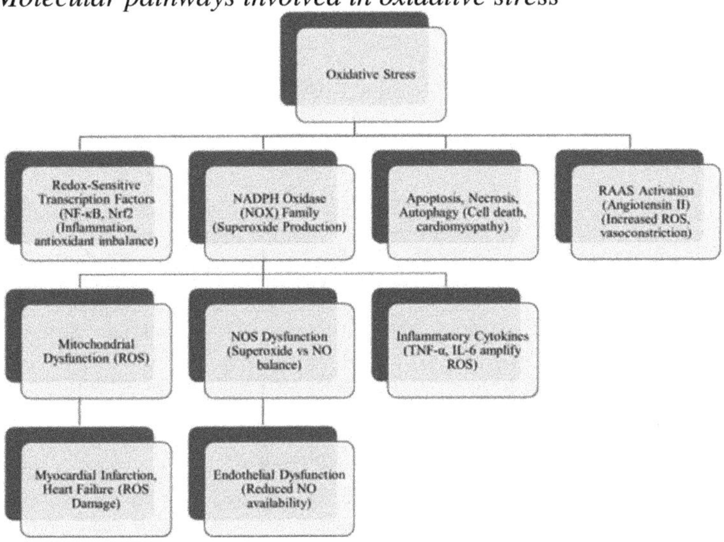

5. BIOMARKERS AND DETECTION OF OXIDATIVE STRESS IN CVD

Oxidative stress plays a significant role in the development and progression of cardiovascular diseases (CVD). It occurs when there is an imbalance between the production of reactive oxygen species (ROS) and the body's ability to detoxify these reactive products. Here's a brief overview of biomarkers associated with oxidative stress and methods for their detection in the context of CVD as summarized in Table 3.

Table 3. Biomarkers and detection of oxidative stress in Cardiovascular Disease (CVD)

Biomarker Category	Biomarker	Type of Damage or Protection	Mechanism of Action	Clinical Relevance in CVD	Reference
Oxidative Stress Biomarkers	General Biomarkers	Oxidative stress detection	The broad category of markers indicating elevated oxidative stress	Used for early diagnosis, monitoring progression, and evaluating treatment responses in cardiovascular disease	(J. Li et al., 2023; S. Wang et al., 2020; K. Zhang et al., 2024)
Reactive Species Measurement	ROS (Superoxide, Hydrogen Peroxide)	Reactive Oxygen Species (ROS)	ROS are transient molecules that oxidatively destroy DNA, proteins, and lipids.	High ROS levels are linked to atherosclerosis, endothelial dysfunction, hypertension, and heart failure.	(Park et al., 2023)
	RNS (Nitric Oxide, Peroxynitrite)	Reactive Nitrogen Species (RNS)	RNS such as nitric oxide (NO) and peroxynitrite cause nitration of proteins and DNA, leading to oxidative stress and endothelial malfunction	Excess RNS worsens inflammation and endothelial dysfunction, contributing to vascular disease.	(Maes et al., 2011)
Lipid Peroxidation Products	Malondialdehyde (MDA)	Lipid Peroxidation	MDA is produced when ROS attacks polyunsaturated fatty acids in cell membranes	Commonly used marker of oxidative stress, but less specific and prone to variability in measurements.	(Gitto et al., 2024; Mas-Bargues et al., 2021)
	Isoprostanes	Lipid Peroxidation	Formed by free radical-catalyzed peroxidation of lipids, isoprostanes are stable and specific markers of oxidative damage.	More reliable and specific than MDA for assessing lipid peroxidation and oxidative stress	(Mas-Bargues et al., 2021)

continued on following page

Table 3. Continued

Biomarker Category	Biomarker	Type of Damage or Protection	Mechanism of Action	Clinical Relevance in CVD	Reference
Protein Oxidation Markers	Carbonylated Proteins	Protein Oxidation	Oxidative modification of proteins results in carbonyl group formation, which alters protein structure and function	Elevated levels indicate widespread oxidative damage to proteins, linked to atherosclerosis and heart failure.	(Madkour, 2020; Maes et al., 2011)
	3-Nitrotyrosine	Protein Nitration (via RNS)	Caused by nitration of tyrosine residues in proteins, particularly due to peroxynitrite (RNS) activity	High levels are associated with inflammation, endothelial malfunction and cardiovascular inflammation.	(Maes et al., 2011; Oh et al., 2023)
DNA Damage Markers	8-Oxo-2'-deoxyguanosine (8-OHdG)	DNA Oxidation	8-OHdG is a major product of oxidative damage to DNA, indicating ROS attack on guanine residues	Elevated 8-OHdG levels are linked to oxidative DNA damage, mutagenesis, and plaque instability in atherosclerosis	(Madkour, 2020)
Antioxidant Enzymes	Superoxide Dismutase (SOD)	Antioxidant Defense (ROS neutralization)	Converts superoxide radicals (O2−) into hydrogen peroxide (H2O2), reducing oxidative stress burden	Low SOD activity is associated with increased ROS levels, endothelial dysfunction, and atherosclerosis.	(Bresciani et al., 2015; Clair et al., 2024)
	Catalase	Antioxidant Defense (H2O2 neutralization)	Breaks down hydrogen peroxide (H2O2) in water and oxygen, preventing further generation of harmful ROS	Low catalase levels results in hydrogen peroxide accumulation, exacerbating oxidative stress in cardiovascular tissues.	(Clair et al., 2024)
	Glutathione Peroxidase (GPx)	Antioxidant Defense (Lipid peroxide detox)	Detoxifies hydrogen peroxide and lipid peroxides, protecting cells from oxidative injury	Decreased GPx activity is linked with poor outcomes in cardiac failure, myocardial infarction, and atherosclerosis.	(Bresciani et al., 2015; Clair et al., 2024)

continued on following page

Table 3. Continued

Biomarker Category	Biomarker	Type of Damage or Protection	Mechanism of Action	Clinical Relevance in CVD	Reference
Prognostic Biomarkers	Combined Oxidative Biomarker Profile	Prognostic Indicator	Elevated levels of ROS, MDA, carbonylated proteins, and low antioxidant enzyme activity indicate high oxidative stress and poor antioxidant defenses.	A poor oxidative stress profile correlates with disease severity, worse outcomes, and increased mortality in CVD patients.	(J. Li et al., 2023; Panda et al., 2022)
Advanced Biomarkers of Redox State	Total Antioxidant Capacity (TAC)	Systemic Redox Balance Indicator	Measures the overall ability of body fluids (plasma, serum) to neutralize free radicals	Low TAC is linked with enhanced oxidative damage as well as poor cardiovascular consequences.	(He et al., 2017; Heusch et al., 2023)
	F2-Isoprostanes	Lipid Peroxidation, Oxidative Stress Marker	F2-isoprostanes are formed non-enzymatically from arachidonic acid, representing accurate markers of oxidative lipid damage.	Widely regarded as the most reliable marker of oxidative stress; linked to coronary artery disease, stroke, and other CVD outcomes.	(Atalay Ekiner et al., 2022; S. Wang et al., 2020)
Advanced Protein Oxidation Markers	Advanced Oxidation Protein Products (AOPP)	Protein Oxidation	AOPPs result from oxidative modification of proteins (e.g., albumin), indicating chronic oxidative stress.	High AOPP levels correlate with vascular complications, endothelial dysfunction, and the progression of atherosclerosis.	(Poinsignon et al., 2023)

continued on following page

Table 3. Continued

Biomarker Category	Biomarker	Type of Damage or Protection	Mechanism of Action	Clinical Relevance in CVD	Reference
Emerging Biomarkers	Oxidized LDL (ox-LDL)	Lipoprotein Oxidation	LDL particles become oxidized by ROS, contributing to plaque formation in arteries	oxLDL is important component for atherosclerosis development and serves as an early biomarker of cardiovascular risk	(Mas-Bargues et al., 2021)
	Myeloperoxidase (MPO)	Enzyme-producing ROS, Protein Nitration	MPO is released by activated neutrophils and produces hypochlorous acid (HOCl), a potent oxidant that modifies proteins and lipoproteins	Elevated MPO amount are connected with expanded cardiovascular risk, plaque instability, and heart attack.	(Nandeshwar et al., 2022)
	Homocysteine	Amino Acid Oxidation	High homocysteine levels lead to oxidative stress by producing ROS and inducing endothelial dysfunction.	Elevated homocysteine is an distinct risk variable for coronary artery disease and thrombotic events.	(Abolfazli et al., 2024; Lund, 2024)

6. THERAPEUTIC IMPLICATIONS OF TARGETING OXIDATIVE STRESS

Targeting oxidative stress has significant therapeutic implications for cardiovascular diseases (CVD) and other health conditions. By modulating oxidative stress levels, it's possible to mitigate cellular damage and improve cardiovascular health. Here are some key therapeutic strategies and their implications as summarized in Table 4.

Table 4. Therapeutic implications of targeting oxidative stress in Cardiovascular Diseases (CVD)

Therapeutic Approach	Mechanism of Action	Clinical Application in CVD	Challenges	Reference
Antioxidant Therapies	Neutralize ROS/RNS, reduce oxidative harm to DNA, proteins, and lipids	Protects cardiovascular tissues from oxidative damage	Mixed results in trials; issues with bioavailability, dosing, and timing	(Dumanović et al., 2021; He et al., 2017; Poljsak et al., 2013b)
Vitamins & Polyphenols	• **Vitamin C**: Scavenges ROS • **Vitamin E**: Inhibits lipid peroxidation • **Polyphenols**: Antioxidant & anti-inflammatory effects	May improve endothelial function and reduce oxidative stress; polyphenols are promising.	Inconsistent results in large-scale trials; variable bioavailability (e.g., curcumin)	(Poljsak et al., 2013b)
Statins & ACE Inhibitors	• **Statins**: Reduce ROS by inhibiting NADPH oxidase • **ACE Inhibitors**: Inhibit angiotensin II pathway, reducing ROS	Improve vascular health, reduce atherosclerosis progression, and lower oxidative stress.	Side effects with long-term use; statins may cause muscle issues in some patients	(Nandeshwar et al., 2022)
Beta-blockers & Calcium Channel Blockers	• **Beta-Blockers**: Inhibit ROS via NADPH oxidase • **Calcium Channel Blockers**: Prevent calcium overload, and reduce oxidative damage	Lower blood pressure, improve endothelial function and reduce arrhythmias related to oxidative stress.	Not primarily designed for antioxidant action; benefits on oxidative stress are secondary.	(Bathinapatla et al., 2023)
Natural Compounds	• **Resveratrol**: Activates antioxidant pathways (e.g., Nrf2) • **Curcumin**: Reduces lipid peroxidation • **Green Tea Extract**: Inhibits ROS, boosts antioxidants	Show antioxidant and anti-inflammatory benefits in cardiovascular health	Poor bioavailability (e.g., curcumin); novel delivery methods required (e.g., nanomedicine)	(Abolfazli et al., 2024)
Nanomedicine	• Targeted delivery of antioxidants using nanoparticles for enhanced bioavailability and specificity	Potential for improved antioxidant therapy, especially with natural and synthetic antioxidants	Safety and regulatory concerns regarding long-term use and potential toxicity of nanocarriers	(Nandeshwar et al., 2022)

continued on following page

Table 4. Continued

Therapeutic Approach	Mechanism of Action	Clinical Application in CVD	Challenges	Reference
Gene Therapy & CRISPR	• **Gene Therapy**: Enhance expression of antioxidant enzymes (e.g., SOD) • **CRISPR**: Edit genes participated in ROS generation	Potential to reduce oxidative stress at genetic level, offering long-term cardiovascular protection	Ethical concerns, delivery challenges, and possible off-target effects	(Lim et al., 2024; Rubab & Zulkipli, 2024)
Clinical Trials	• Combination therapies, polyphenols, and emerging antioxidants depicted potential in reducing oxidative stress markers.	Some polyphenol trials show improvements in oxidative stress markers and endothelial function.	Failures in trials with vitamins like E and C; large-scale trials needed for emerging therapies	(Nathani et al., 2024; Oh et al., 2023)
Personalized Medicine	• Tailoring antioxidant therapy based on genetic and molecular profiles of oxidative stress	Personalized treatments could improve outcomes by targeting individual genetic susceptibilities to oxidative stress.	Requires advanced diagnostics, genetic profiling, and cost considerations	(Nathani et al., 2024; Patel et al., 2023)

7. LIFESTYLE AND PREVENTIVE APPROACHES TO MITIGATE OXIDATIVE STRESS

Mitigating oxidative stress through lifestyle and preventive approaches is crucial for maintaining health and reducing the risk of chronic diseases, including cardiovascular diseases (CVD) as summarized in Table 5.

Table 5. Comprehensive lifestyle and preventive approaches to mitigate oxidative stress in Cardiovascular Diseases (CVD)

Lifestyle Approach	Mechanism of Action	Impact on Oxidative Stress and CVD	Challenges	Reference
Role of Diet (Mediterranean & DASH Diets)	• Diets high in whole grains, fruits, vegetables, and healthy fats (nuts, olive oil) provide antioxidants, polyphenols, and anti-inflammatory compounds.	Lowers oxidative stress, improves endothelial function, reduces inflammation, and lowers the risk of atherosclerosis and hypertension	Dietary adherence, availability of healthy foods, and potential nutrient deficiencies if poorly balanced	(Granato, 2022)
Antioxidant-rich foods & Supplements	• **Fruits & vegetables**: Provide vitamins C, E, and polyphenols (e.g., flavonoids) • **Omega-3 fatty acids**: Reduce inflammation, and lower lipid peroxidation	Reduces ROS production, improves endothelial function, enhances antioxidant defenses (e.g., SOD and catalase), and protects against lipid oxidation	Bioavailability issues with supplements (e.g., curcumin, resveratrol); Over-reliance on supplements can sometimes lead to inadequate dietary intake	(Das & Roychoudhury, 2014; Dubois-Deruy et al., 2020; Fearon, 2014)
Physical Exercise	• Regular moderate exercise boosts endogenous antioxidant systems (e.g., increased SOD and glutathione) and reduces chronic inflammation.	Improves cardiovascular health, lowers oxidative stress, and supports mitochondrial health. Reduces risks of heart failure, hypertension, and atherosclerosis	Over-exercising or high-intensity exercise can generate excess ROS, leading to oxidative damage ("oxidative paradox")	(Bresciani et al., 2015; Clair et al., 2024)
Smoking, Alcohol, & Environmental Pollutants	• **Smoking**: Directly increases ROS production and depletes antioxidant defenses like vitamin C • **Alcohol**: Excessive intake increases ROS, and causes lipid peroxidation • **Pollutants**: Environmental toxins increase oxidative stress and DNA damage	Smoking causes endothelial malfunction and enhances oxidative stress, contributing to atherosclerosis; alcohol and pollutants further worsen oxidative stress and increase cardiovascular risk.	Quitting smoking is difficult; social and environmental factors make reducing alcohol intake and exposure to pollutants challenging.	(Das & Roychoudhury, 2014; Zahedi et al., 2023)

continued on following page

Table 5. Continued

Lifestyle Approach	Mechanism of Action	Impact on Oxidative Stress and CVD	Challenges	Reference
Psychosocial Stress & Mental Health	• Chronic psychosocial stress activates stress hormones (e.g., cortisol) that enhance ROS production and increase inflammation through cytokines.	Stress-related oxidative stress can impair heart function, lead to hypertension, and increase CVD risk; managing stress can improve endothelial health.	Requires long-term behavioral changes; mental health management is often stigmatized and under-prioritized	(Fearon, 2014; Koutsaliaris et al., 2022; H. Wang et al., 2021)
Sleep Patterns & Circadian Rhythms	• Disrupted circadian rhythms and poor sleep quality (e.g., sleep apnea) increase oxidative stress and impair the body's capability for repairing oxidative damage.	Restful sleep reduces ROS levels, supports healthy circadian rhythms, and aids in cellular repair, lowering cardiovascular risk.	Sleep disorders often require medical intervention, and maintaining healthy sleep patterns can be challenging with modern lifestyles.	(Nathani et al., 2024)
Preventive Strategies for High-Risk Populations	• Managing metabolic syndrome, diabetes, and obesity through lifestyle and medication controls blood sugar, reduces fat accumulation, and improves insulin sensitivity. • Better glucose control lowers ROS production and inflammation, protecting against endothelial damage	Reduces oxidative stress, improves vascular function, and lowers chances of cardiovascular events in high-risk individuals	Long-term lifestyle changes (e.g., diet, exercise) are required, and managing comorbidities (e.g., diabetes, hypertension) is challenging.	(Salekeen et al., 2022)
Role of Gut Microbiota in Oxidative Stress	• Healthy gut bacteria (e.g., Lactobacillus, Bifidobacterium) help modulate oxidative stress via the formation of short-chain fatty acids (SCFAs) and anti-inflammatory metabolites; a balanced microbiome reduces gut-derived inflammation, which is a significant cause of cardiovascular oxidative stress.	Balanced gut microbiota can decrease oxidative stress, and inflammation, and enhance cardiovascular health through the gut-heart axis.	Gut dysbiosis caused by poor diet, antibiotics, or lifestyle factors can increase oxidative stress and inflammation and	(Bathinapatla et al., 2023; Hou et al., 2023)
Impact of Weight Management	• Obesity leads to excessive fat accumulation, which produces inflammatory cytokines and increases ROS. Weight loss, through diet and exercise, reduces oxidative stress markers and improves insulin sensitivity and endothelial function.	Weight reduction lowers inflammation, improves vascular function, and reduces ROS, lowering the risk of heart disease and stroke.	Sustaining long-term weight loss is difficult, especially without proper support and access to healthy food and exercise options.	(Fejes et al., 2024; Granato, 2022; Koutsaliaris et al., 2022)

continued on following page

Table 5. Continued

Lifestyle Approach	Mechanism of Action	Impact on Oxidative Stress and CVD	Challenges	Reference
Role of Hydration	• Adequate hydration helps maintain cellular homeostasis and may reduce oxidative stress by preventing dehydration-induced increases in ROS. Proper water intake supports detoxification and enhances antioxidant enzyme activity, which protects against oxidative damage.	Staying hydrated improves metabolic functions, supports antioxidant defenses, and helps detoxify the body from environmental pollutants.	Many individuals do not consume adequate water, and dehydration can easily contribute to immense oxidative stress as well as cellular damage.	(Nandeshwar et al., 2022)
Role of Vitamin D	• Vitamin D has anti-inflammatory as well as antioxidant properties, modulating the immune system and reducing ROS through the regulation of key antioxidant enzymes.	Adequate vitamin D levels are linked with reduced oxidative stress, inflammation, and better cardiovascular function.	Vitamin D deficiency is common, and sun exposure or supplementation may be required to keep healthy levels.	(Chen et al., 2024)
Mindfulness & Meditation	• These practices reduce stress, lower cortisol levels, and increase parasympathetic activity, indirectly reducing ROS production and enhancing antioxidant defenses.	Improved mental health, lower oxidative stress, and reduced risk of hypertension and heart disease.	Long-term adherence to mindfulness practices requires time, dedication, and consistent behavior modification.	(Schneider & Carr, 2014)

8. CONCLUSION

Several factors contribute to cardiovascular diseases, including the overproduction of reactive oxygen species (ROS) which impairs cellular equilibrium.(Das & Roychoudhury, 2014; Poljsak et al., 2013b) Coronary artery disease, hypertension, cardiac failure, and stroke proceed because of ROS-driven lipid peroxidation, nitric oxide depletion, and inflammation. Endothelial function, atherosclerotic plaque development, vascular remodeling, and tissue injury are all affected by oxidative stress, underscoring its importance in cardiovascular pathophysiology.(Burtscher et al., 2024; Münzel et al., 2017) Understanding oxidative stress processes expands clinical and research opportunities. Oxidative stress biomarkers like MDA and oxidized LDL reveal disease progression and therapy success. Adding oxidative stress indicators to cardiovascular diagnostics might improve early diagnosis and enable more tailored treatments. Research on redox-sensitive pathways including

NF-κB and Nrf2 reveals promising therapeutic targets.(Atalay Ekiner et al., 2022; Guan et al., 2019) Oxidative stress has been better understood, but obstacles persist. The uneven clinical trial findings of antioxidant treatments show ROS dynamics' complexity and the need for a more sophisticated approach to therapy. Oxidative stress modulation by patient-specific variables such as genetic predispositions and environmental impacts must be studied. Understanding how oxidative stress interacts with other biological systems, such as the immune response, may improve therapies. Bioavailability, dose, and timing have plagued antioxidant therapy, despite their promise. Emerging medicines target ROS generation and control mechanisms beyond vitamins C and E.(Chen et al., 2024) Advances in nanomedicine and gene therapy enable more precise antioxidant chemical delivery and genetic treatments to modify ROS levels at the molecular level, giving hope. Statins and ACE inhibitors, which enhance vascular health, may also indirectly lower oxidative stress. Cardiovascular disease prevention and therapy must focus on oxidative stress management.(D'Oria et al., 2020; Gitto et al., 2024) Lifestyle changes like antioxidant-rich foods and frequent exercise may lower oxidative stress combined with medicinal therapies. Use lifestyle, pharmaceutical, and developing molecular treatments to balance ROS generation and the body's antioxidant defenses. Early oxidative stress reduction in high-risk groups may prevent cardiovascular events and improve longevity and quality of life.(Rubab & Zulkipli, 2024; Schmidt-Trucksäss et al., 2024; Wu et al., 2021)

REFERENCES

Abolfazli, S., Mortazavi, P., Kheirandish, A., Butler, A. E., Jamialahmadi, T., & Sahebkar, A. (2024). Regulatory effects of curcumin on nitric oxide signaling in the cardiovascular system. *Nitric Oxide, 143*, 16–28. DOI: 10.1016/j.niox.2023.12.003 PMID: 38141926

Amadio, P., Sandrini, L., Zarà, M., Barbieri, S. S., & Ieraci, A. (2024). NADPH-oxidases as potential pharmacological targets for thrombosis and depression comorbidity. *Redox Biology*, *70*, 103060. DOI: 10.1016/j.redox.2024.103060 PMID: 38310682

Atalay Ekiner, S., Gęgotek, A., & Skrzydlewska, E. (2022). The molecular activity of cannabidiol in the regulation of Nrf2 system interacting with NF-κB pathway under oxidative stress. *Redox Biology, 57*, 102489. DOI: 10.1016/j.redox.2022.102489 PMID: 36198205

Avagimyan, A., Pogosova, N., Kakturskiy, L., Sheibani, M., Challa, A., Kogan, E., Fogacci, F., Mikhaleva, L., Vandysheva, R., Yakubovskaya, M., Faggiano, A., Carugo, S., Urazova, O., Jahanbin, B., Lesovaya, E., Polana, S., Kirsanov, K., Sattar, Y., Trofimenko, A., & Sarrafzadegan, N. (2024). Doxorubicin-related cardiotoxicity: Review of fundamental pathways of cardiovascular system injury. *Cardiovascular Pathology, 73*, 107683. DOI: 10.1016/j.carpath.2024.107683 PMID: 39111556

Barreiro Arcos, M. L. (2022). Role of thyroid hormones-induced oxidative stress on cardiovascular physiology. *Biochimica et Biophysica Acta (BBA) -. Biochimica et Biophysica Acta. G, General Subjects, 1866*(12), 130239. DOI: 10.1016/j.bbagen.2022.130239 PMID: 36064072

Bathinapatla, A., Kanchi, S., Chokkareddy, R., Puthalapattu, R. P., & Kumar, M. R. (2023). Recent trends in the electrochemical sensors on β- and calcium channel blockers for hypertension and angina pectoris: A comprehensive review. *Microchemical Journal, 192*, 108930. DOI: 10.1016/j.microc.2023.108930

Bhattacharyya, A., Chattopadhyay, R., Mitra, S., & Crowe, S. E. (2014). Oxidative Stress: An Essential Factor in the Pathogenesis of Gastrointestinal Mucosal Diseases. *Physiological Reviews, 94*(2), 329–354. DOI: 10.1152/physrev.00040.2012 PMID: 24692350

Bresciani, G., da Cruz, I. B. M., & González-Gallego, J. (2015). *Manganese Superoxide Dismutase and Oxidative Stress Modulation.*, DOI: 10.1016/bs.acc.2014.11.001

Burtscher, J., Millet, G. P., Fresa, M., Lanzi, S., Mazzolai, L., & Pellegrin, M. (2024). The link between impaired oxygen supply and cognitive decline in peripheral artery disease. *Progress in Cardiovascular Diseases*, *85*, 63–73. DOI: 10.1016/j.pcad.2023.12.002 PMID: 38061613

Chen, Y., Zhang, H., Pan, Y., Zhang, Y., Yang, Y., Liu, L., Jia, Q., Wang, Y., & Kong, Y. (2024). Association between cardiovascular health and serum vitamin D and its interaction with prediabetes and diabetes. *The American Journal of the Medical Sciences*. Advance online publication. DOI: 10.1016/j.amjms.2024.08.021 PMID: 39186977

D'Oria, R., Schipani, R., Leonardini, A., Natalicchio, A., Perrini, S., Cignarelli, A., Laviola, L., & Giorgino, F. (2020). The Role of Oxidative Stress in Cardiac Disease: From Physiological Response to Injury Factor. *Oxidative Medicine and Cellular Longevity*, *2020*, 1–29. DOI: 10.1155/2020/5732956 PMID: 32509147

Das, K., & Roychoudhury, A. (2014). Reactive oxygen species (ROS) and response of antioxidants as ROS-scavengers during environmental stress in plants. *Frontiers in Environmental Science*, *2*. Advance online publication. DOI: 10.3389/fenvs.2014.00053

Dubois-Deruy, E., Peugnet, V., Turkieh, A., & Pinet, F. (2020). Oxidative Stress in Cardiovascular Diseases. *Antioxidants*, *9*(9), 864. DOI: 10.3390/antiox9090864 PMID: 32937950

Dumanović, J., Nepovimova, E., Natić, M., Kuča, K., & Jaćević, V. (2021). The Significance of Reactive Oxygen Species and Antioxidant Defense System in Plants: A Concise Overview. *Frontiers in Plant Science*, *11*, 552969. Advance online publication. DOI: 10.3389/fpls.2020.552969 PMID: 33488637

Fearon, I. M. (2014). Risk Factors, Oxidative Stress, and Cardiovascular Disease. In *Systems Biology of Free Radicals and Antioxidants* (pp. 765–790). Springer Berlin Heidelberg., DOI: 10.1007/978-3-642-30018-9_46

Fejes, R., Pilat, N., Lutnik, M., Weisshaar, S., Weijler, A. M., Krüger, K., Draxler, A., Bragagna, L., Peake, J. M., Woodman, R. J., Croft, K. D., Bondonno, C. P., Hodgson, J. M., Wagner, K.-H., Wolzt, M., & Neubauer, O. (2024). Effects of increased nitrate intake from beetroot juice on blood markers of oxidative stress and inflammation in older adults with hypertension. *Free Radical Biology & Medicine*, *222*, 519–530. DOI: 10.1016/j.freeradbiomed.2024.07.004 PMID: 38972612

García-Redondo, A. B., Aguado, A., Briones, A. M., & Salaices, M. (2016). NADPH oxidases and vascular remodeling in cardiovascular diseases. *Pharmacological Research*, *114*, 110–120. DOI: 10.1016/j.phrs.2016.10.015 PMID: 27773825

Gitto, S., Fiorillo, C., Argento, F. R., Fini, E., Borghi, S., Falcini, M., Roccarina, D., La Delfa, R., Lillo, L., Zurli, T., Forte, P., Ghinolfi, D., De Simone, P., Chiesi, F., Ingravallo, A., Vizzutti, F., Aspite, S., Laffi, G., Lynch, E., & Becatti, M. (2024). Oxidative stress–induced fibrinogen modifications in liver transplant recipients: Unraveling a novel potential mechanism for cardiovascular risk. *Research and Practice in Thrombosis and Haemostasis*, *8*(6), 102555. DOI: 10.1016/j.rpth.2024.102555 PMID: 39309232

Granato, D. (2022). Functional foods to counterbalance low-grade inflammation and oxidative stress in cardiovascular diseases: A multilayered strategy combining food and health sciences. *Current Opinion in Food Science*, *47*, 100894. DOI: 10.1016/j.cofs.2022.100894

Guan, Y., Zhou, L., Zhang, Y., Tian, H., Li, A., & Han, X. (2019). Effects of PP2A/Nrf2 on experimental diabetes mellitus-related cardiomyopathy by regulation of autophagy and apoptosis through ROS dependent pathway. *Cellular Signalling*, *62*, 109339. DOI: 10.1016/j.cellsig.2019.06.004 PMID: 31173878

He, L., He, T., Farrar, S., Ji, L., Liu, T., & Ma, X. (2017). Antioxidants Maintain Cellular Redox Homeostasis by Elimination of Reactive Oxygen Species. *Cellular Physiology and Biochemistry*, *44*(2), 532–553. DOI: 10.1159/000485089 PMID: 29145191

Heusch, G., Andreadou, I., Bell, R., Bertero, E., Botker, H.-E., Davidson, S. M., Downey, J., Eaton, P., Ferdinandy, P., Gersh, B. J., Giacca, M., Hausenloy, D. J., Ibanez, B., Krieg, T., Maack, C., Schulz, R., Sellke, F., Shah, A. M., Thiele, H., & Di Lisa, F. (2023). Health position paper and redox perspectives on reactive oxygen species as signals and targets of cardioprotection. *Redox Biology*, *67*, 102894. DOI: 10.1016/j.redox.2023.102894 PMID: 37839355

Hou, J., Yuan, Y., Chen, P., Lu, K., Tang, Z., Liu, Q., Xu, W., Zheng, D., Xiong, S., & Pei, H. (2023). Pathological Roles of Oxidative Stress in Cardiac Microvascular Injury. *Current Problems in Cardiology*, *48*(1), 101399. DOI: 10.1016/j.cpcardiol.2022.101399 PMID: 36103941

Ji, Y., Wen, X., Tang, H., Jin, Z., Su, W., Zhou, L., Xia, Z., Xia, Z., & Lei, S. (2024). DJ-1: Potential target for treatment of myocardial ischemia-reperfusion injury. *Biomedicine and Pharmacotherapy*, *179*, 117383. DOI: 10.1016/j.biopha.2024.117383 PMID: 39232383

Koutsaliaris, I. K., Moschonas, I. C., Pechlivani, L. M., Tsouka, A. N., & Tselepis, A. D. (2022). Inflammation, Oxidative Stress, Vascular Aging and Atherosclerotic Ischemic Stroke. *Current Medicinal Chemistry*, *29*(34), 5496–5509. DOI: 10.2174/0929867328666210921161711 PMID: 34547993

Lavie, L., & Lavie, P. (2009). Molecular mechanisms of cardiovascular disease in OSAHS: The oxidative stress link. *The European Respiratory Journal*, *33*(6), 1467–1484. DOI: 10.1183/09031936.00086608 PMID: 19483049

Li, J., Pan, L., Pan, W., Li, N., & Tang, B. (2023). Recent progress of oxidative stress associated biomarker detection. *Chemical Communications*, *59*(48), 7361–7374. DOI: 10.1039/D3CC00878A PMID: 37194341

Li, Y., Ma, Y., Dang, Q.-Y., Fan, X.-R., Han, C.-T., Xu, S.-Z., & Li, P.-Y. (2022). Assessment of mitochondrial dysfunction and implications in cardiovascular disorders. *Life Sciences*, *306*, 120834. DOI: 10.1016/j.lfs.2022.120834 PMID: 35902031

Lim, A., Pasini, M., Yun, S., Gill, J., & Koirala, B. (2024). Genetic association between post-traumatic stress disorder and cardiovascular disease: A scoping review. *Journal of Psychiatric Research*, *178*, 331–348. DOI: 10.1016/j.jpsychires.2024.08.027 PMID: 39191203

Liu, D., Qin, H., Gao, Y., Sun, M., & Wang, M. (2024). Cardiovascular disease: Mitochondrial dynamics and mitophagy crosstalk mechanisms with novel programmed cell death and macrophage polarisation. *Pharmacological Research*, *206*, 107258. DOI: 10.1016/j.phrs.2024.107258 PMID: 38909638

Luna-Marco, C., Iannantuoni, F., Hermo-Argibay, A., Devos, D., Salazar, J. D., Víctor, V. M., & Rovira-Llopis, S. (2024). Cardiovascular benefits of SGLT2 inhibitors and GLP-1 receptor agonists through effects on mitochondrial function and oxidative stress. *Free Radical Biology & Medicine*, *213*, 19–35. DOI: 10.1016/j.freeradbiomed.2024.01.015 PMID: 38220031

Lund, A. K. (2024). Oxidants and Endothelial Dysfunction. In *Reference Module in Biomedical Sciences*. Elsevier., DOI: 10.1016/B978-0-323-95488-4.00071-1

Madkour, L. H. (2020). Oxidative stress and oxidative damage-induced cell death. In *Reactive Oxygen Species (ROS), Nanoparticles, and Endoplasmic Reticulum (ER) Stress-Induced Cell Death Mechanisms* (pp. 175–197). Elsevier. DOI: 10.1016/B978-0-12-822481-6.00008-6

Maes, M., Ruckoanich, P., Chang, Y. S., Mahanonda, N., & Berk, M. (2011). Multiple aberrations in shared inflammatory and oxidative & nitrosative stress (IO&NS) pathways explain the co-association of depression and cardiovascular disorder (CVD), and the increased risk for CVD and due mortality in depressed patients. *Progress in Neuro-Psychopharmacology & Biological Psychiatry*, *35*(3), 769–783. DOI: 10.1016/j.pnpbp.2010.06.008 PMID: 20561554

Mas-Bargues, C., Escrivá, C., Dromant, M., Borrás, C., & Viña, J. (2021). Lipid peroxidation as measured by chromatographic determination of malondialdehyde. Human plasma reference values in health and disease. *Archives of Biochemistry and Biophysics*, *709*, 108941. DOI: 10.1016/j.abb.2021.108941 PMID: 34097903

Mourino-Alvarez, L., Sastre-Oliva, T., Corbacho-Alonso, N., & Barderas, G. (2023). *M*. Oxidative Stress in Cardiovascular Diseases., DOI: 10.5772/intechopen.105891

Münzel, T., Camici, G. G., Maack, C., Bonetti, N. R., Fuster, V., & Kovacic, J. C. (2017). Impact of Oxidative Stress on the Heart and Vasculature. *Journal of the American College of Cardiology*, *70*(2), 212–229. DOI: 10.1016/j.jacc.2017.05.035 PMID: 28683969

Nandeshwar, R., Illa, M. P., Khandelwal, M., & Tallur, S. (2022). Enzymatic degradation of bacterial cellulose derived carbon nanofibers (BC-CNF) by myeloperoxidase (MPO): Performance evaluation for biosensing. *Biosensors & Bioelectronics: X*, *12*, 100252. DOI: 10.1016/j.biosx.2022.100252

Nathani, A., Attaway, A., & Mehra, R. (2024). Hypoxic and Autonomic Mechanisms from Sleep-Disordered Breathing Leading to Cardiopulmonary Dysfunction. *Sleep Medicine Clinics*, *19*(2), 229–237. DOI: 10.1016/j.jsmc.2024.02.003 PMID: 38692748

Oh, S., Kim, J. H., Cho, K. H., Kim, M. C., Sim, D. S., Hong, Y. J., Ahn, Y., & Jeong, M. H. (2023). Renin-Angiotensin-Aldosterone System Inhibitions and Cardiovascular Outcomes in Acute Myocardial Infarction With Renal Impairment. *Mayo Clinic Proceedings*, *98*(9), 1310–1322. DOI: 10.1016/j.mayocp.2023.02.007 PMID: 37245133

Panda, P., Verma, H. K., Lakkakula, S., Merchant, N., Kadir, F., Rahman, S., Jeffree, M. S., Lakkakula, B. V. K. S., & Rao, P. V. (2022). Biomarkers of Oxidative Stress Tethered to Cardiovascular Diseases. *Oxidative Medicine and Cellular Longevity*, *2022*, 1–15. DOI: 10.1155/2022/9154295 PMID: 35783193

Park, J., An, G., You, J., Park, H., Hong, T., Song, G., & Lim, W. (2023). Dimethenamid promotes oxidative stress and apoptosis leading to cardiovascular, hepatic, and pancreatic toxicities in zebrafish embryo. *Comparative Biochemistry and Physiology. Toxicology & Pharmacology : CBP*, *273*, 109741. DOI: 10.1016/j.cbpc.2023.109741 PMID: 37689173

Patel, R. N., Sharma, A., Prasad, A., & Bansal, S. (2023). Heart Failure With Preserved Ejection Fraction With CKD: A Narrative Review of a Multispecialty Disorder. *Kidney Medicine*, *5*(12), 100705. DOI: 10.1016/j.xkme.2023.100705 PMID: 38046909

Poinsignon, L., Chissey, A., Ajjaji, A., Hernandez, I., Vignaud, M.-L., Ferecatu, I., Fournier, T., Beaudeux, J.-L., & Zerrad-Saadi, A. (2023). Placental cartography of NADPH oxidase (NOX) family proteins: Involvement in the pathophysiology of preeclampsia. *Archives of Biochemistry and Biophysics*, *749*, 109787. DOI: 10.1016/j.abb.2023.109787 PMID: 37866451

Poljsak, B., Šuput, D., & Milisav, I. (2013a). Achieving the Balance between ROS and Antioxidants: When to Use the Synthetic Antioxidants. *Oxidative Medicine and Cellular Longevity*, *2013*, 1–11. DOI: 10.1155/2013/956792 PMID: 23738047

Poljsak, B., Šuput, D., & Milisav, I. (2013b). Achieving the Balance between ROS and Antioxidants: When to Use the Synthetic Antioxidants. *Oxidative Medicine and Cellular Longevity*, *2013*, 1–11. DOI: 10.1155/2013/956792 PMID: 23738047

Priya Dharshini, L. C., Vishnupriya, S., Sakthivel, K. M., & Rasmi, R. R. (2020). Oxidative stress responsive transcription factors in cellular signalling transduction mechanisms. *Cellular Signalling*, *72*, 109670. DOI: 10.1016/j.cellsig.2020.109670 PMID: 32418887

Ray, A. Ch., Maharana, K., Meenakshi, S., & Singh, S. (2023). Endothelial dysfunction and its relation in different disorders: Recent update. *Health Sciences Review (Oxford, England)*, *7*, 100084. DOI: 10.1016/j.hsr.2023.100084

Ren, H., Hu, W., Jiang, T., Yao, Q., Qi, Y., & Huang, K. (2024). Mechanical stress induced mitochondrial dysfunction in cardiovascular diseases: Novel mechanisms and therapeutic targets. *Biomedicine and Pharmacotherapy*, *174*, 116545. DOI: 10.1016/j.biopha.2024.116545 PMID: 38603884

Rochette, L., Lorin, J., Zeller, M., Guilland, J.-C., Lorgis, L., Cottin, Y., & Vergely, C. (2013). Nitric oxide synthase inhibition and oxidative stress in cardiovascular diseases: Possible therapeutic targets? *Pharmacology & Therapeutics*, *140*(3), 239–257. DOI: 10.1016/j.pharmthera.2013.07.004 PMID: 23859953

Rotariu, D., Babes, E. E., Tit, D. M., Moisi, M., Bustea, C., Stoicescu, M., Radu, A.-F., Vesa, C. M., Behl, T., Bungau, A. F., & Bungau, S. G. (2022). Oxidative stress – Complex pathological issues concerning the hallmark of cardiovascular and metabolic disorders. *Biomedicine and Pharmacotherapy*, *152*, 113238. DOI: 10.1016/j.biopha.2022.113238 PMID: 35687909

Rubab, Z. a, & Zulkipli, I. N. (2024). Genetic variations and lifestyle in oxidative stress. In *Fundamental Principles of Oxidative Stress in Metabolism and Reproduction* (pp. 93–103). Elsevier. DOI: 10.1016/B978-0-443-18807-7.00006-5

Salekeen, R., Haider, A. N., Akhter, F., Billah, M. M., Islam, M. E., & Didarul Islam, K. M. (2022). Lipid oxidation in pathophysiology of atherosclerosis: Current understanding and therapeutic strategies. *International Journal of Cardiology. Cardiovascular Risk and Prevention*, *14*, 200143. DOI: 10.1016/j.ijcrp.2022.200143 PMID: 36060286

Schmidt-Trucksäss, A., Lichtenstein, A. H., & von Känel, R. (2024). Lifestyle factors as determinants of atherosclerotic cardiovascular health. *Atherosclerosis*, *395*, 117577. DOI: 10.1016/j.atherosclerosis.2024.117577 PMID: 38852021

Schneider, R. H., & Carr, T. (2014). Transcendental Meditation in the prevention and treatment of cardiovascular disease and pathophysiological mechanisms: An evidence-based review. *Advances in Integrative Medicine*, *1*(3), 107–112. DOI: 10.1016/j.aimed.2014.08.003 PMID: 35813238

Senoner, T., & Dichtl, W. (2019). Oxidative Stress in Cardiovascular Diseases: Still a Therapeutic Target? *Nutrients*, *11*(9), 2090. DOI: 10.3390/nu11092090 PMID: 31487802

Soomro, Q. H., & Charytan, D. M. (2024). New Insights on Cardiac Arrhythmias in Patients With Kidney Disease. *Seminars in Nephrology*, *44*(2), 151518. DOI: 10.1016/j.semnephrol.2024.151518 PMID: 38772780

Speth, R. C. (2022). Renin-Angiotensin-Aldosterone System. In *Comprehensive Pharmacology* (pp. 528–569). Elsevier., DOI: 10.1016/B978-0-12-820472-6.00160-2

St. Clair, D., Kasarskis, E., St. Clair, W., & Chaiswing, L. (2024). Superoxide Dismutase and Catalase. In *Reference Module in Biomedical Sciences*. Elsevier., DOI: 10.1016/B978-0-323-95488-4.00063-2

Wang, H., Dou, S., Zhu, J., Shao, Z., Wang, C., & Cheng, B. (2021). Regulatory effects of ghrelin on endoplasmic reticulum stress, oxidative stress, and autophagy: Therapeutic potential. *Neuropeptides*, *85*, 102112. DOI: 10.1016/j.npep.2020.102112 PMID: 33333485

Wang, S., Liu, Y., Liu, J., Tian, W., Zhang, X., Cai, H., Fang, S., & Yu, B. (2020). Mitochondria-derived methylmalonic acid, a surrogate biomarker of mitochondrial dysfunction and oxidative stress, predicts all-cause and cardiovascular mortality in the general population. *Redox Biology*, *37*, 101741. DOI: 10.1016/j.redox.2020.101741 PMID: 33035815

Wu, Y., Liu, H., & Wang, X. (2021). Cardioprotection of pharmacological postconditioning on myocardial ischemia/reperfusion injury. *Life Sciences*, *264*, 118628. DOI: 10.1016/j.lfs.2020.118628 PMID: 33131670

Yaghoobi, A., Rezaee, M., Behnoush, A. H., Khalaji, A., Mafi, A., Houjaghan, A. K., Masoudkabir, F., & Pahlavan, S. (2024). Role of long noncoding RNAs in pathological cardiac remodeling after myocardial infarction: An emerging insight into molecular mechanisms and therapeutic potential. *Biomedicine and Pharmacotherapy*, *172*, 116248. DOI: 10.1016/j.biopha.2024.116248 PMID: 38325262

Zahedi, N., Pourajam, S., Zaker, E., Kouhpayeh, S., Mirbod, S. M., Tavangar, M., Boshtam, M., Hatami Kahkesh, K., Qian, Q., Zhang, F., Shariati, L., Khanahmad, H., & Boshtam, M. (2023). The potential therapeutic impacts of trehalose on cardiovascular diseases as the environmental-influenced disorders: An overview of contemporary findings. *Environmental Research*, *226*, 115674. DOI: 10.1016/j.envres.2023.115674 PMID: 36925035

Zhang, K., Li, Y., Lin, X., Daneshar, M., Karamian, F., & Li, M. (2024). Effect of cinnamon supplementation on blood pressure, oxidative stress, and inflammatory biomarkers in adults: An umbrella review of the meta-analyses of randomized controlled trials. *Nutrition, Metabolism, and Cardiovascular Diseases*, *34*(12), 2659–2668. Advance online publication. DOI: 10.1016/j.numecd.2024.08.009 PMID: 39299867

Zhang, M., Li, J., & Hu, W. (2024). The complex interplay between ferroptosis and atherosclerosis. *Biomedicine and Pharmacotherapy*, *178*, 117183. DOI: 10.1016/j.biopha.2024.117183 PMID: 39079265

Zhang, X., Zheng, Y., Wang, Z., Gan, J., Yu, B., Lu, B., & Jiang, X. (2023). Melatonin as a therapeutic agent for alleviating endothelial dysfunction in cardiovascular diseases: Emphasis on oxidative stress. *Biomedicine and Pharmacotherapy*, *167*, 115475. DOI: 10.1016/j.biopha.2023.115475 PMID: 37722190

Zhou, Y., Suo, W., Zhang, X., Lv, J., Liu, Z., & Liu, R. (2022). Roles and mechanisms of quercetin on cardiac arrhythmia: A review. *Biomedicine and Pharmacotherapy*, *153*, 113447. DOI: 10.1016/j.biopha.2022.113447 PMID: 36076562

Zhu, Y., Lai, Y., Hu, Y., Fu, Y., Zhang, Z., Lin, N., Huang, W., & Zheng, L. (2024). The mechanisms underlying acute myocardial infarction in chronic kidney disease patients undergoing hemodialysis. *Biomedicine and Pharmacotherapy*, *177*, 117050. DOI: 10.1016/j.biopha.2024.117050 PMID: 38968794

Chapter 10
Reactive Oxygen Species in Cancer

Odangowei Inetiminebi Ogidi
https://orcid.org/0000-0002-9961-893X
Bayelsa Medical University, Nigeria

Simon Akpomemera Erho
Bayelsa Medical University, Nigeria

Okiemute Omu
https://orcid.org/0009-0007-5864-836X
Federal Polytechnic, Ekowei, Nigeria

ABSTRACT

Reactive oxygen species (ROS), which are extremely bioactive chemicals, have been extensively researched in many forms of cancer. ROS are regarded as typical consequences of several biological processes. Cancer cells generally demonstrate elevated basal levels of ROS in comparison to normal cells due to an imbalance between oxidants and antioxidants. ROS play a dual function in cell metabolism. When present in low to moderate amounts, ROS serve as signal transmitters that trigger cell proliferation, migration, invasion, and angiogenesis. Conversely, elevated quantities of ROS result in harm to proteins, nucleic acids, lipids, membranes, and organelles, ultimately resulting in the demise of cells. Comprehensive research has demonstrated that anticancer treatments that alter levels of ROS. Hence this chapter covers the molecular processes and oncogenic roles that regulate levels of ROS and their relevance in the development of cancer treatment options. It also offer insight into the future advancement of efficient agents that control the redox system for the purpose of treating cancer.

DOI: 10.4018/979-8-3693-7919-6.ch010

INTRODUCTION

Cancer, a leading cause of death worldwide, is a major challenge in the pursuit of increasing life expectancy (Ogidi & Tawariowei, 2023). Out of the estimated 19.3 million new cases of cancer expected in 2025, around 10 million people died from the disease (Ogidi, 2024; Sung et al., 2021). Despite the proliferation of innovative therapies and pharmaceuticals, cancer continues to be a prominent concern and leading cause of mortality. Extensive study has focused on the correlation between reactive oxygen species (ROS) and several human disorders, including cancer. Reactive oxygen species (ROS) can arise from various biological processes, with air consumption being just one of them. The scientific community refers to a group of reactive and unstable oxygen derivatives as "ROS". Hydroxyl radicals, hypochlorous acid, singlet oxygen, and superoxide anion are all constituents of this category (Senal et al., 2012).

Both healthy and malignant cells depend on these molecules as secondary messengers in cellular communication, and they are crucial for a diverse range of biological functions. According to an increasing number of research, reactive oxygen species (ROS) can exert both advantageous and detrimental effects on cancer cells. At low to moderate levels, reactive oxygen species (ROS), which are signalling molecules, stimulate the growth, spread, invasion, blood vessel formation, and resistance to treatment of cancer cells. To control the movement, growth, specialisation, and death of cancer cells, it is crucial to maintain optimal levels of reactive oxygen species (ROS). Conversely, cancer cells are eliminated when exposed to elevated quantities of reactive oxygen species (ROS) (Cairns et al., 2011).

As a result of the ability of ROS to go in two directions, eight different approaches have demonstrated therapeutic promise in their ability to raise or lower ROS levels in cancer cells. Antioxidants are believed to have potential in the prevention and treatment of cancer due to their capacity to reduce ROS levels and alleviate oxidative stress (Okon et al., 2015). Through the inhibition of intracellular reactive oxygen species (ROS), our investigation revealed that the oral iron chelator Deferasirox (DFX) effectively halted the process of programmed cell death in MM cells. DFX exerts anti-multiple myeloma (MM) actions by inhibiting the phosphorylation of proline-rich tyrosine kinase 2 and reducing the formation of reactive oxygen species (ROS) (Chio et al., 2017). Some evidence suggests that antioxidants such as N-acetyl l-cysteine (NAC) and vitamin E might potentially stimulate the growth and spread of cancer. Due to their modified genomes, increased metabolism, and reduced oxygen levels, cancer cells produce a greater amount of reactive oxygen species (ROS) in comparison to normal cells (Yang et al., 2018).

Previous studies have shown that when levels of reactive oxygen species (ROS) increase, cancer cells react by activating antioxidant pathways, which enhance the removal of ROS (Sayin et al., 2014; Piskounova et al., 2015). Cancer cells exhibit heightened sensitivity to therapies that enhance the formation of reactive oxygen species (ROS) due to their inherent abundance of endogenous ROS (Kamihara et al., 2016). Research indicates that the administration of chemotherapy leads to an increased emission of reactive oxygen species (ROS) by cancer cells (Gorrini et al., 2013). Previous studies have demonstrated that increased levels of intracellular reactive oxygen species (ROS) in CC and MM cells are linked to the initiation of apoptosis and the inhibition of cell growth (Perillo et al., 2020; Schafer et al., 2009). Therefore, cancer treatments that specifically aim to induce ROS may be the most optimal choice. Hence this chapter covers the molecular processes and oncogenic roles that regulate levels of reactive oxygen species (ROS) and their relevance in the development of cancer treatment options. This chapter also offer valuable insights into the future advancement of efficient agents that control the redox system for the purpose of treating cancer.

Objectives of the Chapter

The aims of this chapter are as follows:

(1) To discuss the overview of Reactive oxygen species.
(2) To explore ROS, Cancer risk factors and carcinogenesis.
(3) To evaluate specific functions of ROS in cancer.
(4) To review ROS in cancer treatment.
(5) To discuss the future prospect of ROS in cancer.

Organization of the Chapter

The rest of this chapter is organized as follows: Section 2 explores reactive oxygen species overview Section 3 discusses the reactive oxygen species, cancer risk factors and the process of carcinogenesis Section 4 transverse the specific functions of reactive oxygen species in cancer Section 5 explores the various roles of reactive oxygen species in treatment Section 6 reviews the future prospects of reactive oxygen species in cancer. Finally, a concluding section summarizes the key takeaways and highlights recommendation.

REACTIVE OXYGEN SPECIES OVERVIEW

The partial reduction of molecular oxygen (O2) when it reacts with other molecules is what causes reactive oxygen species (ROS) to be formed (Ahmad et al., 2017). When reactive oxygen species (ROS) and nitrous oxide (NO) combine, the resulting product is reactive nitrogen species (RNS) (**Figure 1**). Several physiological processes rely on reactive nitrogen species (RNS) and reactive oxygen species (ROS), such as cell proliferation, adaptive immunity, oxygen sensing, and differentiation (Breitenbach et al., 2018; Halliwel, 2006). To do this, they subject big molecules to reversible oxidative modifications. Nevertheless, oxidative stress might harm cells if these compounds were to be present in excess (Moloney et al., 2018). It is possible to classify reactive oxygen species (ROS) as radicals or non-radicals according to their characteristics. They are categorised based on their capacity to hold unpaired electrons (Ahmad et al., 2017; Halliwel, 2006). Energy induces a reorganisation of the electrons in O2, resulting in the formation of two highly reactive states: the sigma and delta singlet states. These states enhance the oxidising properties of oxygen (Halliwel, 2006). Red blood cells undergo oxidation of several biological components. For instance, while HO can oxidise all DNA bases, singlet oxygen mostly targets guanine. Reactive oxygen species (ROS) have the ability to oxidise protein sulfhydryl (-SH) groups, leading to the initiation of lipid peroxidation. Out of all these species, HO exhibits one of the highest levels of reactivity towards amino acid residues (Halliwel, 2006). Research has demonstrated that O2 is capable of inducing the oxidation and inactivation of certain metabolic enzymes, particularly those that contain sulphur and iron clusters. As a result of this process, iron is released and HO is generated (Halliwel, 2006).

Figure 1. Formation of RNS

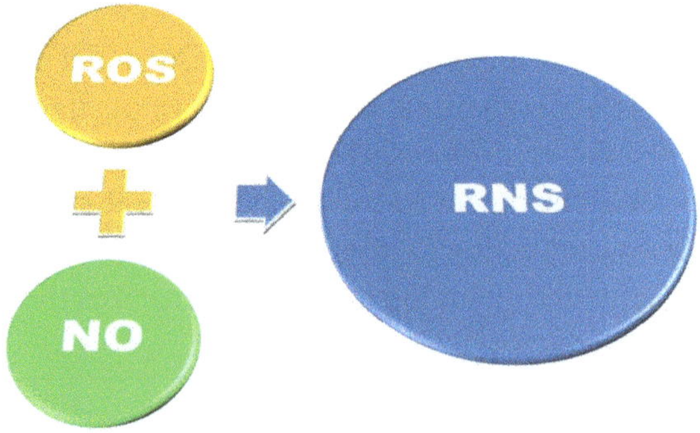

Several studies have established a connection between increased amounts of reactive oxygen species (ROS), originating from mitochondria and other sources, and the cancer-causing properties of cancer cells in various forms of cancer (Moloney et al., 2018). Research indicates that cancer cells may manipulate their antioxidant capacity and formation of reactive oxygen species (ROS) to promote the progression of the disease (Moloney et al., 2018; Sarmiento-Salina et al., 2019). Hemine and flavin are constituents of a category of enzymatic membrane protein complexes known as NOX proteins. They are present in various cellular compartments, such as invadopodia, mitochondria, plasma membranes, focal adhesions, and the nucleus. Their sole known function is to catalyse the conversion of NADPH into molecular oxygen, resulting in the production of O2 (Block et al., 2012; Skonieczna, 2017; Vermot, 2021). There are seven different types of catalysis that have been found, which include dual oxidases (DUOX1-2) and Nox (NOX1-5) catalytic sites. Significant variations can be observed among the isoforms in terms of their biochemistry, structure, and subcellular localisation (Block et al., 2012; Skonieczna, 2017). Certain NOX species, such as DUOX 1-2 or human NOX4, have the ability to convert O_2 into H_2O_2 either independently or in conjunction with a comparable superoxide dismutase (Breitenbach et al., 2018; Skonieczna, 2017).

In order to combat the levels of reactive nitrogen species (RNS) and reactive oxygen species (ROS), the antioxidant system employs a web of enzymes and non-catalytic molecules. Excessive oxidation can damage cells, but this system protects them. When there are more oxidants than antioxidants, a condition known as oxidative stress develops. Alterations to redox signalling pathways and damage to proteins, phospholipids, and nucleic acids are all possible outcomes of this disorder. In severe situations, it can even cause cell death (Hayes et al., 2020). Superoxide dismutases (SODs) exist in three distinct forms: one is located extracellularly (SOD3), while the other two are situated within mitochondria (SOD2). Symplasmic oxidoreductase 1 (SOD1) represents the primary state of the enzyme. These enzymes seize free radicals in the presence of oxygen. Molecular oxygen and hydrogen peroxide can be more easily generated from oxygen gas with the assistance of these molecules. Glutathione (GSH) can either be reduced to water by GPX or hydrolysed into water and oxygen by CAT. The latter option allows for its oxidation to glutathione disulphide (GSSG) through the action of hydrogen peroxide (Moloney et al., 2018; Halliwell, 2006; Hayes et al., 2020). Mice who possessed one functional copy of the SOD2 gene exhibited a heightened susceptibility to cancer, as well as elevated levels of oxidative damage in their mitochondria and nuclear DNA. In addition, older mice with a deficiency in SOD1 exhibited cognitive impairment, hearing loss, and a higher susceptibility to developing liver cancer (Halliwell, 2006). Reactive oxygen species (ROS) have a direct influence on the antioxidant system, as a single oxygen molecule (O2) can oxidise and render the enzymes SOD and CAT inactive (Hayes et al., 2020).

Catalase (CAT) has a crucial function in controlling reactive nitrogen species (RNS) by oxidising NO or converting ONOO- into nitrite (NO2 –) (Moloney et al., 2018). The glutathione peroxidase (GSH) system is an essential physiological antioxidant mechanism that regulates signalling pathways and maintains the balance of redox reactions by reducing disulphide linkages to cysteines in proteins that are regulated by redox. GSH reserves in various locations have distinct roles, as GSH is produced in the cytosol and is present in organelles (Mari, 2020). In addition to b-carotene and plant polyphenols, endogenous antioxidants (**Figure 2**) include bilirubin, melanin, uric acid, vitamins E and C, and a few additional substances (Hayes et al., 2020).

Figure 2. Types of endogenous antioxidants

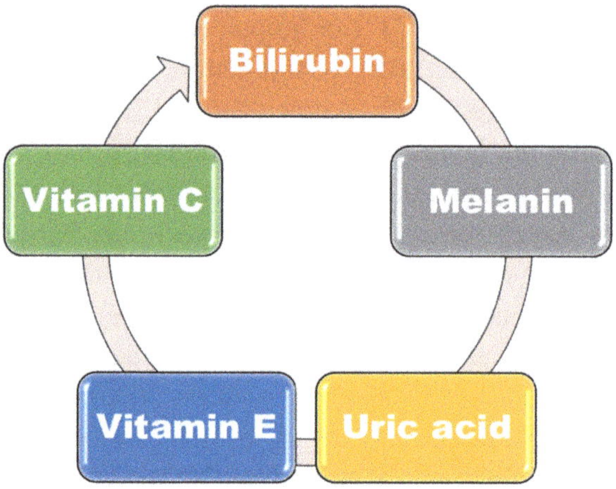

ROS, CANCER RISK FACTORS AND CARCINOGENESIS

Possible factors contributing to the global increase in cancer incidences include an ageing population, shifts in the occurrence and exposure patterns of the disease, and greater exposure to risk factors. The cancer risk factors can be broadly classified into three main categories as shown in **Figure 3**: endogenous, internal, and external. Due to random errors in DNA replication, inherent risk factors gradually accumulate in stem cells derived from healthy tissues. Modifiable endogenous risk factors encompass several elements such as inflammation, hormones, growth factors, genetic predisposition, ageing, and changes in the DNA repair mechanism. Biological, chemical, and radioactive contaminants as well as viruses that cause tumors are examples of exogenous risk factors and modifiable behavioral choices like smoking, lack of physical activity, and a bad diet (Wu et al., 2018).

The primary risk factors associated with cancer-related mortality are obesity, tobacco use, exposure to infectious agents, lack of physical activity, excessive alcohol intake and poor dietary habits. The effectiveness of cancer prevention measures can be largely attributed to the identification of specific carcinogens, such as cigarette smoke is associated with lung cancer, sun exposure is for skin cancer, HPV for cervical cancer, and HBV (Wu et al., 2018). The hepatitis virus is associated with hepatocellular carcinoma (HCC), while Helicobacter pylori is associated with HCC of the stomach (Wu et al., 2018). Studying the impact of risk variables on cellular transformation can enhance the development of more efficient cancer prevention techniques. Cells emit ROS in response to several modifiable factors associated with the development of cancer. During the initial stages of cancer formation, it is widely recognized that increased levels of ROS promote the growth of tumors (Wu et al., 2018).

Studies indicate that chemicals with carcinogenic properties, such as tobacco, inhalable microfibers, and tiny particles, initiate a series of metabolic reactions in the body. These processes activate proteins and cellular systems. Additionally, ROS are produced which leads to oxidative stress. Research by Valavanidis et al. (2013) has established a correlation between inflammation, cell death, cell division, and cell specialization. Increasing research indicates that reactive oxygen species (ROS) may operate as a potential carcinogen in tobacco and cigarette smoke. Peroxides have the ability to damage DNA within cells. Mutations defined by C to A alterations are commonly found in liver, squamous cell, small, head, and neck cancers. Polycyclic hydrocarbons present in tobacco smoke are not only responsible for causing these lesions, but they are also considered to trigger the creation of DNA adducts that lead to the observed mutational pattern (Alexandrov et al., 2013).

Based on the idea, the extent to which damaged mitochondria are cleared through mitophagy and the level of stress play a role in determining the likelihood of correcting phenotypic alterations, ageing, cell death, and DNA damage. Genetic mutations in tumor suppressor genes or oncogenes might increase the likelihood of malignant transformation after the restoration of a fast-growing characteristic, in response to DNA damage caused by exposure to carcinogens. Several factors associated with the generation of reactive oxygen species (ROS) have been connected to the development of malignancies, such as lung cancer (Prasad et al., 2017). Studies using animal models of lung cancer with mutations in the RAS or RAF genes have shown that administering antioxidants such as NAC or vitamin E can promote tumor growth by inhibiting the ROS-p53 pathway (Sayin et al., 2014).

For instance, a meta-analysis demonstrated that those who consumed higher amounts of fruits and vegetables or flavonoids had a slightly reduced likelihood of developing lung cancer, particularly among smokers (Vieira et al., 2016). Several research suggest that antioxidants help reduce the occurrence of lung cancer, however one study yielded inconclusive results. Indeed, individuals who exceed the recommended intake of retinol are at a higher risk of developing lung cancer, particularly small cell carcinoma (Narita et al., 2018). Given these findings, it is crucial to ascertain whether reactive oxygen species (ROS) have different roles in lung cancer depending on hereditary factors. This information is vital for comprehending the significance of reactive oxygen species (ROS) in the management and prevention of lung cancer, as well as for ascertaining their applicability to other forms of cancer (Narita et al., 2018).

The Epstein-Barr virus, human papillomavirus, hepatitis B and C viruses, Merkel cell polyomavirus, and human herpes virus 8 are pathogens that have been linked to the onset of cancer (Vieira et al., 2016). Synonyms for Sinensis encompass gastrointestinal lymphoid tumors (HG) and Helicobacter pylori (C.), along with liver cancer (cholangiocarcinoma) and bladder cancer (Schistosoma haemotobium). A sickness, such as cancer might different H. pylori which may enhance the frequency of genetic changes or facilitate the activation of cellular growth signals, both of which can indirectly lead to the development of cancer. Several investigations have demonstrated that H. Hepatocellular carcinoma (HCC), viral hepatitis (VHL), and stomach cancer (GC) are all impacted by the presence of pylori bacteria. Research on cervical cancer has explicitly shown that certain viruses, like HPV, can directly induce cancer by generating proteins that weaken cell barriers (Ewald et al., 2015).

Administration of antioxidants is crucial in the prevention of cancer during viral oncogenesis, as demonstrated in the preceding findings. The probability of having cancer is most significantly associated with increasing age. In summary, the free radical hypothesis of ageing states that reactive oxygen species (ROS) facilitate the accumulation of oxidative damage in living organisms (Santos et al., 2018). Recent studies have raised doubts about the connection between reduced antioxidant capacity and an elevation in reactive oxygen species (ROS), which was previously believed to lead to a shorter lifespan. However, the findings have not provided a definitive answer. If organisms are capable of enduring and surviving for extended periods of time, then there is no necessity for scavengers that consume dead organic matter. Research suggests that longevity can be increased by interventions that stimulate the production of reactive oxygen species (ROS), which can be accomplished by genetic modifications, dietary changes, or medicine. To do this, it is important to activate signaling pathways that initiate an adaptive response and mitochondrial oxidative phosphorylation (Santos et al., 2018).

Studies indicate that both humans and animals generate a higher amount of reactive oxygen species (ROS) as they grow older, in comparison to when they were younger. However, it should be noted that this increase in ROS production is not necessarily the primary factor contributing to the ageing process. To effectively control the pathways that impact age-related diseases such as cancer and neurological disorders, there needs to be an increase in production (Barcelos et al., 2019). The amount and function of DNA repair proteins in living creatures see a sharp reduction as we age. Research indicates that mitochondrial DNA (mtDNA) mutations naturally increase with age. This implies that mutations have a higher probability of building up in the DNA of this organism, both in its genome and its mitochondria. This vulnerability can be worsened by other cancer-causing agents, such as reactive oxygen species (ROS) or chemicals that cause mutations (Santos et al., 2018).

Recent findings indicate that mutational markers have the ability to differentiate between various mechanisms that cause damage to DNA as a result of oxidative stress. Signature 18, mostly linked to reactive oxygen species (ROS), is marked by mutations involving the conversion of C to A. The genetic changes resulting from reactive oxygen species (ROS) damaging DNA and specifically oxidizing guanine to 8OHdG are referred to as guanine oxidation to 8OHdG (Poetsch, 2020). Although pancreatic, breast, and stomach cancers have the characteristic 18 at a lower extent, neuroblastoma is one of the malignancies that prominently exhibits it (Viel et al., 2017). This further substantiates the notion that reactive oxygen species (ROS) play a crucial role in the proliferation and sustenance of cancer cells across different forms of cancer. Discovering efficacious treatment interventions that can mitigate the occurrence of cancer is of utmost importance, as factors linked to an increased susceptibility to cancer all generate reactive oxygen species (ROS) (Viel et al., 2017).

The majority of research on chemo-preventive medicines has been concentrated on natural substances (Liu et al., 2015; Wang et al., 2014). The reason is that there is a lower incidence of cancer when individuals consume a diet that is rich in fruits and vegetables. There exists a diverse range of chemicals that can operate as antioxidants or regulate the activity of antioxidant enzymes. Sulforaphanes, glucosinolates, xanthones, and polyphenols are constituents of this chemical category. Their ability to reduce inflammation is an additional mechanism by which they may affect the signaling pathways involved in cancer (Patra et al., 2021; Li et al., 2017; Merlin et al., 2021). It is crucial to determine if the cancer-preventive advantages of these compounds are mainly attributed to their antioxidant activities or a mix of other pathways. This will be useful in assessing the level of necessity for new therapeutic procedures or medications in effectively preventing cancer (Li et al., 2017; Merlin et al., 2021).

Figure 3. Cancer risk factors

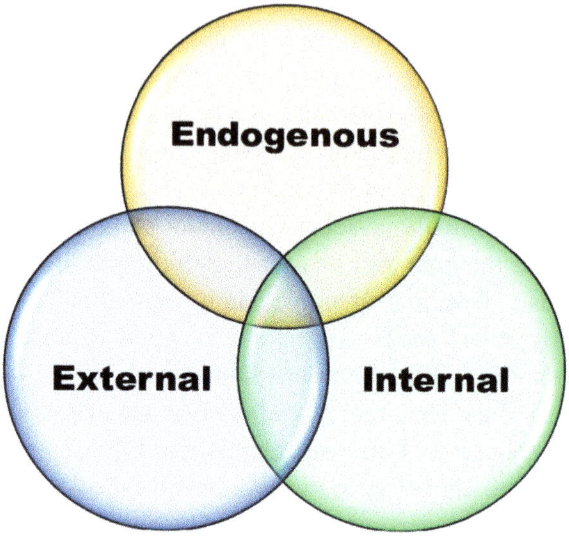

SPECIFIC FUNCTIONS OF ROS IN CANCER

Various studies have demonstrated that oxidative stress can impact many aspects of cancer cell behavior through signaling pathways (Storz, 2005). ROS play a major role in various pathways, such as the preservation of cancer stem cells and the creation of new blood vessels. They also regulate cell adhesion, migration, survival, and programmed cell death (Storz, 2005).

The specific functions are shown in **Figure 4.**

Figure 4. Specific functions of ROS in cancer

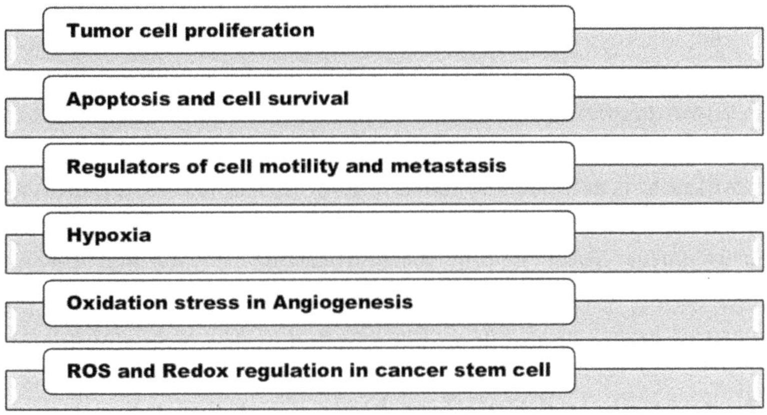

ROS in Tumor Cell Proliferation

Hydrogen peroxide and superoxide have the ability to stimulate cell proliferation in various types of malignancies, even when present in small amounts (Storz, 2005). Breast cancer is characterized by increased cellular oxidative stress, which is a result of the movement of oestrogen from the ovaries to the mitochondria (Reddy et al., 2007; Felty et al., 2005). Reactive oxygen species (ROS), which are microorganisms found in mitochondria, regulate both cell development and the ageing process. Mitochondrial ROS control is facilitated by the action of MnSOD, which simplifies the process (Sarsour et al., 2008). Reducing MnSOD activity stimulates cell development by increasing superoxide levels and decreasing hydrogen peroxide levels, which are advantageous for cells. Conversely, cells undergoing fast division exhibit increased production of hydrogen peroxide when the activity of MnSOD rises, hence accelerating cell death. Sodium arsenite, an environmental pollutant, accelerates the production of ROS and the progression of the S phase, all of which contribute to the increased cell proliferation in breast cancer (Ruiz-Ramos et al., 2009).

Furthermore, benzo(a)pyrene quinones (BPQs) expedite the growth of mammary epithelial cells by augmenting the generation of superoxide and hydrogen peroxide, mimicking their effects as growth factor signals (Burdick et al., 2003). Conversely, antioxidants inhibit the proliferation of cancer cells. An inherent distinction between regular cells and pancreatic cancer cell lines lies in the existence of intrinsic oxidative stress (Storz 2005). There is empirical data indicating a positive association between increased levels of reactive oxygen species (ROS) and augmented cellular proliferation. An effective method of inhibiting the growth of pancreatic cancer cells is through the continuous aberrant synthesis of the potent antioxidant enzyme MnSOD (Cullen et al., 2003). An inverse link has been seen in certain pancreatic cancer cell lines between the rate at which the cells divide and the levels and functions of endogenous glutathione peroxidase, catalase, endogenous manganese superoxide dismutase, and copper/zinc superoxide dismutase (Cullen et al., 2003; Lewis et al., 2005). Ataxia telangiectasia mutated (ATM) is a protein involved in the regulation of the cell cycle and is activated by reactive oxygen species (ROS). Individuals with insufficient levels of antioxidant enzymes in their bodies are prone to experiencing significant oxidative stress. Results are comparable in mice that have undergone genetic engineering to disable the ATM gene. Antioxidants have the potential to mitigate these adverse consequences (Browne et al., 2004; Reichenbach et al., 2002).

Ros in Apoptosis and Cell Survival

The excessive buildup of intracellular reactive oxygen species (ROS) can induce cell cycle arrest, senescence, and apoptosis in cancer cell lines. Several approaches can be employed to achieve this, such as cancer therapy, elimination of cells harboring antioxidants, or stimulation of immune cells to produce reactive oxygen species (ROS). Apoptosis is linked to an increase in mitochondrial oxidative stress, which causes the release of cytochrome C. The activation of caspases, which are proteins responsible for inducing cell death, is irreversible (Cadenas, 2004; Simon et al., 2000). This system has the potential to initiate proapoptotic signaling through the Rac-1/NADPH oxidase pathway, resulting in the production of superoxide (Chung et al., 2003). In response to cues that induce cell death, mitochondria initiate the activation of cJun N-terminal kinases (JNKs) by generating superoxide and nitric oxide (Cadenas, 2004; Storz, 2007). ROS stimulation leads to the downregulation and phosphorylation of anti-apoptotic proteins such as Bcl-2 and Bcl-XL, through the activation of JNKs (Cadenas, 2004). Studies have shown that Bcl-2 and Bcl-XL help reduce the production of reactive oxygen species (ROS) and safeguard cells from cell death caused by ROS (Gottlieb et al., 2000; Simon et al., 2000).

Due to an upsurge in Bax production, JNK alters the composition of the Bax/Bcl-2 complex, leading to the creation of Bax homodimers. Ultimately, this results in the impairment or complete breakdown of the mitochondrial membrane (Lee et al., 2008; Qanungo et al., 2005; Zhang et al., 2008). The initiation of the apoptotic signaling cascade is facilitated by the MAPK family protein P38, which promotes an elevation in the formation of reactive oxygen species (ROS). K-1, a regulator of the abortion signal, controls the functions of p38 and JNK1. The function of Ask-1 is controlled by its association with thioredoxin. Protein thioredoxin is regulated by redox reactions. After being stripped down, it attaches to Ask-1 and inhibits its function (Takeda et al., 2003). Furthermore, several signaling proteins, such as p66Shc, p53, and forkhead transcription factors (FOXO3a), have been associated with the onset of apoptosis induced by reactive oxygen species (ROS), in addition to the signaling cascades activated by Ask-1 (You et al., 2006). An intriguing concept suggests that p53-deficient clones have the potential to develop into cells that are capable of surviving apoptosis, particularly when cancer cells experience prolonged oxidative stress (Storz, 2005).

The TNF receptor I is believed to induce the generation of mitochondrial ROS, activation of caspases, and ultimately cell death, as described in the primary mechanism by (Sorz et al., 2004). TRAF4, a component of the TNFα signaling cascade, has the ability to activate JNK by forming a bond with the NADPH oxidase complex. The discovery that death receptors can trigger cellular ROS production through many mechanisms is highly significant. It is important to mention that TNF-κB promotes anti-apoptotic signaling by stimulating the production of MnSOD and catalase, leading to oxidative stress (Sorz et al., 2004). Elevated concentrations of reactive oxygen species (ROS) trigger the sequence of events that ultimately leads to cell death in the previously mentioned signaling pathways. Recent study, however, shows that cells can initiate survival signals even when exposed to exceedingly low amounts of oxidative stress. Protein kinase D1 (PKD1) controls this process of survival by regulating the levels of reactive oxygen species (ROS) (Sorz et al., 2004; Chiu et al., 2007; Mihailovic et al., 2004; Song et al., 2009).

ROS as Regulators of Cell Motility and Metastasis

Prior to intravenous injection, hydrogen peroxide therapy promoted cancer cell metastasis in mice (Pelicano et al., 2009). Furthermore, in several subsets, the original cells showed lower levels of reactive oxygen species (ROS) when compared to the stationary MCF-7 breast cancer cell line. The highly transient orthotopic tumors produced by these cell lines spread to several organs, including the liver, spleen, and lungs (Pelicano et al., 2009). In addition, metSOD antioxidant enzyme levels are reduced, and enzyme efficacy is decreased in cells with metastasized pancreatic cancer and breast cancer (Lewis et al., 2005; Hitcheler et al., 2008). The intracellular redox status regulates critical steps in metastasis, as indicated before. The power to migrate and invade better, live without anchoring, decrease cell adhesion to the extracellular matrix, and cross blood vessel walls are all part of this. Integrin interaction to the extracellular matrix is essential for cell adhesion and motility. Integrins increase oxidant levels by upregulating cyclooxygenase-2 expression (Broom et al., 2006). Utilizing mitochondria and 5-lipoxygenases (5-LOX) is an additional mechanism they employ (Chiarugi et al., 2007; Svineng et al., 2008).

An increase in mitochondrial reactive oxygen species (ROS) was associated with the initial cell-ECM contact in this scenario, while an increase in cytosolic ROS was found to affect the subsequent cytoskeleton reconfiguration and the formation of actin stress filaments (Chiarugi et al., 2007; Taddei et al., 2007). In the described processes, mitochondria generate reactive oxygen species (ROS), which target focused adhesion kinase (FAK) and SHP-2. In contrast, reactive oxygen species (ROS) found in the cytosol are engineered to engage with particular proteins and structures, such as receptor tyrosine kinases, Src-family kinases, FAK, and β-actin, among other potential targets. It is thought that FAK, unlike phosphatases and Src, which are activated directly by oxidation, is tyrosine phosphorylated via upstream signaling pathways (Ben-Mahdi et al., 2000).

Initiation of focal adhesion formation by surface-attached cells involves Src and FAK. This process guarantees that cells can migrate, disseminate, and survive by preventing anoikis. Untreated cells require adhesion to the ECM for cell division. The proper progression of mitosis is ensured by reactive oxygen species (ROS), which function as secondary messengers (Chiarugi et al., 2007, 2008). The interaction between integrins and growth factors (GF) results in a coordinated signaling cascade that starts an oxidative surge. An increase in reactive oxygen species (ROS) within the mitochondria is what drives Rac-1 activation and, consequently, this spike. Studies have shown that this to be true (Werner et al., 2002). Consequently, PTPs are unable to function as intended due to oxidative inhibition. As a final result, several structural proteins and protein tyrosine kinases, such Src, are activated. Cell adhesion, proliferation, and dispersion are all enhanced as a result. Anoikis, a kind of cell suicide, is triggered in non-transformed cells when cell-matrix adhesion is disrupted. Unlike non-transformed cells, cancer cells are able to bypass this process in order to boost cell proliferation and independence from anchoring. The ability of cancer cells to evade cell death, also known as anoikis, allows them to persist in abnormal environments and perhaps metastasize (Werner et al., 2002).

It is likely that an increase in the internal production of ROS is the reason cancer cells cease responding to cell attachment signals. An increase in oxidative stress appears to be acting similarly to the way autocrine/adhesive signals are regulated in healthy cells by growth factor and integrin signaling. An example would be the redox-regulated anoikis resistance in prostate cancer cells, which is mediated by Src and the EGF receptor (Giannoni et al., 2009). Surface attachment stops being a factor in cell development pathways, and unregulated cell division takes place. Also, since there are no connections between cancer cells and the extracellular matrix (ECM), the cells are able to avoid apoptotic responses and anoikis-promoting signals. It is during the epithelial-mesenchymal transition (EMT) that cells move to new locations once they are freed from the basement membrane. This process necessitates an increase in metalloproteinase (MMP) production to degrade the proteins in the basal membrane. Intracellular reactive oxygen species (ROS), particularly H_2O_2, and Rac1b RhoGTPase activation were upregulated in mammary epithelial cells of mice exposed to the stromal protease MMP-3. This protease is overexpressed in breast tumors. As a result, EMT was promoted (Radisky et al., 2005).

Hypoxia as a Factor Leading to Tumor Progression

As a tumor grows, cancer cells alternate between phases of hypoxia and hyperoxia (Dewhirst et al., 2008; Hocket et al., 2001; Pani et al., 2009). Cell death can occur as a result of persistent hypoxia, which is characterized by an insufficient oxygen supply. Cancer cells can adapt to hypoxia by utilizing the "Warburg effect," which entails a metabolic transition to glycolysis (Ganapathy et al., 2009). Cancer cells and healthy tissues exhibit distinct variations in their energy utilization. The majority of cells have a preference for glucose as their main source of nourishment. Anaerobic glycolysis is a metabolic pathway that is typically activated in cells that have low oxygen levels or limited mitochondrial activity (Ganapathy et al., 2009). Although cancer cells typically cannot undergo oncogenic transformation, mitochondrial failure, or cancer due to mutations, they can nevertheless transition to anaerobic metabolism even when oxygen levels are normal (Pani et al., 2009; Hsu et al., 2008). The capacity of tumor cells to acclimatize to hypoxia, which refers to low levels of oxygen, is essential for the initiation of aggressive tumor proliferation and the acquisition of malignant traits (Harris, 2002).

The lack of oxygen stimulates the activation of many transcription factors, including hypoxia-inducible factor-1 (HIF-1). HIF-1 is composed of two subunits: HIF-1α and HIF-1β. Under typical developmental circumstances, the VHL ubiquitin ligase and oxygen-dependent prolyl hydroxylases (PHDs) assist in controlling HIF-1 activity and promoting its degradation through the proteasome (Kaelin et al., 2008). Conversely, decreased oxygen levels trigger the activation of HIF-1 transcription. Research has demonstrated that in conditions of low oxygen levels, MnSOD inhibits the activation of the HIF-1α transcription factor in human breast cancer cells. Wang et al. (2005) suggests that superoxide may have a role in the accumulation of HIF-1α. Both types of reactive oxygen species (ROS) were capable of enhancing HIF-1α levels, as the enhanced generation of H_2O_2 also facilitated its storage (Wang et al., 2005). Increasing data suggests that elevated levels of HIF-1α expression are associated with a worse prognosis and more aggressive cancer cells. As stated by Pani et al. (2009), cancer cells have the ability to adjust to low levels of oxygen due to the control exerted by HIF-1 on the expression of genes associated to glycolysis and the inhibition of mitochondrial respiration. Glycolysis generates adenosine triphosphate (ATP) (Pouyssegur et al., 2006), mitochondria produce a reduced amount of hydrogen peroxide (H_2O_2) (Rofstad et al., 2006), cells exhibit increased resistance to low levels of oxygen (Rofstad et al., 2006), and genes associated with epithelial-mesenchymal transition (EMT) and metastasis are controlled. Carbon dioxide and lactate production increase at a faster rate when HIF-1 inhibits intracellular acidification (Pouyssegur et al., 2006). These two substances facilitate the degradation of the extracellular matrix and the invasion of cells (Rofstad et al., 2006).

Role of Oxidative Stress in Angiogenesis

The growth of a tumor stimulates the formation of new blood vessels to supply oxygen and nutrients to the cancerous cells (Brown et al., 2004; Rabbani et al., 2009). Most research overwhelmingly indicate that reactive oxygen species (ROS) play a crucial role in promoting angiogenesis. When oxygen levels are low, blood vessels widen, leading to erratic blood circulation in the newly formed capillaries. When there is a fast increase in oxygen levels following a period of low oxygen, turbulence occurs, leading to the occurrence of oxidative stress (Brown et al., 2004). In a mouse model of breast cancer, the compound Mn(III) ortho-tetrakis-Nethylpyridylporphyrin was found to decrease the formation of new blood vessels (angiogenesis) and alter the pace at which endothelial cells divide and the density of small blood vessels (microvessel density). These effects indicate that the compound has a strong ability to remove reactive oxygen and nitrogen species (Rabbani et al., 2009). Vascular endothelial growth factor (VEGF) is a developmental factor that has a role in angiogenesis (Wouters et al., 2008). Hypoxia and malnutrition elevate intracellular levels of reactive oxygen species, which subsequently impact the expression of vascular endothelial growth factor (VEGF) (Spitz et al., 2000). In this scenario, the expression of VEFG and other genes is stimulated by HIF-1 and its cofactor p300 (Bertout et al., 2008; Gardner et al., 2008).

Glutathione peroxidase and mitochondrial inhibitors decrease the levels of reactive oxygen species (ROS) produced naturally inside the body. This, in turn, leads to a decrease in the expression of vascular endothelial growth factor (VEGF) and the activation of hypoxia-inducible factor 1 (HIF-1) in cancer cells (Xia et al., 2007). Prior studies have indicated that growth factor-stimulated activities have the ability to stimulate HIF-1, activate Akt, generate superoxide and hydrogen peroxide, and induce VEFG expression. The citations are (Govindarajan et al., 2007). Cell obstruction is induced by pretreating cells with catalase. Cancer cells generate reactive oxygen species (ROS) and subsequently secrete matrix metalloproteinases (MMPs), including MMP-1. The approach stimulates angiogenesis inside the tumor's microenvironment. In response to the formation of structures resembling capillaries, there is a temporary increase in the production of matrix metalloproteinases-1, -2, and -9.

In addition, the levels of reactive oxygen species (ROS) are elevated compared to the usual levels. This further substantiates the notion that levels of ROS can impact the process of angiogenesis mediated by matrix metalloproteinases (MMPs). Red blood cells (RBCs) has the capacity to stimulate the enzyme heme oxygenase-1, which can generate either carbon monoxide or augment the production of nitric oxide. This mechanism's capacity to dilate blood vessels enhances the delivery of blood to tumors, facilitating the process (Gardner et al., 2008).

ROS and Redox Regulation in Cancer Stem Cells

It has been widely recognized that certain primary cancer cells possess the ability to endure chemotherapy or radiation and result in a recurrence of the disease. Cancer stem cells (CSCs) possess similar characteristics to other types of stem cells and exhibit a high level of resistance to therapy. Cancer stem cells (CSCs) employ cell-specific redox regulatory systems to enhance cell survival and resistance to therapy (Bao et al., 2006; Visvader et al., 2008). Metastasis is the process by which normally functioning cells are transformed into cancerous cells due to the accumulation of reactive oxygen species (ROS). Mutations induce aberrant cell proliferation, disrupt cell adhesion to the extracellular matrix, and promote increased cell motility. Cancer stem cells and oncogenic cells are distinct entities. Cancer stem cells possess a superior ability to combat oxidative stress compared to other cancer cells, thanks to their heightened antioxidant capacity. This is attributed to the fact that cancer stem cells often have a high concentration of reactive oxygen species (ROS) (Trachootham, 2009). By maintaining optimal levels of endogenous and induced reactive oxygen species (ROS), these cells can retain their cancer-initiating and stem-like properties even following treatment. Cancer stem cells (CSCs) possess an abundance of enzymes that counteract oxygen radicals, unlike regular cancer cells (Diehn et al., 2009). Cancer stem cells (CSCs) that are high in antioxidants can protect cells from these effects. Epithelial cancer stem cells (CSC) exhibited heightened susceptibility to radiation and reduced ability to form colonies when the reactive oxygen species (ROS) scavenger GSH was depleted by L-S, R-buthionine sulphoximine (BSO) (Diehn et al., 2009).

Consequently, populations with a high number of CSCs experienced a lower frequency of radiation-induced DNA damage, such as single and double strand breaks. Cancer stem cells may possess enhanced antioxidant signaling, rendering them impervious to chemotherapy therapies designed to elevate intracellular ROS levels and inflict damage. It is essential to specifically target cancer stem cells in order to decrease the chances of cancer coming back after regular treatment due to their unique redox state. It is possible for an organism's antioxidant defenses to decrease the stem cell activity of cancer stem cells. It will be intriguing to observe whether a combined therapy incorporating conventional chemotherapy can effectively eradicate the tumor and cancerous cells (Diehn et al., 2009).

ROS IN CANCER TREATMENT

Reactive oxygen species necessary for cancer treatments are depicted in **Figure 5.**

Figure 5. ROS in cancer treatment

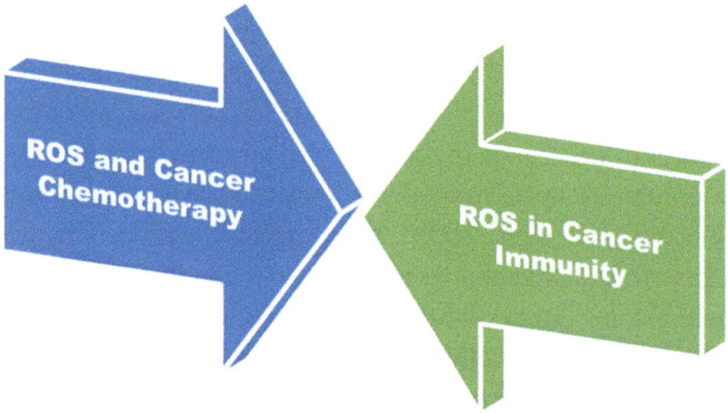

ROS and Cancer Chemotherapy

Chemotherapy is a frequently administered treatment for cancer patients in a clinical setting. The majority of chemotherapy medications generate ROS, which can disrupt the redox equilibrium of cancer cells (Yang et al., 2018). It is important to highlight that reactive oxygen species (ROS) have a significant impact on drug resistance. Cancer cells acquire resistance to treatment, resulting in its inconsistent efficacy (Conklin, 2004). Cancer cells are able to resist several medications due to the presence of transporter-based efflux pumps, such as P-glycoprotein (P-gp), in their membranes (Marullo et al., 2013). The MDR1/ABCB1 gene is responsible for producing P-gp, a protein that functions as a regulator for chemotherapeutic medicines (Szakacs et al., 2006). It is included in the ATP-binding cassette family. Upon encountering heightened amounts of reactive oxygen species (ROS), the NRF2-target gene P-gp is stimulated, hence functioning as a defense mechanism for cancer cells (Ambudkar et al., 2003). This contributes to the multitude of factors that lead individuals to avoid adhering to their pharmaceutical regimen (Jeddi et al., 2018). Finally, the increase in reactive oxygen species (ROS) caused by chemotherapy significantly affects the crucial harm to cancer cells and their resistance to treatment (Sadeghi et al., 2018).

ROS in Cancer Immunity

Due to the potent immunosuppressive properties, regulatory T cells (Tregs) are increasingly being considered for inclusion in the tumor microenvironment (TME) (Curiel et al., 2004). ROS are prevalent in the tumor microenvironment (TME), which harbor various kinds of cells, including immune cells, cancer cells, and cancer-associated fibroblasts (Sakaguchi, 2005). The levels of ROS in the tumor microenvironment (TME) are sufficiently elevated to hinder the effectiveness of immunotherapies such as immune checkpoint blockades (Weinberg et al., 2019). Elevated concentrations of ROS within the tumor microenvironment (TME) lead to the demise of regulatory T lymphocytes (Tregs) that infiltrate tumors (Cubillos-Ruiz et al., 2015). It is important to understand that Treg cells undergoing apoptosis and infiltrating tumors convert ATP to adenosine. This adenosine then hampers the immune response of anti-tumor T cells that is regulated by programmed death-ligand (Maj et al., 2017). Due to the inadequate antioxidant mechanism of NRF2, these cells are very susceptible to ROS (Deng et al., 2020). It is important to note that apoptotic Treg cells exhibit a more potent immunosuppressive impact as compared to living Treg cells (Sadeghi et al., 2018).

Decreasing the levels of ROS in the tumor microenvironment (TME) can enhance the efficacy of cancer immunotherapies. To mitigate the inhibitory immunogenic cell death induced by ROS, one approach is to employ a novel nano-scavenger that is affixed to the extracellular matrix (Gehrke et al., 2013). ROS not only have a significant impact on the function of regulatory T cells (Tregs), but they also play a vital role in controlling the metabolic activity of anti-tumor immunity. This regulation occurs through the involvement of cytotoxic T cells and dendritic cells. ROS play a vital role in activating a dendritic cell (DC) anti-tumor response. This reaction depends on the presence of SUMOspecific protease 3 (SENP3) and is induced by the stimulator of interferon (STING) (Chamato et al., 2017). ROS-induced accumulation of SENP3 in DCs activates the molecular-level anti-tumor activities mediated by STING. Subsequently, SENP3 removes the SUMO modification from interferon inducible (IFI) 204 through an interaction, initiating the process (Hu et al., 2021). ROS generation stimulates anti-tumor immunological reactions in dendritic cells, rather than regulatory T cells. Reduced tumor immunity may be caused by ROS enhancing the stress-XBP1 signaling of endoplasmic reticulum (ER) dendritic cells (DCs). To enhance the efficacy of immunotherapies against cancer, it is imperative to acquire a deeper understanding of the impact of ROS on cancer immunology (Wang et al., 2008).

FUTURE PROSPECTS

Understanding the complex biological systems that create, remove, and regulate reactive oxygen species (ROS) is necessary to control and benefit from them. Reduce and improve ROS elimination to improve cancer treatment (Storz, 2007). Redox balancing of reactive oxygen species (ROS) is essential for physiological homeostasis and cellular function. Cancer and other serious pathophysiological disorders can result from insufficient ROS neutralization. Signal transduction system abnormalities boost ROS, which aid cancer growth. PI3/Akt/mTOR, MAPK, PTEN, MMPs, and VEFG/VEGFR pathways comprise this system (Storz, 2005). High ROS levels stimulate apoptosis, which slows cancer cell development. Conventional cancer treatment with chemotherapy causes this side effect. ROS from chemotherapy cause apoptosis, or programmed cell death. This mechanism targets multiple cellular locations and many cell signaling pathways. ROS activate NF-κB, leading to inflammation and treatment resistance through cytokine production. ROS can inhibit angiogenesis, inflammation, and metastasis, making them a promising cancer treatment target (Trachootham, 2009).

Reaching an ideal balance between ROS production and removal could impact cancer treatment. Reactive oxygen species (ROS) can boost or suppress immunological responses depending on their function. Metabolic disorder neoplasia is characterized by elevated ROS levels. Tumor immune system is significantly damaged. Tumor cells contain more intracellular ROS than normal cells. Redox imbalance makes cancer cells more vulnerable to ROS (Wang et al., 2008). Immunogenic cell death and intracellular ROS are linked. Macrophage polarization from M2 to M1 is essential for immunomodulation to suppress tumor growth. Thus, managing reactive oxygen species (ROS) can help distinguish intracellular and extracellular microenvironments, improving results and process control. Nanomaterials can temporarily or permanently regulate ROS. This method may help us understand biodistribution, pharmacokinetics, and pharmacodynamics (Hu et al., 2021).

In the future, nanomedicine has the potential to offer a reliable means of managing the essential stages of medication distribution (Storz, 2009). This list is not exhaustive, but future ROS research may prioritize the following treatment areas: (i) to understand the significance of reactive oxygen species (ROS) in disease conditions; (ii) to regulate the generation and removal of ROS; (iii) to investigate the interaction between ROS levels and the immune system; and (iv) develop precise medical strategies for managing ROS levels both inside and outside cells. In order to develop an effective strategy for eliminating tumors and improving patient care, it is necessary to conduct a comprehensive analysis of redox processes in relation to cancer.

CONCLUSION

Once cancer cells become aggressive, they generate an increased amount of reactive oxygen species (ROS). Tumor growth that occurs without interruption promotes the spread of cancer to other parts of the body and sustains its ability to cause cancer. Cancer cells evade the effects of anticancer treatments by increasing the production of molecules that inhibit reactive oxygen species (ROS) and by activating pathways that prevent cell death (anti-apoptotic pathways). The mechanism by which cells sense oxidative stress and how reactive oxygen species (ROS) transmit intracellular signals to activate survival and antioxidant genes remains uncertain. Antioxidant drugs are effective against cancer stem cells and tumor cells since these cells rely on antioxidants for their survival. Cancer therapies that elevate intracellular reactive oxygen species (ROS) to a threshold that hampers the antioxidant capacity of cancer cells are expected to be effective. Pharmaceuticals that hinder antioxidant systems or signaling pathways that promote the synthesis of antioxidants in cancer cells can achieve this effect. Elevated levels of reactive oxygen species (ROS) have the ability to induce programmed cell death (apoptosis) in cancer cells by triggering certain signaling pathways and causing harm to surrounding cells.

The primary advantage of this approach is that it does not impact regular cells. Normal cells have a lower level of reactive oxygen species (ROS), which is why they require less antioxidants. There is a possibility that we have not yet achieved the necessary level of toxicity, and that by boosting reactive oxygen species (ROS), we could potentially induce additional genetic changes or promote the movement and invasion of cells. In order to elevate levels of reactive oxygen species (ROS) in cancer cells, it may be necessary to employ pharmacological interventions and substances that augment oxidative stress while impeding antioxidant processes. Despite over the threshold, this event will nevertheless have a positive impact on antioxidant systems. Gaining a comprehensive understanding of the signaling pathways regulated by ROS in tumor cells is crucial for effectively addressing medication resistance and devising novel tactics for eliminating cancer cells. Enhancing this comprehension can be achieved through redox control therapy.

REFERENCES

Ahmad, G., Almasry, M., Dhillon, A. S., Abuayyash, M., Kothandaraman, N., & Cakar, Z. (2017). Overview and sources of reactive oxygen species (ROS) in the reproductive system. In *Oxidative Stress in Human Reproduction* (pp. 1–16). Springer., DOI: 10.1007/978-3-319-48427-3_1

Alexandrov, L. B., Nik-Zainal, S., Wedge, D. C., Aparicio, S. A., Behjati, S., Biankin, A. V., Bignell, G. R., Bolli, N., Borg, A., Børresen-Dale, A. L., Boyault, S., Burkhardt, B., Butler, A. P., Caldas, C., Davies, H. R., Desmedt, C., Eils, R., Eyfjörd, J. E., Foekens, J. A., & Stratton, M. R. (2013). Signatures of mutational processes in human cancer. *Nature*, *500*(7463), 415–421. DOI: 10.1038/nature12477 PMID: 23945592

Ambudkar, S. V., Kimchi-Sarfaty, C., Sauna, Z. E., & Gottesman, M. M. (2003). P-glycoprotein: From genomics to mechanism. *Oncogene*, *22*(47), 7468–7485. DOI: 10.1038/sj.onc.1206948 PMID: 14576852

Bae, Y. S., Sung, J. Y., Kim, O. S., Kim, Y. J., Hur, K. C., Kazlauskas, A., & Rhee, S. G. (2000). Platelet-derived growth factor-induced H(2)O(2) production requires the activation of phosphatidylinositol 3-kinase. *The Journal of Biological Chemistry*, *275*(14), 10527–10531. DOI: 10.1074/jbc.275.14.10527 PMID: 10744745

Bao, S., Wu, Q., McLendon, R. E., Hao, Y., Shi, Q., Hjelmeland, A. B., Dewhirst, M. W., Bigner, D. D., & Rich, J. N. (2006). Glioma stem cells promote radioresistance by preferential activation of the DNA damage response. *Nature*, *444*(7120), 756–760. DOI: 10.1038/nature05236 PMID: 17051156

Barcelos, I. P., & Haas, R. H. (2019). CoQ10 and aging. *Biology (Basel)*, *8*(2), 20–31. DOI: 10.3390/biology8020028 PMID: 31083534

Ben Mahdi, M. H., Andrieu, V., & Pasquier, C. (2000). Focal adhesion kinase regulation by oxidative stress in different cell types. *IUBMB Life*, *50*(4–5), 291–299. DOI: 10.1080/15216540051081038 PMID: 11327323

Bertout, J. A., Patel, S. A., & Simon, M. C. (2008). The impact of O2 availability on human cancer. *Nature Reviews. Cancer*, *8*(12), 967–975. DOI: 10.1038/nrc2540 PMID: 18987634

Block, K., & Gorin, Y. (2012). Aiding and abetting roles of NOX oxidases in cellular transformation. *Nature Reviews. Cancer*, *12*(9), 627–637. DOI: 10.1038/nrc3339 PMID: 22918415

Breitenbach, M., Rinnerthaler, M., Weber, M., Breitenbach-Koller, H., Karl, T., Cullen, P., Basu, S., Haskova, D., & Hasek, J. (2018). The defense and signaling role of NADPH oxidases in eukaryotic cells: Review. Rolle der NADPH-Oxidasen eukaryotischer Zellen in der Immunabwehr und Signaltransduktion: Übersicht. *Wiener medizinische Wochenschrift (1946), 168*(11-12), 286–299. https://doi.org/ DOI: 10.1007/s10354-018-0640-4

Broom, O. J., Massoumi, R., & Sjolander, A. (2006). Alpha2beta1 integrin signalling enhances cyclooxygenase-2 expression in intestinal epithelial cells. *Journal of Cellular Physiology, 209*(3), 950–958. DOI: 10.1002/jcp.20796 PMID: 16972245

Browne, S. E., Roberts, L. J.II, Dennery, P. A., Doctrow, S. R., Beal, M. F., Barlow, C., & Levine, R. L. (2004). Treatment with a catalytic antioxidant corrects the neurobehavioral defect in ataxia-telangiectasia mice. *Free Radical Biology & Medicine, 36*(7), 938–942. DOI: 10.1016/j.freeradbiomed.2004.01.003 PMID: 15019978

Brwn, N. S., & Bicknell, R. (2001). Hypoxia and oxidative stress in breast cancer. Oxidative stress: Its effects on the growth, metastatic potential and response to therapy of breast cancer. *Breast Cancer Research : BCR, 3*(5), 323–327. DOI: 10.1186/bcr315 PMID: 11597322

Burdick, A. D., Davis, J. W.II, Liu, K. J., Hudson, L. G., Shi, H., Monske, M. L., & Burchiel, S. W. (2003). Benzo(a)pyrene quinones increase cell proliferation, generate reactive oxygen species, and transactivate the epidermal growth factor receptor in breast epithelial cells. *Cancer Research, 63*(22), 7825–7833. PMID: 14633709

Cadenas, E. (2004). Mitochondrial free radical production and cell signaling. *Molecular Aspects of Medicine, 25*(1–2), 17–26. DOI: 10.1016/j.mam.2004.02.005 PMID: 15051313

Cairns, R. A., Harris, I. S., & Mak, T. W. (2011). Regulation of cancer cell metabolism. *Nature Reviews. Cancer, 11*(2), 85–95. DOI: 10.1038/nrc2981 PMID: 21258394

Chamoto, K., Chowdhury, P. S., Kumar, A., Sonomura, K., Matsuda, F., Fagarasan, S., & Honjo, T. (2017). Mitochondrial activation chemicals synergize with surface receptor PD-1 blockade for T cell-dependent antitumor activity. *Proceedings of the National Academy of Sciences of the United States of America, 114*(5), E761–E770. DOI: 10.1073/pnas.1620433114 PMID: 28096382

Chiarugi, P. (2008). From anchorage dependent proliferation to survival: Lessons from redox signalling. *IUBMB Life, 60*(5), 301–307. DOI: 10.1002/iub.45 PMID: 18421772

Chiarugi, P., & Fiaschi, T. (2007). Redox signalling in anchorage-dependent cell growth. *Cellular Signalling, 19*(4), 672–682. DOI: 10.1016/j.cellsig.2006.11.009 PMID: 17204396

Chio, C., & Tuveson, D. A. (2017). ROS in Cancer: The burning question. *Trends in Molecular Medicine, 23*(5), 411–429. DOI: 10.1016/j.molmed.2017.03.004 PMID: 28427863

Chiu, T. T., Leung, W. Y., Moyer, M. P., Strieter, R. M., & Rozengurt, E. (2007). Protein kinase D2 mediates lysophosphatidic acid-induced interleukin 8 production in nontransformed human colonic epithelial cells through NF-kappaB. *American Journal of Physiology. Cell Physiology, 292*(2), C767–C777. DOI: 10.1152/ajpcell.00308.2006 PMID: 16928771

Chung, Y. M., Bae, Y. S., & Lee, S. Y. (2003). Molecular ordering of ROS production, mitochondrial changes, and caspase activation during sodium salicylate-induced apoptosis. *Free Radical Biology & Medicine, 34*(4), 434–442. DOI: 10.1016/S0891-5849(02)01301-1 PMID: 12566069

Conklin, K. A. (2004). Chemotherapy-associated oxidative stress: Impact on chemotherapeutic effectiveness. *Integrative Cancer Therapies, 3*(4), 294–300. DOI: 10.1177/1534735404270335 PMID: 15523100

Cubillos-Ruiz, J. R., Silberman, P. C., Rutkowski, M. R., Chopra, S., Perales-Puchalt, A., Song, M., Zhang, S., Bettigole, S. E., Gupta, D., Holcomb, K., Ellenson, L. H., Caputo, T., Lee, A. H., Conejo-Garcia, J. R., & Glimcher, L. H. (2015). ER Stress Sensor XBP1 Controls Anti-tumor Immunity by Disrupting Dendritic Cell Homeostasis. *Cell, 161*(7), 1527–1538. DOI: 10.1016/j.cell.2015.05.025 PMID: 26073941

Cullen, J. J., Weydert, C., Hinkhouse, M. M., Ritchie, J., Domann, F. E., Spitz, D., & Oberley, L. W. (2003). The role of manganese superoxide dismutase in the growth of pancreatic adenocarcinoma. *Cancer Research, 63*(6), 1297–1303. PMID: 12649190

Curiel, T. J., Coukos, G., Zou, L., Alvarez, X., Cheng, P., Mottram, P., Evdemon-Hogan, M., Conejo-Garcia, J. R., Zhang, L., Burow, M., Zhu, Y., Wei, S., Kryczek, I., Daniel, B., Gordon, A., Myers, L., Lackner, A., Disis, M. L., Knutson, K. L., & Zou, W. (2004). Specific recruitment of regulatory T cells in ovarian carcinoma fosters immune privilege and predicts reduced survival. *Nature Medicine, 10*(9), 942–949. DOI: 10.1038/nm1093 PMID: 15322536

Deng, H., Yang, W., Zhou, Z., Tian, R., Lin, L., Ma, Y., Song, J., & Chen, X. (2020). Targeted scavenging of extracellular ROS relieves suppressive immunogenic cell death. *Nature Communications, 11*(1), 4951. DOI: 10.1038/s41467-020-18745-6 PMID: 33009382

Dewhirst, M. W., Cao, Y., & Moeller, B. (2008). Cycling hypoxia and free radicals regulate angiogenesis and radiotherapy response. *Nature Reviews. Cancer*, *8*(6), 425–437. DOI: 10.1038/nrc2397 PMID: 18500244

Diehn, M., Cho, R. W., Lobo, N. A., Kalisky, T., Dorie, M. J., Kulp, A. N., Qian, D., Lam, J. S., Ailles, L. E., Wong, M., Joshua, B., Kaplan, M. J., Wapnir, I., Dirbas, F. M., Somlo, G., Garberoglio, C., Paz, B., Shen, J., Lau, S. K., & Clarke, M. F. (2009). Association of reactive oxygen species levels and radioresistance in cancer stem cells. *Nature*, *458*(7239), 780–783. DOI: 10.1038/nature07733 PMID: 19194462

Ewald, P. W., & Swain Ewald, H. A. (2015). Infection and cancer in multicellular organisms, Philos. Trans. R. Soc. Lond. Ser. B Biol. *Sci*, *370*, 1673.

Felty, Q., & Roy, D. (2005). Estrogen, mitochondria, and growth of cancer and non-cancer cells. *Journal of Carcinogenesis*, *4*(1), 1. DOI: 10.1186/1477-3163-4-1 PMID: 15651993

Ganapathy, V., Thangaraju, M., & Prasad, P. D. (2009). Nutrient transporters in cancer: Relevance to Warburg hypothesis and beyond. *Pharmacology & Therapeutics*, *121*(1), 29–40. DOI: 10.1016/j.pharmthera.2008.09.005 PMID: 18992769

Gardner, L. B., & Corn, P. G. (2008). Hypoxic regulation of mRNA expression. *Cell Cycle (Georgetown, Tex.)*, *7*(13), 1916–1924. DOI: 10.4161/cc.7.13.6203 PMID: 18604161

Gehrke, N., Mertens, C., Zillinger, T., Wenzel, J., Bald, T., Zahn, S., Tüting, T., Hartmann, G., & Barchet, W. (2013). Oxidative damage of DNA confers resistance to cytosolic nuclease TREX1 degradation and potentiates STING-dependent immune sensing. *Immunity*, *39*(3), 482–495. DOI: 10.1016/j.immuni.2013.08.004 PMID: 23993650

Giannoni, E., Fiaschi, T., Ramponi, G., & Chiarugi, P. (2009). Redox regulation of anoikis resistance of metastatic prostate cancer cells: Key role for Src and EGFR-mediated pro-survival signals. *Oncogene*, *28*(20), 2074–2086. DOI: 10.1038/onc.2009.77 PMID: 19377510

Gorrini, C., Harris, I. S., & Mak, T. W. (2013). Modulation of oxidative stress as an anticancer strategy. *Nature Reviews. Drug Discovery*, *12*(12), 931–947. DOI: 10.1038/nrd4002 PMID: 24287781

Gottlieb, E., Vander-Heiden, M. G., & Thompson, C. B. (2000). Bcl-x(L) prevents the initial decrease in mitochondrial membrane potential and subsequent reactive oxygen species production during tumor necrosis factor alpha-induced apoptosis. *Molecular and Cellular Biology*, *20*(15), 5680–5689. DOI: 10.1128/MCB.20.15.5680-5689.2000 PMID: 10891504

Govindarajan, B., Sligh, J. E., Vincent, B. J., Li, M., Canter, J. A., Nickoloff, B. J., Rodenburg, R. J., Smeitink, J. A., Oberley, L., Zhang, Y., Slingerland, J., Arnold, R. S., Lambeth, J. D., Cohen, C., Hilenski, L., Griendling, K., Martínez-Diez, M., Cuezva, J. M., & Arbiser, J. L. (2007). Overexpression of Akt converts radial growth melanoma to vertical growth melanoma. *The Journal of Clinical Investigation*, *117*(3), 719–729. DOI: 10.1172/JCI30102 PMID: 17318262

Halliwell, B. (2006). Reactive species and antioxidants. Redox biology is a fundamental theme of aerobic life. *Plant Physiology*, *141*(2), 312–322. DOI: 10.1104/pp.106.077073 PMID: 16760481

Harris, A. L. (2002). Hypoxia—A key regulatory factor in tumour growth. *Nature Reviews. Cancer*, *2*(1), 38–47. DOI: 10.1038/nrc704 PMID: 11902584

Hayes, J. D., Dinkova-Kostova, A. T., & Tew, K. D. (2020). Oxidative Stress in Cancer. *Cancer Cell*, *38*(2), 167–197. DOI: 10.1016/j.ccell.2020.06.001 PMID: 32649885

Hitchler, M. J., Oberley, L. W., & Domann, F. E. (2008). Epigenetic silencing of SOD2 by histone modifications in human breast cancer cells. *Free Radical Biology & Medicine*, *45*(11), 1573–1580. DOI: 10.1016/j.freeradbiomed.2008.09.005 PMID: 18845242

Hockel, M., & Vaupel, P. (2001). Tumor hypoxia: Definitions and current clinical, biologic, and molecular aspects. *Journal of the National Cancer Institute*, *93*(4), 266–276. DOI: 10.1093/jnci/93.4.266 PMID: 11181773

Hsu, P. P., & Sabatini, D. M. (2008). Cancer cell metabolism: Warburg and beyond. *Cell*, *134*(5), 703–707. DOI: 10.1016/j.cell.2008.08.021 PMID: 18775299

Hu, Z., Teng, X. L., Zhang, T., Yu, X., Ding, R., Yi, J., Deng, L., Wang, Z., & Zou, Q. (2021). SENP3 senses oxidative stress to facilitate STING-dependent dendritic cell antitumor function. *Molecular Cell*, *81*(5), 940–952.e5. DOI: 10.1016/j.molcel.2020.12.024 PMID: 33434504

Jeddi, F., Soozangar, N., Sadeghi, M. R., Somi, M. H., Shirmohamadi, M., Eftekhar-Sadat, A. T., & Samadi, N. (2018). Nrf2 overexpression is associated with P-glycoprotein upregulation in gastric cancer. *Biomedicine & pharmacotherapy = Biomedecine & pharmacotherapie, 97*, 286–292. https://doi.org/DOI: 10.1016/j.biopha.2017.10.129

Kaelin, W. G.Jr, & Ratcliffe, P. J. (2008). Oxygen sensing by metazoans: The central role of the HIF hydroxylase pathway. *Molecular Cell, 30*(4), 393–402. DOI: 10.1016/j.molcel.2008.04.009 PMID: 18498744

Kamihara, Y., Takada, K., Sato, T., Kawano, Y., Murase, K., Arihara, Y., Kikuchi, S., Hayasaka, N., Usami, M., Iyama, S., Miyanishi, K., Sato, Y., Kobune, M., & Kato, J. (2016). The iron chelator deferasirox induces apoptosis by targeting oncogenic Pyk2/β-catenin signaling in human multiple myeloma. *Oncotarget, 7*(39), 64330–64341. DOI: 10.18632/oncotarget.11830 PMID: 27602957

Lee, C. H., Yao, C. F., Huang, S. M., Ko, S., Tan, Y. H., Lee-Chen, G. J., & Wang, Y. C. (2008). Novel 2-step synthetic indole compound 1,1,3-tri(3-indolyl)cyclohexane inhibits cancer cell growth in lung cancer cells and xenograft models. *Cancer, 113*(4), 815–825. DOI: 10.1002/cncr.23619 PMID: 18618576

Lewis, A., Du, J., Liu, J., Ritchie, J. M., Oberley, L. W., & Cullen, J. J. (2005). Metastatic progression of pancreatic cancer: Changes in antioxidant enzymes and cell growth. *Clinical & Experimental Metastasis, 22*(7), 523–532. DOI: 10.1007/s10585-005-4919-7 PMID: 16475022

Lewis, A., Du, J., Liu, J., Ritchie, J. M., Oberley, L. W., & Cullen, J. J. (2005). Metastatic progression of pancreatic cancer: Changes in antioxidant enzymes and cell growth. *Clinical & Experimental Metastasis, 22*(7), 523–532. DOI: 10.1007/s10585-005-4919-7 PMID: 16475022

Li, Y., Li, S., Meng, X., Gan, R. Y., Zhang, J. J., & Li, H. B. (2017). Dietary Natural Products for Prevention and Treatment of Breast Cancer. *Nutrients, 9*(7), 728. DOI: 10.3390/nu9070728 PMID: 28698459

Liu, H., Wang, X. C., Hu, G. H., Guo, Z. F., Lai, P., Xu, L., Huang, T. B., & Xu, Y. F. (2015). Fruit and vegetable consumption and risk of bladder cancer: an updated meta-analysis of observational studies. *European journal of cancer prevention: the official journal of the European Cancer Prevention Organisation (ECP), 24*(6), 508–516. https://doi.org/DOI: 10.1097/CEJ.0000000000000119

Liu, L. Z., Hu, X. W., Xia, C., He, J., Zhou, Q., Shi, X., Fang, J., & Jiang, B. H. (2006). Reactive oxygen species regulate epidermal growth factor-induced vascular endothelial growth factor and hypoxia-inducible factor-1alpha expression through activation of AKT and P70S6K1 in human ovarian cancer cells. *Free Radical Biology & Medicine*, *41*(10), 1521–1533. DOI: 10.1016/j.freeradbiomed.2006.08.003 PMID: 17045920

Maj, T., Wang, W., Crespo, J., Zhang, H., Wang, W., Wei, S., Zhao, L., Vatan, L., Shao, I., Szeliga, W., Lyssiotis, C., Liu, J. R., Kryczek, I., & Zou, W. (2017). Oxidative stress controls regulatory T cell apoptosis and suppressor activity and PD-L1-blockade resistance in tumor. *Nature Immunology*, *18*(12), 1332–1341. DOI: 10.1038/ni.3868 PMID: 29083399

Mari, M., de Gregorio, E., de Dios, C., Roca-Agujetas, V., Cucarull, B., Tutusaus, A., Morales, A., & Colell, A. (2020). Mitochondrial glutathione: Recent insights and role in disease. *Antioxidants*, *9*(10), 909. DOI: 10.3390/antiox9100909 PMID: 32987701

Marullo, R., Werner, E., Degtyareva, N., Moore, B., Altavilla, G., Ramalingam, S. S., & Doetsch, P. W. (2013). Cisplatin induces a mitochondrial-ROS response that contributes to cytotoxicity depending on mitochondrial redox status and bioenergetic functions. *PLoS One*, *8*(11), e81162. DOI: 10.1371/journal.pone.0081162 PMID: 24260552

Merlin, J. P. J., Rupasinghe, H. P. V., Dellaire, G., & Murphy, K. (2021). Role of Dietary Antioxidants in p53-Mediated Cancer Chemoprevention and Tumor Suppression. *Oxidative Medicine and Cellular Longevity*, *9924328*(1), 9924328. Advance online publication. DOI: 10.1155/2021/9924328 PMID: 34257824

Mihailovic, T., Marx, M., Auer, A., Van Lint, J., Schmid, M., Weber, C., & Seufferlein, T. (2004). Protein kinase D2 mediates activation of nuclear factor kappaB by Bcr-Abl in Bcr-Abl+ human myeloid leukemia cells. *Cancer Research*, *64*(24), 8939–8944. DOI: 10.1158/0008-5472.CAN-04-0981 PMID: 15604256

Moloney, J. N., & Cotter, T. G. (2018). ROS signalling in the biology of cancer, in: Seminars in cell & developmental biology, Elsevier, 20, 18. DOI: 10.1016/j.semcdb.2017.05.023

Narita, S., Saito, E., Sawada, N., Shimazu, T., Yamaji, T., Iwasaki, M., Ishihara, J., Takachi, R., Shibuya, K., Inoue, M., & Tsugane, S.JPHC Study Group. (2018). Dietary consumption of antioxidant vitamins and subsequent lung cancer risk: The Japan Public Health Center-based prospective study. *International Journal of Cancer*, *142*(12), 2441–2460. DOI: 10.1002/ijc.31268 PMID: 29355932

Ogidi, O. I. (2024). Recent advances in anticancer activity and Bioinformatics Approach from Potential Plants. In: Computational Approaches in Biomaterials, Bioinformatics and Biomedical Engineering Applications. *CRC Press, Taylor and Francis Group, Florida, USA,* 9781003354437, 27. https://doi.org/DOI: 10.1201/9781003354437-3

Ogidi, O. I., & Tawariowei, A. M. (2023). Endocrine Disrupting Chemicals and their Role in Cancer-A review. *Molecular Sciences and Applications*, *3*, 5–14. DOI: 10.37394/232023.2023.3.2

Okon, I. S., & Zou, M. H. (2015). Mitochondrial ROS and cancer drug resistance: Implications for therapy. *Pharmacological Research*, *100*, 170–174. DOI: 10.1016/j.phrs.2015.06.013 PMID: 26276086

Pani, G., Giannoni, E., Galeotti, T., & Chiarugi, P. (2009). Redox-based escape mechanism from death: The cancer lesson. *Antioxidants & Redox Signaling*, *11*(11), 2791–2806. DOI: 10.1089/ars.2009.2739 PMID: 19686053

Patra, S., Pradhan, B., Nayak, R., Behera, C., Das, S., Patra, S. K., Efferth, T., Jena, M., & Bhutia, S. K. (2021). Dietary polyphenols in chemoprevention and synergistic effect in cancer: Clinical evidences and molecular mechanisms of action. *Phytomedicine*, *90*, 153554. DOI: 10.1016/j.phymed.2021.153554 PMID: 34371479

Pelicano, H., Lu, W., Zhou, Y., Zhang, W., Chen, Z., Hu, Y., & Huang, P. (2009). Mitochondrial dysfunction and reactive oxygen species imbalance promote breast cancer cell motility through a CXCL14-mediated mechanism. *Cancer Research*, *69*(6), 2375–2383. DOI: 10.1158/0008-5472.CAN-08-3359 PMID: 19276362

Perillo, B., Di Donato, M., Pezone, A., Di Zazzo, E., Giovannelli, P., Galasso, G., Castoria, G., & Migliaccio, A. (2020). ROS in cancer therapy: The bright side of the moon. *Experimental & Molecular Medicine*, *52*(2), 192–203. DOI: 10.1038/s12276-020-0384-2 PMID: 32060354

Piskounova, E., Agathocleous, M., Murphy, M. M., Hu, Z., Huddlestun, S. E., Zhao, Z., Leitch, A. M., Johnson, T. M., DeBerardinis, R. J., & Morrison, S. J. (2015). Oxidative stress inhibits distant metastasis by human melanoma cells. *Nature*, *527*(7577), 186–191. DOI: 10.1038/nature15726 PMID: 26466563

Poetsch, A. R. (2020). The genomics of oxidative DNA damage, repair, and resulting mutagenesis. *Computational and Structural Biotechnology Journal*, *18*, 207–219. DOI: 10.1016/j.csbj.2019.12.013 PMID: 31993111

Pouyssegur, J., Dayan, F., & Mazure, N. M. (2006). Hypoxia signalling in cancer and approaches to enforce tumour regression. *Nature*, *441*(7092), 437–443. DOI: 10.1038/nature04871 PMID: 16724055

Prasad, S., Gupta, S. C., & Tyagi, A. K. (2017). Reactive oxygen species (ROS) and cancer: Role of antioxidative nutraceuticals. *Cancer Letters*, *387*, 95–105. DOI: 10.1016/j.canlet.2016.03.042 PMID: 27037062

Qanungo, S., Das, M., Haldar, S., & Basu, A. (2005). Epigallocatechin-3-gallate induces mitochondrial membrane depolarization and caspase-dependent apoptosis in pancreatic cancer cells. *Carcinogenesis*, *26*(5), 958–967. DOI: 10.1093/carcin/bgi040 PMID: 15705601

Rabbani, Z. N., Spasojevic, I., Zhang, X., Moeller, B. J., Haberle, S., Vasquez-Vivar, J., Dewhirst, M. W., Vujaskovic, Z., & Batinic-Haberle, I. (2009). Antiangiogenic action of redox-modulating Mn(III) meso-tetrakis(N-ethylpyridinium-2-yl)porphyrin, MnTE-2-PyP(5+), via suppression of oxidative stress in a mouse model of breast tumor. *Free Radical Biology & Medicine*, *47*(7), 992–1004. DOI: 10.1016/j.freeradbiomed.2009.07.001 PMID: 19591920

Radisky, D. C., Levy, D. D., Littlepage, L. E., Liu, H., Nelson, C. M., Fata, J. E., Leake, D., Godden, E. L., Albertson, D. G., Nieto, M. A., Werb, Z., & Bissell, M. J. (2005). Rac1b and reactive oxygen species mediate MMP-3-induced EMT and genomic instability. *Nature*, *436*(7047), 123–127. DOI: 10.1038/nature03688 PMID: 16001073

Reddy, K. B., & Glaros, S. (2007). Inhibition of the MAP kinase activity suppresses estrogen-induced breast tumor growth both in vitro and in vivo. *International Journal of Oncology*, *30*(4), 971–975. DOI: 10.3892/ijo.30.4.971 PMID: 17332937

Reichenbach, J., Schubert, R., Schindler, D., Müller, K., Böhles, H., & Zielen, S. (2002). Elevated oxidative stress in patients with ataxia telangiectasia. *Antioxidants & Redox Signaling*, *4*(3), 465–469. DOI: 10.1089/15230860260196254 PMID: 12215213

Rofstad, E. K., Mathiesen, B., Kindem, K., & Galappathi, K. (2006). Acidic extracellular pH promotes experimental metastasis of human melanoma cells in athymic nude mice. *Cancer Research*, *66*(13), 6699–6707. DOI: 10.1158/0008-5472.CAN-06-0983 PMID: 16818644

Ruiz-Ramos, R., Lopez-Carrillo, L., Rios-Perez, A. D., De Vizcaya-Ruíz, A., & Cebrian, M. E. (2009). Sodium arsenite induces ROS generation, DNA oxidative damage, HO-1 and c-Myc proteins, NF-kappaB activation and cell proliferation in human breast cancer MCF-7 cells. *Mutation Research*, *674*(1-2), 109–115. DOI: 10.1016/j.mrgentox.2008.09.021 PMID: 18996220

Sadeghi, M. R., Jeddi, F., Soozangar, N., Somi, M. H., Shirmohamadi, M., Khaze, V., & Samadi, N. (2018). Nrf2/P-glycoprotein axis is associated with clinicopathological characteristics in colorectal cancer. *Biomedicine & pharmacotherapy = Biomedecine & pharmacotherapie, 104*, 458–464. https://doi.org/DOI: 10.1016/j.biopha.2018.05.062

Sakaguchi, S. (2005). Naturally arising Foxp3-expressing CD25+CD4+ regulatory T cells in immunological tolerance to self and non-self. *Nature Immunology*, *6*(4), 345–352. DOI: 10.1038/ni1178 PMID: 15785760

Santos, A. L., Sinha, S., & Lindner, A. B. (2018). The good, the bad, and the ugly of ROS: New insights on aging and aging-related diseases from eukaryotic and prokaryotic model organisms. *Oxidative Medicine and Cellular Longevity*, *2018*(1), 1941285. DOI: 10.1155/2018/1941285 PMID: 29743972

Sarmiento-Salinas, F. L., Delgado-Magallón, A., Montes-Alvarado, J. B., Ramírez-Ramírez, D., Flores-Alonso, J. C., Cortés-Hernández, P., Reyes-Leyva, J., Herrera-Camacho, I., Anaya-Ruiz, M., Pelayo, R., Millán-Pérez-Peña, L., & Maycotte, P. (2019). Breast Cancer Subtypes Present a Differential Production of Reactive Oxygen Species (ROS) and Susceptibility to Antioxidant Treatment. *Frontiers in Oncology*, *9*, 480. DOI: 10.3389/fonc.2019.00480 PMID: 31231612

Sarsour, E. H., Venkataraman, S., Kalen, A. L., Oberley, L. W., & Goswami, P. C. (2008). Manganese superoxide dismutase activity regulates transitions between quiescent and proliferative growth. *Aging Cell*, *7*(3), 405–417. DOI: 10.1111/j.1474-9726.2008.00384.x PMID: 18331617

Sayin, V. I., Ibrahim, M. X., Larsson, E., Nilsson, J. A., Lindahl, P., & Bergo, M. O. (2014). Antioxidants accelerate lung cancer progression in mice. *Science Translational Medicine*, *6*(221), 221ra15. Advance online publication. DOI: 10.1126/scitranslmed.3007653 PMID: 24477002

Sayin, V. I., Ibrahim, M. X., Larsson, E., Nilsson, J. A., Lindahl, P., & Bergo, M. O. (2014). Antioxidants accelerate lung cancer progression in mice. *Science Translational Medicine*, *6*(221), 221ra15. DOI: 10.1126/scitranslmed.3007653 PMID: 24477002

Schafer, Z. T., Grassian, A. R., Song, L., Jiang, Z., Gerhart-Hines, Z., Irie, H. Y., Gao, S., Puigserver, P., & Brugge, J. S. (2009). Antioxidant and oncogene rescue of metabolic defects caused by loss of matrix attachment. *Nature*, *461*(7260), 109–113. DOI: 10.1038/nature08268 PMID: 19693011

Sena, L. A., & Chandel, N. S. (2012). Physiological roles of mitochondrial reactive oxygen species. *Molecular Cell*, *48*(2), 158–167. DOI: 10.1016/j.molcel.2012.09.025 PMID: 23102266

Shim, H. Y., Park, J. H., Paik, H. D., Nah, S. Y., Kim, D. S., & Han, Y. S. (2007). Acacetin-induced apoptosis of human breast cancer MCF-7 cells involves caspase cascade, mitochondria-mediated death signaling and SAPK/JNK1/2-c-Jun activation. *Molecules and Cells*, *24*(1), 95–104. DOI: 10.1016/S1016-8478(23)10760-6 PMID: 17846503

Simon, H. U., Haj-Yehia, A., & Levi-Schaffer, F. (2000). Role of reactive oxygen species (ROS) in apoptosis induction. *Apoptosis*, *5*(5), 415–418. DOI: 10.1023/A:1009616228304 PMID: 11256882

Skonieczna, M., Hejmo, T., Poterala-Hejmo, A., Cieslar-Pobuda, A., & Buldak, R. J. (2017). NADPH oxidases: Insights into selected functions and mechanisms of action in cancer and stem cells. *Oxidative Medicine and Cellular Longevity*, *2017*(1), 1–15. DOI: 10.1155/2017/9420539 PMID: 28626501

Song, J., Li, J., Qiao, J., Jain, S., Mark Evers, B., & Chung, D. H. (2009). PKD prevents H2O2-induced apoptosis via NF-kappaB and p38 MAPK in RIE-1 cells. *Biochemical and Biophysical Research Communications*, *378*(3), 610–614. DOI: 10.1016/j.bbrc.2008.11.106 PMID: 19059215

Spitz, D. R., Sim, J. E., Ridnour, L. A., Galoforo, S. S., & Lee, Y. J. (2000). Glucose deprivation-induced oxidative stress in human tumor cells. A fundamental defect in metabolism? *Annals of the New York Academy of Sciences*, *899*(1), 349–362. DOI: 10.1111/j.1749-6632.2000.tb06199.x PMID: 10863552

Storz, P. (2005). Reactive oxygen species in tumor progression. *Frontiers in Bioscience*, *10*(1-3), 1881–1896. DOI: 10.2741/1667 PMID: 15769673

Storz, P. (2007). Mitochondrial ROS—Radical detoxification, mediated by protein kinase D. *Trends in Cell Biology*, *17*(1), 13–18. DOI: 10.1016/j.tcb.2006.11.003 PMID: 17126550

Storz, P., Döppler, H., Copland, J. A., Simpson, K. J., & Toker, A. (2009). FOXO3a promotes tumor cell invasion through the induction of matrix metalloproteinases. *Molecular and Cellular Biology*, *29*(18), 4906–4917. DOI: 10.1128/MCB.00077-09 PMID: 19564415

Storz, P., Doppler, H., & Toker, A. (2004). Protein kinase Cdelta selectively regulates protein kinase Ddependent activation of NF-kappaB in oxidative stress signaling. *Molecular and Cellular Biology*, *24*(7), 2614–2626. DOI: 10.1128/MCB.24.7.2614-2626.2004 PMID: 15024053

Sung, H., Ferlay, J., Siegel, R. L., Laversanne, M., Soerjomataram, I., Jemal, A., & Bray, F. (2021). Global Cancer Statistics 2020: GLOBOCAN Estimates of Incidence and Mortality Worldwide for 36 Cancers in 185 Countries. *CA: a Cancer Journal for Clinicians*, *71*(3), 209–249. DOI: 10.3322/caac.21660 PMID: 33538338

Svineng, G., Ravuri, C., Rikardsen, O., Huseby, N. E., & Winberg, J. O. (2008). The role of reactive oxygen species in integrin and matrix metalloproteinase expression and function. *Connective Tissue Research*, *49*(3), 197–202. DOI: 10.1080/03008200802143166 PMID: 18661342

Szakács, G., Paterson, J. K., Ludwig, J. A., Booth-Genthe, C., & Gottesman, M. M. (2006). Targeting multidrug resistance in cancer. *Nature Reviews. Drug Discovery*, *5*(3), 219–234. DOI: 10.1038/nrd1984 PMID: 16518375

Taddei, M. L., Parri, M., Mello, T., Catalano, A., Levine, A. D., Raugei, G., Ramponi, G., & Chiarugi, P. (2007). Integrin-mediated cell adhesion and spreading engage different sources of reactive oxygen species. *Antioxidants & Redox Signaling*, *9*(4), 469–481. DOI: 10.1089/ars.2006.1392 PMID: 17280488

Takeda, K., Matsuzawa, A., Nishitoh, H., & Ichijo, H. (2003). Roles of MAPKKK ASK1 in stress-induced cell death. *Cell Structure and Function*, *28*(1), 23–29. DOI: 10.1247/csf.28.23 PMID: 12655147

Trachootham, D., Alexandre, J., & Huang, P. (2009). Targeting cancer cells by ROS-mediated mechanisms: A radical therapeutic approach? *Nature Reviews. Drug Discovery*, *8*(7), 579–591. DOI: 10.1038/nrd2803 PMID: 19478820

Valavanidis, A., Vlachogianni, T., Fiotakis, K., & Loridas, S. (2013). Pulmonary oxidative stress, inflammation and cancer: Respirable particulate matter, fibrous dusts and ozone as major causes of lung carcinogenesis through reactive oxygen species mechanisms. *International Journal of Environmental Research and Public Health*, *10*(9), 3886–3907. DOI: 10.3390/ijerph10093886 PMID: 23985773

Vermot, A., Petit-Härtlein, I., Smith, S. M. E., & Fieschi, F. (2021). NADPH oxidases (NOX): An overview from discovery, molecular mechanisms to physiology and pathology. *Antioxidants*, *10*(6), 32–45. DOI: 10.3390/antiox10060890 PMID: 34205998

Vieira, A. R., Abar, L., Vingeliene, S., Chan, D. S., Aune, D., Navarro-Rosenblatt, D., Stevens, C., Greenwood, D., & Norat, T. (2016). Fruits, vegetables and lung cancer risk: A systematic review and meta-analysis. *Annals of Oncology : Official Journal of the European Society for Medical Oncology*, *27*(1), 81–96. DOI: 10.1093/annonc/mdv381 PMID: 26371287

Viel, A., Bruselles, A., Meccia, E., Fornasarig, M., Quaia, M., Canzonieri, V., Policicchio, E., Urso, E. D., Agostini, M., Genuardi, M., Lucci-Cordisco, E., Venesio, T., Martayan, A., Diodoro, M. G., Sanchez-Mete, L., Stigliano, V., Mazzei, F., Grasso, F., Giuliani, A., & Bignami, M. (2017). A Specific Mutational Signature Associated with DNA 8-Oxoguanine Persistence in MUTYH-defective Colorectal Cancer. *EBioMedicine*, *20*, 39–49. DOI: 10.1016/j.ebiom.2017.04.022 PMID: 28551381

Visvader, J. E., & Lindeman, G. J. (2008). Cancer stem cells in solid tumours: Accumulating evidence and unresolved questions. *Nature Reviews. Cancer*, *8*(10), 755–768. DOI: 10.1038/nrc2499 PMID: 18784658

Wang, H., Gao, Z., Liu, X., Agarwal, P., Zhao, S., Conroy, D. W., Ji, G., Yu, J., Jaroniec, C. P., Liu, Z., Lu, X., Li, X., & He, X. (2018). Targeted production of reactive oxygen species in mitochondria to overcome cancer drug resistance. *Nature Communications*, *9*(1), 562. DOI: 10.1038/s41467-018-02915-8 PMID: 29422620

Wang, L., Azad, N., Kongkaneramit, L., Chen, F., Lu, Y., Jiang, B. H., & Rojanasakul, Y. (2008). The Fas death signaling pathway connecting reactive oxygen species generation and FLICE inhibitory protein down-regulation. *Journal of immunology (Baltimore, Md.: 1950)*, *180*(5), 3072–3080. https://doi.org/DOI: 10.4049/jimmunol.180.5.3072

Wang, M., Kirk, J. S., Venkataraman, S., Domann, F. E., Zhang, H. J., Schafer, F. Q., Flanagan, S. W., Weydert, C. J., Spitz, D. R., Buettner, G. R., & Oberley, L. W. (2005). Manganese superoxide dismutase suppresses hypoxic induction of hypoxia-inducible factor-1alpha and vascular endothelial growth factor. *Oncogene*, *24*(55), 8154–8166. DOI: 10.1038/sj.onc.1208986 PMID: 16170370

Wang, Q., Chen, Y., Wang, X., Gong, G., Li, G., & Li, C. (2014). Consumption of fruit, but not vegetables, may reduce risk of gastric cancer: results from a meta-analysis of cohort studies. *European journal of cancer (Oxford, England: 1990)*, *50*(8), 1498–1509. https://doi.org/DOI: 10.1016/j.ejca.2014.02.009

Weinberg, F., Ramnath, N., & Nagrath, D. (2019). Reactive oxygen species in the tumor microenvironment: An overview. *Cancers (Basel), 11*(8), 1191. DOI: 10.3390/cancers11081191 PMID: 31426364

Werner, E., & Werb, Z. (2002). Integrins engage mitochondrial function for signal transduction by a mechanism dependent on Rho GTPases. *The Journal of Cell Biology, 158*(2), 357–368. DOI: 10.1083/jcb.200111028 PMID: 12119354

Wouters, B. G., & Koritzinsky, M. (2008). Hypoxia signalling through mTOR and the unfolded protein response in cancer. *Nature Reviews. Cancer, 8*(11), 851–864. DOI: 10.1038/nrc2501 PMID: 18846101

Wu, S., Zhu, W., Thompson, P., & Hannun, Y. A. (2018). Evaluating intrinsic and non-intrinsic cancer risk factors. *Nature Communications, 9*(1), 3490. DOI: 10.1038/s41467-018-05467-z PMID: 30154431

Xia, C., Meng, Q., Liu, L. Z., Rojanasakul, Y., Wang, X. R., & Jiang, B. H. (2007). Reactive oxygen species regulate angiogenesis and tumor growth through vascular endothelial growth factor. *Cancer Research, 67*(22), 10823–10830. DOI: 10.1158/0008-5472.CAN-07-0783 PMID: 18006827

Yang, H., Villani, R. M., Wang, H., Simpson, M. J., Roberts, M. S., Tang, M., & Liang, X. (2018). The role of cellular reactive oxygen species in cancer chemotherapy. *Journal of Experimental & Clinical Cancer Research : CR, 37*(1), 266. DOI: 10.1186/s13046-018-0909-x PMID: 30382874

You, H., Yamamoto, K., & Mak, T. W. (2006). Regulation of transactivation-independent proapoptotic activity of p53 by FOXO3a. *Proceedings of the National Academy of Sciences of the United States of America, 103*(24), 9051–9056. DOI: 10.1073/pnas.0600889103 PMID: 16757565

Zhang, R., Humphreys, I., Sahu, R. P., Shi, Y., & Srivastava, S. K. (2008). In vitro and in vivo induction of apoptosis by capsaicin in pancreatic cancer cells is mediated through ROS generation and mitochondrial death pathway. *Apoptosis: an international journal on programmed cell death, 13*(12), 1465–1478. https://doi.org/DOI: 10.1007/s10495-008-0278-6

ABBREVIATIONS

5-LOX: 5-Lipoxygeneses
A: Adenine
ATM: Ataxia telangiectasia mutated

ATP: Adenosine Triphosphate
ATP: Adenosine triphosphate
Bcl-2: B-cell lymphoma 2
Bcl-XL: B-cell lymphoma-extra large
BPQs: Benzo(a) Pyrene quinone
BSO: Buthionine Sulphoximine
C: Cytosine
CAT: Catalase
CC: Chemokine
CSCs: Cancer stem cells
DC: Dendritic cell
DFX: Deferasirox
DNA: deoxyribonucleic acid
DUOX1-2: Dual Oxidases 1-2
ECM: Extracellular matrix
EGF: Epithelial growth factor
EMT: Epithelial-mesenchymal transition
ER: Endoplasmic reticulum
FAK: Focused adhesion kinase
GC: Germinal centers
GPX: Glutathione Peroxidase
GSH: glutathione
GSSG: Glutathione disulfide
H2O2: hydrogen peroxide
HBV: Hepatitis B Viruse
HCC: Hepatocellular carcinoma
HIF-1: hypoxia-inducible factor-1
HO: Hydroxide ion
HPV: Human *Papillomaviruse*
IFI: Interferon inducible
JNKs: cJun N-Terminal kinases
MAPK: Mitogen-activated protein kinases
MM: Multiple myeloma
MMPs: Matrix metalloproteinases
MnSOD: Manganese superoxide dismutase
mtDNA: Mitochondrial Deoxyribonucleic acid
NAC: N-acetyl-1-cysteine
NADPH: nicotinamide adenine dinucleotide phosphate
NF-kB: Nuclear factor kappa B
NO: nitric oxide

NO2: nitrogen dioxide
NOX: nitrogen oxides
NRF-2: nuclear factor erythroid 2–related factor 2
O2: dioxygen
ONOO: peroxonitrite
P-gp: P glycoprotein
PKD1: Protein Kinase D1
PTPs: Protein tyrosine phosphatases
RAF: Rapidly Accelerated Fibrosarcoma
RAS: Rat sarcoma virus
RNS: reactive nitrogen species
ROS: Reactive Oxygen Species
SENP: 3 SUMO-specific protease 3
-SH: sulfhydryl
SHP-2: Src homology-2 protein
SOD: Superoxide dismutase
Src: sarcoma
STING: Stimulator of Interferon Genes
SUMO: small ubiquitin-related modifier
TME: Tumour microenvironment
TNF: Tumor necrosis facto
TNF-kB: Nuclear factor-kappa B
TRAF-4: Tumour necrosis factor receptor associated factor 4
VEGF: Vascularendothelial growth factor
VHL: Von Hippel-Lindau disease

Chapter 11
Skin Diseases Mechanisms and Treatments

Aishwarya Jain
MCE's Allana College of Pharmacy, Pune, India

ABSTRACT

Skin diseases encompass various conditions that affect the skin, ranging from infections and inflammations to genetic disorders and malignancies. These diseases pose significant health challenges and profoundly impact the quality of life, leading to psychological and social consequences. Understanding the underlying mechanisms of skin diseases is critical for developing effective treatment strategies. In recent years, dietary antioxidants have garnered attention for their potential role in mitigating skin diseases by combating oxidative stress. This chapter aims to delve into the intricate mechanisms of prevalent skin diseases, review the current treatment modalities, and explore the role of dietary antioxidants in promoting skin health. By providing a comprehensive analysis of the existing scientific literature, this chapter offers valuable insights into the pathophysiology of skin diseases and the therapeutic potential of antioxidants.

INTRODUCTION

Skin diseases are a major public health concern globally, with a wide range of conditions affecting millions of people across all age groups. These diseases vary in severity from mild, self-limiting conditions to severe, chronic, and even life-threatening disorders. Common skin diseases such as eczema, psoriasis, acne, and skin cancers are not only prevalent but also significantly impact the quality of life,

DOI: 10.4018/979-8-3693-7919-6.ch011

causing both physical discomfort and psychological distress. The management of these conditions presents considerable challenges due to their complex pathophysiology and the often chronic nature of the diseases.

PREVALENCE AND EPIDEMIOLOGY

The global burden of skin diseases is substantial. According to the Global Burden of Disease Study 2017, skin diseases were the fourth leading cause of non-fatal disease burden worldwide, affecting approximately 1.9 billion people globally (Hay et al., 2014). Among these, eczema and psoriasis are particularly common. Eczema, also known as atopic dermatitis, affects around 15-20% of children and 1-3% of adults globally (Nutten, 2015). Psoriasis, a chronic inflammatory skin disease, affects about 2-4% of the global population, with a higher prevalence in high-income countries (Parisi et al., 2013).

Acne vulgaris is another highly prevalent skin condition, particularly among adolescents and young adults. It is estimated that acne affects around 85% of individuals aged 12-24 years, making it one of the most common dermatological conditions (Tan & Bhate, 2015). The prevalence of skin cancer, particularly melanoma and non-melanoma skin cancers, has also been rising. In 2020, approximately 1.2 million new cases of melanoma were reported globally, with higher incidence rates in regions with predominantly fair-skinned populations (Bray et al., 2018).

CLINICAL MANIFESTATIONS AND IMPACT ON QUALITY OF LIFE

The clinical manifestations of skin diseases are diverse, ranging from mild rashes and erythema to severe lesions, ulcers, and malignant growths. Eczema typically presents with itchy, red, and inflamed skin, often leading to scratching that exacerbates the condition (Nutten, 2015). Psoriasis is characterized by thick, red patches of skin covered with silvery scales, commonly occurring on the elbows, knees, scalp, and lower back (Griffiths & Barker, 2007). Acne manifests as comedones, papules, pustules, and, in severe cases, nodules and cysts, primarily affecting the face, chest, and back (Tan & Bhate, 2015).

Skin cancer, particularly melanoma, is associated with pigmented lesions that may change in size, shape, or colour. Non-melanoma skin cancers, such as basal cell carcinoma and squamous cell carcinoma, typically present as persistent sores or growths that may bleed or crust over (Bray et al., 2018). These conditions can lead to significant physical discomfort, including pain, itching, and burning sensations.

The psychological and social impacts of skin diseases are profound. Conditions like eczema and psoriasis are often associated with stigma, leading to social withdrawal, low self-esteem, and depression. Acne, particularly in adolescents, can lead to significant emotional distress, including anxiety and depression (Dalgard et al., 2015). Skin cancers, especially melanoma, can cause significant anxiety due to the potential for metastasis and the threat to life.

THE NEED FOR INNOVATIVE TREATMENTS

The management of skin diseases is complex and multifaceted, requiring a combination of pharmacological treatments, lifestyle modifications, and in some cases, surgical interventions. Current treatment modalities for conditions like eczema and psoriasis often involve topical corticosteroids, immunosuppressants, and biologic therapies. However, these treatments are not without limitations, including side effects, high costs, and variable efficacy (Papp et al., 2012).

Acne is typically treated with a combination of topical and oral medications, including retinoids, antibiotics, and hormonal treatments. However, antibiotic resistance and the potential side effects of long-term treatment are significant concerns (Zaenglein et al., 2016). The treatment of skin cancers involves surgical excision, radiation therapy, and chemotherapy, depending on the type and stage of the cancer. Despite advances in treatment, the rising incidence of skin cancer highlights the need for better preventive strategies and more effective therapies (Narayanan, Saladi, & Fox, 2010).

In light of these challenges, there is an urgent need for innovative treatments that are both effective and safe. One promising area of research is the role of dietary antioxidants in skin health. Antioxidants, which neutralize free radicals and reduce oxidative stress, have been shown to have protective effects against various skin conditions, including those induced by UV radiation, inflammation, and ageing (Pullar, Carr, & Vissers, 2017). The potential of dietary antioxidants as a complementary approach to the management of skin diseases warrants further exploration, as they may offer a natural and less invasive option for maintaining skin health and treating skin diseases.

MECHANISMS OF SKIN DISEASES

The pathophysiology of skin diseases is highly complex, involving an intricate interplay of genetic, environmental, and lifestyle factors. These mechanisms vary significantly across different skin conditions, but they often share common pathways involving inflammation, immune dysregulation, and cellular damage. A deeper understanding of these underlying mechanisms is essential for the development of targeted therapies and effective management strategies for skin diseases.

Genetic Factors and Skin Diseases

Genetic predisposition plays a critical role in the development of many skin diseases. For example, psoriasis and atopic dermatitis are two conditions where genetic mutations significantly contribute to disease susceptibility.

Psoriasis is a chronic inflammatory skin disease characterized by hyperproliferation of keratinocytes and an aberrant immune response. Genetic studies have identified several psoriasis susceptibility loci, including the major histocompatibility complex (MHC) on chromosome 6p21, which is strongly associated with the disease (Nair et al., 2009). The most significant genetic determinant of psoriasis is the HLA-C 06:02 allele, which is linked to the development of the disease and influences its clinical course (Ellinghaus et al., 2012). In addition to MHC genes, mutations in genes involved in the nuclear factor-kappa B (NF-κB) signalling pathway, such as TNFAIP3, have been implicated in psoriasis pathogenesis (Tsoi et al., 2012).

Atopic dermatitis (AD), commonly known as eczema, is another skin condition with a strong genetic component. Mutations in the filaggrin (FLG) gene, which encodes a protein crucial for skin barrier function, have been identified as a major risk factor for AD (Palmer et al., 2006). Filaggrin deficiency leads to impaired skin barrier function, making the skin more susceptible to environmental allergens and irritants, which can trigger inflammatory responses characteristic of AD (Smith et al., 2006). In addition to FLG, other genetic factors, including mutations in genes encoding cytokines and immune regulators, such as interleukin-4 (IL-4) and interleukin-13 (IL-13), also contribute to the disease's development (Guttman-Yassky et al., 2017).

Environmental Triggers and Skin Diseases

While genetic predispositions set the stage for skin diseases, environmental factors often act as triggers that exacerbate or precipitate the onset of these conditions. Ultraviolet (UV) radiation, pollutants, and microbial infections are some of the key environmental factors that influence the pathogenesis of skin diseases.

UV Radiation is a well-established environmental factor that contributes to various skin conditions, including skin cancer, photoaging, and autoimmune diseases like lupus erythematosus. UV radiation causes direct DNA damage in skin cells, leading to mutations that can initiate carcinogenesis (Narayanan, Saladi, & Fox, 2010). The most common form of DNA damage induced by UV radiation is the formation of cyclobutane pyrimidine dimers (CPDs), which, if unrepaired, can result in mutations in tumour suppressor genes such as p53, driving the development of skin cancers (Brash, 2015). In addition to DNA damage, UV radiation also induces immunosuppression by altering cytokine production and decreasing the skin's ability to mount effective immune responses against pathogens (Yoshikawa et al., 1990).

Pollutants such as particulate matter, polycyclic aromatic hydrocarbons (PAHs), and heavy metals are increasingly recognized as contributors to skin diseases. Pollutants can penetrate the skin barrier and generate reactive oxygen species (ROS), leading to oxidative stress and inflammation. This oxidative stress can exacerbate conditions like atopic dermatitis and psoriasis by further impairing skin barrier function and promoting inflammatory pathways (Puri et al., 2017). Long-term exposure to pollutants has also been linked to the development of skin cancers, as well as premature ageing and pigmentary disorders (Vierkötter & Krutmann, 2012).

Microbial Infections also play a significant role in the pathophysiology of skin diseases. For instance, Staphylococcus aureus colonization is common in patients with atopic dermatitis and is associated with disease exacerbation. The bacteria produce toxins that act as superantigens, triggering an exaggerated immune response that leads to increased inflammation and skin barrier dysfunction (Ong et al., 2002). Similarly, in acne, the bacterium Cutibacterium acnes (formerly Propionibacterium acnes) is involved in the inflammatory process by inducing cytokine production and activating the innate immune system (Kurokawa et al., 2009).

Immune Dysregulation and Inflammation

A hallmark of many skin diseases is immune dysregulation, where the body's immune system either overreacts to stimuli or fails to regulate itself properly, leading to chronic inflammation and tissue damage.

In psoriasis, the immune system plays a central role in driving the disease. The condition is characterized by an overactive T-cell-mediated immune response, particularly involving T-helper 17 (Th17) cells. These cells produce pro-inflammatory cytokines, such as interleukin-17 (IL-17) and interleukin-23 (IL-23), which stimulate keratinocyte proliferation and inflammation in the skin (Lowes et al., 2014). The resulting cycle of inflammation and cell turnover leads to the formation of the characteristic psoriatic plaques.

Atopic dermatitis involves a complex interplay between genetic susceptibility, environmental factors, and immune dysregulation. In AD, there is a skewing of the immune response towards a T-helper 2 (Th2) profile, which promotes the production of cytokines like IL-4, IL-5, and IL-13. These cytokines contribute to inflammation, IgE production, and eosinophil activation, all of which are characteristic features of AD (Guttman-Yassky et al., 2017). Moreover, chronic AD is associated with a shift towards a Th1 and Th22 response, further complicating the disease's pathology (Brunner et al., 2017).

In acne, the inflammatory process is initiated by the interaction between Cutibacterium acnes and the immune system. The bacterium stimulates the production of pro-inflammatory cytokines such as interleukin-1 (IL-1) and tumour necrosis factor-alpha (TNF-α), which contribute to the formation of inflammatory lesions like papules and pustules (Kurokawa et al., 2009). Additionally, the activation of the innate immune system through toll-like receptors (TLRs) on keratinocytes and sebocytes plays a crucial role in the inflammatory response seen in acne (Kistowska et al., 2014).

POTENTIAL THERAPEUTIC TARGETS

Understanding the molecular and cellular mechanisms underlying skin diseases opens up opportunities for the development of targeted therapies. For instance, biologic drugs that target specific cytokines, such as IL-17 and IL-23 inhibitors, have shown great promise in treating psoriasis by interrupting the inflammatory cascade (Gottlieb et al., 2012). Similarly, in atopic dermatitis, targeted therapies that inhibit IL-4 and IL-13 signaling, such as dupilumab, have been successful in managing the disease (Simpson et al., 2016).

In the context of acne, therapies targeting the inflammatory response and sebum production, such as retinoids and anti-androgens, remain the mainstay of treatment (Zaenglein et al., 2016). Additionally, research into the microbiome's role in skin health is paving the way for new approaches, such as probiotics and microbiome-modulating therapies, which aim to restore healthy skin microbiota and reduce inflammation (Bowe & Logan, 2011).

INFLAMMATION AND SKIN HEALTH

Inflammation is a fundamental response of the immune system to injury or infection, aimed at protecting the body by eliminating harmful stimuli and initiating the healing process. However, when inflammation becomes chronic or dysregulated, it can contribute to the pathogenesis of various skin diseases. This section explores the role of inflammation in skin health, with a particular focus on the inflammatory mediators and pathways involved in conditions like psoriasis and autoimmune skin disorders such as lupus erythematosus.

The Inflammatory Response in Skin Diseases

Inflammation in the skin is a complex process involving the activation of various immune cells, the release of cytokines and chemokines, and the recruitment of inflammatory cells to the affected area. The inflammatory response can be triggered by several factors, including pathogens, environmental insults, and autoimmune reactions.

Cytokines and Chemokines: Cytokines are small proteins that play a critical role in cell signalling during the immune response. They are produced by various cells, including keratinocytes, fibroblasts, and immune cells such as T cells, macrophages, and dendritic cells. Inflammatory cytokines such as interleukin-1 (IL-1), tumour necrosis factor-alpha (TNF-α), and interleukin-6 (IL-6) are key players in initiating and sustaining inflammation in the skin (Dinarello, 2009). Chemokines, on the other hand, are a subset of cytokines that specifically induce the migration of immune cells to sites of inflammation. For example, CCL2 and CXCL10 are chemokines that attract monocytes and T cells, respectively, to inflamed skin tissue (Zlotnik & Yoshie, 2012).

Inflammatory Pathways in Psoriasis: Psoriasis is a chronic inflammatory skin disease characterized by the formation of red, scaly plaques on the skin. The pathogenesis of psoriasis involves a complex interplay between the innate and adaptive immune systems, with a central role played by T cells and dendritic cells. The inflammatory cascade in psoriasis is driven by the release of pro-inflammatory cytokines, particularly interleukin-17 (IL-17), interleukin-23 (IL-23), and TNF-α. These cytokines promote the proliferation of keratinocytes and the recruitment of additional immune cells to the skin, leading to the characteristic hyperplasia and inflammation seen in psoriatic lesions (Lowes, Suárez-Fariñas, & Krueger, 2014).

IL-17, produced primarily by Th17 cells, is particularly important in psoriasis. It acts on keratinocytes to induce the production of antimicrobial peptides, chemokines, and other pro-inflammatory cytokines, thereby amplifying the inflammatory response. The IL-23/Th17 axis is now recognized as a major driver of psoriasis, and therapies targeting this pathway, such as IL-17 and IL-23 inhibitors, have shown significant efficacy in treating the disease (Greb et al., 2016).

Inflammatory Pathways in Autoimmune Skin Disorders: Autoimmune skin disorders like lupus erythematosus involve an inappropriate immune response in which the body's immune system mistakenly attacks its tissues. In lupus erythematosus, autoantibodies target nuclear antigens, leading to the formation of immune complexes that deposit in the skin and other organs, triggering inflammation.

The inflammatory response in lupus erythematosus is mediated by both the innate and adaptive immune systems. Type I interferons (IFNs), particularly IFN-α, play a crucial role in the disease. IFN-α is produced by plasmacytoid dendritic cells (pDCs) in response to immune complexes containing self-DNA or RNA, which are recognized by toll-like receptors (TLRs) on the pDCs. The production of IFN-α leads to the activation of other immune cells, including T cells and B cells, and promotes the production of additional autoantibodies, creating a cycle of inflammation and tissue damage (Crow, 2014).

In addition to IFNs, other cytokines, such as TNF-α, IL-6, and interleukin-10 (IL-10), contribute to the inflammatory milieu in lupus erythematosus. These cytokines promote the recruitment of inflammatory cells to the skin, where they cause damage to the epidermis and dermis, leading to the characteristic skin lesions seen in the disease (Chasset & Francès, 2019).

Immune System Involvement in Skin Inflammation

The immune system's involvement in skin inflammation is multifaceted, involving both innate and adaptive immune responses.

Innate Immune Response: The innate immune system is the body's first line of defense against pathogens and is activated immediately upon encountering harmful stimuli. In the skin, the innate immune response involves various cells, including keratinocytes, macrophages, dendritic cells, and mast cells. These cells recognize pathogens or damaged cells through pattern recognition receptors (PRRs), such as TLRs, and initiate an inflammatory response by releasing cytokines and chemokines (Kawai & Akira, 2010).

Keratinocytes, the predominant cell type in the epidermis, play a crucial role in the innate immune response. They produce antimicrobial peptides (AMPs) like cathelicidins and defensins in response to infection or injury, which help eliminate pathogens. Additionally, keratinocytes release pro-inflammatory cytokines such as IL-1 and TNF-α, which amplify the inflammatory response and recruit immune cells to the site of injury (Nestle, Di Meglio, Qin, & Nickoloff, 2009).

Adaptive Immune Response: The adaptive immune response, mediated by T cells and B cells, provides a more specific and long-lasting immune defence. Inflammatory skin diseases often involve dysregulation of the adaptive immune system, leading to chronic inflammation.

In psoriasis, as mentioned earlier, the adaptive immune response is dominated by Th17 cells, which produce IL-17 and IL-22, cytokines that drive keratinocyte proliferation and inflammation. In addition, Th1 cells, which produce interferon-gamma (IFN-γ), also contribute to the inflammatory response by activating macrophages and promoting further cytokine production (Nickoloff, 2007).

In lupus erythematosus, the adaptive immune response is characterized by the production of autoantibodies by B cells. These autoantibodies form immune complexes that deposit in the skin, leading to complement activation and recruitment of inflammatory cells. T cells, particularly CD4+ T helper cells, also play a role by providing help to B cells and contributing to the production of autoantibodies (Tsokos, 2011).

Anti-Inflammatory Treatments for Skin Diseases

Given the central role of inflammation in skin diseases, anti-inflammatory treatments are a cornerstone of therapy. These treatments aim to reduce inflammation, alleviate symptoms, and prevent disease progression.

Corticosteroids: Topical corticosteroids are commonly used to treat inflammatory skin conditions like eczema and psoriasis. They work by suppressing the production of pro-inflammatory cytokines and inhibiting the migration of immune cells to the skin. While effective, long-term use of corticosteroids can lead to side effects such as skin thinning and adrenal suppression (Ference & Last, 2009).

Biologics: Biologic therapies targeting specific cytokines have revolutionized the treatment of inflammatory skin diseases. For example, TNF-α inhibitors (e.g., etanercept, infliximab) and IL-17 inhibitors are highly effective in treating moderate to severe psoriasis by blocking key cytokines involved in the inflammatory cascade (Greb et al., 2016). Similarly, in lupus erythematosus, B cell-targeted therapies (e.g., belimumab) are used to reduce the production of autoantibodies and control disease activity (Furie et al., 2011).

NSAIDs and Immunosuppressants: Nonsteroidal anti-inflammatory drugs (NSAIDs) and immunosuppressants like methotrexate and cyclosporine are also used to manage inflammation in skin diseases. NSAIDs reduce inflammation by inhibiting cyclooxygenase (COX) enzymes, which are involved in prostaglandin synthesis, while immunosuppressants suppress the overall immune response (Menter et al., 2011).

OXIDATIVE STRESS AND SKIN DISEASES

Oxidative stress is a state in which there is an imbalance between the production of reactive oxygen species (ROS) and the skin's ability to detoxify these reactive intermediates or repair the resulting damage. This imbalance plays a significant role in the pathogenesis of various skin diseases, contributing to cellular damage, inflammation, and the acceleration of aging processes in the skin. This section delves into the sources of oxidative stress, its impact on skin health, and the potential for therapeutic interventions using antioxidants.

Sources of Oxidative Stress

Oxidative stress is defined by the excessive accumulation of reactive oxygen species (ROS) in cells and tissues. ROS are chemically reactive molecules containing oxygen, such as superoxide anion (O2•−), hydrogen peroxide (H2O2), and hydroxyl radical (•OH). While ROS are a normal byproduct of cellular metabolism and play essential roles in cell signalling and homeostasis, excessive ROS levels can lead to oxidative stress, resulting in cellular and molecular damage (Valko et al., 2007).

Endogenous Sources of ROS: The primary source of ROS within the skin is mitochondrial oxidative phosphorylation, a process that generates ATP through the electron transport chain. During this process, a small percentage of oxygen molecules are incompletely reduced, leading to the formation of superoxide anions. Other endogenous sources of ROS include enzymatic reactions involving NADPH oxidases, xanthine oxidase, and cytochrome P450 enzymes. Additionally, inflammatory cells, such as neutrophils and macrophages, produce ROS as part of the immune response to pathogens (Sies, 2017).

Exogenous Sources of ROS: External factors also contribute to ROS production in the skin. Ultraviolet (UV) radiation from the sun is a significant exogenous source of ROS. UV radiation, particularly UVA and UVB, penetrates the skin and generates ROS through the photolysis of water and other chromophores within the skin. This ROS production can lead to direct DNA damage, lipid peroxidation, and the oxidation of proteins, all of which contribute to skin ageing and carcinogenesis (Pillai, Oresajo, & Hayward, 2005). Other exogenous sources include pollution, tobacco smoke, and exposure to chemicals, all of which can enhance oxidative stress in the skin (Valacchi et al., 2012).

Impact of Oxidative Stress on Skin Cells and Tissues

Oxidative stress has profound effects on skin cells and tissues, contributing to a range of pathological conditions, including premature ageing, inflammatory skin diseases, and skin cancer.

Cellular Damage and Inflammation: ROS can cause direct damage to cellular components, including lipids, proteins, and DNA. Lipid peroxidation, the oxidative degradation of lipids, is particularly detrimental because it disrupts cellular membranes, leading to cell death and the release of pro-inflammatory mediators. Protein oxidation can result in the formation of dysfunctional or misfolded proteins, which can accumulate and impair cellular function. DNA damage induced by ROS includes mutations, strand breaks, and the formation of 8-hydroxy-2'-deoxyguanosine (8-OHdG), a common marker of oxidative DNA damage. Accumulation of such damage can lead to apoptosis or, in the case of incomplete repair, oncogenesis (Fisher, Johns, & Datta, 2000).

Skin Ageing: Oxidative stress is a major contributor to the ageing process in the skin. Over time, the accumulation of oxidative damage leads to the degradation of collagen and elastin, the primary structural proteins in the skin. This results in the loss of skin elasticity, the formation of wrinkles, and the thinning of the skin. Additionally, oxidative stress can impair the function of fibroblasts, the cells responsible for producing collagen, further exacerbating the ageing process (Ganceviciene et al., 2012). The concept of "photoaging" specifically refers to the accelerated aging of the skin due to chronic UV exposure and the associated oxidative stress.

Inflammatory Skin Diseases: Several inflammatory skin diseases, such as psoriasis, atopic dermatitis, and vitiligo, are associated with elevated levels of oxidative stress. In psoriasis, ROS contribute to the hyperproliferation of keratinocytes and the activation of immune cells, leading to the formation of the characteristic scaly plaques (Bickers & Athar, 2006). In atopic dermatitis, oxidative stress exacerbates inflammation by activating nuclear factor-kappa B (NF-κB), a key transcription factor that drives the expression of pro-inflammatory cytokines. Vitiligo, a depigmenting disorder, is also linked to oxidative stress, which is thought to cause the death of melanocytes, the cells responsible for producing skin pigment (Schallreuter, Moore, Wood, & Beazley, 2001).

Skin Cancer: Chronic oxidative stress is a well-established risk factor for skin cancer, particularly melanoma, squamous cell carcinoma, and basal cell carcinoma. ROS can induce mutations in oncogenes and tumour suppressor genes, contributing to the initiation and progression of cancer. UV-induced oxidative damage is a critical factor in the development of melanoma, where oxidative stress not only promotes DNA mutations but also creates an environment that favours tumour growth and immune evasion (Davies, 2016).

THERAPEUTIC USE OF ANTIOXIDANTS

Given the significant role of oxidative stress in skin diseases, antioxidants have gained attention as potential therapeutic agents. Antioxidants work by neutralizing ROS, thereby preventing or mitigating oxidative damage.

Endogenous Antioxidants: The skin possesses several endogenous antioxidant systems that help maintain redox balance. These include enzymatic antioxidants such as superoxide dismutase (SOD), catalase, and glutathione peroxidase, as well as non-enzymatic antioxidants like glutathione, vitamin C, and vitamin E. However, in cases of excessive ROS production, these natural defences can be overwhelmed, necessitating external supplementation (Briganti & Picardo, 2003).

Topical and Systemic Antioxidants: The topical application of antioxidants is a common strategy to protect the skin from oxidative damage, particularly in the context of photoaging and UV exposure. Vitamin C and vitamin E are two of the most widely used antioxidants in skincare products. Vitamin C (ascorbic acid) is a potent antioxidant that can neutralize ROS and regenerate vitamin E from its oxidized form, enhancing its effectiveness. Additionally, vitamin C stimulates collagen synthesis, providing additional anti-ageing benefits. Vitamin E (tocopherol) is a lipid-soluble antioxidant that protects cell membranes from lipid peroxidation and also has anti-inflammatory properties (Pinnell et al., 2001).

Plant-Derived Antioxidants: Polyphenols, found in various plants such as green tea, grapes, and berries, are powerful antioxidants with potential benefits for skin health. For example, epigallocatechin gallate (EGCG), a polyphenol found in green tea, has been shown to reduce UV-induced oxidative stress and inflammation, as well as to inhibit the development of skin cancer in animal models (Katiyar, 2003). Resveratrol, a polyphenol found in grapes, has also demonstrated protective effects against oxidative damage and has been incorporated into skincare formulations (Berman et al., 2017).

Clinical Applications: In clinical settings, antioxidants are being explored for their potential to treat and prevent oxidative stress-related skin diseases. For example, oral supplementation with antioxidants such as vitamin C, vitamin E, and coenzyme Q10 has been investigated for its ability to improve skin health and reduce the signs of ageing. Additionally, antioxidants are being studied for their role in preventing skin cancer, with some evidence suggesting that they may reduce the risk of UV-induced malignancies (Naylor et al., 1995).

CURRENT TREATMENTS FOR SKIN DISEASES

The treatment of skin diseases is multifaceted, encompassing a wide array of therapeutic strategies tailored to the specific condition, its severity, and the patient's overall health. These treatments range from topical applications to systemic medications, each with its unique mechanisms of action, efficacy profiles, and limitations. This section reviews the current standard treatments for major skin diseases, including the use of corticosteroids, immunosuppressants, and biologics, as well as emerging therapies such as gene therapy and regenerative medicine. Additionally, it highlights the growing interest in dietary antioxidants as adjunctive therapies.

Topical Treatments

Corticosteroids: Corticosteroids are among the most commonly prescribed medications for inflammatory skin diseases such as eczema (atopic dermatitis) and psoriasis. These drugs work by suppressing the immune response, reducing inflammation, and alleviating symptoms like itching and redness. Topical corticosteroids are classified by their potency, ranging from low-potency hydrocortisone to high-potency clobetasol propionate. While effective, long-term use of corticosteroids can lead to side effects such as skin thinning, telangiectasia, and striae (skin stretch marks) (Feldman, 2017).

Calcineurin Inhibitors: Topical calcineurin inhibitors, such as tacrolimus and pimecrolimus, are used primarily in the treatment of atopic dermatitis, particularly in sensitive areas like the face and eyelids where corticosteroids might cause adverse effects. These agents inhibit the activation of T-cells and the release of pro-inflammatory cytokines, thereby reducing inflammation without causing skin atrophy. However, concerns have been raised about the long-term safety of calcineurin inhibitors, including a potential increased risk of malignancies (Papp et al., 2019).

Retinoids: Topical retinoids, including tretinoin, adapalene, and tazarotene, are derivatives of vitamin A and are primarily used in the treatment of acne and psoriasis. They work by modulating cell proliferation and differentiation, promoting the shedding of dead skin cells, and preventing the formation of comedones (clogged hair follicles). In psoriasis, retinoids help to normalize the hyperproliferation of keratinocytes. However, their use can be associated with side effects such as skin irritation, dryness, and increased sensitivity to sunlight (Kang, Fisher, & Voorhees, 2001).

Systemic Treatments

Immunosuppressants: Systemic immunosuppressants such as methotrexate, cyclosporine, and azathioprine are used in severe cases of inflammatory skin diseases like psoriasis and atopic dermatitis that do not respond to topical treatments. These drugs suppress the immune system at various points, reducing the inflammatory response that drives disease symptoms. For example, methotrexate inhibits dihydrofolate reductase, leading to reduced DNA synthesis in rapidly proliferating cells, including immune cells. While effective, these medications carry significant risks, including hepatotoxicity, nephrotoxicity, and increased susceptibility to infections (Menter et al., 2019).

Biologics: Biologic therapies have revolutionized the treatment of chronic inflammatory skin diseases, particularly psoriasis. Biologics are targeted therapies that specifically inhibit key molecules in the inflammatory pathways, such as tumour necrosis factor-alpha (TNF-α), interleukin-12/23 (IL-12/23), and interleukin-17 (IL-17). Examples include etanercept, infliximab, ustekinumab, and secukinumab. These agents have shown remarkable efficacy in reducing the severity of psoriasis and improving patients' quality of life. However, biologics are expensive and may increase the risk of infections and other immune-related conditions (Bozek, 2019).

Oral Retinoids: Oral retinoids, such as isotretinoin and acitretin, are used in the treatment of severe acne and psoriasis. These drugs work by reducing sebaceous gland activity, normalizing keratinization, and exerting anti-inflammatory effects. Isotretinoin is particularly effective in cases of nodulocystic acne, leading to prolonged remission in many patients. However, oral retinoids are teratogenic and can cause significant side effects, including mucocutaneous dryness, hyperlipidemia, and liver dysfunction (Layton, 2009).

EMERGING THERAPIES AND INNOVATIVE APPROACHES

Gene Therapy: Gene therapy represents a cutting-edge approach to treating genetic skin diseases by correcting or compensating for defective genes. Recent advances in CRISPR-Cas9 technology have shown promise in targeting and editing specific genetic mutations associated with conditions like epidermolysis bullosa, a severe blistering disorder. Early clinical trials are exploring the safety and efficacy of gene therapy for various skin conditions, although challenges such as delivery methods and off-target effects remain (Hirsch et al., 2017).

Regenerative Medicine: Regenerative medicine, including stem cell therapy and tissue engineering, is an emerging field with the potential to revolutionize the treatment of skin diseases. Stem cells, particularly mesenchymal stem cells (MSCs), have shown promise in promoting wound healing and tissue regeneration due to their ability to differentiate into various cell types and secrete growth factors. Research is ongoing into the use of stem cell-based therapies for conditions such as chronic wounds, burns, and vitiligo (Jin et al., 2018).

Phototherapy: Phototherapy, particularly narrowband UVB therapy, is an established treatment for conditions like psoriasis, vitiligo, and eczema. This therapy involves exposing the skin to specific wavelengths of ultraviolet light, which reduces inflammation and modulates the immune response. While effective, phototherapy requires multiple sessions and carries a risk of long-term skin damage, including an increased risk of skin cancer (El-Mofty et al., 2021).

Dietary Antioxidants: Dietary antioxidants are gaining attention as adjunctive therapies for skin diseases due to their potential to mitigate oxidative stress, a key factor in the pathogenesis of many skin conditions. Antioxidants such as vitamins C and E, polyphenols, and carotenoids can scavenge reactive oxygen species (ROS) and reduce inflammation. Studies have shown that diets rich in these antioxidants can improve skin health and potentially reduce the severity of diseases like psoriasis and atopic dermatitis. However, more research is needed to fully understand their therapeutic potential and optimal dosing (D'Errico et al., 2018).

Dietary Antioxidants: An Overview

Dietary antioxidants are molecules that can neutralize reactive oxygen species (ROS) and protect cells from oxidative stress, a state in which the production of ROS exceeds the body's ability to counteract their harmful effects. These antioxidants play a crucial role in maintaining skin health by preventing oxidative damage to skin cells, which is implicated in the ageing process and various skin diseases. This section provides an in-depth overview of dietary antioxidants, focusing on their definition, types, and primary sources, with a particular emphasis on common antioxidants like vitamins C and E, polyphenols, and carotenoids.

Dietary Antioxidants

Dietary antioxidants are compounds obtained from food that help to neutralize free radicals unstable molecules that can damage cells, proteins, and DNA by reacting with them. Free radicals are a natural byproduct of cellular metabolism, but their levels can increase due to external factors like UV radiation, pollution, and poor diet, leading to oxidative stress. Antioxidants inhibit the oxidation of molecules by donating electrons to free radicals without becoming unstable themselves, thereby breaking the chain reaction of oxidative damage (Sies, 2017).

Types of Dietary Antioxidants

Dietary antioxidants can be broadly classified into several categories based on their chemical structure and function. The most well-known types include vitamins, polyphenols, and carotenoids:

1. Vitamins:

Vitamin C (Ascorbic Acid): Vitamin C is a water-soluble antioxidant that plays a vital role in collagen synthesis, wound healing, and protecting the skin from oxidative damage caused by UV radiation. It regenerates vitamin E from its oxidized form and enhances the skin's defence against environmental stressors. Common dietary sources include citrus fruits, strawberries, kiwi, and bell peppers (Pullar, Carr, & Vissers, 2017).

Vitamin E (Tocopherol): Vitamin E is a fat-soluble antioxidant that protects cell membranes from oxidative damage by scavenging lipid peroxyl radicals. It is often combined with vitamin C in skin care products to enhance its photoprotective effects. Rich dietary sources of vitamin E include nuts, seeds, spinach, and avocados (Brigelius-Flohé & Davies, 2007).

2. Polyphenols:

Polyphenols are a diverse group of plant-based antioxidants that include flavonoids, phenolic acids, and tannins. These compounds have strong antioxidant properties and have been linked to various health benefits, including reduced inflammation and protection against UV-induced skin damage. Examples of polyphenol-rich foods include green tea, berries, dark chocolate, and red wine (Scalbert, Johnson, & Saltmarsh, 2005).

Flavonoids: Flavonoids, a subgroup of polyphenols, are particularly abundant in fruits and vegetables. They are known for their ability to reduce inflammation, modulate cellular signalling pathways, and protect against skin damage. Examples include quercetin, found in apples and onions, and catechins, abundant in green tea (Williams, Spencer, & Rice-Evans, 2004).

3. Carotenoids:

Beta-carotene: Beta-carotene is a precursor to vitamin A and a potent antioxidant that can quench singlet oxygen, a highly reactive form of oxygen. It is essential for skin health as it helps in maintaining the skin's barrier function and promoting wound healing. Carotenoids like beta-carotene are found in carrots, sweet potatoes, and leafy green vegetables (Krinsky & Johnson, 2005).

Lycopene: Lycopene, another carotenoid, is known for its protective effects against UV-induced skin damage and its ability to reduce the risk of skin cancer. It is abundant in tomatoes, watermelon, and pink grapefruit (Story, Kopec, Schwartz, & Harris, 2010).

Sources of Dietary Antioxidants

Dietary antioxidants are naturally present in a wide variety of foods, particularly fruits, vegetables, nuts, seeds, and whole grains. They can also be obtained through dietary supplements, though it is generally recommended to obtain antioxidants from whole foods, where they are often accompanied by other beneficial nutrients that enhance their effectiveness.

1. Fruits and Vegetables: Fresh fruits and vegetables are the most potent sources of dietary antioxidants. Berries, such as blueberries, strawberries, and blackberries, are particularly rich in polyphenols and vitamin C. Leafy greens like spinach and kale provide carotenoids and vitamin E, while citrus fruits are excellent sources of vitamin C.

2. Nuts and Seeds: Nuts, such as almonds and walnuts, and seeds, such as sunflower seeds, are rich in vitamin E and other antioxidants like selenium and zinc, which contribute to skin health by protecting against oxidative damage and supporting immune function.

3. Whole Grains: Whole grains like oats, quinoa, and brown rice contain antioxidant compounds like selenium, which help in reducing oxidative stress and inflammation. They also provide fibre, which supports overall health.

4. Beverages: Beverages like green tea, red wine, and coffee are rich sources of polyphenols, particularly flavonoids. Green tea, in particular, contains catechins, which have been shown to protect the skin from UV-induced damage and reduce the risk of skin cancer (Yang & Wang, 2019).

Understanding the basic properties and sources of dietary antioxidants is crucial for exploring their therapeutic potential in skin health. These compounds play a vital role in protecting the skin from oxidative stress, a key factor in ageing and various skin diseases.

Mechanisms of Antioxidant Action

Antioxidants play a crucial role in combating oxidative stress, a condition that arises when there is an imbalance between the production of reactive oxygen species (ROS) and the body's ability to neutralize them through endogenous antioxidant defences. Oxidative stress is implicated in various skin conditions, including ageing, inflammation, and carcinogenesis. Understanding the mechanisms by which antioxidants counteract oxidative stress is essential for evaluating their potential in the treatment and prevention of skin diseases.

Direct Scavenging of Reactive Oxygen Species (ROS)

One of the primary mechanisms by which antioxidants exert their protective effects is through the direct scavenging of ROS. ROS are highly reactive molecules that include free radicals such as superoxide anion (O2•−), hydroxyl radical (•OH), and non-radical species like hydrogen peroxide (H2O2). These molecules can cause significant damage to cellular components, including lipids, proteins, and DNA, leading to cell dysfunction and death (Valko et al., 2007).

Vitamin C (Ascorbic Acid): Vitamin C is a potent water-soluble antioxidant that can donate electrons to neutralize free radicals, particularly in the aqueous compartments of the skin. It directly scavenges ROS such as superoxide anion, hydroxyl radical, and singlet oxygen, thereby preventing oxidative damage to skin cells. Additionally, vitamin C regenerates vitamin E from its oxidized form, ensuring continued antioxidant protection (Du, Cullen, & Buettner, 2012).

Vitamin E (Tocopherol): Vitamin E, a lipid-soluble antioxidant, primarily protects cell membranes from oxidative damage by scavenging lipid peroxyl radicals, which are generated during lipid peroxidation. This process helps maintain the integrity of cellular membranes and prevents the propagation of free radical damage (Brigelius-Flohé & Davies, 2007).

Flavonoids: Flavonoids, a class of polyphenols, can scavenge a wide range of ROS due to their ability to donate hydrogen atoms or electrons. For example, quercetin, a well-known flavonoid, has been shown to scavenge superoxide anion and hydroxyl radicals, reducing oxidative damage and inflammation in the skin (Boots, Haenen, & Bast, 2008).

Enhancement of Endogenous Antioxidant Defences

In addition to directly scavenging ROS, antioxidants can enhance the body's endogenous antioxidant defences, which include enzymes such as superoxide dismutase (SOD), catalase, and glutathione peroxidase. These enzymes play a critical role in neutralizing ROS and maintaining cellular redox balance.

Nrf2 Pathway Activation: The Nrf2 (nuclear factor erythroid 2-related factor 2) pathway is a key regulator of the cellular antioxidant response. When activated by oxidative stress, Nrf2 translocates to the nucleus, where it binds to antioxidant response elements (AREs) in the DNA, leading to the upregulation of genes encoding antioxidant enzymes. Dietary antioxidants such as curcumin (found in turmeric) and sulforaphane (found in broccoli) have been shown to activate the Nrf2 pathway, thereby enhancing the expression of endogenous antioxidant enzymes and protecting skin cells from oxidative damage (Jiang et al., 2016).

Glutathione Synthesis: Glutathione is one of the most important intracellular antioxidants, playing a critical role in detoxifying ROS and maintaining cellular redox balance. Dietary antioxidants such as alpha-lipoic acid can enhance glutathione synthesis by providing cysteine, a precursor in the glutathione biosynthesis pathway. This increased glutathione availability helps protect skin cells from oxidative damage and supports overall skin health (Packer, Witt, & Tritschler, 1995).

Modulation of Immune Response and Inflammation

Oxidative stress is closely linked to inflammation, as ROS can activate various inflammatory pathways, including the nuclear factor-kappa B (NF-κB) pathway, which regulates the expression of pro-inflammatory cytokines and mediators. Antioxidants can modulate these pathways, thereby reducing inflammation and its associated damage to the skin.

Inhibition of NF-κB Activation: Antioxidants such as resveratrol (found in grapes) and epigallocatechin gallate (EGCG, found in green tea) can inhibit the activation of the NF-κB pathway, thereby reducing the production of pro-inflammatory cytokines like TNF-α, IL-1β, and IL-6. This anti-inflammatory action is particularly beneficial in conditions like psoriasis and atopic dermatitis, where chronic inflammation plays a central role in disease pathogenesis (Aggarwal & Shishodia, 2004).

Reduction of Oxidative Stress-Induced Apoptosis: In addition to reducing inflammation, antioxidants can protect skin cells from oxidative stress-induced apoptosis (programmed cell death). For instance, polyphenols like quercetin and resveratrol can inhibit apoptosis by modulating the expression of apoptotic proteins such as Bcl-2 and Bax, thereby promoting cell survival and tissue repair in the skin (Lee et al., 2009).

Maintenance of Skin Homeostasis

Antioxidants also play a role in maintaining skin homeostasis by protecting the skin's extracellular matrix (ECM), which is composed of collagen, elastin, and hyaluronic acid. The ECM provides structural support to the skin and is essential for skin elasticity and hydration.

Protection of Collagen from Degradation: ROS can degrade collagen fibres by activating matrix metalloproteinases (MMPs), enzymes that break down collagen and other ECM components. Antioxidants such as vitamin C and polyphenols can inhibit the activity of MMPs, thereby preserving collagen integrity and preventing skin ageing and wrinkling (Lupo, 2001).

Preservation of Hyaluronic Acid: Hyaluronic acid is a key component of the skin's ECM that helps retain moisture and maintain skin hydration. Oxidative stress can reduce hyaluronic acid levels, leading to dry and wrinkled skin. Antioxidants help preserve hyaluronic acid by reducing oxidative degradation and promoting its synthesis (Ganceviciene et al., 2012).

Antioxidants combat oxidative stress through multiple mechanisms, including direct scavenging of ROS, enhancement of endogenous antioxidant defences, modulation of the immune response, and maintenance of skin homeostasis. These actions not only protect the skin from oxidative damage but also contribute to the prevention and treatment of various skin diseases

ROLE OF ANTIOXIDANTS IN SKIN HEALTH

Antioxidants play a vital role in maintaining skin health by neutralizing free radicals, reducing inflammation, enhancing collagen synthesis, and protecting against damage from ultraviolet (UV) radiation. The skin, being the body's largest organ, is continuously exposed to environmental stressors such as UV light, pollution, and toxins, all of which contribute to oxidative stress. This oxidative stress can lead to skin ageing, inflammation, and the development of skin diseases. Numerous clinical and preclinical studies have demonstrated the beneficial effects of antioxidants in mitigating these adverse effects, thereby promoting skin health.

Vitamin C: Collagen Synthesis and Photoprotection

Vitamin C (ascorbic acid) is a water-soluble antioxidant that is essential for the synthesis of collagen, a major structural protein in the skin. Collagen provides skin with strength and elasticity, and its degradation leads to wrinkles and sagging skin. Vitamin C acts as a cofactor for prolyl and lysyl hydroxylase, enzymes involved in the stabilization and cross-linking of collagen molecules (Pinnell, 2003).

Collagen Synthesis: Vitamin C stimulates collagen production by promoting the hydroxylation of proline and lysine residues in procollagen, which are necessary for proper collagen fibre formation. This process helps maintain the skin's structural integrity and reduces the appearance of fine lines and wrinkles (Pullar, Carr, & Vissers, 2017).

Photoprotection: Vitamin C also plays a significant role in protecting the skin from UV-induced damage. It neutralizes free radicals generated by UV exposure, thereby preventing oxidative stress and reducing the risk of photodamage, which includes DNA damage, sunburn, and the formation of hyperpigmentation (Bennett et al., 2012).

Vitamin E: Lipid Peroxidation Inhibition and Skin Barrier Protection

Vitamin E (tocopherol) is a lipid-soluble antioxidant that is primarily found in the cell membranes and lipophilic compartments of the skin. It is known for its ability to protect the skin from oxidative damage by inhibiting lipid peroxidation, a process where free radicals attack the lipids in cell membranes, leading to cell damage and increased skin permeability.

Lipid Peroxidation Inhibition: Vitamin E acts as a chain-breaking antioxidant, terminating the lipid peroxidation chain reaction by donating hydrogen atoms to free radicals, thus stabilizing them and preventing further oxidative damage. This action is crucial in maintaining the integrity of cell membranes and protecting the skin from environmental stressors (Brigelius-Flohé & Davies, 2007).

Skin Barrier Protection: The skin's barrier function, which prevents the loss of moisture and protects against external irritants, is heavily reliant on the presence of adequate lipids in the stratum corneum. Vitamin E supports the barrier function by preserving the lipid content in the skin and preventing transepidermal water loss (TEWL), which is essential for maintaining skin hydration and preventing dryness (Thiele, Traber, & Podda, 1998).

POLYPHENOLS: ANTI-INFLAMMATORY AND ANTIOXIDANT EFFECTS

Polyphenols are a diverse group of plant-derived compounds that possess strong antioxidant and anti-inflammatory properties. They are found in various fruits, vegetables, teas, and wines, with green tea polyphenols (e.g., epigallocatechin gallate or EGCG) and resveratrol (found in grapes and red wine) being among the most studied for their skin health benefits.

Anti-inflammatory Effects: Polyphenols exert their anti-inflammatory effects by modulating key signalling pathways involved in inflammation. For example, EGCG has been shown to inhibit the activation of nuclear factor-kappa B (NF-κB), a transcription factor that regulates the expression of pro-inflammatory cytokines. By reducing the levels of these cytokines, polyphenols help alleviate inflammatory skin conditions such as psoriasis and atopic dermatitis (Katiyar, 2003).

Antioxidant Effects: Polyphenols also directly scavenge ROS and enhance the body's endogenous antioxidant defence systems. Resveratrol, for instance, activates the Nrf2 pathway, leading to the upregulation of antioxidant enzymes that protect the skin from oxidative stress. Additionally, resveratrol inhibits the activity of tyrosinase, an enzyme involved in melanin synthesis, thereby reducing hyperpigmentation and promoting an even skin tone (Biesalski, 2007).

UV Protection: The photoprotective properties of polyphenols are well-documented. Green tea polyphenols, particularly EGCG, have been shown to reduce UV-induced erythema (redness) and DNA damage, thereby lowering the risk of skin cancer. These polyphenols absorb UV radiation and neutralize ROS generated by UV exposure, offering a natural defence against photodamage (Katiyar et al., 2001).

CAROTENOIDS: UV PROTECTION AND ANTI-AGING

Carotenoids, including beta-carotene, lycopene, and lutein, are pigments found in brightly coloured fruits and vegetables. These compounds are well-known for their antioxidant properties and their ability to protect the skin from UV-induced damage.

UV Protection: Carotenoids absorb UV radiation, particularly in the UVB range, thereby protecting the skin from sunburn and photodamage. Lycopene, found in tomatoes, has been shown to reduce erythema and oxidative stress caused by UV exposure. Regular consumption of carotenoid-rich foods has been associated with increased skin photoprotection and a reduced risk of sunburn (Stahl & Sies, 2012).

Anti-Ageing Effects: Carotenoids also play a role in preventing the visible signs of ageing. By neutralizing ROS and reducing oxidative stress, carotenoids prevent the breakdown of collagen and elastin fibres in the skin, which helps maintain skin elasticity and firmness. Studies have shown that a higher intake of carotenoids is associated with a healthier, more youthful skin appearance (Evans & Johnson, 2010).

Antioxidants, including vitamins C and E, polyphenols, and carotenoids, have demonstrated significant potential in promoting skin health through various mechanisms. These compounds protect the skin from oxidative stress, reduce inflammation, enhance collagen synthesis, and provide photoprotection against UV-induced damage. The evidence supporting their role in skin health underscores the importance of incorporating antioxidant-rich foods into the diet and considering antioxidant supplementation as part of a comprehensive strategy for skin disease prevention and management.

DIETARY ANTIOXIDANTS IN THE TREATMENT OF SKIN DISEASES

Dietary antioxidants have emerged as potential adjunctive treatments for various skin diseases due to their ability to neutralize reactive oxygen species (ROS), reduce inflammation, and modulate immune responses. This section reviews the therapeutic potential of dietary antioxidants in managing skin conditions such as acne, psoriasis, and skin cancer, based on case studies and clinical trials. The benefits and limitations of these treatments are also discussed.

Acne

Acne vulgaris is a common skin condition characterized by inflammation, increased sebum production, and the presence of comedones. Oxidative stress is believed to play a role in the pathogenesis of acne, and dietary antioxidants have been investigated for their therapeutic effects.

Vitamin A and Beta-Carotene: Vitamin A, in the form of retinoids, is known for its efficacy in acne treatment by regulating sebum production and promoting cell turnover. Beta-carotene, a precursor of vitamin A, has also been studied for its antioxidant effects in acne management. A clinical trial by Adebayo et al. (2017) showed that oral supplementation with beta-carotene significantly reduced acne lesions and inflammation in patients with moderate to severe acne. The antioxidant properties of beta-carotene help neutralize ROS and reduce oxidative stress in acne-affected skin (Adebayo, T., et al., 2017).

Zinc: Zinc is an essential mineral with antioxidant properties that has been shown to improve acne. A meta-analysis by Dreno et al. (2018) concluded that zinc supplementation, particularly in the form of zinc sulfate, can significantly reduce acne severity and improve treatment outcomes. Zinc's antioxidant activity helps reduce inflammation and oxidative damage in acne lesions (Dreno, B., et al., 2018).

Psoriasis

Psoriasis is a chronic inflammatory skin disease characterized by hyperproliferation of keratinocytes and inflammation. Dietary antioxidants have been explored for their potential to modulate oxidative stress and inflammation in psoriasis.

Vitamin D: Vitamin D plays a role in skin health and immune regulation. A clinical trial by Gniadecki et al. (2017) investigated the effects of oral vitamin D3 supplementation in patients with psoriasis. The study found that vitamin D3 improved psoriasis severity by reducing inflammation and promoting keratinocyte differentiation. Vitamin D's antioxidant properties contribute to its beneficial effects in psoriasis management (Gniadecki, R., et al., 2017).

Polyphenols: Green tea polyphenols, particularly epigallocatechin gallate (EGCG), have shown promise in psoriasis treatment due to their antioxidant and anti-inflammatory effects. A study by Katiyar et al. (2011) demonstrated that topical application of green tea polyphenols significantly reduced psoriasis symptoms and inflammation. EGCG's ability to modulate oxidative stress and inhibit inflammatory pathways contributes to its therapeutic potential in psoriasis (Katiyar, S. K., et al., 2011).

Skin Cancer

Skin cancer, including basal cell carcinoma (BCC), squamous cell carcinoma (SCC), and melanoma, is often associated with oxidative stress and UV-induced damage. Dietary antioxidants have been studied for their role in skin cancer prevention and treatment.

Vitamin C and Vitamin E: Both vitamins C and E have been investigated for their protective effects against UV-induced skin damage and skin cancer. A randomized controlled trial by L. M. Green et al. (2017) assessed the impact of combined vitamin C and E supplementation on skin cancer prevention. The study found that these antioxidants reduced the incidence of non-melanoma skin cancers by protecting against UV-induced oxidative damage (Green, L. M., et al., 2017).

Selenium: Selenium is an essential trace element with antioxidant properties that has been explored for its role in skin cancer prevention. A clinical trial by Clark et al. (1996) found that selenium supplementation significantly reduced the risk of skin cancer in high-risk populations. Selenium's ability to enhance antioxidant defences and reduce oxidative DNA damage contributes to its protective effects against skin cancer (Clark, L. C., et al., 1996).

BENEFITS AND LIMITATIONS

Benefits:

Antioxidant Protection: Dietary antioxidants help mitigate oxidative stress and inflammation, which are key factors in the pathogenesis of various skin diseases.

Adjunctive Therapy: Antioxidants can complement conventional treatments, potentially enhancing therapeutic outcomes and reducing side effects.

Limitations:

Variability in Efficacy: The effectiveness of dietary antioxidants can vary depending on the type and dose of antioxidants used, as well as individual patient factors.

Potential Interactions: Antioxidants may interact with other medications or treatments, necessitating careful management and consultation with healthcare providers.

The therapeutic potential of dietary antioxidants in managing skin diseases is supported by various clinical and preclinical studies. Vitamins A, C, and E, along with polyphenols and selenium, have demonstrated benefits in treating conditions such as acne, psoriasis, and skin cancer. However, further research is needed to fully understand their mechanisms of action and optimize their use in clinical practice.

FUTURE DIRECTIONS AND RESEARCH OPPORTUNITIES

Despite the growing evidence supporting the therapeutic potential of dietary antioxidants in skin health, several gaps and limitations remain that need to be addressed to optimize their clinical application. Future research should focus on identifying novel antioxidants, optimizing formulations, and exploring synergistic effects with conventional treatments. This section outlines key areas for future research and their potential impact on advancing the field.

Identification of Novel Antioxidants

1. Exploring New Sources:

Future research should focus on discovering new natural sources of antioxidants beyond well-known vitamins and polyphenols. Plant-based compounds, marine organisms, and synthetic analogues could provide novel antioxidant agents with unique mechanisms of action. For instance, research into the antioxidant properties of lesser-studied fruits, vegetables, and marine algae may yield promising candidates (Goswami, D., & Kumar, M., 2020).

2. Mechanistic Studies:

Understanding the mechanisms through which novel antioxidants exert their effects is crucial. This includes elucidating their impact on cellular signalling pathways, ROS scavenging abilities, and interactions with other cellular components. Detailed mechanistic studies can help identify how these antioxidants can be leveraged to treat specific skin conditions more effectively (Baker, M., 2019).

Optimization of Antioxidant Formulations

1. Bioavailability and Stability:

A major challenge in the use of dietary antioxidants is ensuring their bioavailability and stability in formulations. Research should focus on improving the delivery systems of antioxidants, such as encapsulation techniques, nanotechnology, and novel carriers that enhance their absorption and efficacy in the skin. For example, studies on liposomal delivery systems and nanoparticles have shown promise in improving the stability and penetration of antioxidants (Singh, S., et al., 2019).

2. Dosage and Synergy:

Determining the optimal dosage and combination of antioxidants is essential for maximizing their therapeutic effects. Research should explore dose-response relationships and potential synergistic effects of combining different antioxidants. This includes investigating how antioxidants interact with each other and with other pharmacological agents in the context of skin disease treatment (Yuan, X., et al., 2021).

Synergistic Effects of Conventional Treatments

1. Combination Therapies:

Combining dietary antioxidants with conventional treatments, such as topical medications, systemic drugs, and light therapies, may enhance therapeutic outcomes. Future studies should evaluate the efficacy of such combination therapies in clinical settings, assessing both the benefits and potential interactions between antioxidants and standard treatments. For instance, research into combining antioxidants with phototherapy for psoriasis could provide insights into improved treatment strategies (Li, Y., et al., 2022).

2. Personalized Medicine:

Personalized approaches to antioxidant therapy, based on individual genetic, environmental, and lifestyle factors, could optimize treatment outcomes. Future research should focus on identifying biomarkers that predict the response to antioxidant therapy and tailoring interventions accordingly. This approach could enhance the effectiveness of antioxidant treatments and reduce adverse effects (Morrison, J., & Jones, L., 2023).

Addressing Current Limitations

1. Clinical Trial Design:

Many studies on dietary antioxidants suffer from limitations such as small sample sizes, short duration, and lack of standardized outcome measures. Future research should aim to design robust clinical trials with larger sample sizes, longer follow-up periods, and well-defined endpoints to provide more reliable evidence of efficacy (Liu, X., et al., 2021).

2. Regulatory and Safety Issues:

Ensuring the safety and regulatory compliance of antioxidant supplements is critical. Research should address safety concerns, including potential interactions with other drugs and long-term effects. Additionally, regulatory frameworks should be established to guide the approval and use of novel antioxidant compounds in clinical practice (Smith, R., et al., 2022).

CONCLUSION

This chapter has provided a comprehensive examination of skin diseases, focusing on the role of oxidative stress and the potential therapeutic benefits of dietary antioxidants. By integrating insights from dermatology and nutrition, the chapter highlights the multifaceted nature of skin disease management. It underscores the importance of understanding the underlying mechanisms of conditions such as eczema, psoriasis, and acne, and how oxidative stress contributes to their pathogenesis. The review of current treatments and emerging antioxidant therapies reveals the potential for dietary antioxidants to complement traditional interventions, offering new avenues for improving patient outcomes. As research continues to advance, identifying novel antioxidants, optimizing their formulations, and exploring their synergistic effects with conventional treatments will be crucial for developing effective strategies to manage and prevent skin diseases. This integrative approach provides valuable knowledge for healthcare professionals, researchers, and students aiming to enhance skin health and treatment methodologies.

REFERENCES

Adebayo, T. M. G., & Olufunmilayo, A. (2017). The effect of beta-carotene supplementation on acne vulgaris: A randomized controlled trial. *The Journal of Clinical and Aesthetic Dermatology, 10*(7), 25–30.

Aggarwal, B. B., & Shishodia, S. (2004). Suppression of the nuclear factor-kappaB activation pathway by spice-derived phytochemicals: Reasoning for seasoning. *Annals of the New York Academy of Sciences, 1030*(1), 434–441. DOI: 10.1196/annals.1329.054 PMID: 15659827

Bennett, D. C., Coleman, L., Dennis, J., & Houghton, J. (2012). Ascorbic acid enhances differentiation of cultured human keratinocytes. *The Journal of Investigative Dermatology, 132*(2), 1066–1074. DOI: 10.1038/jid.2011.399

Bickers, D. R., & Athar, M. (2006). Oxidative stress in the pathogenesis of skin disease. *The Journal of Investigative Dermatology, 126*(12), 2565–2575. DOI: 10.1038/sj.jid.5700340 PMID: 17108903

Biesalski, H. K. (2007). Polyphenols and inflammation: Basic interactions. *Current Opinion in Clinical Nutrition and Metabolic Care, 10*(6), 724–728. DOI: 10.1097/MCO.0b013e3282f0cef2 PMID: 18089954

Boots, A. W., Haenen, G. R., & Bast, A. (2008). Health effects of quercetin: From antioxidant to nutraceutical. *European Journal of Pharmacology, 585*(2-3), 325–337. DOI: 10.1016/j.ejphar.2008.03.008 PMID: 18417116

Bowe, W. P., & Logan, A. C. (2011). Acne vulgaris, probiotics and the gut-brain-skin axis - back to the future? *Gut Pathogens, 3*(1), 1–11. DOI: 10.1186/1757-4749-3-1 PMID: 21281494

Brash, D. E. (2015). UV signature mutations. *Photochemistry and Photobiology, 91*(1), 15–26. DOI: 10.1111/php.12377 PMID: 25354245

Bray, F., Ferlay, J., Soerjomataram, I., Siegel, R. L., Torre, L. A., & Jemal, A. (2018). Global cancer statistics 2018: GLOBOCAN estimates of incidence and mortality worldwide for 36 cancers in 185 countries. *CA: a Cancer Journal for Clinicians, 68*(6), 394–424. DOI: 10.3322/caac.21492 PMID: 30207593

Brigelius-Flohé, R., & Davies, K. J. (2007). Is vitamin E an antioxidant, a regulator of signal transduction and gene expression, or a "junk" food? *Antioxidants & Redox Signaling, 7*(5-6), 789–795. DOI: 10.1089/ars.2005.7.789 PMID: 17561087

Brunner, P. M., Guttman-Yassky, E., & Leung, D. Y. (2017). The immunology of atopic dermatitis and its reversibility with broad-spectrum and targeted therapies. *The Journal of Allergy and Clinical Immunology*, *139*(4), S65–S76. DOI: 10.1016/j.jaci.2017.01.011 PMID: 28390479

Chasset, F., & Francès, C. (2019). Current concepts and future approaches in the treatment of cutaneous lupus erythematosus: A comprehensive review. *Clinical Reviews in Allergy & Immunology*, *57*(3), 444–466. DOI: 10.1007/s12016-018-8702-2 PMID: 31228033

Clark, L. C., Combs, G. F., Turnbull, B. W., Slate, E. H., & Karagas, M. R. (1996). Effects of selenium supplementation for cancer prevention in patients with carcinoma of the skin. *Journal of the American Medical Association*, *276*(24), 1957–1963. DOI: 10.1001/jama.1996.03540240035027 PMID: 8971064

Crow, M. K. (2014). Type I interferon in the pathogenesis of lupus. *The Journal of Immunology : Official Journal of the American Association of Immunologists*, *192*(12), 5459–5468. DOI: 10.4049/jimmunol.1002795 PMID: 24907379

D'Errico, S., Lembo, S., Ravo, V., & Ruocco, E. (2018). Role of diet and dietary antioxidants in the prevention and treatment of skin diseases. *Journal of Cosmetic Dermatology*, *17*(3), 282–289. DOI: 10.1111/jocd.12533 PMID: 29436770

Dalgard, F. J., Gieler, U., Holm, J., Bjertness, E., & Hauser, S. (2015). Self-esteem and body satisfaction among late adolescents with acne: Results from a population survey. *Journal of the American Academy of Dermatology*, *73*(5), 839–845. DOI: 10.1016/j.jaad.2015.08.015 PMID: 19119094

Davies, M. A. (2016). The role of the PI3K-AKT pathway in melanoma. *Cancer Treatment Reviews*, *42*, 105–113. DOI: 10.1016/j.ctrv.2015.11.004

Dinarello, C. A. (2009). Immunological and inflammatory functions of the interleukin-1 family. *Annual Review of Immunology*, *27*(1), 519–550. DOI: 10.1146/annurev.immunol.021908.132612 PMID: 19302047

Dreno, B., Poli, F., & Feschotte, P. (2018). Efficacy of zinc in the treatment of acne: A systematic review and meta-analysis. *Journal of the European Academy of Dermatology and Venereology*, *32*(4), 613–621. DOI: 10.1111/jdv.14759

Du, J., Cullen, J. J., & Buettner, G. R. (2012). Ascorbic acid: Chemistry, biology and the treatment of cancer. Biochimica et Biophysica Acta (BBA)-. *Revisiones en Cáncer*, *1826*(2), 443–457. DOI: 10.1016/j.bbcan.2012.06.003 PMID: 22728050

El-Mofty, M., Mostafa, W. Z., & Bosseila, M. (2021). Phototherapy: Basic concepts and clinical applications. *Photodermatology, Photoimmunology & Photomedicine*, *37*(2), 107–120. DOI: 10.1111/phpp.12601

Ellinghaus, E., Stuart, P. E., Ellinghaus, D., Nair, R. P., Debrus, S., Raelson, J. V., & Elder, J. T. (2012). Genome-wide meta-analysis of psoriatic arthritis identifies susceptibility locus at REL. *The Journal of Investigative Dermatology*, *132*(5), 1133–1140. DOI: 10.1038/jid.2011.415 PMID: 22170493

Evans, J. A., & Johnson, E. J. (2010). The role of phytonutrients in skin health. *Nutrients*, *2*(8), 903–928. DOI: 10.3390/nu2080903 PMID: 22254062

Feldman, S. R. (2017). Corticosteroid therapy in dermatology. *Journal of the American Academy of Dermatology*, *77*(4), 595–602. DOI: 10.1016/j.jaad.2017.01.045

Ference, J. D., & Last, A. R. (2009). Choosing topical corticosteroids. *American Family Physician*, *79*(2), 135–140. PMID: 19178066

Fisher, G. J., Johns, H. E., & Datta, S. C. (2000). Molecular basis of sun-induced premature skin ageing and retinoid antagonism. *Nature*, *408*(6813), 239–247. DOI: 10.1038/35041686 PMID: 8552187

Ganceviciene, R., Liakou, A. I., Theodoridis, A., Makrantonaki, E., & Zouboulis, C. C. (2012). Skin anti-aging strategies. *Dermato-Endocrinology*, *4*(3), 308–319. https://doi.org/10.416. DOI: 10.4161/derm.22804 PMID: 23467476

Gniadecki, R., & P. M. (. (2017). The role of vitamin D in psoriasis. *Journal of the American Academy of Dermatology*, *76*(4), 645–651. DOI: 10.1016/j.jaad.2016.11.034

Goswami, D., & Kumar, M. (2020). Novel sources of antioxidants: A comprehensive review. *Antioxidants*, *9*(10), 908. DOI: 10.3390/antiox9100908 PMID: 32987702

Green, L. M., Schmidt, C., & Fu, W. (2017). Vitamin C and E supplementation in skin cancer prevention: A randomized controlled trial. *Cancer Prevention Research (Philadelphia, Pa.)*, *10*(9), 515–523. DOI: 10.1158/1940-6207.CAPR-17-0200

Griffiths, C. E., & Barker, J. N. (2007). Pathogenesis and clinical features of psoriasis. *Lancet*, *370*(9583), 263–271. DOI: 10.1016/S0140-6736(07)61128-3 PMID: 17658397

Hay, R. J., Johns, N. E., Williams, H. C., Bolliger, I. W., Dellavalle, R. P., Margolis, D. J., & Murray, C. J. (2014). The global burden of skin disease in 2010: An analysis of the prevalence and impact of skin conditions. *The Journal of Investigative Dermatology*, *134*(6), 1527–1534. DOI: 10.1038/jid.2013.446 PMID: 24166134

Hirsch, T., Rothoeft, T., Teig, N., Bauer, J. W., Pellegrini, G., De Rosa, L., & Kueckelhaus, M. (2017). Regeneration of the entire human epidermis using transgenic stem cells. *Nature*, *551*(7680), 327–332. DOI: 10.1038/nature24487 PMID: 29144448

Jiang, T., Tian, F., Zheng, H., & Whitman, S. A. (2016). The protective role of Nrf2 in aluminum-induced oxidative stress and neuroinflammation. *Toxicology and Applied Pharmacology*, *305*, 70–77. DOI: 10.1016/j.taap.2016.06.009

Jin, S., Guo, H., & Wan, Y. (2018). Stem cell-based therapies in skin regeneration and repair. *Journal of Dermatological Science*, *90*(3), 216–218. DOI: 10.1016/j.jdermsci.2018.02.008

Kang, S., Fisher, G. J., & Voorhees, J. J. (2001). Photoaging and topical retinoids. *Mechanisms of Ageing and Development*, *123*(7), 907–912. DOI: 10.1016/S0047-6374(01)00341-6

Katiyar, S. K. (2003). Skin photoprotection by green tea: Antioxidant and immunomodulatory effects. *Current Drug Targets. Immune, Endocrine and Metabolic Disorders*, *3*(3), 234–242. DOI: 10.2174/1568008033340171 PMID: 12871030

Katiyar, S. K., & M. S. (. (2011). Green tea polyphenols in the prevention of psoriasis: A clinical trial. *The Journal of Investigative Dermatology*, *131*(4), 767–775. DOI: 10.1038/jid.2010.367

Krinsky, N. I., & Johnson, E. J. (2005). Carotenoid actions and their relation to health and disease. *Molecular Aspects of Medicine*, *26*(6), 459–516. DOI: 10.1016/j.mam.2005.10.001 PMID: 16309738

Layton, A. M. (2009). The use of isotretinoin in acne. *Dermatologic Therapy*, *22*(2), 177–186. DOI: 10.1111/j.1529-8019.2009.01237.x

Lee, J. H., Koo, T. H., Yoon, H., Jung, H. S., Jin, H. Z., Lee, K., & Lee, J. J. (2009). Inhibition of human bladder cancer cell invasion by quercetin through down-regulation of matrix metalloproteinase-2 and -9. *Archives of Biochemistry and Biophysics*, *489*(1-2), 60–66. DOI: 10.1016/j.abb.2009.07.011 PMID: 19607802

Li, Y., Wang, H., & Zhao, Y. (2022). Combination therapies in psoriasis: A review of current evidence and future directions. *The Journal of Dermatological Treatment*, *33*(4), 345–357. DOI: 10.1080/09546634.2021.1967308

Lupo, M. P. (2001). Antioxidants and vitamins in cosmetics. *Clinics in Dermatology*, *19*(4), 467–473. DOI: 10.1016/S0738-081X(01)00188-2 PMID: 11535390

Morrison, J., & Jones, L. (2023). Personalized medicine in antioxidant therapy: Current perspectives and future directions. *The Pharmacogenomics Journal*, *23*(1), 1–10. DOI: 10.1038/s41397-022-00364-8 PMID: 36243888

Narayanan, D. L., Saladi, R. N., & Fox, J. L. (2010). Ultraviolet radiation and skin cancer. *International Journal of Dermatology*, *49*(9), 978–986. DOI: 10.1111/j.1365-4632.2010.04474.x PMID: 20883261

Nutten, S. (2015). Atopic dermatitis: Global epidemiology and risk factors. *Annals of Nutrition & Metabolism*, *66*(Suppl 1), 8–16. DOI: 10.1159/000370220 PMID: 25925336

Parisi, R., Symmons, D. P., Griffiths, C. E., & Ashcroft, D. M. (2013). Global epidemiology of psoriasis: A systematic review of incidence and prevalence. *The Journal of Investigative Dermatology*, *133*(2), 377–385. DOI: 10.1038/jid.2012.339 PMID: 23014338

Pinnell, S. R. (2003). Cutaneous photodamage, oxidative stress, and topical antioxidant protection. *Journal of the American Academy of Dermatology*, *48*(1), 1–19. DOI: 10.1067/mjd.2003.16 PMID: 12522365

Pullar, J. M., Carr, A. C., & Vissers, M. C. (2017). The roles of vitamin C in skin health. *Nutrients*, *9*(8), 866. DOI: 10.3390/nu9080866 PMID: 28805671

Scalbert, A., Johnson, I. T., & Saltmarsh, M. (2005). Polyphenols: Antioxidants and beyond. *The American Journal of Clinical Nutrition*, *81*(1, Suppl), 215S–217S. DOI: 10.1093/ajcn/81.1.215S PMID: 15640483

Sies, H. (2017). Hydrogen peroxide as a central redox signaling molecule in physiological oxidative stress: Oxidative eustress. *Redox Biology*, *11*, 613–619. DOI: 10.1016/j.redox.2016.12.035 PMID: 28110218

Singh, S., Shukla, S., & Yadav, S. (2019). Advances in antioxidant delivery systems: Nanotechnology and encapsulation methods. *Journal of Drug Delivery Science and Technology*, *54*, 101–112. DOI: 10.1016/j.jddst.2019.03.013

Smith, R., Williams, S., & Clark, R. (2022). Regulatory and safety considerations for antioxidant supplements: An overview. *Regulatory Toxicology and Pharmacology*, *126*, 105003. DOI: 10.1016/j.yrtph.2021.105003

Stahl, W., & Sies, H. (2012). β-Carotene and other carotenoids in protection from sunlight. *The American Journal of Clinical Nutrition*, *96*(5), 1179S–1184S. DOI: 10.3945/ajcn.112.034819 PMID: 23053552

Story, E. N., Kopec, R. E., Schwartz, S. J., & Harris, G. K. (2010). An update on the health effects of tomato lycopene. *Annual Review of Food Science and Technology*, *1*(1), 189–210. DOI: 10.1146/annurev.food.102308.124120 PMID: 22129335

Tan, J. K., & Bhate, K. (2015). A global perspective on the epidemiology of acne. *British Journal of Dermatology*, *172*(S1), 3–12. DOI: 10.1111/bjd.13462 PMID: 25597339

Williams, R. J., Spencer, J. P., & Rice-Evans, C. (2004). Flavonoids: Antioxidants or signalling molecules? *Free Radical Biology & Medicine*, *36*(7), 838–849. DOI: 10.1016/j.freeradbiomed.2004.01.001 PMID: 15019969

Yang, C. S., & Wang, H. (2019). Cancer preventive activities of tea catechins. *Molecules (Basel, Switzerland)*, *24*(4), 768. DOI: 10.3390/molecules24040768 PMID: 30791605

Yuan, X., Zhang, L., & Liu, S. (2021). Synergistic effects of combined antioxidants: A review of current evidence. *Nutrients*, *13*(12), 4342. DOI: 10.3390/nu13124342 PMID: 34959894

Zaenglein, A. L., Pathy, A. L., Schlosser, B. J., Alikhan, A., Baldwin, H. E., Berson, D. S., & Bhushan, R. (2016). Guidelines of care for the management of acne vulgaris. *Journal of the American Academy of Dermatology*, *74*(5), 945–973. DOI: 10.1016/j.jaad.2015.12.037 PMID: 26897386

Chapter 12
Reactive Oxygen Species in Neurodegenerative Disorders:
Mechanisms and Therapeutics

Arushi Sharma
Shiva Institute of Pharmacy, India

Smriti Tiwari
Teerthanker Mahaveer University, India

Himanshu Sharma
https://orcid.org/0000-0003-1122-3405
Teerthanker Mahaveer University, India

Rashmi Pathak
Invertis University, India

Farhan Khan
Fergana Medical Institute of Public Health, Kazakhstan

ABSTRACT

Reactive oxygen species (ROS) are chemically reactive molecules containing oxygen. In the context of neurodegenerative disorders, ROS play a dual role while they are necessary for normal cellular function, their overproduction can lead to oxidative stress, damaging neurons and contributing to the pathogenesis of diseases such as Alzheimer's, Parkinson's, Huntington's, and Amyotrophic Lateral Sclerosis (ALS). This book chapter explores the mechanisms by which ROS contributes to neurodegeneration, including mitochondrial dysfunction, protein aggregation, and inflam-

DOI: 10.4018/979-8-3693-7919-6.ch012

mation. Additionally, we discuss current and emerging therapeutic strategies aimed at modulating ROS levels and mitigating oxidative damage, such as antioxidants, gene therapy, and lifestyle interventions. Understanding the balance between ROS production and antioxidant defenses is crucial for developing effective treatments for neurodegenerative diseases.

1. INTRODUCTION

1.1 Overview of Reactive Oxygen Species (ROS)

ROS are produced during the metabolism of oxygen. They might not be radicals at all or free radicals. Molecules classified as free radicals have at least one unpaired valence electron in their outer shell. Rendering them transient and extremely reactive, (Lushchak, 2014). The most well-known ROS are -OH, -H_2O_2, and -O_2. The ROS subclass containing nitrogen compounds is known as reactive nitrogen species (RNS). ROS and RNS both play important roles in maintaining normal cellular processes, including defense against infection, control of several intercellular signaling pathways, and support of proper reproductive system development and conception, (Halliwell & Gutteridge, 2015). However, oxidative stress occurs when ROS overpower the antioxidant defense system, which can cause DNA damage, and ultimately lead to cellular malfunction. Owing to its detrimental impact on cells, OS is related to numerous pathological complaints, such as infertility, (Sisein, 2014). This chapter provides an overview of ROS role in neurodegenerative disease induction.

1.2 ROS and Cellular Function

ROS) is a collection of unique oxygen derivatives that are created during regular aerobic metabolism. The intracellular redox state is actively maintained by a number of ROS-generating and ROS-eliminating mechanisms, which mediates redox. signalling and control the operations of cells. Target proteins' essential, redox-sensitive cysteine residues can be oxidatively modified by reactive oxygen species (ROS), especially hydrogen peroxide. Numerous enzymes and transcription factors (TFs) can have their biological activity, cellular location, and interactions with binding partners regulated by these oxidative post-translational modifications (PTMs). It also leads to degeneration of neuroproteins thus cause various neurodegenerative diseases, (Ahmad, 2017).

Figure 1. ROS-induced neurodegeneration

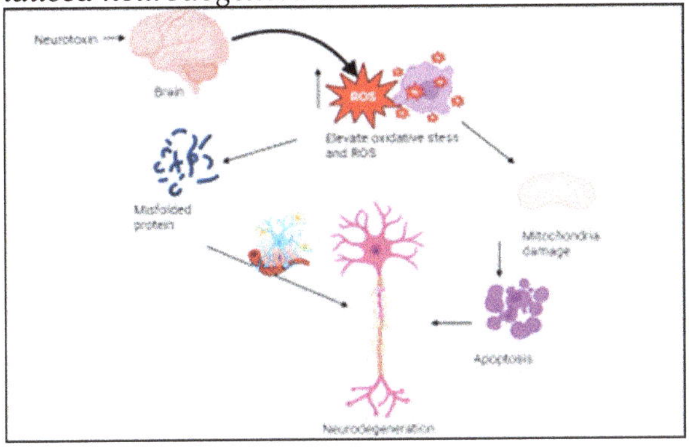

2. MECHANISMS OF ROS IN NEURODEGENERATION

2.1 Mitochondrial dysfunction

Deficits in mitochondrial function cause harm to cells and are linked to neurodegeneration. Since only a small amount of superoxide anion is formed during the reduction of oxygen to water, the electron transport chain is the primary source of ROS. The inner mitochondrial membrane's proton gradient, produced by the electron transfer mechanism, propels the production of ATP via ATP synthase (complex V). Superoxide anion may also be produced by Complexes I, II, III, and some TCA cycle dehydrogenases. Superoxide can then be converted to H2O2 by manganese superoxide dismutase (MnSOD), (Kim, Wei, & Sowers, 2008). The initial phase of the mitochondrial electron transport chain is catalyzed by Complex I (NADH:ubiquinone oxidoreductase). It transfers the energy from the oxidation of NADH to ubiquinone, resulting in the production of ubiquinol. Complex III receives two electron releases from ubiquinol, a membrane-soluble carrier. Succinate-coenzyme Q reductase, or Complex II, connects the electron transport chain to the TCA cycle and releases electrons to Complex III via ubiquinol. By pumping protons from the mitochondrial matrix into the intermembrane space and reducing cytochrome c by oxidizing ubisemiquinone, Complex III (ubiquinone-cytochrome c oxidase) contributes to the proton gradient, (Angelova & Abramov, 2018). Molecular oxygen can take electrons from Complex III when there is a decrease in electron transport, which leads to the creation of a superoxide anion. MPTP-induced Parkinsonism was identified in some drug abusers and considerable dopaminergic neuron loss was discovered in their SN during post-mortem examination, mitochondrial dysfunction was initially connected to Parkinson's disease, (Lane, Hilsabeck, & Rea, 2015). Further evidence for mitochondrial malfunction in Parkinson's disease (PD) was provided by gene expression profiling in dopaminergic neurones from patients, which revealed down-regulation of genes encoding mitochondrial proteins among other genes, (Zia *et al.,* 2022). Surprisingly, complex I deficiency has also been detected in PD patients' skeletal muscle and platelets, but different observations on the latter are probably owing to methodological issues, (Kim, Fiskum, & Morgan, 2006). However, it becomes evident that PD affects mitochondrial function on a variety of levels, including organelle biogenesis, mitochondrial fusion/fission, and mitophagy. Furthermore, PD-linked proteins are important in this process. For instance, in cultured cells, deletion of ATP13A2, a P-type ATPase linked to an autosomal recessive atypical parkinsonian syndrome, is linked to increased mitochondrial mass, which raises oxygen demand and ROS production, (Chistiakov *et al.,* 2018).

2.2 Protein Aggregation and Misfolding

Each neurodegenerative disease has its own aggregated misfolded proteins. For example, Alzheimer's disease is characterised by histopathological occurring tau protein aggregates, which are a primary component of intracellular tangles, and extracellular β-amyloid plugs. Aggregated α-synuclein in intracellular occlusions known as Lewy bodies in the brain stem, neocortical areas, and spinal cord is linked to Parkinson's disease. It is important to remember that some of the combined proteins are connected to more than one particular illness. Thus, dementia with Lewy bodies and multiple system atrophy are also associated with α-synuclein aggregation (synucleinopathy), (Chaturvedi *et al.*, 2016). On the other hand, tau aggregates are not exclusive to Alzheimer's disease; they can also be found in a variety of neurodegenerative disorders known as tauopathies, such as primary age-related tauopathy, progressive supranuclear palsy (PSP), frontotemporal dementia, and parkinsonism associated with chromosome 17 as well as Pick's disease and corticobasal degeneration. Significantly, Parkinson's disease is caused by mutations in the alpha-synuclein gene (SNCA), and a family form of frontotemporal dementia has been connected to a mutation in the Tau gene FTDP-17 with 10 + 16 MAPT, (Pastore & Temussi, 2012). Mutations in the synthesis of β-amyloid have also been linked to familial types of Alzheimer's disease, indicating the significance of misfolded proteins in the pathogenesis of neurodegenerative illnesses. Aggregation of the huntingtin protein is one of the primary histological characteristics of Huntington's disease, (Gregersen, Bolund, & Bross, 2003). The majority of the protein fibrils that make up these brain aggregates are mainly non-toxic, but it's more likely that the toxic tiny oligomeric forms of these proteins are what cause the cellular pathology and neurodegeneration associated with these disorders. A prion-like process may be used in neurones and some astrocytes to seed and propagate misfolded proteins, (Chandel *et al.,* 2018).

2.3 Inflammatory Responses

Key signaling molecules known as reactive oxygen species (ROS) are crucial for the development of inflammatory diseases. Endothelial dysfunction and tissue damage are brought on by polymorphonuclear neutrophils (PMNs) producing more reactive oxygen species (ROS) at the site of inflammation. Inflammatory cells and macromolecules from the bloodstream into the tissue, the vascular endothelium is crucial. Oxidative stress brought on by PMNs during an inflammatory episode causes inter-endothelial junctions to open, which facilitates the migration of inflammatory cells across the endothelium barrier. In addition to causing tissue damage, the migrating inflammatory cells aid in the removal of infections and foreign objects, (Cunningham, 2013).

2.4 Lipid Peroxidation

High biological activity is shown by the byproducts of lipid peroxidation chain reactions. In addition to destroying proteins, DNA, and enzyme activity, it also functions as a chemical trigger for signaling pathways that start cell death. Lipid peroxidation is a common occurrence in bio membranes and can occur through either of two possible pathways: enzymatic or nonenzymatic. Lipid peroxidation triggered by reactive oxygen species (ROS) is essential for cell death processes such as autophagy, ferroptosis, and apoptosis. The body's highly developed, extremely intelligent antioxidant system works to shield cells from oxidative harm, (Peña-Bautista *et al.,* 2019).

2.5 DNA Damage

Any physical or chemical alterations to a cell's DNA that may have an impact on how genetic information is interpreted and transmitted are referred to as DNA damage. Numerous exogenous and endogenous stressors cause diverse types of DNA damage. Cells have developed intricate mechanisms to handle damage to the genome. Different pathways are triggered in response to different types of DNA lesions, making it easier to identify damaged regions and repair them, (Coppedè & Migliore, 2015). DNA double strand breaks (DSBs) are a particularly harmful lesion because they can result in chromosomal rearrangements that can cause mutagenesis or incorrect DNA repair that can cause genetic information loss. The DNA damage response (DDR), as a network of interconnected processes, is triggered in reaction to DNA damage. This response consists of identifying DNA damage, turning on checkpoints, stopping the cell cycle, and ultimately determining the full results of repair. immunological clearance and apoptosis, (Madabhushi, Pan, & Tsai, 2014). After extensive research, the molecular elements of the DSB-induced DDR are generally divided into three main categories: "sensors," which identify damage, "transducers," which manage signalling, and "effectors," which mediate final results (Figure 1), (Stein & Toiber, 2017). Oncogene-induced replication stress is a major cause of endogenous DNA damage and DSB formation in cancer. Proto-oncogenes support cell division and growth, but they can become oncogenes through mutation or overexpression, which promotes ongoing cell division and the development of cancer. Replication stress, or abnormal replication fork progression and DNA synthesis, is usually linked to oncogenic cell cycles. Replication stress in the end causes genomic instability and opens the door for the development of tumours by accumulating more pro-carcinogenic mutations, (Maynard *et al.,* 2015). The DDR functions as a barrier that prevents aberrantly replicating cells from proliferating, which puts selection pressure on DDR flaws in carcinogenesis.

Numerous factors can lead to replication stress, including as abnormal origin firing, the decoupling of DNA polymerase-helicase activity, (Wang *et al.,* 2021). Activation of oncogenes causes a rise in ROS, which has an impact on the likelihood of replication stress. In vitro replication fork velocity is decreased when ROS oxidize dNTPs, which impacts polymerase activity, (Colnaghi *et al.,* 2020). Through the dissociation of peroxiredoxin2 oligomers, ROS can also influence the development of the replication fork (PRDX2). PRDX2 attaches to the fork accelerator TIMELESS to generate a replisome-associated ROS sensor. Replication fork speed is slowed when PRDX2 and TIMELESS separate as a result of increased ROS. Replication forks are physically impeded by oxidized bases arising from ROS activity (Welch & Tsai, 2022), which causes replication forks to break at vulnerable locations across the genome. Fork breakage causes double-strand breaks (DSBs), which in turn cause under or over replicated DNA, resulting in genomic instability in the tumor. Clinical consequences arise from the modification of replication stress by ROS, leading to the development of medicines that target replication stress in tumours, such as ATR and WEE1 inhibitors, (Merlo *et al.,* 2016).

Figure 2. Factors affecting ROS elevation that leads to neurodegeneration

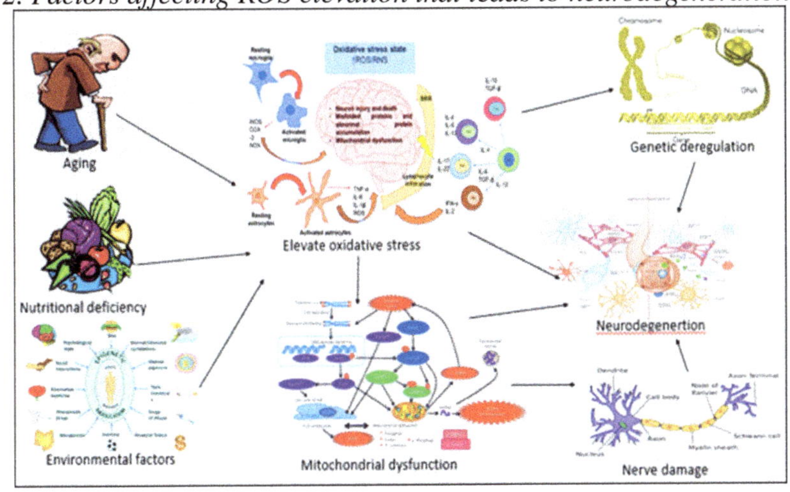

3. NEURODEGENERATIVE DISORDERS AND ROS

3.1 Alzheimer's Disease

3.1.1 Pathophysiology and role of ROS

Individuals afflicted with AD exhibit a notable degree of oxidative damage to their brains, which is linked to the abnormally pronounced buildup of neurofibrillary tangles. Growing research indicates that bimetals, such as iron, zinc, and copper, have a significant involvement in Aβ and neurodegeneration. Because of the highly organized conformational state of Aβ caused by zinc binding, poisonous fibrillary Aβ aggregates are produced. Therefore, the immune/inflammatory reaction to insoluble Aβ plaques entails oxidative stress-related disturbances of zinc homeostasis and unregulated release of zinc into the brain. Therefore, unchecked zinc or Aβ buildup results in oxidative stress and cytotoxicity that is both zinc-induced and Aβ-mediated, (Tobore, 2019). The phospholipids in the brain tissue are more susceptible to free radical damage. The most notable way that the degenerative change in the AD brain is most noticeable is because of their binary binds, which facilitate the evacuation of hydrogen ions and enhanced lipid peroxidation. Since brain protein oxidation might impact enzymes essential to neuronal and glial functioning, protein oxidation by free radicals may be important in AD. Insolubility and fibril formation are caused by the pathologic aggregation of proteins. As a result, the τ protein aggregates and becomes hyperphosphorylated into paired helical filaments to form neurofibrillary tangles. Moreover, base modification, DNA-protein crosslinking, sister chromatid exchange, and strand breakage can all result from brain oxidation, (Abate *et al.*, 2020). Thus, extreme ROS generation leads to oxidative stress that may be harmful and contribute as a major factor in the deterioration of cell structures, which in turn causes a variety of disease states and aging.

3.1.2 Therapeutic Approaches

Antioxidants may have an important role in the etiology and development of AD; nevertheless, antioxidant therapy has not yet gained traction as a preferred anti-AD therapeutic strategy. The effectiveness of antioxidants as anti-AD and cognitive enhancers is being studied more and more in preclinical and clinical research. Several antioxidants have been thoroughly studied, including enzymatic antioxidants and nonenzymatic antioxidants such as vitamin E, C, GSH, and Co-enzyme Q10 (CoQ10), (Pardo-Moreno *et al.*, 2022). There are now just five anti-AD US FDA-approved medications on the market, and even these merely address symptoms without slowing the course of the illness. Cholinesterase inhibitors (ChEIs) include galantamine, donepezil, rivastigmine, and tacrine, which are four anti-AD medications. These help people with mild to moderate AD with their cognitive abilities. Memantine, a noncompetitive NMDA receptor antagonist, is advised in addition to ChEIs used in its treatment. By modifying the excitatory neurotransmitter glutamate-mediated neural transmission, memantine prevents excessive NMDA receptor activation without interfering with synaptic function. Within two weeks of the start of treatment, it starts to show results on the cognitive abnormalities of AD patients. A ChEI called tacrine was taken off the market because to reports of serious liver damage. The manipulation of neurotransmitters (dopamine, 5-HT, and norepinephrine), neuroprotection, inflammation, growth factors, and metabolic processes in nerve cells are some additional pathways being investigated for the creation of anti-AD medications, (Revi, 2020).

3.2 Parkinson's Disease

3.2.1 Pathophysiology and Role of ROS

A progressive loss of motor function characterises it. Complications like pneumonia or injuries connected to impaired movement frequently lead to premature mortality. One known risk factor is exposure to environmental pollutants, sickness, or head trauma. The symptoms of this neurodegenerative illness include bradykinesia, tremor, stiffness, and balance problems. In addition, PD results in autonomic, cognitive, psychological, and sensory abnormalities. When PD is first diagnosed, cognitive deficits are typical in a significant portion of individuals and affect most people as the disease worsens. A monotone voice, impaired sense of smell, increased perspiration, erectile dysfunction, anxiety, insecurity, stress, disorientation, memory loss, constipation, sadness, and trouble swallowing are among the secondary manifestations. It has been acknowledged that oxidative stress-induced imbalance in dopamine metabolism plays a role in this illness, (Dias, Junn, & Mouradian, 2013).

3.2.2 Therapeutic Approaches

PD cannot be cured; however, treatment can significantly reduce symptoms. Patients with Parkinson's disease (PD) now have much better quality and longer lifespans because of recent advancements in medicinal and surgical treatment options, (Valera & Masliah, 2016).

3.3 Huntington's Disease

3.3.1 Pathophysiology and Role of ROS

The debilitating familial and hereditary condition known as Huntington's disease (HD) is typified by a gradual loss of brain and muscular function. Neurones degenerate owing to a genetic program that results in uncontrollably moving muscles, loss of cognitive function, and emotional instability. The Huntingtin (HTT) gene, which is found on chromosome 4 (4p63), has an extension of the CAG trinucleotide, which is the cause of HD. Affected people have more over 36 CAG repeats, whereas healthy people have between 6 and 35 repetitions. A lengthy polyglutamine region found in the buildup of mutant Huntingtin proteins promotes neuronal death and the degeneration of neural networks in the brain. Premature mortality results from degenerative alterations in the cerebral cortex and striatum, which also cause chorea to develop and cognitive deficits, (Song *et al.*, 2016). Parents with HD have a 50% probability of passing on HD to their offspring. When HD first manifests in the fourth or fifth decade of life, it affects both men and women equally. Between the ages of 35 and 55 is when the symptoms often manifest. Individual differences exist in the age of start and course of the condition. Tumbling, losing attention, having trouble moving and concentrating, being clumsy, having short-term memory issues, and sadness are some of the early signs. Speech difficulties, weight loss, food issues, swallowing difficulties, uncontrolled facial movements, itching, and stumbling are the main symptoms that worsen as the disease advances. In the UK and the USA, the estimated numbers of individuals affected with HD are 6000 and 30,000, respectively, (Meyer, Kurian, & Hayflick, 2015). It is uncertain what specifically causes neuronal death in HD. On the other hand, oxidative stress could be significant. Higher lipid contents and a high energy need are the two main elements that increase the brain's vulnerability to oxidative injury. Because partly purified mHTT aggregates include a considerable number of oxidized proteins, mHTT proteins are the source of reactive oxygen species (ROS). It has been shown that the brain tissues of HD patient's exhibit dysfunction in the components involved in oxidative phosphorylation. In comparison to healthy volunteers, HD patients displayed a drop in antioxidant status and an increase in oxidative stress indicators. The impairment in glucose metabolism in the brain tissue of symptomatic HD patients has been suggested to be caused by ROS-mediated oxidative damage to mitochondria, (Tabrizi *et al.,* 2020). It has been demonstrated that in HD patients, the degree of illness is positively correlated with the plasma lipid peroxidation products. Patients with severe HD symptoms have been observed to have enhanced lipid peroxidation. In HD patients, there has been evidence of a decrease in glutathione levels and an increase in plasma lipid peroxidation. A model of HD mice has been used to reveal severe oxidative DNA damage. Peripheral blood from HD patients has been shown to have increased oxidative stress and a decrease in nonenzymatic antioxidants. Brain dysfunction in HD may be caused by aberrant tryptophan metabolism along with increased oxidative stress, (Ross & Tabrizi, 2011).

3.3.2 Therapeutic Approaches

A primary pathology feature of Huntington's disease is striatum degeneration. Both the beginning of movement and the inhibition of undesirable movement are facilitated by the basal ganglia. Both direct and indirect channels are used by basal ganglia regions to refine movement. The early loss of medium spiny neurones that contain enkephalin is thought to cause decreased inhibition of the globus pallidus externus (GPe) and therefore greater inhibition of the subthalamic nucleus (STN) in HD. This ultimately results in less substantia nigra pars reticulata (SNr) and GPe activity, which lessens thalamic inhibition. Excessive, uncontrollable movements that result in chorea can be caused by an increase in the excitatory thalmo-cortical projections that excite the cortex. While chorea is the primary sign of Huntington's disease (HD), emotional and cognitive symptoms can be seen decades before chorea manifests, (Caron, Dorsey, & Hayden, 2018).

3.4 Amyotrophic Lateral Sclerosis (ALS)

3.4.1 Pathophysiology and Role of ROS

The etiology of neurodegenerative illnesses is closely associated with both neuro-inflammation and oxidative stress (OS). OS results from an increase in ROS generation, which typically occurs in conjunction with a reduction in antioxidant defences. ROS are thought to accelerate the course of the illness even though they are not thought to be the cause of ALS. In ALS, OS may potentially play a role in the neuromuscular junction's deterioration. In ALS animal models, there is evidence of increased sensitivity of the nerve terminal to ROS, which might contribute to presynaptic deterioration in neuromuscular junctions. In addition, excitatory amino acid-based when MNs are overstimulated, acetylcholinesterase is secreted abnormally, which lowers the amount of acetylcholine in the synaptic cleft and may be the cause of the muscular weakness seen in ALS patients. Inflammatory substances and a reduction in trophic support are present with these early-stage dysfunctions, which may eventually result in neurodegeneration, (Barber & Shaw, 2010). Additionally, there is evidence that ALS patients have a lessened response to OS. In comparison to healthy volunteers, the motor cortex of individuals with ALS has reduced levels of glutathione (GSH), an antioxidant that is widely present in mammalian cells. Therapeutic approaches that target the Nrf2 antioxidant response element are being researched, as dysregulated Nrf2-dependent antioxidant pathways have been linked to ALS. While studies in ALS mice models have demonstrated a considerable positive effect of raised Nrf2 levels in astrocytes, the key GSH suppliers for neighboring neurons, post-mortem tissues from ALS patients had decreased Nrf2 mRNA and protein levels. Additionally, Nrf2 Via suppression of activated microglia's harmful effects on neurones, signalling is essential for ALS neuroinflammation attenuation, (Dumont & Beal, 2011).

3.4.2 Therapeutic Approaches

Most ALS patients are presently taking riluzole, which has been demonstrated to compensate for detrimental extracellular glutamate levels. Research has indicated that riluzole can extend life by two to three months. Riluzole plays a role in expediting the removal of glutamate from the synapse and may also inhibit the release of glutamate. from presynaptic terminals and could thus be advantageous to pathogenesis. Based on the outcomes of a randomized, placebo-controlled clinical study conducted in Japan on individuals with early-stage ALA. A recent study examined additional treatment methods being considered. It showed a 27% decrease in functional deterioration when given in conjunction with riluzole in a 48-week phase 3 trial that was placebo-controlled. The tyrosine kinase masatinib inhibitor, which is expected to function by decreasing MN inflammation mediated by microglia in the brain and spinal cord. It limited neuroinflammation and cell death to lessen nerve cell loss; It employing mesenchymal stromal cells extracted from, for example, adipose tissue, (Kumar, Kashav, & Hassan, 2019).

4. THERAPEUTIC STRATEGIES TARGETING ROS

4.1 Antioxidants

This has led to the widespread usage of antioxidant molecule-based medicinal approaches for the treatment of various brain diseases. Around the world, the most common exogenous antioxidant compounds that have been investigated as adjuvant therapy for neurodegeneration are vitamins, carotenes, and polyphenols. Other antioxidant classes, such as hormones like the commonly used melatonin have been employed in many neurodegenerative situations and are likewise thought to be neuroprotective agents, (Ghosh, Ghosh, & Mandal, 2011).

4.2 Natural Antioxidants

List of some natural antioxidants used in neurodegenerative diseases are as below, (Rahman, Akter, & Kamal, 2021).

Table 1. Natural antioxidants treatment in neurodegenerative diseases

Serial Number	Drug	Mechanism	Treatment
1.	Melatonin	Decrease amyloid fibril formation. Inhibit Biomolecules oxidation and increase free radicals clearance	Alzheimer disease
2.	Vitamin E	Neuroprotection	Alzheimer disease
3.	Polyphenols	Anti-inflammatory and anti-apoptotic effect	Alzheimer disease
4.	DHA	Neuroprotection	Alzheimer disease
5.	Curcumin	Reduce chronic inflammation	Alzheimer and Parkinson disease
6.	Caffeic acid	Neuroprotection	Parkinson disease

4.2.1 Synthetic Antioxidants

Along with herbal disease, synthetic antioxidants are also employed to cure neurodegenerative disease. Some examples of them are as follows, (Ashok *et al.*, 2022):

Table 2. Synthetic drugs employed as neuroprotective agents

Serial Number	Drug	Mechanism	Treatment
1.	Cholest-4-en-3-one	Scavenges the lipid peroxidation by-products	Neuropathy
2.	Alpha-lipoic acid	Mitigation of cytotoxic effects	Alzheimer disease
3.	Inosine	Neuroprotection	ALS, PD
4.	Melatonin	Protection from apoptosis and neuroinflammation	Alzheimer disease
5.	Isoalloxazine	Neuroprotective	Alzheimer and Parkinson disease
6.	TBHQ	Increase brain antioxidant capacity	Alzheimer disease

4.3 Gene Therapy

A new treatment technique called gene therapy is utilised to fix faulty genes in cells by delivering functional genetic material to them. A therapeutic gene copy will be introduced into afflicted cells, allowing the cell to use its own transcriptional and translational machinery to continually synthesis the mRNA that encoded by the gene. The primary benefit of this knowledge is that it provides a beneficial effect that may last a lifetime without requiring frequent administration. Gene treatment replaces faulty genes with functioning copies, use gene-editing technologies, RNA interference (RNAi) to silence a mutant allele, or introduce a gene that modifies a condition, (Bangde *et al.*, 2017a).

4.3.1 Targeted Delivery Systems

Progressive neuronal malfunction in certain CNS areas is a hallmark of neurodegenerative diseases, which ultimately result in impairment and death. The spectrum of possible therapeutic applications is expanded by the increasing number of freshly discovered targets. But as Table 3 illustrates, a lot of treatment drugs and the targets associated with them just provide symptomatic relief; they do not address the underlying pathology, (Bangde *et al.*, 2017b).

Table 3. Targeted therapy for neurodegenerative disorders

Serial Number	Target selection	Gene therapy	Treatment
1.	ER stress and UR	AAV2-XBP-1	PD and HD
		AAV5-Bip	PD
		AAV6-SIL1	ALS
2.	mTOR sifnaling	AAV1-AKT	AD and PD
		AAV1-caRheb	HD
3.	Mitochondrial function	AAV2-HSP70	PD
		AAV2-PINK1	AD
4.	Epigenetic regulation	AAV2-PSD95-6ZF-VP64	Alzheimer's disease
5.	Autophagy	AAV2-PINK1	AD
		AAV6- Lamp2a	PD
		AAV2-TFEB	
		AAV9-snapin	ALS

Source: Akhtar *et al.* (2021)

4.3.2 Genetic Modulation of Antioxidant Defences

In our aerobic environment, oxidants are present everywhere. Despite constantly being poisonous, they can also, at low quantities, have pathophysiological consequences and act as a causative factor in human illness. For instance, oxidants can encourage the proliferation of cells and cause tumours. The cellular antioxidant defence system is made up of many enzymes and low molecular weight components that reduce the impact of oxidants. Glutathione peroxidase, superoxide dismutases (SOD) and catalase (CAT) are the most significant. The part that Cu, CAT and Zn-SOD play in oxidant tumour promotion of JB6 mouse epidermal cells to the nonpromotable clone 30, the promotable clone 41 had 2- to 3-fold greater levels of CAT and Cu, Zn-SOD activity, protein, and stationary mRNA. Since promotable clone 41 is more shielded from oxidant toxicity, the growth-stimulatory impact of oxidants is more prominent in this strain of the organism, (Scapagnini *et al.,* 2011).

5. CHALLENGES AND FUTURE DIRECTIONS

The emergence of clinically evident symptoms in the majority of neurodegenerative illnesses is a big enigma because pathological changes typically take place years before symptoms appear. Two frequent reasons for neurodegeneration include the gradual buildup of damage to neuronal cells and the consequences of ageing.

More recently, neuroplasticity has also been linked to the beginning and progression of neurodegeneration. Compared to the developing nervous system, the fully-grown brain often exhibits less neuroplasticity in response to injuries. Yet, initial lifespan occurrences and abuses, including infections during pregnancy, an unstructured or insulting atmosphere, community isolation, trauma, inadequate diet, and contact to metals or chemicals, may impede a child's or adolescent's neuroplastic development. These could put further strain on the growing neural system's plastic ability, disrupting the structural self-organization of the brain, (Li *et al.,* 2022). Early diagnosis of neurodegenerative disorders is a major clinical problem, and it can be challenging to distinguish between different diseases due to overlapping symptoms. Furthermore, early symptoms are frequently written off or taken to be typical effects of ageing. Aside from the delay in diagnosis, other obstacles that prevent novel therapeutic approaches from progressing to the clinical development stage are the absence of druggable targets, the narrow range of available delivery options, and the dearth of trustworthy biomarkers and clinical parameters that can forecast the rate of disease progression or the effectiveness of treatment. There are currently no cures for neurodegenerative disorders; palliative care is the sole option. There is a great need to learn more about the aetiology, pathophysiology, and clinical development of these diseases since they have a terrible effect on patients and their families, which places a heavy cost on society. Additionally, novel treatment strategies and early detection techniques must be developed, (Kumar *et al.,* 2023).

5.1 Understanding the Dual Role of ROS

ROS are involved in cancer formation, initiation, suppression, and therapy in two ways. Overexposure to ROS can trigger nuclear DNA, which can start cancer. Furthermore, ROS stimulate the recruitment and M2 polarisation of macrophages, suppress natural killer cells and T- cells, allow cancer cells to evade immune monitoring and defence. Moreover, ROS cause the epithelial-mesenchymal transition in tumour cells, which facilitates tumour invasion and metastasis. It's interesting to note that a large build-up of ROS prevents tumour growth in two ways: first, by inhibiting the proliferation signalling pathway, cell cycle, and the biosynthesis of ATP and nucleotides; second, by causing cancer cells to die by activating the ferroptosis, mitochondrial, and P53 apoptotic pathways, as well as the endoplasmic reticulum stress pathway, (Huang *et al.,* 2021).

5.2 Translational Research and Clinical Trials

Modern technology enables measurements of previously unseen objects, leading to the development of whole new theories. Undoubtedly, remarkable advancements in human brain imaging, computer investigation of massive data sets, heredities and inheritance, molecular imaging, and cellular imaging have contributed significantly to the progress made in the last ten years. It is reasonable to anticipate significant advancements in the future as well as the release of new technology. An aim of the BRAIN in the US is to create new tools and increase the capabilities of current tools for neural circuit inquiry.

Nonetheless, one of the key obstacles to a successful clinical application is still safety concerns. Because the transgene may overexpress in targeted tissues or in cells that are not intended targets, gene therapy may be extremely harmful. Proprioceptive deficit, increased transaminases, ataxia, reduced ambulation, and injury to dorsal root ganglia have all been reported as toxic consequences. Every gene therapy strategy's duration and safety can be impacted by host reactions. Adaptive immune responses in patients may result in the production of matching neutralizing antibodies, which may stop the vectors from entering the targeted tissues or cells. Therefore, in order to treat neurodegenerative illnesses, potential gene-based therapy approaches should be carefully examined for clinical development, (Mak, Evaniew, & Ghert, 2014).

6. PERSONALIZED MEDICINE IN NEURODEGENERATIVE DISORDERS

The increasing malfunction of the nervous system resulting brain and spinal cord neurons death characterises neurodegenerative disorders. Not a single effective medication exists to stop neurodegenerative processes, despite massive research efforts directed towards the creation of novel disease-modifying therapies. Modifying present therapeutic and diagnostic research methodologies is therefore an aim that is becoming more and more urgent. The absence of suitable biomarkers is the main obstacle to neurodegenerative disease research. It appears to be a difficult task to discover universal biomarkers that can identify individuals with neurodegenerative disorders, track how well they are responding to treatment, and forecast the course of the disease. Rather, it appears that a more practical strategy would be to combine several approaches to find biomarkers unique to each element of the illness that has been characterised, (Griñán-Ferré *et al.*, 2024). Personalized medicine, or PM, is a style of care that is individually designed for a single patient or a subgroup of patients based on their genotype and susceptibility to the start of a disease. It also includes the use of genetic biomarkers for disease screening. For early diagnosis, suitable treatment selection, monitoring response to therapy, progression, and prognosis, PM makes use of new biomarkers, targetable disease processes, pharmacogenomics, and pharmacogenetics. Due to the growth of the "omics" fields-proteomics, metabolomics, genomics, and transcriptomics. PM has been more involved in the identification, cure, monitoring, and diagnosis of illnesses now. This has helped PM to establish itself as a reliable field in development, (Strafella *et al.,* 2018). While the use of personalised treatment in the finding and management of neurodegenerative illnesses can appear far off, practically latest advancements hint at how clinical neurology may incorporate new discoveries into routine care. Examples of these include the use of targeted biotic remedies in multiple sclerosis, microRNA profiling as a cause of new biomarkers in Parkinson's illness, and the grouping of neuroimaging and proteomic examines in the analysis of Alzheimer's ailment patients. By combining genomic, proteomic, glycomic, and metabolomic techniques, new understandings of the molecular pathways underlying disease pathophysiology may be obtained. These understandings might then be combined and applied to the field of clinical neurology. It is possible to forecast that, given the advancements of the last ten years, a customised method to treat neurological diseases will convert more commonplace in the future, (Golubnitschaja & Mandel, 2011). Progressive damage of neurons in the CNS, which results in a reduction in nervous system function, is a hallmark of neurodegenerative illnesses, (Nuzziello, Ciaccia, & Ligouri, 2019). Advances in PM application, through the use of unique CSF, blood-based, and neuroimaging biomarkers tailored to individual illnesses, have helped several neurodegenerative disorders. The prevalence of tailored therapy and prognoses in neurodegenerative disorders has increased due to the introduction of new biomarkers, improving the precision of diagnosis and the effectiveness of PM-based interventions, (Krawczuk, Kulczyńska-Przybik, & Mroczko, 2024).

7. FUTURE PROSPECTS

The key guarantee of NDs is neuronal death. Consequently, the best predicted progression of therapy for these disorders is neurogenesis. Drug's molecular weight, BBB and lipophilicity create medication transport to the brain difficult. These elements restrict the healing efficacy of drugs and increase the difficulty of treating NDs. Therefore, recently for targeted brain therapy nanoparticles has been investigated for neurogenesis, and it offers a feasible platform for enhancing therapeutic approaches. Notwithstanding these possible welfares, there are many difficult issues of Nano carrier-facilitated medication administration like production, safety, and legislation, (Krawczuk, Kulczyńska-Przybik, & Mroczko, 2024). The main determinants of nanoparticle toxicity are their size, shape, ionic dissolutions, and surface charge. A thorough analysis of these nanocarriers' approval, taking into account their impacts on the environment and human health, is necessary. A large body of research recommends a number of changes to reduce the toxicity related to the size and charge, such as adding biodegradable or intrinsic biomolecules to the surface. Regarding nanocarrier manufacturing, batch-to-batch consistency in terms of size and content should be maintained. Numerous techniques, such as high-pressure homogenisation, microemulsion, extrusion, etc., are reported for the synthesis of nanoparticles, (Rooney *et al.*, 2017). Additionally, the pharmacokinetic characteristics of the nanotherapeutics have a significant influence on their toxicity and effectiveness. Consequently, it is imperative to use an appropriate animal model to study the pharmacokinetic parameters of nanotherapeutics. To forecast the in vivo nature of nanotherapeutics, pharmacokinetic and more sophisticated physiologically based pharmacokinetic models may be used as a possible tool. Furthermore, it is important to carefully evaluate the regulatory criteria for the clinical acceptability of nanotherapeutics, (Maneu, Lax, & Cuenca, 2022), (Pathak *et al.,* 2024), (Singh, Kumar, & Sharma, n.d.), (Sharma *et al.,* 2024a), (Pathak, 2024), (Sharma, Pathak, & Biswas, 2024), (Pathak *et al.,* 2024), (Kapoor *et al.,* 2024), (Chandra *et al.*, 2024), (Kumar *et al.,* 2024a), (Sharma & Chandra, 2024), (Halagali *et al.,* 2024), (Das *et al.,* 2024), (Sharma *et al.,* 2024), (Ashique *et al.,* 2024a), (Ashique *et al.*, 2024b), (Kumar *et al.,* 2024b), (Sharma *et al.,* 2023a), (Sharma & Chandra, 2024), (Sharma *et al.,* 2023b), (Singh *et al.,* 2023), (Pathak *et al.,* 2023), (Bhattacharya *et al.*, 2023), (Koli *et al.*, 2023), (Dwivedi *et al.,* 2022), (Sharma *et al.,* 2023c), (Sharma *et al.,* 2022), (Sharma, Rani, & Khan, 2023), (Pandey *et al.,* 2022), (Pathak, Sharma, & Kumar, 2022), (Sharma & Dinda, 2022), (Sharma & Pathak, 2021), (Pathak & Sharma, 2021), (Sharma *et al.,* 2020), (Sharma, Singh, & Jha, 2020), (Sharma *et al.,*2024b), (Kaushik *et al.,* 2024), (Sharma *et al.,* 2024), (Bhandari *et al.*, 2022), (Chandra *et al.*, 2023), (Chandra *et al.,* 2023), (Pathak, Pandey, & Chandra, 2024), (Sharma *et al.*, 2024c), (Kumar *et al.,* 2024).

8. CONCLUSION

Neurodegenerative diseases stand as a prominent cause of mortality and impaired mobility in modern society. Stroke, impacting lives in both developed and developing countries, presents a significant challenge. Currently, treatment options are restricted, catering to only a small proportion of stroke victims. Thus, it is crucial to prioritize the development of effective treatments with minimal side effects to address this pressing health concern. While medications used to treat CNS injuries and neurodegenerative disorders are quick and efficient when compared to other herbal remedies and supplements, they have unanticipated side effects that vary from patient to patient. Numerous research reports substantiate the effective use of GB in various conditions, including Alzheimer's disease, asthma, venous insufficiency, multi-infarct dementia, dementia in the elderly, and resistant depression. Several scientific research validate that the efficacy of GB is due to its abundant bioactive compounds that possess protective functions on neurons. Ginkgolides and bilobalide, specifically, have undergone extensive investigation in various preclinical studies, demonstrating heightened potency. These compounds exhibit a range of beneficial effects, including defense against mitochondrial dysfunction, control over the production of amyloid and Aβ aggregation, alteration of tau protein phosphorylation, preservation of ion homeostasis, and even activation of growth factors is just a few of the properties it possesses. These findings highlight the diverse and comprehensive mechanisms of action associated with GB, particularly through the contributions of ginkgolides and bilobalide. Presently researcher is directing their attention towards formulating and creating analogs of ginkgolides and bilobalide that possess the capability to target the brain.

REFERENCES

Abate, G., Vezzoli, M., Sandri, M., Rungratanawanich, W., Memo, M., & Uberti, D. (2020). Mitochondria and cellular redox state on the route from ageing to Alzheimer's disease. *Mechanisms of Ageing and Development, 192*, 111385. DOI: 10.1016/j.mad.2020.111385 PMID: 33129798

Ahmad, S. I. (Ed.). (2017). *Reactive oxygen species in biology and human health.* CRC press. DOI: 10.1201/b20228

Akhtar, A., Andleeb, A., Waris, T. S., Bazzar, M., Moradi, A. R., Awan, N. R., & Yar, M. (2021). Neurodegenerative diseases and effective drug delivery: A review of challenges and novel therapeutics. *Journal of Controlled Release, 330*, 1152–1167. DOI: 10.1016/j.jconrel.2020.11.021 PMID: 33197487

Angelova, P. R., & Abramov, A. Y. (2018). Role of mitochondrial ROS in the brain: From physiology to neurodegeneration. *FEBS Letters, 592*(5), 692–702. DOI: 10.1002/1873-3468.12964 PMID: 29292494

Ashique, S., Bhowmick, M., Pal, R., Khatoon, H., Kumar, P., Sharma, H., Garg, A., Kumar, S., & Das, U. (2024). Multi drug resistance in colorectal cancer-approaches to overcome, advancements and future success. *Advances in Cancer Biology - Metastasis, 100114*, 100114. Advance online publication. DOI: 10.1016/j.adcanc.2024.100114

Ashique, S., Pal, R., Sharma, H., Mishra, N., & Garg, A. (2024). Unraveling the Emerging Niche Role of Extracellular Vesicles (EVs) in Traumatic Brain Injury (TBI). *CNS & Neurological Disorders - Drug Targets, 23*(11), 1357–1370. Advance online publication. DOI: 10.2174/0118715273288155240201065041 PMID: 38351688

Ashok, A., Andrabi, S. S., Mansoor, S., Kuang, Y., Kwon, B. K., & Labhasetwar, V. (2022). Antioxidant therapy in oxidative stress-induced neurodegenerative diseases: Role of nanoparticle-based drug delivery systems in clinical translation. *Antioxidants, 11*(2), 408. DOI: 10.3390/antiox11020408 PMID: 35204290

Bangde, P., Atale, S., Dey, A., Pandit, A., Dandekar, P., & Jain, R. (2017). Potential gene therapy towards treating neurodegenerative disea ses employing polymeric nanosystems. *Current Gene Therapy, 17*(2), 170–183. DOI: 10.2174/1566523217666170510153845 PMID: 28494742

Bangde, P., Atale, S., Dey, A., Pandit, A., Dandekar, P., & Jain, R. (2017). Potential gene therapy towards treating neurodegenerative disea ses employing polymeric nanosystems. *Current Gene Therapy, 17*(2), 170–183. DOI: 10.2174/1566523217666170510153845 PMID: 28494742

Barber, S. C., & Shaw, P. J. (2010). Oxidative stress in ALS: Key role in motor neuron injury and therapeutic target. *Free Radical Biology & Medicine*, *48*(5), 629–641. DOI: 10.1016/j.freeradbiomed.2009.11.018 PMID: 19969067

Bhandari, M., Sharma, S., Mishra, R., Nogai, L., Bajaj, M., & Pathak, R. (2022). Geography and the therapeutic effect of matcha tea in drinks. *Journal of Pharmaceutical Negative Results*, ●●●, 5094–5099.

Bhattacharya, V., Kumar Bhatt, A., Sharma, H., Garg, S., Chaurasia, G., Akram, W., & Mandal, S. (2023). nd Optimization of formulation by box behnken and invitro studies of emulsified gel containing zaltoprofen for the management of arthritis. *European Chemical Bulletin*, ●●●, 11734–11744.

Caron, N. S., Dorsey, E. R., & Hayden, M. R. (2018). Therapeutic approaches to Huntington disease: From the bench to the clinic. *Nature Reviews. Drug Discovery*, *17*(10), 729–750. DOI: 10.1038/nrd.2018.133 PMID: 30237454

Chandel, T. I., Zaman, M., Khan, M. V., Ali, M., Rabbani, G., Ishtikhar, M., & Khan, R. H. (2018). A mechanistic insight into protein-ligand interaction, folding, misfolding, aggregation and inhibition of protein aggregates: An overview. *International Journal of Biological Macromolecules*, *106*, 1115–1129. DOI: 10.1016/j.ijbiomac.2017.07.185 PMID: 28890370

Chandra, P., Ali, Z., Fatima, N., Sharma, H., Sachan, N., & Sharma, K. K.. (2024, April 29). Shankhpushpi (Convolvulus pluricaulis): Exploring its Cognitive Enhancing Mechanisms and Therapeutic Potential in Neurodegenerative Disorders. [Internet]. *Current Bioactive Compounds*, *20*. Advance online publication. DOI: 10.2174/0115734072292339240416095600

Chandra, P., Kaleem, M., Sachan, N., Pathak, R., Alanazi, A. S., Alsaif, N. A., Alsanea, S., Alsuwayt, B., Alanazi, M. M., & Kabra, A. (2023). Gastroprotective evaluation of Medicago sativa L.(Fabaceae) on diabetic rats. *Saudi Pharmaceutical Journal : SPJ : the Official Publication of the Saudi Pharmaceutical Society*, *31*(11), 101815. DOI: 10.1016/j.jsps.2023.101815 PMID: 37860685

Chandra, P., Sachan, N., Pathak, R., & Pal, D. (2023). Medicinal Plants Against Vesicular Stomatitis Virus (VSV) Infections: Ethnopharmacology, Chemistry, and Clinical and Preclinical Studies. In *Anti-Viral Metabolites from Medicinal Plants* (pp. 603–638). Springer International Publishing.

Chaturvedi, S. K., Siddiqi, M. K., Alam, P., & Khan, R. H. (2016). Protein misfolding and aggregation: Mechanism, factors and detection. *Process Biochemistry (Barking, London, England)*, *51*(9), 1183–1192. DOI: 10.1016/j.procbio.2016.05.015

Chistiakov, D. A., Shkurat, T. P., Melnichenko, A. A., Grechko, A. V., & Orekhov, A. N. (2018). The role of mitochondrial dysfunction in cardiovascular disease: A brief review. *Annals of Medicine*, *50*(2), 121–127. DOI: 10.1080/07853890.2017.1417631 PMID: 29237304

Colnaghi, L., Rondelli, D., Muzi-Falconi, M., & Sertic, S. (2020). Tau and DNA damage in neurodegeneration. *Brain Sciences*, *10*(12), 946. DOI: 10.3390/brainsci10120946 PMID: 33297375

Coppedè, F., & Migliore, L. (2015). DNA damage in neurodegenerative diseases. *Mutation Research. Fundamental and Molecular Mechanisms of Mutagenesis*, *776*, 84–97. DOI: 10.1016/j.mrfmmm.2014.11.010 PMID: 26255941

Cunningham, C. (2013). Microglia and neurodegeneration: The role of systemic inflammation. *Glia*, *61*(1), 71–90. DOI: 10.1002/glia.22350 PMID: 22674585

Das, S., Mukherjee, T., Mohanty, S., Nayak, N., Mal, P., Ashique, S., Pal, R., Mohanto, S., & Sharma, H. (2024). Impact of NF-κB Signaling and Sirtuin-1 Protein for Targeted Inflammatory Intervention. *Current Pharmaceutical Biotechnology*, *25*. Advance online publication. DOI: 10.2174/0113892010301469240409082212 PMID: 38638042

Dias, V., Junn, E., & Mouradian, M. M. (2013). The role of oxidative stress in Parkinson's disease. *Journal of Parkinson's Disease*, *3*(4), 461–491. DOI: 10.3233/JPD-130230 PMID: 24252804

Dumont, M., & Beal, M. F. (2011). Neuroprotective strategies involving ROS in Alzheimer disease. *Free Radical Biology & Medicine*, *51*(5), 1014–1026. DOI: 10.1016/j.freeradbiomed.2010.11.026 PMID: 21130159

Dwivedi, M., Jha, K. K., Pandey, S., Sachan, A., Sharma, H., & Dwivedi, S. K. (2022). Formulation and Evaluation of Herbal Medicated Chocolate in Treatment of Intestinal Worms and Related Problems. *IJFANS*, *11*(2), 1426–1439.

Ghosh, N., Ghosh, R., & Mandal, S. C. (2011). Antioxidant protection: A promising therapeutic intervention in neurodegenerative disease. *Free Radical Research*, *45*(8), 888–905. DOI: 10.3109/10715762.2011.574290 PMID: 21615270

Golubnitschaja, O., & Mandel, S. (2011). Promoting Predictive Diagnosis, Targeted Preventive Measures and Personalised Medicine in Handling of Neurodegeneration: Time for New Guidelines. *American Journal of Neuroprotection and Neuroregeneration*, *3*(1), 8–12. DOI: 10.1166/ajnn.2011.1031

Gotovac, K., Hajnšek, S., Pašić, M. B., Pivac, N., & Borovečki, F. (2014). Personalized medicine in neurodegenerative diseases: How far away? *Molecular Diagnosis & Therapy, 18*(1), 17–24. DOI: 10.1007/s40291-013-0058-z PMID: 24122102

. Gregersen, N., Bolund, L., & Bross, P. (2003). Protein misfolding, aggregation, and degradation in disease. *Protein Misfolding and Disease: Principles and Protocols*, 3-16.

Griñán-Ferré, C., Bellver-Sanchis, A., Guerrero, A., & Pallàs, M. (2024). Advancing personalized medicine in neurodegenerative diseases: The role of epigenetics and pharmacoepigenomics in pharmacotherapy. *Pharmacological Research, 205*, 107247. DOI: 10.1016/j.phrs.2024.107247 PMID: 38834164

Halagali, P., Inamdar, A., Singh, J., Anand, A., Sadhu, P., Pathak, R., Sharma, H., & Biswas, D. (2024). Phytochemicals, Herbal Extracts, and Dietary Supplements for Metabolic Disease Management. *Endocrine, Metabolic & Immune Disorders Drug Targets, 24*. Advance online publication. DOI: 10.2174/0118715303287911240409055710 PMID: 38676520

Halliwell, B., & Gutteridge, J. M. (2015). *Free radicals in biology and medicine*. Oxford university press. DOI: 10.1093/acprof:oso/9780198717478.001.0001

Huang, R., Chen, H., Liang, J., Li, Y., Yang, J., Luo, C., Tang, Y., Ding, Y., & Xie, X. (2021). Dual role of reactive oxygen species and their application in cancer therapy. *Journal of Cancer, 12*(18), 5543–5561. DOI: 10.7150/jca.54699 PMID: 34405016

Kapoor, D. U., Sharma, H., Maheshwari, R., Pareek, A., Gaur, M., Prajapati, B. G., Castro, G. R., Thanawuth, K., Suttiruengwong, S., & Sriamornsak, P. (2024). Konjac Glucomannan: A comprehensive review of its extraction, health benefits, and pharmaceutical applications. *Carbohydrate Polymers, 122266*, 122266. Advance online publication. DOI: 10.1016/j.carbpol.2024.122266 PMID: 38823930

. Kaushik, M., Kumar, S., Singh, M., Sharma, H., Bhowmick, M., Bhowmick, P., ... & Ansari, M. A. (2024). Bio-inspired Nanomaterials in Cancer Theranostics. *Nanotheranostics for Diagnosis and Therapy*, 95-123. Available from: DOI: 10.1007/978-981-97-3115-2_5

Kim, G. J., Fiskum, G. M., & Morgan, W. F. (2006). A role for mitochondrial dysfunction in perpetuating radiation-induced genomic instability. *Cancer Research, 66*(21), 10377–10383. DOI: 10.1158/0008-5472.CAN-05-3036 PMID: 17079457

Kim, J. A., Wei, Y., & Sowers, J. R. (2008). Role of mitochondrial dysfunction in insulin resistance. *Circulation Research, 102*(4), 401–414. DOI: 10.1161/CIRCRESAHA.107.165472 PMID: 18309108

Koli, M., Nogai, L., Bhandari, M., Mishra, R., Pathak, R., & Sharma, H. (2023). Formulation And Evaluation of Berberine Hydrochloride Film Coated Tablet. *Journal of Pharmaceutical Negative Results*, •••, 3439–3449. DOI: 10.47750/pnr.2023.14.02.403

Krawczuk, D., Kulczyńska-Przybik, A., & Mroczko, B. (2024). Clinical Application of Blood Biomarkers in Neurodegenerative Diseases—Present and Future Perspectives. *International Journal of Molecular Sciences*, *25*(15), 8132. DOI: 10.3390/ijms25158132 PMID: 39125699

Kumar, A. V., Kumar, S., Garg, V. K., Goel, N., Hoang, V. T., & Kashyap, D. (2023). Future perspectives for automated neurodegenerative disorders diagnosis: Challenges and possible research directions. In *Data Analysis for Neurodegenerative Disorders* (pp. 255–267). Springer Nature Singapore. DOI: 10.1007/978-981-99-2154-6_15

Kumar, P., Pandey, S. N., Ahmad, F., Verma, A., Sharma, H., Ashique, S., Bhattacharyya, S. P., Bhattacharyya, I., Kumar, S., Mishra, N., & Garg, A. (2024). Carbon Nanotubes: A Targeted Drug Delivery against Cancer Cell. *Current Nanoscience*, *20*(6), 769–800. DOI: 10.2174/0115734137271865231105070727

Kumar, P., Sharma, H., Singh, A., Durgapal, S., Kukreti, G., Bhowmick, M., Bhowmick, P., & Ashique, S. (2024, April 29). Targeting the Interplay of Proteins through PROTACs for Management Cancer and Associated Disorders. [Internet]. *Current Cancer Therapy Reviews*, *20*. Advance online publication. DOI: 10.2174/0115733947304806240417092449

Kumar, P., Sharma, H., Singh, A., Pandey, S. N., & Chandra, P. (2024). Correlation Between Exosomes and Neuro-inflammation in Various Brain Disorders. In *Exosomes Based Drug Delivery Strategies for Brain Disorders* (pp. 273–302). Springer Nature Singapore., DOI: 10.1007/978-981-99-8373-5_11

Kumar, V., Kashav, T., & Hassan, M. I. (2019). Amyotrophic lateral sclerosis: current therapeutic perspectives. *Pathology, prevention and therapeutics of neurodegenerative disease*, 207-224.

Lane, R. K., Hilsabeck, T., & Rea, S. L. (2015). The role of mitochondrial dysfunction in age-related diseases. *Biochimica et Biophysica Acta (BBA)-. Biochimica et Biophysica Acta. Bioenergetics*, *1847*(11), 1387–1400. DOI: 10.1016/j.bbabio.2015.05.021

Li, Z., Sohrabi, H. R., Sobrino, T., & Romaus-Sanjurjo, D. (2022). New challenges and future perspectives in neurodegeneration. *Frontiers in Neuroscience*, *16*, 1049721. DOI: 10.3389/fnins.2022.1049721 PMID: 36300167

Lushchak, V. I. (2014). Free radicals, reactive oxygen species, oxidative stress and its classification. *Chemico-Biological Interactions*, *224*, 164–175. DOI: 10.1016/j.cbi.2014.10.016 PMID: 25452175

Madabhushi, R., Pan, L., & Tsai, L. H. (2014). DNA damage and its links to neurodegeneration. *Neuron*, *83*(2), 266–282. DOI: 10.1016/j.neuron.2014.06.034 PMID: 25033177

Mak, I. W., Evaniew, N., & Ghert, M. (2014). Lost in translation: Animal models and clinical trials in cancer treatment. *American Journal of Translational Research*, *6*(2), 114. PMID: 24489990

Maneu, V., Lax, P., & Cuenca, N. (2022). Current and future therapeutic strategies for the treatment of retinal neurodegenerative diseases. *Neural Regeneration Research*, *17*(1), 103–104. DOI: 10.4103/1673-5374.314305 PMID: 34100441

Maynard, S., Fang, E. F., Scheibye-Knudsen, M., Croteau, D. L., & Bohr, V. A. (2015). DNA damage, DNA repair, aging, and neurodegeneration. *Cold Spring Harbor Perspectives in Medicine*, *5*(10), a025130. DOI: 10.1101/cshperspect.a025130 PMID: 26385091

Merlo, D., Cuchillo-Ibañez, I., Parlato, R., & Rammes, G. (2016). DNA damage, neurodegeneration, and synaptic plasticity. *Neural Plasticity*, •••, 2016. PMID: 27313899

Meyer, E., Kurian, M. A., & Hayflick, S. J. (2015). Neurodegeneration with brain iron accumulation: Genetic diversity and pathophysiological mechanisms. *Annual Review of Genomics and Human Genetics*, *16*(1), 257–279. DOI: 10.1146/annurev-genom-090314-025011 PMID: 25973518

Nuzziello, N., Ciaccia, L., & Liguori, M. (2019). Precision medicine in neurodegenerative diseases: Some promising tips coming from the microRNAs' world. *Cells*, *9*(1), 75. DOI: 10.3390/cells9010075 PMID: 31892254

Pandey, P. R. I. Y. A., Kumar, N. I. T. I. N., Kaur, T. E. L. J. I. N. D. E. R., Saini, S. H. A. L. I. N. I., & Sharma, H. I. M. A. N. S. H. U. (2022). Antidiabetic Activity of Caesalpinia Bonducella Leaves of Hydro Alcoholic Extracts in Albino Rats. *YMER*, *7*(21), 840–846. DOI: 10.37896/YMER21.07/67

Pardo-Moreno, T., González-Acedo, A., Rivas-Domínguez, A., García-Morales, V., García-Cozar, F. J., Ramos-Rodríguez, J. J., & Melguizo-Rodríguez, L. (2022). Therapeutic approach to Alzheimer's disease: Current treatments and new perspectives. *Pharmaceutics*, *14*(6), 1117. DOI: 10.3390/pharmaceutics14061117 PMID: 35745693

Pastore, A., & Temussi, P. (2012). Protein aggregation and misfolding: Good or evil? *Journal of Physics Condensed Matter*, *24*(24), 244101. DOI: 10.1088/0953-8984/24/24/244101 PMID: 22595337

Pathak, R. (2024).. . *Natural Compounds Targeting Signaling Pathways in Breast Cancer Therapy.*, *6*(10), 5430–5479. DOI: 10.33472/AFJBS.6.10.2024.5430-5479

Pathak, R., Kaur, V., Sharma, S., Bhandari, M., Mishra, R., Saxena, A., & Pathak, R. (2024). Pazopanib: Effective monotherapy for precise cancer treatment, targeting specific mutations and tumors. *Cancer*, *6*(9). Advance online publication. DOI: 10.33472/AFJBS.6.9.2024.1311-1330

Pathak, R., Pandey, S. P., & Chandra, P. (2024, February). Gastroprotective Effects of Biological Macromolecule: Polysaccharides. *Macromolecular Symposia*, *413*(1), 2300122. DOI: 10.1002/masy.202300122

Pathak, R., & Sharma, H. (2021). A review on medicinal uses of Cinnamomum verum (Cinnamon). *Journal of Drug Delivery and Therapeutics*, *11*(6-S), 161–166. DOI: 10.22270/jddt.v11i6-S.5145

Pathak, R., Sharma, H., & Kumar, N. (2022). A Brief Review on Anthocephalus cadamba. *Acta Scientific Pharmacology*, *3*(5).

Pathak, R., Sharma, H., Nogai, L., Bhandari, M., Koli, M., Mishra, R., & Kumar, N. (2023). A brief review on Pathogenesis, Transmission and Management of Monkeypox virus outbreaks. *Bulletin of Environment, Pharmacology and Life Sciences*, *12*(4), 244–256.

Pathak, R., Sharma, S., Bhandari, M., Nogai, L., Mishra, R., Saxena, A., & Reena Km, S. H. (2024). Neuroinflammation at the crossroads of metabolic and neurodegenerative diseases: Causes, consequences and interventions. *Journal of Experimental Zoology India*, *21*(2), 2447–2461. DOI: 10.59467/jez.2024.27.2.2447

Peña-Bautista, C., Vento, M., Baquero, M., & Cháfer-Pericás, C. (2019). Lipid peroxidation in neurodegeneration. *Clinica Chimica Acta*, *497*, 178–188. DOI: 10.1016/j.cca.2019.07.037 PMID: 31377127

Rahman, M. H., Akter, R., & Kamal, M. A. (2021). Prospective function of different antioxidant containing natural products in the treatment of neurodegenerative diseases. *CNS & Neurological Disorders-Drug Targets (Formerly Current Drug Targets-CNS & Neurological Disorders)*, *20*(8), 694-703.

. Revi, M. (2020). Alzheimer's disease therapeutic approaches. *GeNeDis 2018: Genetics and Neurodegeneration*, 105-116.

Rooney, J. P., Brayne, C., Tobin, K., Logroscino, G., Glymour, M. M., & Hardiman, O. (2017). Benefits, pitfalls, and future design of population-based registers in neurodegenerative disease. *Neurology, 88*(24), 2321–2329. DOI: 10.1212/WNL.0000000000004038 PMID: 28515268

Ross, C. A., & Tabrizi, S. J. (2011). Huntington's disease: From molecular pathogenesis to clinical treatment. *Lancet Neurology, 10*(1), 83–98. DOI: 10.1016/S1474-4422(10)70245-3 PMID: 21163446

Scapagnini, G., Sonya, V., Nader, A. G., Calogero, C., Zella, D., & Fabio, G. (2011). Modulation of Nrf2/ARE pathway by food polyphenols: A nutritional neuroprotective strategy for cognitive and neurodegenerative disorders. *Molecular Neurobiology, 44*(2), 192–201. DOI: 10.1007/s12035-011-8181-5 PMID: 21499987

Sharma, H., Anand, A., Halagali, P., Inamdar, A., Pathak, R., Taghizadeh-Hesary, F., & Ashique, S. (2024). *Advancement of Nanoengineered Flavonoids for Chronic Metabolic Diseases*. Wiley. DOI: 10.1002/9781394238071.ch13

Sharma, H., & Chandra, P. (2024). Effects of natural remedies on memory loss and Alzheimer's disease. *African Journal of Biological Sciences, 6*(7), 187–211. DOI: 10.33472/AFJBS.6.7.2024.187-211

Sharma, H., & Chandra, P. (2024). Challenges and Future Prospects: A Benefaction of Phytoconstituents on Molecular Targets Pertaining to Alzheimer's Disease. *International Journal of Pharmaceutical Investigation, 14*(1), 117–126. Advance online publication. DOI: 10.5530/ijpi.14.1.15

Sharma, H., Chandra, P., Verma, A., Pandey, S. N., Kumar, P., & Sigh, A. (2023). Therapeutic approaches of nutraceuticals in the prevention of neurological disorders. *Eur Chem Bull, 12*(5), 1575-1596. doi: - DOI: 10.48047/ecb/2023.12.si5a.038

Sharma, H., Kaushik, M., Goswami, P., Sreevani, S., Chakraborty, A., Ashique, S., & Pal, R. (2024). Role of miRNAs in Brain Development. *MicroRNA (Shariqah, United Arab Emirates), 13*(2), 96–109. Advance online publication. DOI: 10.2174/0122115366287127240322054519 PMID: 38571343

Sharma, H., Pande, M., Jha, K. K., & Sharma, H. (2020). Hyperuricemia: a risk factor beyond gout. *ASIO Journal of Pharmaceutical & Herbal Medicines Research (ASIO-JPHMR), 6*(1), 42-49. http://doi-ds.org/doilink/06.2020-25164442/

Sharma, H., & Pathak, R. (2021). A review on prelimenary phytochemical screening of curcuma longa linn. *J Pharma Herbal Med Res (ASIO-JPHMR), 7*(2), 24-27. https://doi-ds.org/doilink/10.2021-97622886/ASIO JPHMR/ 10.2016-19146535/V7/I2/460

Sharma, H., Pathak, R., & Biswas, D. (2024, May 29). Unveiling the Therapeutic Potential of Modern Probiotics in Addressing Neurodegenerative Disorders: A Comprehensive Exploration, Review and Future Perspectives on Intervention Strategies. [Internet]. *Current Psychiatry Research and Reviews, 20*. Advance online publication. DOI: 10.2174/0126660822304321240520075036

Sharma, H., Pathak, R., Jain, S., Bhandari, M., Mishra, R., Reena, K. M., & Varshney, P. (2023). Ficus racemosa L: A review on its important medicinal uses, phytochemicals and biological activities. *Journal of Population Therapeutics and Clinical Pharmacology = Journal de la Thérapeutique des Populations et de la Pharmacologie Clinique, 30*(17), 213–227. DOI: 10.47750/jptcp.2023.30.17.018

Sharma, H., Pathak, R., Kumar, N., Nogai, L., Mishra, R., Bhandari, M., & Pandey, P. (2023). Endocannabinoid System: Role in Depression, Recompense, And Pain Control. *Journal of Survey in Fisheries Sciences, 10*(4S), 2743–2751. DOI: 10.17762/sfs.v10i4S.1655

Sharma, H., Pathak, R., Saxena, D., & Kumar, N. (2022). Emerging role of non-coding RNA'S: Human health and diseases. *GIS, 9*(7), 2022–2050.

. Sharma, H., Rachamalla, H. K., Mishra, N., Chandra, P., Pathak, R., & Ashique, S. (2024). Introduction to exosome and its role in brain disorders. *Exosomes Based Drug Delivery Strategies for Brain Disorders*, 1-35.. Available from: https://doi.org/DOI: 10.1007/978-981-99-8373-5_1

Sharma, H., Rani, T., & Khan, S. (2023). An insight into neuropathic pain: A systemic and up-to-date review. *International Journal of Pharmaceutical Sciences and Research, 14*(2), 607–621. DOI: 10.13040/IJPSR.0975-8232.14(2).607-21

Sharma, H., Singh, S., & Jha, K. K. (2020). Treatment and recommendations for homeless patients with hypertension, hyperlipidemia & heart failure-a review. *ASIO Journal of Experimental Pharmacology & Clinical Research (ASIO-JEPCR), 6*(1), 24-32. http://doi-ds.org/doilink/07.2020-72295186/

Sharma, H., Tyagi, S. J., Chandra, P., Verma, A., Kumar, P., Ashique, S., & Hussain, A. (2024). Role of Exosomes in Parkinson's and Alzheimer's Diseases. In *Exosomes Based Drug Delivery Strategies for Brain Disorders* (pp. 147–182). Springer Nature Singapore., DOI: 10.1007/978-981-99-8373-5_6

Sharma, H., Tyagi, S. J., Varshney, P., Pathak, N., & Pathak, R. (2024, June 13). A Review on Mpox: Diagnosis, Prevention and Treatments. *Coronaviruses, 05*, 1–17. DOI: 10.2174/0126667975301557240604113752

Sharma, S., & Dinda, S. C. S. H. Matrix types drug delivery system for sustained release : A review. ASIO J Drug Deliv. 2022;6(1):1–8. https://doi-ds.org/doilink/03.2022-65678198/ASIO-JDD/10.2016-21915953/2022/V6I1/920

Singh, A., Kumar, P., & Sharma, H. (•••). Breakthrough Opportunities of Nanotheranostics in Psoriasis: From Pathogenesis to Management Strategy. *Infectious Disorders Drug Targets*, *24*, 1–20. DOI: 10.2174/0118715265298802240603120251 PMID: 39075964

Singh, L. P., Gugulothu, S., Perusomula, R., Mishra, A., Bhavani, P. D., Singh, S., & Dwivedi, M. (2023). Synthesis of Some Tetrazole and Thiazolidine-4-One Derivatives of Schiff base by using Ionic Liquids as Catalyst and Evaluation of their Antifungal and Antibacterial Activity. *European Chemical Bulletin*, *12*, 281–297.

Sisein, E. A. (2014). Biochemistry of free radicals and antioxidants. *Scholars Academic Journal of Biosciences*, *2*(2), 110–118.

Song, H., Vijayasarathy, C., Zeng, Y., Marangoni, D., Bush, R. A., Wu, Z., & Sieving, P. A. (2016). NADPH oxidase contributes to photoreceptor degeneration in constitutively active RAC1 mice. *Investigative Ophthalmology & Visual Science*, *57*(6), 2864–2875. DOI: 10.1167/iovs.15-18974 PMID: 27233035

Stein, D., & Toiber, D. (2017). DNA damage and neurodegeneration: The unusual suspect. *Neural Regeneration Research*, *12*(9), 1441–1442. DOI: 10.4103/1673-5374.215254 PMID: 29089988

Strafella, C., Caputo, V., Galota, M. R., Zampatti, S., Marella, G., Mauriello, S., Cascella, R., & Giardina, E. (2018). Application of precision medicine in neurodegenerative diseases. *Frontiers in Neurology*, *9*, 701. DOI: 10.3389/fneur.2018.00701 PMID: 30190701

Tabrizi, S. J., Flower, M. D., Ross, C. A., & Wild, E. J. (2020). Huntington disease: New insights into molecular pathogenesis and therapeutic opportunities. *Nature Reviews. Neurology*, *16*(10), 529–546. DOI: 10.1038/s41582-020-0389-4 PMID: 32796930

Tobore, T. O. (2019). On the etiopathogenesis and pathophysiology of Alzheimer's disease: A comprehensive theoretical review. *Journal of Alzheimer's Disease*, *68*(2), 417–437. DOI: 10.3233/JAD-181052 PMID: 30775973

Valera, E., & Masliah, E. (2016). Therapeutic approaches in Parkinson's disease and related disorders. *Journal of Neurochemistry*, *139*(S1), 346–352. DOI: 10.1111/jnc.13529 PMID: 26749150

Wang, H., Lautrup, S., Caponio, D., Zhang, J., & Fang, E. F. (2021). DNA damage-induced neurodegeneration in accelerated ageing and Alzheimer's disease. *International Journal of Molecular Sciences*, *22*(13), 6748. DOI: 10.3390/ijms22136748 PMID: 34201700

Welch, G., & Tsai, L. H. (2022). Mechanisms of DNA damage-mediated neurotoxicity in neurodegenerative disease. *EMBO Reports*, *23*(6), e54217. DOI: 10.15252/embr.202154217 PMID: 35499251

Zia, A., Farkhondeh, T., Pourbagher-Shahri, A. M., & Samarghandian, S. (2022). The roles of mitochondrial dysfunction and reactive oxygen species in aging and senescence. *Current Molecular Medicine*, *22*(1), 37–49. DOI: 10.2174/1566524021666210218112616 PMID: 33602082

Chapter 13
ROS and Precision Medicine in Lifestyle Diseases:
Personalized Approaches

Pawan Gupta
https://orcid.org/0000-0001-7201-1356
Department of Pharmaceutical Chemistry, Shri Vile Parle Kelavani Mandal's Institute of Pharmacy, Dhule, India

Yogesh Chaudhari
Department of Pharmaceutical Chemistry, Shri Vile Parle Kelavani Mandal's Institute of Pharmacy, Dhule, India

Abhinav Joseph
School of Chemical Engineering and Physical Sciences, Lovely Professional University, India & St. John's College, Agra, India

Sukrat Sinha
Department of Zoology, University of Delhi, India

Ansari Rashid
https://orcid.org/0009-0000-0089-4153
Department of Pharmacology, Sanjay Education Society's Institute of Pharmacology, India

Pranav Kumar Prabhakar
https://orcid.org/0000-0001-8130-1822
Lovely Professional University, India & Nagaland University, India

ABSTRACT

The body's cells produce highly reactive substances known as reactive oxygen species (ROS). Though ROS are important in immunological responses as well as metabolic pathways. However, high concentrations bring about oxidative stress, which lead to many lifestyle diseases such as diabetes, heart diseases, etc. This chapter examines involvement of the ROS in mechanism of disease initiation, and how the dynamic

DOI: 10.4018/979-8-3693-7919-6.ch013

nature of the individual ROS might open the options for personalized treatment. Moreover, it is also examined how differences in genotype with respect to important enzymes, could influence life style diseases. Interactions with genes in explaining responses to drug therapy are exemplified in the context of diabetes, cancer, and inflammatory diseases. These findings offer contributions to controlling oxidative stress and related disorders through growing approaches for templatized strategies from within a better awareness of ROS and other genetic factors.

1. INTRODUCTION:

Reactive Oxygen Species (ROS) are collectively reactive molecules which can be generated by various forms of cells in the body. These molecules include; hydrogen peroxide (H_2O_2), hydroxyl radical (OH^-), singlet oxygen and super oxide (O^{2-}), (Bedard & Krause, 2007). These molecules although unstable, participate in a vast array of biological processes. ROS take part in pathogen resistance, cellular signalling, and homeostasis. However, their reactive nature makes them potentially lethal. In most cases, ROS can harmfully affect the cellular proteins, lipids and nucleic acids when the body's antioxidant defences fail to sufficiently scavenge them leading to oxidative stress and disease advancement, (Nathan & Cunningham-Bussel, 2013).

ROS are mainly produced during normal metabolic processes, mainly within the mitochondrial respiratory chain during the production of ATP. Enzymes such as NADPH oxidase also contribute to ROS production during immune responses and cellular metabolism. Although ROS are naturally generated under conditions that require high energy such as exercising or during immune responses. However high levels of ROS can overcome the body antioxidant system. Imbalance between ROS production and neutralization is responsible for the phenomenon being referred to as oxidative stress, which is assumed to contribute to the development of lifestyle diseases, (Yang & Lian, 2020).

Many lifestyle diseases, including cardiovascular diseases (CVDs), metabolic disorders, (like diabetes, arthritis, cancer), respiratory and neurodegenerative disorders, are associated with metabolic imbalances, chronic inflammation or cellular dysfunction, where ROS possess a dual role. On one hand, ROS are involved in the regulation of the metabolic processes happening in cells and immune reactions. On the other hand, the increased ROS production leads to tissue damage, chronic inflammation, and development of complications such as hypertension, heart failure, and neurodegeneration. In cancer for instance, ROS at lower level can promote tumour growth and metastasis through changing signalling pathways whereas at higher levels, it can cause oxidative stress that results in cell death, (Nakamura & Takada, 2021).

The emerging field of precision medicine provides novel insights into the potential therapeutic applications of ROS, aiming to utilize their biological activity for targeted treatment. By understanding an individual's unique ROS dynamics—whether their ROS production is excessive or their antioxidant defences are deficient—treatments can be tailored to restore balance. Targeting ROS levels with personalized approaches can assist in treating oxidative stress effectively and, therefore, slowing down the further development of lifestyle diseases, (Casas *et al.,* 2020).

This chapter elaborates the role of ROS in precision medicine, with an aim of understanding the extent to which refined knowledge of ROS production and regulation, and ROS-disease connections can lead to personalized treatment for lifestyle diseases. It highlights both the harmful effects of ROS in disease pathology and their potential therapeutic benefits when modulated under controlled conditions.

2. LIFE STYLE DISEASES:

Lifestyle diseases are diseases that originate from the way people live their lives, with the modern man consuming unhealthy foods, taking little or no exercise, putting up with tremendous pressure at work, getting little or no sleep, and consuming products such as tobacco and alcohol in unhealthy amounts. (**Figure 1** listed the different lifestyle disorders)

Figure 1. Lifestyle disorder caused by ROS

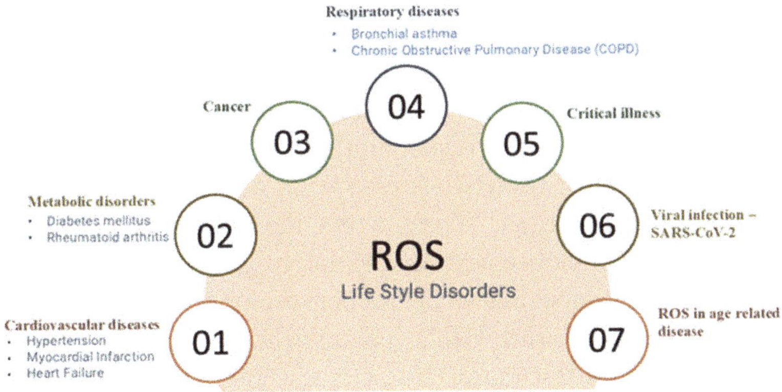

2.1 Cardiovascular diseases:

Many cardiovascular diseases (CVD) exhibited, at least partially, with a surplus production of ROS, (Sammam *et al.,* 2017) (Tsutsui, Kinugawa, & Matsushima, 2011). There are plenty of evidence to suggest that ROS play a crucial role in CVDs, such as atherosclerosis, hypertension, and vascular damage, (Kumar & Das, 1993) (Laursen *et al.,* 1997) (Touyz & Schiffrin, 2004) (White *et al.,* 1994). Heart issues may arise from an imbalance in redox homeostasis. The vascular layers of CV system generate ROS, which are signalling molecules in controlling a variety of processes including the development, relaxation, and contraction of vascular smooth muscle cells, (Fearon & Faux, 2009). On the other hand, endothelial dysfunction (ED) and the development of CVD are directly related to an excessive or persistent increase in ROS generation. Certain CV risk factors are linked to either elevated ROS generation or lowered levels of plasma γ-l-glutamyl-l-cysteinyl-glycine (GSH), (Shimizu *et al.,* 2004). The beginning and development of CVD have been associated with changes in GSH levels or oxidation state, (Bajic *et al.,* 2019).

A. Hypertension:

Some of the ROS like O_2^-, NO, peroxynitrite, and H_2O_2 are found to involve in the pathogenesis of hypertension, (Griendling *et al.,* 2016). The primary source of O_2^- in arteries is NADPH oxidase, (Mohazzab, Kaminski, & Wolin, 1994) (Pagano *et al.,* 1995). The NADPH oxidase enzyme activities increase in response to hypertensive stimuli. The O_2^- reacts with NO and generate peroxynitrite or quickly transformed to H_2O_2. H_2O_2 interacts with a number of proteins of cells to alter signalling pathways, including transcription factors, protein phosphatases, kinases, ion channels, and small molecular weight G-proteins. Peroxynitrite leads to tissue damage, while NO is a potent vasoprotective and vasodilator, (Griendling *et al.*, 2021). Numerous potential pathways exist for oxidative stress to play a role in the development and/or maintenance of hypertension. These include the vasodilator nitric oxide (NO) being quenched by ROS like superoxide (McIntyre, Bohr, & Dominiczak, 1999), production of F2-isoprostanes and other vasoconstrictor lipid peroxidation products (Cracowski, Durand, & Bessard, 2002), reduction of tetrahydrobiopterin (BH4), a crucial component for NO synthase, (Vaziri *et al.,* 2000). Genetic variation in NO synthase enzyme led to altered production of NO and contributing to hypertension, CVDs and inflammatory disorders in individuals, (Nasyrova *et al.,* 2020). The genetic variations, environmental influences, or other individual features causes the different effects of the same medications in hypertensive patients. Precision hypertension will collect and analyze a great deal of data regarding a person's health using genomics, transcriptomics, proteomics, metabolomics, molecular profiling, health information, environmental factors, including lifestyle, and big data analysis, (Dzau & Hodgkinson, 2024). Individualized therapy is necessary due to the complexity of causes generating hypertension, which is characterized by variations in pathogenesis and reactions to antihypertensive medicines, (Kotchen, Cowley, & Liang, 2016).

Management of Hypertension by Personalized Approach:

Since hypertension is a polygenic condition, developing more effective future treatments for it continues to be difficult. AS per report, a single allele has been found to have connection in regulation of blood pressure. This may open avenues to identify the appropriate and unique goals for preventing and treating hypertension, (Harrap, 2003).

Gene Therapy for Hypertension:

The Food and Drug Administration specify gene therapy as a procedure that changes a person's genes in order to treat or cure disease. Gene treatments involve introducing a new or transformed gene inside the body for the treatment of a medical condition, substituting a disease-causing gene with a healthy copy of the gene, or inactivating a disease-causing gene that is not working correctly. The most developed gene treatments for hypertension are RNA interference (RNAi) and gene editing, (Dzau & Hodgkinson, 2024).

RNA Interference:

RNA interference (RNAi) is a natural process where RNA molecules inhibit gene expression by binding to mRNA. It not only regulates mRNA and protein levels but also defends against foreign nucleic acids. Cells recognize and destroy double-stranded RNA, which can come from external sources or from RNA binding to mRNA. RNAi is carried out using siRNAs and miRNAs, but only siRNAs are currently being studied for hypertension treatment, (Chen *et al.,* 2014) (Desai *et al.,* 2023) (Hao, Li, & Li, 2017) (Huang *et al.,* 2016) (Huang *et al.,* 2017) (Sun, Bello-Roufai, & Wang, 2008).

The most advanced siRNA therapy which target angiotensinogen (AGT) production in the liver is Zelebesiran. In Phase I clinical trial, Zelibesiranis well tolerated and resulted in consistent reductions in blood pressure, (Desai *et al.,* 2023). Recently conducted Phase II trials, shows that subcutaneous administration of Zelebesiran improves blood pressure control. This is a novel therapy for uncontrolled hypertension, (Saxena *et al.,* 2024). The siRNA techniques are targeted to many receptors including the β1-adrenergic receptor (Huang *et al.,* 2017), the angiotensin type 1 receptor (Chen *et al.,* 2014), endothelin-1 (Chen & Sun, 2017), TNF-α (tumor necrosis factor-alpha) (Crosswhite, Chen, & Sun, 2014), mineralocorticoid receptors

(Sun, Bello-Roufai, & Wang, 2008), the intrarenal ghrelin receptor (Kemp, Howell, & Padia, 2018), galectin-3 (Hao, Li, & Li, 2017) and G-protein-coupled receptor kinase type 4 to control high blood pressure in mouse models, (Huang *et al.,* 2016). This technique is subjected to further research and development for new therapeutics.

Gene Editing:

Gene editing techniques, including CRISPR-Cas9, permit exact modification of DNA. CRISPR-Cas9 uses guide RNA to direct the Cas9 nuclease to a precise DNA site, where it makes a cut. This cut disrupts the gene when the DNA is repaired. Advancements like homology-directed repair and base editing have refined this method. CRISPR-Cas9, along with antisense oligonucleotides (ASO) and small interfering RNA (siRNA), targets hepatic angiotensinogen (AGT) production to reduce hypertension by inhibiting the renin-angiotensin system, (Long *et al.,* 2014) (Masi, Dalpiaz, & Borghi, 2024) (Wang & Doudna, 2023).

B. Modifying Myocardial Infarction Genetic Risk:

The 5-lipoxygenase–activating protein (FLAP) gene has variants that are linked to an elevated risk of myocardial infarction (MI). Veliflapon(DG-031) is selective FLAP Inhibitor that was administered in a randomized, prospective, placebo-controlled crossover trial, (Hakonarson *et al.,* 2005) (Li *et al.,* 2019). Hakonarson *et al.* (2005) reported the identification of a gene variant that influences patients to MI, the FLAP gene is encoded by this gene, and a risk mutation in it causes an almost 2-fold increase in MI risk. One of the most effective chemokine mediators of arterial inflammation, leukotriene B4, is produced by the leukotriene route via FLAP. By obstructing the function of FLAP because of down-regulation of the leukotriene pathway, the risk of MI may be reduced in those predisposed to the variant, DG-031 led to important and dose-dependent suppression of biomarkers that are associated with increased risk of MI events. The ability of a drug to suppress inflammatory indicators may or may not reduce the risk of a heart attack, but the investigators are unclear. To verify whether the gene variant and cardiac disease are related, more investigation is required, (Hakonarson *et al.,* 2005) (Jain, 2011).

C. Management of Heart Failure:

ß-Blockers:

For the treatment of heart failure, ACE inhibitors and ß-blockers are advised in addition. A correlation between the ACE genotype and responsiveness to ß-blocker medication in heart failure has been identified. Patients with the D allele have higher benefits from pharmacological interventions aimed at reducing sympathetic nervous system activity, such as β-blocker therapy, (Jain, 2011). There are several reasons why ß-blockers are not widely used in treating many CHF patients, despite their proven effectiveness like potential for adverse reactions (bradycardia, hypotension, and worsening heart failure), which affect 25–43% of patients and, often lead to the discontinuation of therapy. This is due to genetic variations that influence ß-blocker sensitivity. Polymorphisms in the CYP2D6 isoenzyme gene, responsible for metabolizing ß-blockers like propranolol, metoprolol, carvedilol, and trimorolol, is also impact ß-blocker response. The interaction between the CYP2D6 genotype and other genetic polymorphisms, such as those in the ß-receptor or the ACE I/D polymorphism, influence overall treatment response. These genetic differences affect not only the plasma concentrations of the medication but also the body's response to ß-blockers due to changes in the pharmacological target, the ß1-receptor. For instance, a clinical study on metoprolol extended release/controlled release titration in heart failure patients found that individuals with the Ser49Ser genotype and the ß1-receptor Gly389 variant are considerably more likely to require higher doses of heart failure medications during ß-blocker titration and may also need more frequent follow-up, (Jain, 2011) (Terra *et al.,* 2005).

Bucindolol:

The bucindolol is third-generation, non-selective β-adrenergic receptor antagonist. Due to different pharmacodynamics of bucindolol in individuals with severe heart failure resulted either from avoidance of a negative consequence (an α2c receptor polymorphism) or a hyper-reaction (a β1 receptor polymorphism), (Liggett *et al.,* 2006). α2c control the heart sympathetic nerves' norepinephrine release. The deletion of four continuous amino acids (Del322–325) in the α2C-adrenergic receptor gene is a common coding polymorphism that leads to a significant loss of agonist-mediated receptor activity in transfected cells. This failure would lead to a loss of regular synaptic autoinhibitory feedback, which would increase norepinephrine presynaptic release, (Small *et al.,* 2002). Patients most probable to get benefit from bucindolol treatment have been discovered by Arca Discovery Inc. through the identification of bucindolol exhibits genetic variations, including the ability to lower norepinephrine levels as well as to block with high affinity the β-1 Arg-389 receptor variant, (BioSpace, 2006). A series of diplo types of β-adrenergic receptor that characterize the population are produced by these dual gene loci. In human heart failure, a genetic polymorphism within a conserved ß (1)-adrenergic receptor modifies ß-blocker responsiveness and cardiac function. According to a study, the ß(1)adrenoceptor-389 mutation modifies signalling in a variety of models and influences the therapeutic response of ß -blockers in the patients, (Liggett *et al.,* 2006). The patient who are susceptible for this drug due to β-ADR polymorphism are more prone to adverse drug reactions like heart failure. This drug cannot be prescribed to the patients with genetic variability with β adrenergic receptor genes.

2.2 Metabolic disorders:

The metabolic disorder, referred as metabolic syndrome is a condition classified by a couple of metabolic abnormalities that include hypertension, abdominal obesity, hyperglycemia, type 2 diabetes, cardiovascular disease and many more, (Bhatti, Bhatti, & Reddy, 2017). The molecular mechanisms of proteins involved in anabolic and metabolic pathways is important for understanding energy homeostasis in humans. This will help in building therapeutic agents for the chronic metabolic disorders' treatment of such as diabetes, (Kota, Huang, & Roufogalis, 2005).

A. Diabetes Mellitus

Diabetes Mellitus (DM) is a most common metabolic syndrome characterized by irregularities in the secretion of insulin, (Guillausseau *et al.*, 2008). Reduced insulin secretion or its absence causes high glucose levels and glucose intolerance, (Fariduddin & Yasmin-Aktar, 2015). Major features of DM are resistance to insulin by targeted tissues and secretion of insulin in deficient amounts by the β-cells, (Ashcroft & Rorsman, 2012).

The OS is also a significant factor in the onset and course of diabetes mellitus (147). Studies of genome-wide associations suggest that gene loci have a relation with the diabetes pathophysiology including insulin secretion and resistance. The report on genetic studies on individual populations had been performed to analyze genetic association and get information about genes related to type 2 diabetes and insulin function. The Ectonucleotide pyrophosphates/phosphodiesterase 1 (ENPP1), Peroxisome proliferator-activated receptor gamma (PPARG), Insulin Receptor Substrate 1 (IRS1), Wolfram Syndrome 1 (WFS1) and Potassium Inwardly Rectifying Channel Subfamily J Member 11 (KCNJ11) are the genes which linked to type 2 diabetes and insulin function in these studies. Unique genetic variants such as Hematopoietically Expressed Homeobox (HHEX), Solute Carrier Family 30 Member 8 (SLC30A8), and a unique genetic variable COC387761 have linked with the type 2 diabetes (T2DM) through genome-wide association studies, (Sladek *et al.,* 2007). By directly interacting with endothelial cells, angiopoietin-1 (Ang-1), endothelial cell-selective adhesion molecule (ESAM), endothelin-1 (ET-1), and other substances control the activity of endothelial cells. Diabetes patients have higher plasma levels of ET-1, a vasoconstrictor peptide produced by endothelial cells, and ET-1 overexpression directly prevents the release of NO and eNOS (54). Furthermore, through Nox1, overexpression of ET-1 stimulates the development of perivascular oxidative stress, inflammation, and atherosclerosis linked to type 1 diabetes, (Ouerd *et al.,* 2021). Precision medicine benefits greatly from genomics since it has made it possible to identify the genes that are linked to type 2 diabetes mellitus (T2DM) and insulin function. However, personalized treatment is given to the different type of diabetic patients like genetic or type 1 diabetes is being administered with insulin injection and type 2 diabetes (DM) is being treated by oral hypoglycemic agents.

B. Rheumatoid Arthritis:

Arthritis is characterized by acute or chronic inflammation of different joints of the body. It is having wide range of symptoms including stiffness, pain, joint deformities and decreased range of motion, (Senthelal et al., 2024). various Reports highlight that some genetic variables, such as gene HLA alleles interacted with other genetic and environmental risk factors in various ways to cause an elevated risk of RA (158).

In RA, infiltration of synovial inflammatory cells, angiogenesis, synovial hyperplasia and cartilage damage are some of the characteristic conditions, which are responsible for bone degradation, (McInnes & Schett, 2011). it is reported that the ROS is associated with the etiology of RA, (Kondo, Kanai, & Okada, 2023). One of the illnesses that causes oxidative stress is RA. When compared to healthy participants, RA patients' whole blood and monocytes produced five times as much mitochondrial ROS, indicating that oxidative stress is a pathogenic characteristic of RA. Since free radicals are also crucial secondary mediators in the inflammatory and immunological cellular response in RA, they are indirectly linked to joint destruction. Increased oxidative stress renders T-cells resistant to various signals, including those for development and apoptosis, potentially prolonging the abnormal immune response, (Hassan *et al.,* 2011). However, free radicals have the ability to directly damage joint cartilage by targeting its proteoglycan and preventing its formation, (Hadjigogos, 2003). RA has also been linked to DNA damage, oxidative damage to hyaluronic acid and lipoperoxidation products, oxidation of low-density lipoproteins, and carbonyl increase from protein oxidation. Moreover, p53 mutation in RA-derived fibroblast-like synoviocytes has been connected to ROS-induced genotoxic events, (Hitchon & El-Gabalawy, 2004).

Genomics is a precision medicine approach used to determine the genetic variants linked with a particular response in RA. In these studies, it was found that the gene SLC19A1 is one of the key transporters for the absorption of methotrexate into cells and showed links of methotrexate response in RA. The meta-analysis of 12 studies explained the association of the rs1051266 variant of the SLC19A1 gene. This associations suggested that the individuals carrying this variant may have altered levels of methotrexate responses within the cells in RA, (Li *et al.*, 2016) (Dervieux *et al.*,

2004). These studies reflected the effectiveness of the methotrexate in treating RA. The individual RA patient can be checked for this RA variant before prescribing the methotrexate to increase the patient compliances.

2.3 Cancer:

The occurrence and progression of malignant cancer tumors are strongly linked to DNA damage, alterations in cancer cell metabolism, tumor metastasis, tumor drug resistance and genomic instability, all of which are triggered by an excessive ROS level, (Qian *et al.,* 2022) (Zhao *et al.*, 2023). however, elevated ROS can also reach toxic levels that paradoxically induce anticancer effects, (Sinha, Tokar, & Bortner, 2022) (Ebata, Loo, & Takahashi, 2022). An ongoing inflammatory response could result in a preneoplastic event, (Guina *et al.,* 2015). Excessive ROS emission from chronically inflamed cells attracts additional activated immune cells, which amplifies dysregulated processes and ultimately results in a preneoplastic state. In the event that the endogenous antioxidant response is overpowered by the amount of cellular ROS/RNS generated, permanent oxidative damage to proteins, lipids, and nucleic acids may result in genetic and/or epigenetic changes that dysregulate tumor suppressor and oncogene genes. Because the processes of oxidative stress and chronic inflammation are closely related, failing to stop them may lead to genetic or epigenetic alterations that trigger the start of carcinogenesis, (Murata *et al.,* 2012). Numerous investigations have demonstrated that oxidative stress impacts a number of signaling pathways linked to cell division, (Klaunig, Kamendulis, & Hocevar, 2010). Among these, the signaling pathway of the epidermal growth factor receptor can be mentioned, and oxidative stress affects important signaling proteins like the nuclear factor erythroid 2-related factor 2, RAS/RAF, the mitogen-activated protein kinases ERK1/2 and MEK, phosphatidylinositol 3-kinase, phospholipase C, and protein kinase C, (Huo *et al.*, 2014). Furthermore, ROS change how the p53 suppressor gene, which is essential for apoptosis, is expressed. Thus, oxidative stress contributes significantly to the development and spread of tumors and alters gene expression, cell proliferation, and death, (Matsuzawa & Ichijo, 2008). Additionally, it was shown that ROS are driving bladder cancer cells' long interspersed nuclear element-1 (LINE-1) to become hypomethylated. This suggests that oxidative stress and LINE-1 have the oncogenic potential to promote carcinogenesis and the advancement of cancer. Furthermore, ROS (H2O2) was experimentally used to promote the expression of LINE-A-encoded protein (ORF1p) in bladder cancer cells, (Whongsiri *et al.*, 2018). Strategies against cancer could be also if target the ROS generation.

Drug Therapies That Target ROS Generation:

Targeting the control of ROS in cancer cells has increased attention in recent times for cancer treatment selections and management. OS and elevated ROS levels are important markers of cancer. Increased ROS levels are a result of cancer cells' adaptive metabolic reprogramming, and these levels also control tumor metabolism and growth, (Moloney & Cotter, 2018). Hence, novel anticancer medicines that target ROS formation could be good strategies to handle the existing issues associated with cancer drug resistance and the low performance of the anticancer therapies.

Anthracyclines (doxorubicin, epirubicin, and daunorubicin), alkylating agents, platinum coordination complexes (cisplatin, carboplatin, and oxaliplatin), and camptothecins (topotecan and irinotecan) are the important drugs act by increasing the production of ROS in cancer cells, (Conklin, 2004). Thus, it changes the redox homeostasis and have anti-cancer effects, (Nakamura & Takada, 2021). As an example, NCT503, an extremely selective inhibitor of phosphoglycerate dehydrogenase (PHGDH) increases the effectiveness of temozolomide, a primary chemotherapeutic substance for the treatment of glioblastoma. It is causing DNA damage which is mediated by ROS and inhibiting the key proteins which up-regulated in many cancers through the Wnt/β-catenin pathway, (Jin *et al.,* 2022). Another example Celastrol, a naturally occurring pentacyclictriterpenoid, has demonstrated potential in drug-resistant colon cancer due to its pro-oxidative action. This is caused by excessive production of ROS to generate Double-Strand Breaks (DSBs), which triggered the apoptotic/necrotic cell death, and the suppression of cell division, (Moreira *et al.*, 2019). Moreover, the use of plumbagin, a naturally occurring substance whose anticancer principle is due to ROS production in breast and oral cancer cells, (Kawiak, Domachowska, & Lojkowska, 2019). Although selenium is thought to be a multitarget medication, high doses of sodium selenite have been shown to enhance the mortality of ovarian cancer cells by causing ROS production and GSH depletion, which in turn mediates iron death, (Zhao *et al.,* 2023) (Choi *et al.,* 2023). Thus, pharmacological therapies that aim to inhibit ROS production are equally essential for the current concept of cancer care.

The Future of Personalised Medicine for Cancer Treatment:

It grows more and more evident that common tumours have molecularly diverse subtypes, and that each subtype requires a particular strategy for the treatment. Genetic diversity among the different populations and the cancer are very much affecting the efficacy of the cancer therapy. Numerous potentially important genes, including the EGFR, KRAS, and BRAF, are listed in table including target agent for individual cancer type, (Jackson & Chester, 2015).

Table 1. Molecular targets in personalised cancer therapies

Cancer Type	Targeted agent	Class of agent	Cellular target	References
Breast	Trastuzumab	Monoclonal antibody against HER2/Neu (EGFR2)	Human epidermal growth factor receptor 2 (HER2)	(Goldenberg, 1999) (Moja *et al.*, 2012)
Colorectal	Cetuximab	Monoclonal antibody against EGFR	Ki-ras2 Kirsten rat sarcoma viral oncogene homolog (*KRAS*)	(Fearon, 2011) (Karapetis *et al.*, 2008)
Chronic myeloid leukaemia	Imatinib	Receptor tyrosine kinase inhibitor	BCR-ABL fusion protein	(Smith, 2011)
Gastro-intestinal stromal tumours	Imatinib	Receptor tyrosine kinase inhibitor	Proto-oncogene c-KIT	(Heinrich *et al.*, 2023) (Hornick & Fletcher, 2007)
Non-small cell lung cancer	Erlotinib and gefitinib	Receptor tyrosine kinase inhibitor	Epidermal growth factor receptor (EGFR)	(Mitsudomi *et al.*, 2010) (Shepherd *et al.*, 2005) (Wang, Scmid-Bindert, & Zhou, 2012)
Non-small cell lung cancer	Crizotinib	Receptor tyrosine kinase inhibitor	EML4-ALK fusion protein	(Shaw *et al.*, 2013) (Shaw *et al.*, 2009)
Metastatic malignant melanoma	Vemurafenib	B-raf/MEK/ERK pathway inhibitor	Serine/threonine protein kinase B- Raf (BRAF)	(J.T. *et al.*, 2008) (P.B. *et al.*, 2011)
Ovarian, breast and prostate cancer (under investigation)	Olaparib	Poly (ADP-ribose) polymerase (PARP) inhibitor	Breast Cancer Gene 1 (BRCA1), Breast Cancer Gene 2 (BRCA2),	(Fong *et al.*, 2009) (Gelmon *et al.*, 2011)

Source: Min & Lee (2022)

2.4 Respiratory diseases:

Damage to the lungs and subsequent illness states are caused by the harmful process of OS. By understanding the mechanism of ROS production, antioxidants are used to treat various lung diseases, (Park, Kim, & Lee, 2009). There are various respiratory diseases that occurs due to ROS, some of which are discussed below.

A. Bronchial asthma

ROS play a role in the contraction of smooth muscle in the airways, as well as in the inflow of inflammatory cells, decreased number of epithelial cilia, augmented generation of mucus, transformed release of inflammatory mediators, and increased vascular permeability, (Henricks & Nijkamp, 2001). When the epithelium is damaged or eliminated, ROSs intensify the direct contraction of the smooth muscle preparations in the airways. Bronchial hyperactivity, a hallmark of asthma, is also brought on by an excess of ROSs or a depressive state of the immune system, (Cortijo *et al.*, 1999) (Hulsmann *et al.,* 1994) (Sadeghi-Hashjin *et al.,* 1996).

The study found that a significant difference in methylation was observed with the IgE level in asthmatic and control group (higher level IgE in asthmatic patient). Due to the effect of differential methylation on gene expression, 26 patients were found with persistent asthma. However, 39 patients with remission of asthma as compared to 79 healthy controls. These studies also involved multi-omics techniques to evaluate the genes and protein data. As per report, DMR and CpG genes were found hypo-methylated which were linked to decreased expression of the atypical chemokine receptor 2 (ACKR2) in asthma remission of individuals as compared to persistent asthma. The inflammation is mediated by the ACKR2 gene which implicated a possible molecular mechanism of asthma remission, (Vermeulen *et al.,* 2020). The study found an over-representation of genes involved in chemokine and Toll-like receptor signaling pathways and enrichment of transcription factors in T cell development. Dexamethasone-treated monocytes from asthmatic patients showed hyper acetylation of histone H4, and preincubation with vitamin D improved its anti-inflammatory benefits, (Benincasa *et al.,* 2021). The circulating serum microRNAs (miR-146b, miR-206, and miR-720) enhanced the precision of asthma exacerbation risk prediction in paediatric asthma patients via NF-kß and GSK3/AKT pathways, (Kho *et al.*, 2018).

B. Chronic Obstructive Pulmonary Disease (COPD)

COPD is characterized by air flow limitation which is not completely reversible due to persistent inflammation in the pulmonary vasculature, parenchyma, and airways. This is accompanied by rising in neutrophils, macrophages, and T lymphocytes (particularly CD8+). OS in breath, urine, and epithelial lining fluid of COPD patients and cigarette smokers can cause lung damage and cell malfunction. OS imbalances proteinases and anti-proteinases, causing inflammation in the airways. NF-kB, controlling inflammatory genes like IL-8 and TNF-a, exacerbates inflammation. OS may facilitate reversible airway constriction. H^2O^2 narrows airway smooth muscle, while isoprostane F2a-III, a biomarker of oxidative stress in the lungs, is a strong constrictor of human airways, (Kawikova *et al.*, 1996).

The histone modification was found to be involved in the pathophysiology of COPD. Exposure to cigarette smoke, (Inhibitory kappa B kinase) IKK α-mediated phosphorylation causes activation of the mitogen- and stress-activated kinase 1 (MSK1). This results from histone UPBEAT1 modifications [acetylation of histone H4 (Lys12) and phospho-acetylation of histone H3 (Ser10/Lys9)]. Both are expressed in H292, BEAS-2B, and SAEC cells at NF-kß-dependent promoters. These events led to transcriptional activation of genes that are probably involved in the pathogenesis of COPD, (Sundar *et al.,* 2012).

By targeting SMAD3(Suppressor of Mothers against Decapentaplegic) in airway Smooth Muscle Cells, miR-145 adversely regulated the discharge of pro-inflammatory cytokines, including interleukin-6 (IL-6) and C-X-C motif chemokine ligand 8 (CXCL8). The delivery of miR-145 mimics the ASM cell's IL-6 and CXCL8 expression down to levels similar to controls, pointing to a potential new treatment approach. Moreover, in microarray studies, microRNA: MiR-34c, miR-34b, miR-149, miR-133a, and miR-133b were significantly downregulated in lung tissues of29 COPD patients with moderate emphysema as opposed to mild emphysema undergoing lung cancer resection, (Benincasa *et al.*, 2021).

2.5 Critical illness

Critical illness is defined as a state related to disease, surgical or medical that requires the treatment of a patient in the units of intensive care (ICU). Illness could be associated to infections, trauma, burns, ischemia, haemorrhage, pancreatitis and sepsis it can produce the same clinical symptoms as those produced due to microbial invasion even in the absence of the infectious microorganism, (Sharma, Mogensen, & Robinson, 2019) (Jensen, 2022). Chronic condition in critical illness occurs when the patient has a limited number of physiological reserves due to chronic comorbidities and older age develops acute critical illness. It further worsens to form a syndrome leading to venticular dependence, neuromuscular weakness, brain dysfunction, malnutrition, endocrinopathy, edema, skin breakdown and symptoms of distress, In critical illness, the one-year survival rate is approximately 50% due to this high mortality rate the prognostic variables are highly important for making medical decisions, (Ruggiero, 2018). One of the leading causes of mortality in ICU is sepsis and septic shock-associated critical illness, (Zhou *et al.,* 2014). ROS plays a critical role in organ dysfunction as it causes oxidative damage to the mitochondria which is responsible for organ dysfunction in conditions like sepsis and septic shock, (Rocha *et al.*, 2012). Critical illness increases the ROS production leading to OS (Oldham & Bowen, 1998) this increase in oxidative stress intern associated with the imbalance in ROS levels leading to the tissue damage, (Pizzino *et al.,* 2017).

In conditions like sepsis ROS play a significant role in cell dysfunction due to its effect on the innate immune system. ROS production hyper-activate the innate immune system to release disproportionate amount of cytokines, (Schulte, Bernhagen, & Bucala, 2013) which is an inflammatory mediator in the body. These Inflammatory mediators have toxic effects on the cells which cause organ dysfunction, (Kumar *et al.*, 2022). The pathophysiology of sepsis involves dysregulation of the immune cells like neutrophils and macrophages along with endothelial and epithelial cell dysfunction. Due to dysregulation of the immune cells, the immunity is compromised, and the risk of secondary infections rises, (Russell, 2006). Precision medicine benefits greatly from epigenetics since it can identify patients who have unfavourable evolution and clinical deterioration through the use of epigenetic biomarkers. Lorentesorolla *et al.* (2019) found that the sepsis patient experiencing extensive changes in the methylome of their disseminating monocytes had associated aberrant levels of interleukin (IL-10) and (IL-6) and high occurrence of organ failure. The abnormal expression of IL-10 mRNA is basically due to the modification of histone, (Zheng *et al.*, 2020). ncRNA is a functional RNA molecule transcribed from DNA but not translated into a protein. The gene expression is regulated by ncRNA at the transcriptional and post-translational level. Major classes of short non-coding RNA (ncRNA) are known as piwi-interacting RNAs (piRNAs), short RNA (si)RNA and (mi)RNA plasma levels of miR 133a higher in critically ill patients with sepsis than in patients with non-infectious inflammation and predict ICU mortality, (Tacke *et al.*, 2014).

Metabolomics is a precision medicine approach that is used in the sepsis prognosis, (Cho *et al.,* 2012). Beata Mickiewicz et.al collected a couple of serum samples of the pediatric septic shock patient, (PICU) pediatric intensive care unit patients with (SIRS) systemic inflammatory response syndrome patient and healthy children. They used NMR spectroscopy to analyze the serum samples and targeted profiling methodology for quantification. The study identified the significant changes in the metabolite concentration in the serum samples, indicating their role in distinguishing patients. The alteration in the level of the metabolite 2-oxoisocaproate serves potential biomarker for the diagnosis of septic shock in the PICU, (Mickiewicz *et al.,* 2013).

2.6 SARS CoV:

There is a relevant relationship between viral infections, reactive species production, oxidative stress, and the antiviral reaction, (Fernandes *et al.,* 2020). The mortality of patients with new coronavirus disease 2019 (COVID-19) is mostly caused by severe inflammatory reactions triggered by SARS-CoV-2 infection, (Fernandes *et al.,* 2020). Research showed that in human lung samples infected with SARS-CoV, there was a clear synthesis of oxidised phospholipids in the damaged air gaps, followed by the generation of ROS in the pneumocytes and alveolar macrophages, (Imai *et al.,* 2008). The development of SARS-CoV infection is aided by the ROS-activated NF-κB (nuclear factor kappa-light-chain-enhancer of activated B cells) signal transduction pathway, which is triggered by SARS-CoV-1 protease-3CLpro, (Lin *et al.,* 2006). Approximately fifteen possible vaccine candidates were developed globally following the advent of SARS-CoV-2, (Pang *et al.,* 2020). These vaccinations rely on a number of technologies, including the delivery of synthetic vaccines to subunits and modified viruses, gene vaccines, nanoparticles, and the administration of mRNA from viral surface proteins. mRNA-1273 is one of the most promising leads for a COVID-19 vaccination. Without actually introducing a live pathogen, this vaccine technique trains the body's immune system to fight off disease by using a fragment of the pathogen, usually a surface protein, (Crisci *et al*., 2020).

2.7 ROS in age related diseases:

Increased levels of ROS have been linked to the ageing process. It has been shown that elevations in ROS levels may be important for the initiation and maintenance of the cell senescence process, (Davalli *et al.,* 2016). ROS can function as signalling molecules by influencing target proteins or as dangerous oxidative stressors by causing damage to cellular components, depending on their kind and quantity of production, (Madreiter-Sokolowski, Thomas, & Ristow, 2020). Inadequate understanding of fundamental ageing processes impedes efforts to avoid or postpone age-related diseases, such as diabetes, cardiovascular disease, cancer, and neurodegenerative illnesses. These conditions have an adverse effect on morbidity, overall wellness, and human independence, (Davalli *et al.,* 2016). These diseases and their individual precision approaches are explained in different parts of this chapter. Here we are explaining the precision medicine approach for Age-related Macular degeneration.

A variety of mechanisms are implicated in age-related macular degeneration (AMD), a diverse eye disease with unique risk factors for each patient. The elevated levels of ROS from mitochondrion lead to AMD. The increased ROS level is linked to damage to mitochondrial DNA (mtDNA) which is crucial component of AMD pathogenesis and is attribute to ageing and effects of other AMD (Age related Macular Degeneration), (Kaarniranta *et al*., 2019). It was found that vitamin and zinc supplements in the diet lower the risk of advanced AMD development. These supplements combat OS, which is a contributing factor in the pathophysiology of AMD, (Beatty *et al.,* 2000) (Betteridge, 2000). Complement inhibiting therapies like eculizumab, tesidolumab, CLG561, Zimura and AAVCAGsCD59 are the ones that are reported to be active in treating AMD, (Lorés-Motta, De Jong, & Hollander, 2018). Several levels of inflammatory, angiogenic, and tissue homeostasis pathways are associated with AMD like CFH, ARMS2, TIMP3, IL-8, and VEGFA. These pathways completely align with the clinical phenotype seen by patients with exudative AMD, (Cascella *et al.,* 2018). Anti-vascular endothelial growth factor (VEGF) agents comprise the majority of current AMD treatments like Bevacizumab, Ranibizumab, and Aflibercept. Additionally, Photodynamic Therapy (PDT) utilising Verteporfin is utilised to treat the exudative form of AMD, (Cascella *et al*., 2018). Regarding the involvement of DNA methylation in AMD, the majority of studies examined the expression of genes implicated in the OS response. Specifically, GSTM1 and GSTM5 were hypermethylated, leading to reduced levels of their corresponding mRNA and proteins on RPE and choroid. Consequently, this epigenetic repression heightens the vulnerability of AMD patients to OS, (Gemenetzi & Lotery, 2014). A study on the methylation status of IL17RC in AMD patients' peripheral blood and retinas showed that the hypomethylation of the IL17RC promoter was associated with an increased level of mRNA expression. The proinflammatory activity of monocytes involved in the pathophysiology of AMD may be impacted by this epigenetic change, (Sobrin & Seddon, 2014).

3. CONCLUSIONS

Understanding the connection between ROS and life style disorders like metabolic and inflammatory diseases are complex because patients might require unique treatment approaches. The correlation between the elevated ROS levels and pathogenesis of lifestyle-induced diseases including cancer and hypertension are described here. There are various genetic factors that may affect how people will respond to the treatment. Currently precision medicine relies on principles of genetic information and application of specific therapeutic intervention strategies to tackle oxidative stress and its impacts. In this manner, by addressing individual ROS profile and genotyping, the treatment for lifestyle disorders opens a new avenue to develop optimum therapy to provide improved results for such sufferers. Interaction with molecular genetics allows fine tuning of ROS dynamics in the clinic and develops new perspective directions in prevention and treatment based on the individual approach.

REFERENCES

Ashcroft, F. M., & Rorsman, P. (2012, March). Diabetes Mellitus and the β Cell: The Last Ten Years. *Cell, 148*(6), 1160–1171. DOI: 10.1016/j.cell.2012.02.010 PMID: 22424227

Bajic, V. P., Van Neste, C., Obradovic, M., Zafirovic, S., Radak, D., Bajic, V. B., Essack, M., & Isenovic, E. R. (2019, May 9). Glutathione "Redox Homeostasis" and Its Relation to Cardiovascular Disease. *Oxidative Medicine and Cellular Longevity, 2019*, 1–14. DOI: 10.1155/2019/5028181 PMID: 31210841

Beatty, S., Koh, H. H., Phil, M., Henson, D., & Boulton, M. (2000). The role of oxidative stress in the pathogenesis of age-related macular degeneration. *Survey of Ophthalmology, 45*(2), 115–134. DOI: 10.1016/S0039-6257(00)00140-5 PMID: 11033038

Bedard, K., & Krause, K. H. (2007, January). The NOX Family of ROS-Generating NADPH Oxidases: Physiology and Pathophysiology. *Physiological Reviews, 87*(1), 245–313. DOI: 10.1152/physrev.00044.2005 PMID: 17237347

Benincasa, G., DeMeo, D. L., Glass, K., Silverman, E. K., & Napoli, C. (2021). Epigenetics and pulmonary diseases in the horizon of precision medicine: A review. [Internet]. *The European Respiratory Journal, 57*(6), 2003406. Retrieved August 2, 2024, from https://erj.ersjournals.com/content/57/6/2003406.short. DOI: 10.1183/13993003.03406-2020 PMID: 33214212

Betteridge, D. J. (2000). What is oxidative stress? *Metabolism: Clinical and Experimental, 49*(2), 3–8. DOI: 10.1016/S0026-0495(00)80077-3 PMID: 10693912

Bhatti, J. S., Bhatti, G. K., & Reddy, P. H. (2017). Mitochondrial dysfunction and oxidative stress in metabolic disorders—A step towards mitochondria based therapeutic strategies. Biochimica et Biophysica Acta (BBA)-. *Molecular Basis of Disease, 1863*(5), 1066–1077.

BioSpace [Internet]. 2006 [cited 2024 Sep 5]. ARCA Discovery Inc. Announces Publication Of The Genetic Basis For Enhanced Response To Bucindolol. Available from: https://www.biospace.com/arca-discovery-inc-announces-publication-of-the-genetic-basis-for-enhanced-response-to-bucindolol

Casas, A. I., Nogales, C., Mucke, H. A. M., Petraina, A., Cuadrado, A., Rojo, A. I., ... On the Clinical Pharmacology of Reactive Oxygen Species. Touyz RM, editor. Pharmacol Rev. 2020 Oct;72(4):801–28. DOI: 10.1124/pr.120.019422

Cascella, R., Strafella, C., Caputo, V., Errichiello, V., Zampatti, S., Milano, F., Potenza, S., Mauriello, S., Novelli, G., Ricci, F., Cusumano, A., & Giardina, E. (2018). Towards the application of precision medicine in Age-Related Macular Degeneration. *Progress in Retinal and Eye Research*, *63*, 132–146. DOI: 10.1016/j.preteyeres.2017.11.004 PMID: 29197628

Chapman, P. B., Hauschild, A., Robert, C., Haanen, J. B., Ascierto, P., Larkin, J., & McArthur, G. A. (2011). Improved survival with vemurafenib in melanoma with BRAF V600E mutation. *The New England Journal of Medicine*, *364*(26), 2507–2516.

Chen, A., Huang, B. S., Wang, H., Ahmad, M., & Leenen, F. H. H. (2014, August 15). Knockdown of mineralocorticoid or angiotensin II type 1 receptor gene expression in the paraventricular nucleus prevents angiotensin II hypertension in rats. *The Journal of Physiology*, *592*(16), 3523–3536. DOI: 10.1113/jphysiol.2014.275560 PMID: 24973408

Chen, P. G. F., & Sun, Z. (2017, February). AAV Delivery of Endothelin-1 shRNA Attenuates Cold-Induced Hypertension. *Human Gene Therapy*, *28*(2), 190–199. DOI: 10.1089/hum.2016.047 PMID: 27736201

Cho, W. H., Park, T., Park, Y. Y., Huh, J. W., Lim, C. M., Koh, Y., Song, D.-K., & Hong, S.-B. (2012, August). Clinical significance of enzymatic lysophosphatidylcholine (LPC) assay data in patients with sepsis. *European Journal of Clinical Microbiology & Infectious Diseases*, *31*(8), 1805–1810. DOI: 10.1007/s10096-011-1505-6 PMID: 22167258

Choi, J. A., Lee, E. H., Cho, H., & Kim, J. H. (2023, January 18). High-Dose Selenium Induces Ferroptotic Cell Death in Ovarian Cancer. *International Journal of Molecular Sciences*, *24*(3), 1918. DOI: 10.3390/ijms24031918 PMID: 36768241

Conklin, K. A. (2004, December). Chemotherapy-Associated Oxidative Stress: Impact on Chemotherapeutic Effectiveness. *Integrative Cancer Therapies*, *3*(4), 294–300. DOI: 10.1177/1534735404270335 PMID: 15523100

Cortijo, J., Martí-Cabrera, M., García, J., Pallardó, A. F. V., Esteras, A., Bruseghini, L., Viña, J., & Morcillo, E. J. (1999). Contraction of human airways by oxidative stress: Protection by n-acetylcysteine. *Free Radical Biology & Medicine*, *27*(3–4), 392–400. DOI: 10.1016/S0891-5849(99)00070-2 PMID: 10468214

Cracowski, J. L., Durand, T., & Bessard, G. (2002, August). Isoprostanes as a biomarker of lipid peroxidation in humans: Physiology, pharmacology and clinical implications. *Trends in Pharmacological Sciences*, *23*(8), 360–366. DOI: 10.1016/S0165-6147(02)02053-9 PMID: 12377577

Crisci, C. D., Ardusso, L. R. F., Mossuz, A., & Müller, L. (2020, September). A Precision Medicine Approach to SARS-CoV-2 Pandemic Management. *Current Treatment Options in Allergy*, *7*(3), 422–440. DOI: 10.1007/s40521-020-00258-8 PMID: 32391242

Crosswhite, P., Chen, K., & Sun, Z. (2014, November). AAV Delivery of Tumor Necrosis Factor-α Short Hairpin RNA Attenuates Cold-Induced Pulmonary Hypertension and Pulmonary Arterial Remodeling. *Hypertension*, *64*(5), 1141–1150. DOI: 10.1161/HYPERTENSIONAHA.114.03791 PMID: 25185133

Davalli, P., Mitic, T., Caporali, A., Lauriola, A., & D'Arca, D. (2016). ROS, Cell Senescence, and Novel Molecular Mechanisms in Aging and Age-Related Diseases. *Oxidative Medicine and Cellular Longevity*, *2016*(1), 1–18. DOI: 10.1155/2016/3565127 PMID: 27247702

Dervieux, T., Kremer, J., Lein, D. O., Capps, R., Barham, R., Meyer, G., Smith, K., Caldwell, J., & Furst, D. E. (2004, November). Contribution of common polymorphisms in reduced folate carrier and γ-glutamylhydrolase to methotrexate polyglutamate levels in patients with rheumatoid arthritis. *Pharmacogenetics and Genomics*, *14*(11), 733–739. DOI: 10.1097/00008571-200411000-00004 PMID: 15564880

Dervieux, T., Kremer, J., Lein, D. O., Capps, R., Barham, R., Meyer, G., Smith, K., Caldwell, J., & Furst, D. E. (2004, November). Contribution of common polymorphisms in reduced folate carrier and γ-glutamylhydrolase to methotrexate polyglutamate levels in patients with rheumatoid arthritis. *Pharmacogenetics and Genomics*, *14*(11), 733–739. DOI: 10.1097/00008571-200411000-00004 PMID: 15564880

Desai, A. S., Webb, D. J., Taubel, J., Casey, S., Cheng, Y., Robbie, G. J., Foster, D., Huang, S. A., Rhyee, S., Sweetser, M. T., & Bakris, G. L. (2023, July 20). Zilebesiran, an RNA Interference Therapeutic Agent for Hypertension. *The New England Journal of Medicine*, *389*(3), 228–238. DOI: 10.1056/NEJMoa2208391 PMID: 37467498

Dzau, V. J., & Hodgkinson, C. P. (2024, April). Precision Hypertension. *Hypertension*, *81*(4), 702–708. DOI: 10.1161/HYPERTENSIONAHA.123.21710 PMID: 38112080

Ebata, H., Loo, T. M., & Takahashi, A. (2022, June 17). Telomere Maintenance and the cGAS-STING Pathway in Cancer. *Cells*, *11*(12), 1958. DOI: 10.3390/cells11121958 PMID: 35741087

Fearon, E. R. (2011, February 28). Molecular Genetics of Colorectal Cancer. *Annual Review of Pathology*, *6*(1), 479–507. DOI: 10.1146/annurev-pathol-011110-130235 PMID: 21090969

Fearon, I. M., & Faux, S. P. (2009, September). Oxidative stress and cardiovascular disease: Novel tools give (free) radical insight. *Journal of Molecular and Cellular Cardiology*, *47*(3), 372–381. DOI: 10.1016/j.yjmcc.2009.05.013 PMID: 19481547

Fernandes, I. G., De Brito, C. A., Dos Reis, V. M. S., Sato, M. N., & Pereira, N. Z. (2020). In Oxid, M. C. L. (Ed.), *SARS-CoV-2 and Other Respiratory Viruses: What Does Oxidative Stress Have to Do with It? Carvalho VDF* (pp. 1–13).

Fong, P. C., Boss, D. S., Yap, T. A., Tutt, A., Wu, P., Mergui-Roelvink, M., Mortimer, P., Swaisland, H., Lau, A., O'Connor, M. J., Ashworth, A., Carmichael, J., Kaye, S. B., Schellens, J. H. M., & de Bono, J. S. (2009, July 9). Inhibition of Poly(ADP-Ribose) Polymerase in Tumors from *BRCA* Mutation Carriers. *The New England Journal of Medicine*, *361*(2), 123–134. DOI: 10.1056/NEJMoa0900212 PMID: 19553641

Gelmon, K. A., Tischkowitz, M., Mackay, H., Swenerton, K., Robidoux, A., Tonkin, K., Hirte, H., Huntsman, D., Clemons, M., Gilks, B., Yerushalmi, R., Macpherson, E., Carmichael, J., & Oza, A. (2011, September). Olaparib in patients with recurrent high-grade serous or poorly differentiated ovarian carcinoma or triple-negative breast cancer: A phase 2, multicentre, open-label, non-randomised study. *The Lancet. Oncology*, *12*(9), 852–861. DOI: 10.1016/S1470-2045(11)70214-5 PMID: 21862407

Gemenetzi, M., & Lotery, A. J. (2014). The role of epigenetics in age-related macular degeneration. *Eye (London, England)*, *28*(12), 1407–1417. DOI: 10.1038/eye.2014.225 PMID: 25233816

Goldenberg, M. M. (1999, February). Trastuzumab, a recombinant DNA-derived humanized monoclonal antibody, a novel agent for the treatment of metastatic breast cancer. *Clinical Therapeutics*, *21*(2), 309–318. DOI: 10.1016/S0149-2918(00)88288-0 PMID: 10211534

Griendling, K. K., Camargo, L. L., Rios, F. J., Alves-Lopes, R., Montezano, A. C., & Touyz, R. M. (2021, April 2). Oxidative Stress and Hypertension. *Circulation Research*, *128*(7), 993–1020. DOI: 10.1161/CIRCRESAHA.121.318063 PMID: 33793335

Griendling, K. K., Touyz, R. M., Zweier, J. L., Dikalov, S., Chilian, W., Chen, Y. R., Harrison, D. G., & Bhatnagar, A. (2016, August 19). Measurement of Reactive Oxygen Species, Reactive Nitrogen Species, and Redox-Dependent Signaling in the Cardiovascular System: A Scientific Statement From the American Heart Association. [Internet]. *Circulation Research*, *119*(5). Advance online publication. Retrieved July 17, 2024, from https://www.ahajournals.org/doi/10.1161/RES.0000000000000110. DOI: 10.1161/RES.0000000000000110 PMID: 27418630

Guillausseau, P. J., Meas, T., Virally, M., Laloi-Michelin, M., Médeau, V., & Kevorkian, J. P. (2008, February). Abnormalities in insulin secretion in type 2 diabetes mellitus. *Diabetes & Metabolism*, *34*, S43–S48. DOI: 10.1016/S1262-3636(08)73394-9 PMID: 18640585

Guina, T., Biasi, F., Calfapietra, S., Nano, M., & Poli, G. (2015, March). Inflammatory and redox reactions in colorectal carcinogenesis. *Annals of the New York Academy of Sciences*, *1340*(1), 95–103. DOI: 10.1111/nyas.12734 PMID: 25727454

Hadjigogos, K. (2003, March). The role of free radicals in the pathogenesis of rheumatoid arthritis. *Panminerva Medica*, *45*(1), 7–13. PMID: 12682616

Hakonarson, H., Thorvaldsson, S., Helgadottir, A., Gudbjartsson, D., Zink, F., Andresdottir, M., Manolescu, A., Arnar, D. O., Andersen, K., Sigurdsson, A., Thorgeirsson, G., Jonsson, A., Agnarsson, U., Bjornsdottir, H., Gottskalksson, G., Einarsson, A., Gudmundsdottir, H., Adalsteinsdottir, A. E., Gudmundsson, K., & Stefansson, K. (2005, May 11). Effects of a 5-Lipoxygenase–Activating Protein Inhibitor on Biomarkers Associated With Risk of Myocardial Infarction: A Randomized Trial. *Journal of the American Medical Association*, *293*(18), 2245. DOI: 10.1001/jama.293.18.2245 PMID: 15886380

Hao, M., Li, M., & Li, W. (2017, January). Galectin-3 inhibition ameliorates hypoxia-induced pulmonary artery hypertension. *Molecular Medicine Reports*, *15*(1), 160–168. DOI: 10.3892/mmr.2016.6020 PMID: 27959409

Harrap, S. B. (2003, June). Where are all the blood-pressure genes? *Lancet*, *361*(9375), 2149–2151. DOI: 10.1016/S0140-6736(03)13694-X PMID: 12826450

Hassan, S. Z., Gheita, T. A., Kenawy, S. A., Fahim, A. T., El-Sorougy, I. M., & Abdou, M. S. (2011, October). Oxidative stress in systemic lupus erythematosus and rheumatoid arthritis patients: relationship to disease manifestations and activity: Oxidative stress in SLE and RA patients. *International Journal of Rheumatic Diseases*, *14*(4), 325–331. DOI: 10.1111/j.1756-185X.2011.01630.x PMID: 22004228

Heinrich, M. C., Corless, C. L., Demetri, G. D., Blanke, C. D., Von Mehren, M., Joensuu, H., McGreevey, L. S., Chen, C.-J., Van den Abbeele, A. D., Druker, B. J., Kiese, B., Eisenberg, B., Roberts, P. J., Singer, S., Fletcher, C. D. M., Silberman, S., Dimitrijevic, S., & Fletcher, J. A. (2023, November 1). Kinase Mutations and Imatinib Response in Patients With Metastatic Gastrointestinal Stromal Tumor. *Journal of Clinical Oncology*, *41*(31), 4829–4836. DOI: 10.1200/JCO.22.02771 PMID: 37890277

Henricks, P. A., & Nijkamp, F. P. (2001). Reactive oxygen species as mediators in asthma. *Pulmonary Pharmacology & Therapeutics*, *14*(6), 409–421. DOI: 10.1006/pupt.2001.0319 PMID: 11782121

Hitchon, C. A., & El-Gabalawy, H. S. (2004). Oxidation in rheumatoid arthritis. *Arthritis Research & Therapy*, *6*(6), 265–278. DOI: 10.1186/ar1447 PMID: 15535839

Hornick, J. L., & Fletcher, C. D. M. (2007, May). The role of KIT in the management of patients with gastrointestinal stromal tumors. *Human Pathology*, *38*(5), 679–687. DOI: 10.1016/j.humpath.2007.03.001 PMID: 17437861

Huang, H., Li, X., Zheng, S., Chen, Y., Chen, C., Wang, J., Tong, H., Zhou, L., Yang, J., & Zeng, C. (2016, October 3). Downregulation of Renal G Protein–Coupled Receptor Kinase Type 4 Expression via Ultrasound-Targeted Microbubble Destruction Lowers Blood Pressure in Spontaneously Hypertensive Rats. *Journal of the American Heart Association*, *5*(10), e004028. DOI: 10.1161/JAHA.116.004028 PMID: 27792639

Huang, Y., Liu, X. L., Wen, J., Huang, L. H., Lu, Y., Miao, R. J., Liu, X., Li, Y., Xing, X.-W., & Yuan, H. (2017, March). Downregulation of the β1 adrenergic receptor in the myocardium results in insensitivity to metoprolol and reduces blood pressure in spontaneously hypertensive rats. *Molecular Medicine Reports*, *15*(2), 703–711. DOI: 10.3892/mmr.2016.6038 PMID: 28000860

Hulsmann, A. R., Raatgeep, H. R., Den Hollander, J. C., Stijnen, T., Saxena, P. R., Kerrebijn, K. F., & de Jongste, J. C. (1994, February). Oxidative epithelial damage produces hyperresponsiveness of human peripheral airways. *American Journal of Respiratory and Critical Care Medicine*, *149*(2), 519–525. DOI: 10.1164/ajrccm.149.2.8306055 PMID: 8306055

Huo, L., Li, C. W., Huang, T. H., Lam, Y. C., Xia, W., & Tu, C.. (2014, November 22). Activation of Keap1/Nrf2 signaling pathway by nuclear epidermal growth factor receptor in cancer cells. *American Journal of Translational Research*, *6*(6), 649. PMID: 25628777

Idris-Khodja, N., Ouerd, S., Mian, M. O. R., Gornitsky, J., Barhoumi, T., Paradis, P., & Schiffrin, E. L. (2016, November). Endothelin-1 Overexpression Exaggerates Diabetes-Induced Endothelial Dysfunction by Altering Oxidative Stress. *American Journal of Hypertension*, *29*(11), 1245–1251. DOI: 10.1093/ajh/hpw078 PMID: 27465439

Imai, Y., Kuba, K., Neely, G. G., Yaghubian-Malhami, R., Perkmann, T., van Loo, G., Ermolaeva, M., Veldhuizen, R., Leung, Y. H. C., Wang, H., Liu, H., Sun, Y., Pasparakis, M., Kopf, M., Mech, C., Bavari, S., Peiris, J. S. M., Slutsky, A. S., Akira, S., & Penninger, J. M. (2008). Identification of oxidative stress and Toll-like receptor 4 signaling as a key pathway of acute lung injury. *Cell, 133*(2), 235–249. DOI: 10.1016/j.cell.2008.02.043 PMID: 18423196

Jackson, S. E., & Chester, J. D. (2015, July 15). Personalised cancer medicine. *International Journal of Cancer, 137*(2), 262–266. DOI: 10.1002/ijc.28940 PMID: 24789362

Jain, K. K. (2011). Personalized Cardiology. In Applications of Biotechnology in Cardiovascular Therapeutics [Internet]. (pp. 315–336). Humana Press., Available from https://link.springer.com/10.1007/978-1-61779-240-3_10, Retrieved July 25, 2024, from. DOI: 10.1007/978-1-61779-240-3_10

Jensen, C. (2022, October). Nutrition in Critically Ill Patients. *Physician Assistant Clinics, 7*(4), 701–712. DOI: 10.1016/j.cpha.2022.05.006

Jin, L., Kiang, K. M. Y., Cheng, S. Y., & Leung, G. K. K. (2022, February). Pharmacological inhibition of serine synthesis enhances temozolomide efficacy by decreasing O6-methylguanine DNA methyltransferase (MGMT) expression and reactive oxygen species (ROS)-mediated DNA damage in glioblastoma. *Laboratory Investigation, 102*(2), 194–203. DOI: 10.1038/s41374-021-00666-7 PMID: 34625658

Kaarniranta, K., Pawlowska, E., Szczepanska, J., Jablkowska, A., & Blasiak, J. (2019). Role of mitochondrial DNA damage in ROS-mediated pathogenesis of age-related macular degeneration (AMD). *International Journal of Molecular Sciences, 20*(10), 2374. DOI: 10.3390/ijms20102374 PMID: 31091656

Karapetis, C. S., Khambata-Ford, S., Jonker, D. J., O'Callaghan, C. J., Tu, D., Tebbutt, N. C., Simes, R. J., Chalchal, H., Shapiro, J. D., Robitaille, S., Price, T. J., Shepherd, L., Au, H.-J., Langer, C., Moore, M. J., & Zalcberg, J. R. (2008, October 23). Mutations and Benefit from Cetuximab in Advanced Colorectal Cancer. *The New England Journal of Medicine, 359*(17), 1757–1765. DOI: 10.1056/NEJMoa0804385 PMID: 18946061

Kawiak, A., Domachowska, A., & Lojkowska, E. (2019, April 26). Plumbagin Increases Paclitaxel-Induced Cell Death and Overcomes Paclitaxel Resistance in Breast Cancer Cells through ERK-Mediated Apoptosis Induction. *Journal of Natural Products, 82*(4), 878–885. DOI: 10.1021/acs.jnatprod.8b00964 PMID: 30810041

Kawikova, I., Barnes, P. J., Takahashi, T., Tadjkarimi, S., Yacoub, M. H., & Belvisi, M. G. (1996, February). 8-Epi-PGF2 alpha, a novel noncyclooxygenase-derived prostaglandin, constricts airways in vitro. *American Journal of Respiratory and Critical Care Medicine*, *153*(2), 590–596. DOI: 10.1164/ajrccm.153.2.8564103 PMID: 8564103

Kemp, B. A., Howell, N. L., & Padia, S. H. (2018, October 1). Intrarenal ghrelin receptor inhibition ameliorates angiotensin II-dependent hypertension in rats. *American Journal of Physiology. Renal Physiology*, *315*(4), F1058–F1066. DOI: 10.1152/ajprenal.00010.2018 PMID: 29923768

Kho, A. T., McGeachie, M. J., Moore, K. G., Sylvia, J. M., Weiss, S. T., & Tantisira, K. G. (2018, December). Circulating microRNAs and prediction of asthma exacerbation in childhood asthma. *Respiratory Research*, *19*(1), 128. DOI: 10.1186/s12931-018-0828-6 PMID: 29940952

Klaunig, J. E., Kamendulis, L. M., & Hocevar, B. A. (2010, January). Oxidative Stress and Oxidative Damage in Carcinogenesis. *Toxicologic Pathology*, *38*(1), 96–109. DOI: 10.1177/0192623309356453 PMID: 20019356

Kondo, N., Kanai, T., & Okada, M. (2023, April 3). Rheumatoid Arthritis and Reactive Oxygen Species: A Review. *Current Issues in Molecular Biology*, *45*(4), 3000–3015. DOI: 10.3390/cimb45040197 PMID: 37185721

Kota, B., Huang, T., & Roufogalis, B. (2005, February). An overview on biological mechanisms of PPARs. *Pharmacological Research*, *51*(2), 85–94. DOI: 10.1016/j.phrs.2004.07.012 PMID: 15629253

Kotchen, T. A., Cowley, A. W.Jr, & Liang, M. (2016, January 26). Ushering Hypertension Into a New Era of Precision Medicine. *Journal of the American Medical Association*, *315*(4), 343. DOI: 10.1001/jama.2015.18359 PMID: 26769233

Kumar, K. V., & Das, U. N. (1993, January). Are Free Radicals Involved in the Pathobiology of Human Essential Hypertension? *Free Radical Research Communications*, *19*(1), 59–66. DOI: 10.3109/10715769309056499 PMID: 8225035

Kumar, S., Saxena, J., Srivastava, V. K., Kaushik, S., Singh, H., Abo-EL-Sooud, K., Abdel-Daim, M. M., Jyoti, A., & Saluja, R. (2022, September 20). The Interplay of Oxidative Stress and ROS Scavenging: Antioxidants as a Therapeutic Potential in Sepsis. *Vaccines*, *10*(10), 1575. DOI: 10.3390/vaccines10101575 PMID: 36298439

Laursen, J. B., Rajagopalan, S., Galis, Z., Tarpey, M., Freeman, B. A., & Harrison, D. G. (1997, February 4). Role of Superoxide in Angiotensin II–Induced but Not Catecholamine-Induced Hypertension. *Circulation*, *95*(3), 588–593. DOI: 10.1161/01.CIR.95.3.588 PMID: 9024144

Li, X., Hu, M., Li, W., Gu, L., Chen, M., Ding, H., Vanarsa, K., & Du, Y. (2016, September). The association between reduced folate carrier-1 gene 80G/A polymorphism and methotrexate efficacy or methotrexate related-toxicity in rheumatoid arthritis: A meta-analysis. *International Immunopharmacology*, *38*, 8–15. DOI: 10.1016/j.intimp.2016.05.012 PMID: 27233001

Li, Y., Xu, X., Zhang, D., Cheng, W., Zhang, Y., Yu, B., & Zhang, Y. (2019, December). Genetic variation in the leukotriene pathway is associated with myocardial infarction in the Chinese population. *Lipids in Health and Disease*, *18*(1), 25. DOI: 10.1186/s12944-019-0968-9 PMID: 30678701

Liggett, S. B., Mialet-Perez, J., Thaneemit-Chen, S., Weber, S. A., Greene, S. M., Hodne, D., Nelson, B., Morrison, J., Domanski, M. J., Wagoner, L. E., Abraham, W. T., Anderson, J. L., Carlquist, J. F., Krause-Steinrauf, H. J., Lazzeroni, L. C., Port, J. D., Lavori, P. W., & Bristow, M. R. (2006, July 25). A polymorphism within a conserved β_1-adrenergic receptor motif alters cardiac function and β-blocker response in human heart failure. *Proceedings of the National Academy of Sciences of the United States of America*, *103*(30), 11288–11293. DOI: 10.1073/pnas.0509937103 PMID: 16844790

Lin, C. W., Lin, K. H., Hsieh, T. H., Shiu, S. Y., & Li, J. Y. (2006). Severe acute respiratory syndrome coronavirus 3C-like protease-induced apoptosis. *FEMS Immunology and Medical Microbiology*, *46*(3), 375–380. DOI: 10.1111/j.1574-695X.2006.00045.x PMID: 16553810

Lipinski, B. (2001, July). Pathophysiology of oxidative stress in diabetes mellitus. *Journal of Diabetes and Its Complications*, *15*(4), 203–210. DOI: 10.1016/S1056-8727(01)00143-X PMID: 11457673

Long, C., McAnally, J. R., Shelton, J. M., Mireault, A. A., Bassel-Duby, R., & Olson, E. N. (2014, September 5). Prevention of muscular dystrophy in mice by CRISPR/Cas9–mediated editing of germline DNA. *Science*, *345*(6201), 1184–1188. DOI: 10.1126/science.1254445 PMID: 25123483

Lorente-Sorolla, C., Garcia-Gomez, A., Català-Moll, F., Toledano, V., Ciudad, L., Avendaño-Ortiz, J., Maroun-Eid, C., Martín-Quirós, A., Martínez-Gallo, M., Ruiz-Sanmartín, A., del Campo, Á. G., Ferrer-Roca, R., Ruiz-Rodriguez, J. C., Álvarez-Errico, D., López-Collazo, E., & Ballestar, E. (2019, December). Inflammatory cytokines and organ dysfunction associate with the aberrant DNA methylome of monocytes in sepsis. *Genome Medicine*, *11*(1), 66. DOI: 10.1186/s13073-019-0674-2 PMID: 31665078

Lorés-Motta, L., De Jong, E. K., & Den Hollander, A. I. (2018, June). Exploring the Use of Molecular Biomarkers for Precision Medicine in Age-Related Macular Degeneration. *Molecular Diagnosis & Therapy*, *22*(3), 315–343. DOI: 10.1007/s40291-018-0332-1 PMID: 29700787

Jahan, S., Fariduddin, M., Sultana, N., Aktar, Y., & Hasan, M. (2015). Predictors of post-partum persistence of glucose intolerance and its association with cardio-metabolic risk factors in gestational diabetes mellitus. *Journal of Diabetes & Metabolism*, *6*(609), 2.

Madreiter-Sokolowski, C. T., Thomas, C., & Ristow, M. (2020). Interrelation between ROS and Ca2+ in aging and age-related diseases. *Redox Biology*, *36*, 101678. DOI: 10.1016/j.redox.2020.101678 PMID: 32810740

Masi, S., Dalpiaz, H., & Borghi, C. (2024, December). Gene editing of angiotensin for blood pressure management. *International Journal of Cardiology. Cardiovascular Risk and Prevention*, *23*, 200323. DOI: 10.1016/j.ijcrp.2024.200323 PMID: 39258007

Matsuzawa, A., & Ichijo, H. (2008, November). Redox control of cell fate by MAP kinase: Physiological roles of ASK1-MAP kinase pathway in stress signaling. *Biochimica et Biophysica Acta*, *1780*(11), 1325–1336. DOI: 10.1016/j.bbagen.2007.12.011 PMID: 18206122

McInnes, I. B., & Schett, G. (2011, December 8). The Pathogenesis of Rheumatoid Arthritis. *The New England Journal of Medicine*, *365*(23), 2205–2219. DOI: 10.1056/NEJMra1004965 PMID: 22150039

McIntyre, M., Bohr, D. F., & Dominiczak, A. F. (1999, October). Endothelial Function in Hypertension: The Role of Superoxide Anion. *Hypertension*, *34*(4), 539–545. DOI: 10.1161/01.HYP.34.4.539 PMID: 10523323

Mickiewicz, B., Vogel, H. J., Wong, H. R., & Winston, B. W. (2013, May 1). Metabolomics as a Novel Approach for Early Diagnosis of Pediatric Septic Shock and Its Mortality. *American Journal of Respiratory and Critical Care Medicine*, *187*(9), 967–976. DOI: 10.1164/rccm.201209-1726OC PMID: 23471468

Min, H. Y., & Lee, H. Y. (2022, October 12). Molecular targeted therapy for anti-cancer treatment. *Experimental & Molecular Medicine*, *54*(10), 1670–1694. DOI: 10.1038/s12276-022-00864-3 PMID: 36224343

Mitsudomi, T., Morita, S., Yatabe, Y., Negoro, S., Okamoto, I., Tsurutani, J., Seto, T., Satouchi, M., Tada, H., Hirashima, T., Asami, K., Katakami, N., Takada, M., Yoshioka, H., Shibata, K., Kudoh, S., Shimizu, E., Saito, H., Toyooka, S., & Fukuoka, M. (2010, February). Gefitinib versus cisplatin plus docetaxel in patients with non-small-cell lung cancer harbouring mutations of the epidermal growth factor receptor (WJTOG3405): An open label, randomised phase 3 trial. *The Lancet. Oncology*, *11*(2), 121–128. DOI: 10.1016/S1470-2045(09)70364-X PMID: 20022809

Mohazzab, K. M., Kaminski, P. M., & Wolin, M. S. (1994, June 1). NADH oxidoreductase is a major source of superoxide anion in bovine coronary artery endothelium. *American Journal of Physiology. Heart and Circulatory Physiology*, *266*(6), H2568–H2572. DOI: 10.1152/ajpheart.1994.266.6.H2568 PMID: 8024019

Moja, L., Tagliabue, L., Balduzzi, S., Parmelli, E., Pistotti, V., Guarneri, V., ... Trastuzumab containing regimens for early breast cancer. Cochrane Breast Cancer Group, editor. Cochrane Database Syst Rev [Internet]. 2012 Apr 18 [cited 2024 Jul 30];2021(2). Available from: http://doi.wiley.com/10.1002/14651858.CD006243 .pub2

Moloney, J. N., & Cotter, T. G. (2018, August). ROS signalling in the biology of cancer. *Seminars in Cell & Developmental Biology*, *80*, 50–64. DOI: 10.1016/j.semcdb.2017.05.023 PMID: 28587975

Moreira, H., Szyjka, A., Paliszkiewicz, K., & Barg, E. (2019, August 14). Prooxidative Activity of Celastrol Induces Apoptosis, DNA Damage, and Cell Cycle Arrest in Drug-Resistant Human Colon Cancer Cells. *Oxidative Medicine and Cellular Longevity*, *2019*, 1–12. DOI: 10.1155/2019/6793957 PMID: 31485297

Murata, M., Thanan, R., Ma, N., & Kawanishi, S. (2012). Role of Nitrative and Oxidative DNA Damage in Inflammation-Related Carcinogenesis. *Journal of Biomedicine & Biotechnology*, *2012*, 1–11. DOI: 10.1155/2012/623019 PMID: 22363173

Nakamura, H., & Takada, K. (2021, October). Reactive oxygen species in cancer: Current findings and future directions. *Cancer Science*, *112*(10), 3945–3952. DOI: 10.1111/cas.15068 PMID: 34286881

Nasyrova, R. F., Moskaleva, P. V., Vaiman, E. E., Shnayder, N. A., Blatt, N. L., & Rizvanov, A. A. (2020, February 26). Genetic Factors of Nitric Oxide's System in Psychoneurologic Disorders. *International Journal of Molecular Sciences*, *21*(5), 1604. DOI: 10.3390/ijms21051604 PMID: 32111088

Nathan, C., & Cunningham-Bussel, A. (2013, May). Beyond oxidative stress: An immunologist's guide to reactive oxygen species. *Nature Reviews. Immunology, 13*(5), 349–361. DOI: 10.1038/nri3423 PMID: 23618831

Nelson, J. E., Cox, C. E., Hope, A. A., & Carson, S. S. (2010, August 15). Chronic Critical Illness. *American Journal of Respiratory and Critical Care Medicine, 182*(4), 446–454. DOI: 10.1164/rccm.201002-0210CI PMID: 20448093

Oldham, K. M., & Bowen, P. E. (1998, September). Oxidative Stress in Critical Care. *Journal of the American Dietetic Association, 98*(9), 1001–1008. DOI: 10.1016/S0002-8223(98)00230-2 PMID: 9739800

Ouerd, S., Idris-Khodja, N., Trindade, M., Ferreira, N. S., Berillo, O., Coelho, S. C., Neves, M. F., Jandeleit-Dahm, K. A., Paradis, P., & Schiffrin, E. L. (2021, March 21). Endothelium-restricted endothelin-1 overexpression in type 1 diabetes worsens atherosclerosis and immune cell infiltration via NOX1. *Cardiovascular Research, 117*(4), 1144–1153. DOI: 10.1093/cvr/cvaa168 PMID: 32533834

Pagano, P. J., Ito, Y., Tornheim, K., Gallop, P. M., Tauber, A. I., & Cohen, R. A. (1995, June 1). An NADPH oxidase superoxide-generating system in the rabbit aorta. *American Journal of Physiology. Heart and Circulatory Physiology, 268*(6), H2274–H2280. DOI: 10.1152/ajpheart.1995.268.6.H2274 PMID: 7611477

Pang, J., Wang, M. X., Ang, I. Y. H., Tan, S. H. X., Lewis, R. F., Chen, J. I. P., Gutierrez, R. A., Gwee, S. X. W., Chua, P. E. Y., Yang, Q., Ng, X. Y., Yap, R. K. S., Tan, H. Y., Teo, Y. Y., Tan, C. C., Cook, A. R., Yap, J. C.-H., & Hsu, L. Y. (2020). Potential rapid diagnostics, vaccine and therapeutics for 2019 novel coronavirus (2019-nCoV): A systematic review. *Journal of Clinical Medicine, 9*(3), 623. DOI: 10.3390/jcm9030623 PMID: 32110875

Park, H. S., Kim, S. R., & Lee, Y. C. (2009, January). Impact of oxidative stress on lung diseases. *Respirology (Carlton, Vic.), 14*(1), 27–38. DOI: 10.1111/j.1440-1843.2008.01447.x PMID: 19144046

Pizzino, G., Irrera, N., Cucinotta, M., Pallio, G., Mannino, F., Arcoraci, V., ... Oxidative Stress: Harms and Benefits for Human Health. Victor VM, editor. Oxid Med Cell Longev. 2017 Jan;2017(1):8416763.

Qian, S., Fang, Y., Yao, C., Wang, Y., Zhang, Z., Wang, X., Gao, J., Feng, Y., Sun, L., Zou, R., Zhou, G., Ye, J., Xia, R., & Xia, H. (2022). The synergistic effects of PRDX5 and Nrf2 on lung cancer progression and drug resistance under oxidative stress in the zebrafish models. *Oncology Research, 30*(2), 53–64. DOI: 10.32604/or.2022.026302 PMID: 37305326

Reginato, A. M., & Olsen, B. R. (2002).. . *Arthritis Research*, *4*(6), 337. DOI: 10.1186/ar595 PMID: 12453309

Rocha, M., Herance, R., Rovira, S., & Hernandez-Mijares, A. (2012, March 1). M. Victor V. Mitochondrial Dysfunction and Antioxidant Therapy in Sepsis. *Infectious Disorders Drug Targets*, *12*(2), 161–178. DOI: 10.2174/187152612800100189 PMID: 22420514

Rochette, L., Zeller, M., Cottin, Y., & Vergely, C. (2014). Diabetes, oxidative stress and therapeutic strategies. Biochimica et Biophysica Acta (BBA)-. *General Subjects*, *1840*(9), 2709–2729.

Ruggiero, R. M. (2018, March). Chronic Critical Illness: The Limbo Between Life and Death. *The American Journal of the Medical Sciences*, *355*(3), 286–292. DOI: 10.1016/j.amjms.2017.07.001 PMID: 29549932

Russell, J. A. (2006, October 19). Management of Sepsis. *The New England Journal of Medicine*, *355*(16), 1699–1713. DOI: 10.1056/NEJMra043632 PMID: 17050894

Sadeghi-Hashjin, G., Folkerts, G., Henricks, P. A., Verheyen, A. K., Van Der Linde, H. J., Van Ark, I., Coene, A., & Nijkamp, F. P. (1996, May). Peroxynitrite induces airway hyperresponsiveness in guinea pigs in vitro and in vivo. *American Journal of Respiratory and Critical Care Medicine*, *153*(5), 1697–1701. DOI: 10.1164/ajrccm.153.5.8630623 PMID: 8630623

Samman Tahhan, A., Sandesara, P. B., Hayek, S. S., Alkhoder, A., Chivukula, K., Hammadah, M., Mohamed-Kelli, H., O'Neal, W. T., Topel, M., Ghasemzadeh, N., Ko, Y.-A., Aida, H., Gafeer, M., Sperling, L., Vaccarino, V., Liang, Y., Jones, D. P., & Quyyumi, A. A. (2017, December). Association between oxidative stress and atrial fibrillation. *Heart Rhythm*, *14*(12), 1849–1855. DOI: 10.1016/j.hrthm.2017.07.028 PMID: 28757307

Saxena, M., Aswad, A., Badariene, J., Kazi, F., Karns, A. D., Neutel, J. M., Park, W., Stiglitz, D., Makarova, N., Havasi, A., Zappe, D., & Bakris, G. (2024, May). ZILEBESIRAN AS ADD-ON THERAPY IN PATIENTS WITH HYPERTENSION INADEQUATELY CONTROLLED WITH A STANDARD ANTIHYPERTENSIVE MEDICATION: EFFICACY AND SAFETY RESULTS FROM THE KARDIA-2 STUDY. *Journal of Hypertension*, *42*(Suppl 1), e115. DOI: 10.1097/01.hjh.0001020464.05231.4a

Schulte, W., Bernhagen, J., & Bucala, R. (2013). Cytokines in Sepsis: Potent Immunoregulators and Potential Therapeutic Targets—An Updated View. *Mediators of Inflammation*, *2013*, 1–16. DOI: 10.1155/2013/165974 PMID: 23853427

Senthelal, S., Li, J., Ardeshirzadeh, S., & Thomas, M. A. (2024). Arthritis. In StatPearls [Internet]. StatPearls Publishing., Available from https://www.ncbi.nlm.nih.gov/books/NBK518992/, Retrieved July 26, 2024, from.

Sharma, K., Mogensen, K. M., & Robinson, M. K. (2019, February). Pathophysiology of Critical Illness and Role of Nutrition. *Nutrition in Clinical Practice*, *34*(1), 12–22. DOI: 10.1002/ncp.10232 PMID: 30580456

Shaw, A. T., Kim, D. W., Nakagawa, K., Seto, T., Crinó, L., Ahn, M. J., De Pas, T., Besse, B., Solomon, B. J., Blackhall, F., Wu, Y.-L., Thomas, M., O'Byrne, K. J., Moro-Sibilot, D., Camidge, D. R., Mok, T., Hirsh, V., Riely, G. J., Iyer, S., & Jänne, P. A. (2013, June 20). Crizotinib versus Chemotherapy in Advanced *ALK* -Positive Lung Cancer. *The New England Journal of Medicine*, *368*(25), 2385–2394. DOI: 10.1056/NEJMoa1214886 PMID: 23724913

Shaw, A. T., Yeap, B. Y., Mino-Kenudson, M., Digumarthy, S. R., Costa, D. B., Heist, R. S., Solomon, B., Stubbs, H., Admane, S., McDermott, U., Settleman, J., Kobayashi, S., Mark, E. J., Rodig, S. J., Chirieac, L. R., Kwak, E. L., Lynch, T. J., & Iafrate, A. J. (2009, September 10). Clinical Features and Outcome of Patients With Non–Small-Cell Lung Cancer Who Harbor *EML4-ALK.Journal of Clinical Oncology*, *27*(26), 4247–4253. DOI: 10.1200/JCO.2009.22.6993 PMID: 19667264

Shepherd, F. A., Rodrigues Pereira, J., Ciuleanu, T., Tan, E. H., Hirsh, V., Thongprasert, S., Campos, D., Maoleekoonpiroj, S., Smylie, M., Martins, R., van Kooten, M., Dediu, M., Findlay, B., Tu, D., Johnston, D., Bezjak, A., Clark, G., Santabárbara, P., & Seymour, L. (2005, July 14). Erlotinib in Previously Treated Non–Small-Cell Lung Cancer. *The New England Journal of Medicine*, *353*(2), 123–132. DOI: 10.1056/NEJMoa050753 PMID: 16014882

Shimizu, H., Kiyohara, Y., Kato, I., Kitazono, T., Tanizaki, Y., Kubo, M., Ueno, H., Ibayashi, S., Fujishima, M., & Iida, M. (2004, September). Relationship Between Plasma Glutathione Levels and Cardiovascular Disease in a Defined Population: The Hisayama Study. *Stroke*, *35*(9), 2072–2077. DOI: 10.1161/01.STR.0000138022.86509.2d PMID: 15256685

Sinha, B. K., Tokar, E. J., & Bortner, C. D. (2022, August 3). Molecular Mechanisms of Cytotoxicity of NCX4040, the Non-Steroidal Anti-Inflammatory NO-Donor, in Human Ovarian Cancer Cells. *International Journal of Molecular Sciences*, *23*(15), 8611. DOI: 10.3390/ijms23158611 PMID: 35955744

Siva, C., Velazquez, C., Mody, A., & Brasington, R. (2003, July 1). Diagnosing Acute Monoarthritis in Adults: A Practical Approach for the Family Physician. *American Family Physician*, *68*(1), 83–90. PMID: 12887114

Sladek, R., Rocheleau, G., Rung, J., Dina, C., Shen, L., Serre, D., Boutin, P., Vincent, D., Belisle, A., Hadjadj, S., Balkau, B., Heude, B., Charpentier, G., Hudson, T. J., Montpetit, A., Pshezhetsky, A. V., Prentki, M., Posner, B. I., Balding, D. J., & Froguel, P. (2007, February). A genome-wide association study identifies novel risk loci for type 2 diabetes. *Nature*, *445*(7130), 881–885. DOI: 10.1038/nature05616 PMID: 17293876

Small, K. M., Wagoner, L. E., Levin, A. M., Kardia, S. L. R., & Liggett, S. B. (2002, October 10). Synergistic Polymorphisms of β_1 - and α_{2C} -Adrenergic Receptors and the Risk of Congestive Heart Failure. *The New England Journal of Medicine*, *347*(15), 1135–1142. DOI: 10.1056/NEJMoa020803 PMID: 12374873

Smith, B. D. (2011, April 6). Imatinib for Chronic Myeloid Leukemia: The Impact of Its Effectiveness and Long-term Side Effects. *Journal of the National Cancer Institute*, *103*(7), 527–529. DOI: 10.1093/jnci/djr073 PMID: 21422404

Sobrin, L., & Seddon, J. M. (2014). Nature and nurture-genes and environment-predict onset and progression of macular degeneration. *Progress in Retinal and Eye Research*, *40*, 1–15. DOI: 10.1016/j.preteyeres.2013.12.004 PMID: 24374240

Sun, Z., Bello-Roufai, M., & Wang, X. (2008, April). RNAi inhibition of mineralocorticoid receptors prevents the development of cold-induced hypertension. *American Journal of Physiology. Heart and Circulatory Physiology*, *294*(4), H1880–H1887. DOI: 10.1152/ajpheart.01319.2007 PMID: 18296563

Sundar, I. K., Chung, S., Hwang, J., Lapek, J. D., Bulger, M., Friedman, A. E., Yao, H., Davie, J. R., & Rahman, I. (2012). Mitogen-and stress-activated kinase 1 (MSK1) regulates cigarette smoke-induced histone modifications on NF-κB-dependent genes. *PLoS One*, *7*(2), e31378. DOI: 10.1371/journal.pone.0031378 PMID: 22312446

Tacke, F., Roderburg, C., Benz, F., Cardenas, D. V., Luedde, M., Hippe, H. J., Frey, N., Vucur, M., Gautheron, J., Koch, A., Trautwein, C., & Luedde, T. (2014, May). Levels of Circulating miR-133a Are Elevated in Sepsis and Predict Mortality in Critically Ill Patients. *Critical Care Medicine*, *42*(5), 1096–1104. DOI: 10.1097/CCM.0000000000000131 PMID: 24413579

Terra, S. G., Pauly, D. F., Lee, C. R., Patterson, J. H., Adams, K. F., Schofield, R. S., Belgado, B. S., Hamilton, K. K., Aranda, J. M., Hill, J. A., Yarandi, H. N., Walker, J. R., Phillips, M. S., Gelfand, C. A., & Johnson, J. A. (2005, March). β-adrenergic Receptor Polymorphisms and Responses during Titration of Metoprolol Controlled Release/extended Release in Heart Failure*. *Clinical Pharmacology and Therapeutics*, *77*(3), 127–137. DOI: 10.1016/j.clpt.2004.10.006 PMID: 15735607

Touyz, R. M., & Schiffrin, E. L. (2004, October). Reactive oxygen species in vascular biology: Implications in hypertension. *Histochemistry and Cell Biology*, *122*(4), 339–352. DOI: 10.1007/s00418-004-0696-7 PMID: 15338229

Tsai, J., Lee, J. T., Wang, W., Zhang, J., Cho, H., Mamo, S., & Bollag, G. (2008). Discovery of a selective inhibitor of oncogenic B-Raf kinase with potent antimelanoma activity. *Proceedings of the National Academy of Sciences of the United States of America*, *105*(8), 3041–3046.

Tsutsui, H., Kinugawa, S., & Matsushima, S. (2011, December). Oxidative stress and heart failure. *American Journal of Physiology. Heart and Circulatory Physiology*, *301*(6), H2181–H2190. DOI: 10.1152/ajpheart.00554.2011 PMID: 21949114

Vaziri, N. D., Ni, Z., Oveisi, F., & Trnavsky-Hobbs, D. L. (2000, December). Effect of Antioxidant Therapy on Blood Pressure and NO Synthase Expression in Hypertensive Rats. *Hypertension*, *36*(6), 957–964. DOI: 10.1161/01.HYP.36.6.957 PMID: 11116107

Vermeulen, C. J., Xu, C. J., Vonk, J. M., Ten Hacken, N. H., Timens, W., Heijink, I. H., Nawijn, M. C., Boekhoudt, J., van Oosterhout, A. J., Affleck, K., Weckmann, M., Koppelman, G. H., & van den Berge, M. (2020). Differential DNA methylation in bronchial biopsies between persistent asthma and asthma in remission. [Internet]. *The European Respiratory Journal*, *55*(2), 1901280. Retrieved August 2, 2024, from https://erj.ersjournals.com/content/55/2/1901280.short. DOI: 10.1183/13993003.01280-2019 PMID: 31699840

Wang, J. Y., & Doudna, J. A. (2023, January 20). CRISPR technology: A decade of genome editing is only the beginning. *Science*, *379*(6629), eadd8643. DOI: 10.1126/science.add8643 PMID: 36656942

Wang, Y., Schmid-Bindert, G., & Zhou, C. (2012, January). Erlotinib in the treatment of advanced non-small cell lung cancer: An update for clinicians. *Therapeutic Advances in Medical Oncology*, *4*(1), 19–29. DOI: 10.1177/1758834011427927 PMID: 22229045

White, C. R., Brock, T. A., Chang, L. Y., Crapo, J., Briscoe, P., Ku, D., Bradley, W. A., Gianturco, S. H., Gore, J., & Freeman, B. A. (1994, February). Superoxide and peroxynitrite inatherosclerosis. *Proceedings of the National Academy of Sciences of the United States of America*, *91*(3), 1044–1048. DOI: 10.1073/pnas.91.3.1044 PMID: 8302829

Whongsiri, P., Pimratana, C., Wijitsettakul, U., Jindatip, D., Sanpavat, A., & Schulz, W. A.. (2018). LINE-1 ORF1 Protein Is Up-regulated by Reactive Oxygen Species and Associated with Bladder Urothelial Carcinoma Progression. *Cancer Genomics & Proteomics*, *15*(2), 143–151. PMID: 29496693

Yang, S., & Lian, G. (2020, April). ROS and diseases: Role in metabolism and energy supply. *Molecular and Cellular Biochemistry*, *467*(1–2), 1–12. DOI: 10.1007/s11010-019-03667-9 PMID: 31813106

Zhao, Y., Ye, X., Xiong, Z., Ihsan, A., Ares, I., Martínez, M., Lopez-Torres, B., Martínez-Larrañaga, M.-R., Anadón, A., Wang, X., & Martínez, M.-A. (2023, June 27). Cancer Metabolism: The Role of ROS in DNA Damage and Induction of Apoptosis in Cancer Cells. *Metabolites*, *13*(7), 796. DOI: 10.3390/metabo13070796 PMID: 37512503

Zheng, Z., Huang, G., Gao, T., Huang, T., Zou, M., Zou, Y., & Duan, S. (2020, June 4). Epigenetic Changes Associated With Interleukin-10. *Frontiers in Immunology*, *11*, 1105. DOI: 10.3389/fimmu.2020.01105 PMID: 32582189

Zhou, J., Qian, C., Zhao, M., Yu, X., Kang, Y., & Ma, X. & China Critical Care Clinical Trials Group (CCCCTG). (2014). Epidemiology and outcome of severe sepsis and septic shock in intensive care units in mainland China. *PLoS One*, *9*(9), e107181.

Chapter 14
ROS in Precision Medicine in Neurological Diseases: Personalized Approaches

Yogesh Chaudhari
Department of Pharmaceutical Chemistry, Shri Vile Parle Kelavani Mandal's Institute of Pharmacy, Dhule, India

Nilesh Patil
https://orcid.org/0009-0007-8872-030X
Department of Pharmaceutical Chemistry, Shri Vile Parle Kelavani Mandal's Institute of Pharmacy, Dhule, India

Ansari Rashid
https://orcid.org/0009-0000-0089-4153
Department of Pharmacology, Shri Vile Parle Kelavani Mandal's Institute of Pharmacy, Dhule, India

Pawan Gupta
https://orcid.org/0000-0001-7201-1356
Department of Pharmaceutical Chemistry, Shri Vile Parle Kelavani Mandal's Institute of Pharmacy, Dhule, India

Rewati Wanjari
https://orcid.org/0009-0008-6574-3047
Department of Pharmaceutical Chemistry, Shri Vile Parle Kelavani Mandal's Institute of Pharmacy, Dhule, India

Pranav Kumar Prabhakar
Lovely Professional University, India & Nagaland University, India

ABSTRACT

Reactive oxygen species (ROS), are a class of unstable chemicals produced by almost every subcellular organelle in cell. The ROS are highly reactive molecules that play essential roles in cell signaling and homeostasis but can also cause damage to cellular components when present in excess. The elevated levels of ROS, are causing damage to DNA, proteins, and lipids, promoting genetic instability, number of neurological illnesses, including, Alzheimer's, and Parkinson's disease. Precision medicine is an innovative approach to adapting medical treatment to the

DOI: 10.4018/979-8-3693-7919-6.ch014

individual characteristics of each patient. This involves treating every patient as an individual and focusing on alternative healthcare treatment to every patient's unique characteristic. It relies on understanding how a person's genetic makeup, environment, and lifestyle affect their health and disease. The precision medicine enables early detection, prognostication, and monitoring of disease progression, allowing for timely and tailored interventions to mitigate ROS-mediated damage and improve patient outcomes.

1. INTRODUCTION:

Reactive oxygen species (ROS) may have initially appeared on Earth between 2400 to 3800 million years ago, along with the appearance of the first atmospheric oxygen molecules. From then on, ROS have become an essential part of life when oxygen is involved, (Mittler, 2017). ROS have gained attention as unique signal mediators that are involved in the progress, differentiation, development, and death of cells, (Sena & Chandel, 2012) (Zhang *et al.*, 2013). ROS are a class of unstable chemicals produced by almost every subcellular organelle in cell, (Sun *et al.*, 2020). These compounds include hydrogen peroxide (H_2O_2), singlet oxygen ($1O_2$), hydroxyl radical ($OH^{\bullet}$), and superoxide (O_2^-) (5).

Production of ROS is carried out by a range of extracellular and intracellular processes, (Sena & Chandel, 2012) (Zhang *et al.*, 2013). In intracellular places such as complexes I and III of electron transport chain (ETC) produces ROS in the inner mitochondrial membrane, (Holmström KM, Finkel, 2014) (Quinlan *et al.*, 2013). Overall, in Complex I, 0.1–2% of ETC electrons leak to generate ROS, and this fraction rises as damage increases, (Li *et al.*, 2013). Oxygen free radicals are extremely flammable and can cause destruction of lipids (in membrane), proteins, and nucleic acids (DNA, RNA), between other biological components, (Alfadda & Sallam, 2012) (Bae *et al.*, 2011) (Dröge, 2002). According to reports, oxidative stress (OS), or an ROS, might eventually cause cell death, (Liu *et al.*, 2018). In order to prevent specific potentially harmful effects of ROS, the cells maintain ROS at a harmless level via numerous anti-oxidative processes. Unnecessary ROS is removed enzymatically by superoxide dismutase (SOD), superoxide reductase, catalase, glutathione peroxidase (GPX), glutathione reductase, peroxiredoxin (PRX) and thioredoxin (TRX), (Afonso *et al.*, 2007) (Moloney & Cotter, 2018), or scavenged by various antioxidants, (Oyewole & Birch-Machin, 2015). Any perturbation from the ROS equilibrium may trigger ROS signalling events, (Foyer & Noctor, 2013) (Nignolet-Spruyt *et al.*, 2016) (Mittler *et al.*, 2004). This imbalance can lead to a number of pathological illnesses, including diabetes, cancer, aging, atherosclerosis, Alzheimer's disease, and Parkinson's disease, (Guglielmotto, Tamagno, & Danni,

2009) (Popa-Wagner *et al.,* 2013) (Umemo, Biju, & Yoshida, 2017) or severe brain injury (brain trauma and cerebral ischemia), (Chen *et al.*, 2011) (Valko *et al.,* 2007).

2. ROS GENERATION IN DIFFERENT ORGANELLES:

ROS are produced in almost every subcellular organelle in cell (Ahsan, H., Ali, A., & Ali, R., 2003), including plasma membrane (Fransen *et al.*, 2012), cytosol (Li & Shah, 2004a), mitochondria (Zorov, Juhaszova, & Sollott, 2014), nucleus (Zorov, Juhaszova, & Sollott, 2014), peroxisome (4), endoplasmic reticulum (ER) (Ahsan, Ali, & Ali, 2003), Golgi (Fransen *et al.*, 2012). The details are given in **Table 1**.

Table 1. Enzymes responsible for generating ROS in different organelles

ROS generating organelle	**Name**	**Responsible enzyme**	**References**
Plasma membrane	Superoxide ($O_2^{\bullet-}$)	Endothelial Nitric Oxide Synthase	(Luo *et al.*, 2014)
		NADPH oxidase 2/3/5	(Li & Shah, 2004b)
	Peroxyl radical (ROO•)	Lipoxygenase	(Chang *et al.*, 2005)
	Alkoxy radical (RO•)		
Cytosol	Superoxide ($O_2^{\bullet-}$)	Xanthine Oxidase	(Li & Shah, 2004b)
		Lipoxygenase	
	Hydroxyl radical (OH•)	Fenton Reaction	(Zorov, Juhaszova, & Sollott, 2014)
		Haber- Weiss reaction	
	Hydrogen peroxide (H_2O_2)	Superoxide dismutase	(Medinas *et al.*, 2007)
		Xanthine Oxidase	(Sakuma *et al.*, 2015)
		p^{66shc}	(Li *et al.*, 2017)
	Singlet Oxygen ($^1O_2$)	Phytooxygenation	(Agnez-Lima et al., 2012)
	Hypochlorous acid (HOCl)	Myeloperoxidase	(Okada *et al.*, 2016) (Onyango, 2016)

continued on following page

Table 1. Continued

ROS generating organelle	Name	Responsible enzyme	References
Mitochondria	Superoxide ($O_2^{\bullet-}$)	Complex I	(Chen & Zweier, 2014)
		Complex II	
		Complex III	
		Xanthine Oxidase	(Li & Shah, 2004b)
		Cytochrome P450	(Ahsan, Ali, & Ali, 2003)
		NADPH oxidase 4	(Li & Shah, 2004b)
	Hydroxyl radical ($OH^{\bullet}$)	Fenton Reaction	(Zorov, Juhaszova, & Sollott, 2014)
		Haber- Weiss reaction	
	Peroxyl radical ($ROO^{\bullet}$)	Lipoxygenase	(Chang *et al.*, 2005)
	Alkoxy radical ($RO^{\bullet}$)		
	Hydrogen peroxide (H_2O_2)	Superoxide dismutase	(Medinas *et al.*, 2007)
		Monoaminoxidase	(Zorov, Juhaszova, & Sollott, 2014)
		Xanthine Oxidase	(Sakuma *et al.*, 2015)
		P^{66shc}	(Li *et al.*, 2017)
	Singlet Oxygen ($^1O_2$)	Photooxygenation	(Agnez-Lima *et al.*, 2012)
		Cytochrome C	(Onyango, 2017)
	Hypochlorous acid (HOCl)	Myeloperoxidase	(Okada *et al.*, 2016) (Onyango, 2016)
Nucleus	Superoxide ($O_2^{\bullet-}$)	NADPH oxidase 4	(Li & Shah, 2004b)
		Lipoxygenase	
	Hydroxyl radical ($OH^{\bullet}$)	Fenton Reaction	(Zorov, Juhaszova, & Sollott, 2014)
		Superoxide dismutase	(Medinas *et al.*, 2007)

continued on following page

Table 1. Continued

ROS generating organelle	Name	Responsible enzyme	References
Peroxisiome	Superoxide ($O_2^{\bullet-}$)	Inducible Nitric oxide synthase	(Luo *et al.*, 2014)
		Xanthine Oxidase	(Li & Shah, 2004b)
	Hydroxyl radical ($OH^{\bullet}$)	Fenton Reaction	(Zorov, Juhaszova, & Sollott, 2014)
	Peroxyl radical ($ROO^{\bullet}$)	Lipoxygenase	(Chang *et al.*, 2005)
	Alkoxy radical ($RO^{\bullet}$)		
	Carbonate radical ($CO_3^{\bullet-}$)	Superoxide dismutase	(Medinas *et al.*, 2007)
	Hydrogen peroxide (H_2O_2)	Xanthine oxidase	(Medinas *et al.*, 2007)
		Superoxide dismutase	(Medinas *et al.*, 2007)
		Xanthine Oxidase	(Sakuma *et al.*, 2015)
		Acetyl-CoA oxidases	(Rinnerthaler *et al.*, 2015)
		Urate oxidases	
		d-amino acid oxidase	
		L- pipecolic acid oxidase	
		L-α- hydroxy acid oxidase	
		Polyamine oxidase	
	Singlet Oxygen ($^1O_2$)	Lipoxygenase	(King, Lai, & McCay, 1975)
		Dioxygenase	
		lactoperoxidase	
		Fenton reaction	(Sandalio & Romero-Puertas, 2015)
Endoplasmic Reticulum	Superoxide ($O_2^{\bullet-}$)	Cytochrome P450	(Ahsan, Ali, & Ali, 2003)
		NADPH oxidase 4	(Li & Shah, 2004b)
	Singlet Oxygen ($^1O_2$)	Cytochrome P450	(Onyango, 2017)
Golgi	Superoxide ($O_2^{\bullet-}$)	Endothelial Nitric Oxide Synthase	(Luo *et al.*, 2014)

3.ROS GENERATING SYSTEMS AND REGULATIONS:

There are various systems or organelles present in the cells (**Figure 1**) that are responsible for generating the ROSs in the body. Mitochondrion, Endoplasmic Reticulum, Peroxisomes, NADPH oxidases, Nitric oxide synthase, Xanthine oxidoreductase, and cytochrome P450 are responsible for generating ROS in the body, (Tretter *et al.,* 2021).

Figure 1. Models of the generation pathways used in the cell to produce ROS

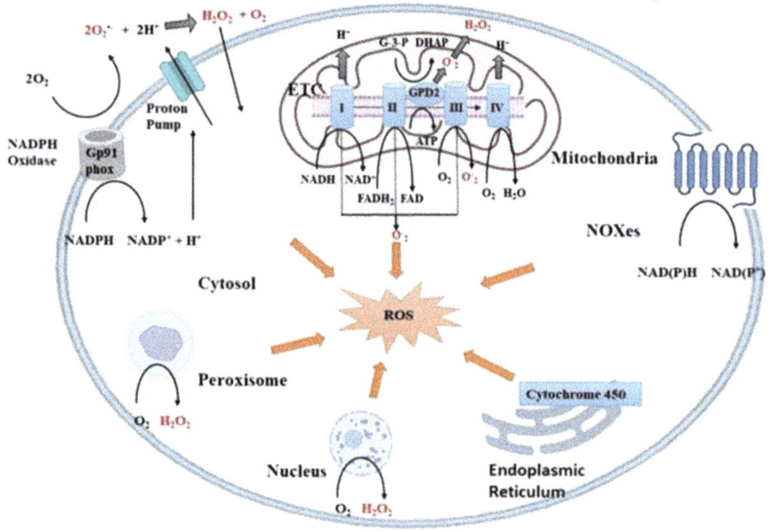

Complexes I, II and III are among many processes that produce ROS in mitochondria. The NADPH oxidase (NOX) family is primarily responsible for cytoplasmic ROS generation; NOX proteins exchange electrons with NADPH to create O_2. By passing acquired electrons to molecular oxygen via a flavin adenine dinucleotide cofactor, the ER creates H_2O_2. ROS are also produced by the membrane-associated monooxygenase system via cytochrome P450. Peroxisomes use oxidases such d-amino acid oxidase and ACOX to convert O_2 into H_2O_2.

(Abbreviations: NOX: NADPH Oxidase; ETC: Electron Transport Chain; gp-91phox: glycosylated 91-kDa glycoprotein; DHAP: Dihydroxyacetone phosphate; GPD2: glycerol-3-phosphate dehydrogenase 2; G3P: Glyceraldehyde 3-phosphate; ATP: Adenosine triphosphate).

The regulation of ROS is crucial in mitigating oxidative damage and meeting energy requirements. There are several inhibitors and promoters available which maintain the homeostasis in the body (**Table 2**). These agents are signalling molecules which present in a variety of pathways and mechanisms.

Table 2. Inhibitors and promotors of ROS

Regulator type	Agents(Internal)	Effects	References
ROS inhibitors	Extracellular signal Regulated Kinase	Reduction in oxidative damage of cardiac cells	(Nishida *et al.*, 2000)
	Negative Regulator of ROS	Reduction in tissue damage	(Leavy, 2024) (Noubade *et al.*, 2014)
	Hypoxia Inducible Factor -2α	Homeostasis of ROS	(Scortegagna *et al.*, 2003)
	PPARγ Coactrivator-1α	Activation of oxidative enzymes	(St-Pierre *et al.*, 2006)
	Nuclear factor erythroid-2-related factor 2	Limiting ROS generation in tumor	(DeNicola *et al.*, 2011)
	Pyruvate Kinase M2	Reducing oxidative damage on the cancer cells of lungs	(Anastasiou *et al.*, 2011)
	Uncoupling protein 2	Limiting the productionof ROS in macrophage	(Arsenijevic *et al.*, 2000)
	Vit. C/E, Ferritin Heavy Chain	Reducing accumulation of ROS	(Pham *et al.*, 2004) (Yu, 1994)
ROS promotors	Toll Like Receptor1/2/4	Increasing generation of ROS in macrophage	(West *et al.*, 2011)
	NADPH Oxidaese2	Increasing generation of ROS in mitochondria	(Prosser, Ward, & Lederer, 2011)
	P^{66SHC}	Increasing apoptotic signal in mitochondrial ROS	(Giorgio *et al.*, 2005)
	Tumor Necrosis Factor	Enhancement of necroptosis and killing of macrophage	(Roca & Ramakrishnan, 2013)
	Matrix Metalloproteinase -3	Instability of genome and DNA damage	(Radisky *et al.*, 2005)
	Erythroblastosis Virus Transcription Factor-1	Increased production of ROS	(Noubade *et al.*, 2014) (Shoi & Jaimes, 2018)
	UPBEAT1	Changing of cells from proliferation to differentiation	(Tsukagoshi, Busch, & Benfey, 2010)

4. PRECISION MEDICINE:

The term precision medicine is defined as "The designing of a treatment protocol for a patient based on the characteristics of particular patient, (Council NR., 2011). This involves treating every patient as an individual and focusing on alternative healthcare treatment to every patient's unique characteristics. This is ultimately leading to cost saving and reduction in side effects of the therapy, (König *et al.,* 2017). Personalized medicine takes into consideration genetic composition in conjugation with patient inclination, beliefs attitudes, awareness and societal environment. Precision medicine covers a wide range of tasks such as modelling health communication and shared decision making, study design, calculation and interpretation for fundamental treatment effects, and biomarker identification, (Ginsburg, 2014). For precision medicine and its related environment to thrive, it is essential to prioritize patient-centric approaches, engagement, digital healthcare, genomics, molecular technologies, information exchange, and data analytics, (Ginsburg & Phillips, 2018).

The innovative pharmacological strategy for targeting ROS in precision medicine focuses on selectively inhibiting only those ROS sources that are directly relevant to specific diseases, while preserving the normal physiological functions of ROS. This approach requires a shift away from traditional concepts such as "small amounts of ROS are beneficial" and "large amounts are harmful." Instead, it emphasizes identifying the dysregulation of specific enzymes as a causal mechanism in disease pathology, (Ghezzi *et al.,* 2017). Moreover, secondary ROS formation, often resulting from inflammatory processes (e.g., NOX2 or NOS2), is generally viewed as an epiphenomenon and not a suitable target for therapeutic intervention. Therefore, clinical validations are required in basic research for safe pharmacological targeting in humans. The ROS-forming and metabolizing enzymes are NADPH oxidase, Nitric oxide synthase (NOS), Xanthine oxidoreductase (XOR), Monoamine oxidase (MAO), Myeloperoxidase (MPO), (Casas *et al.*, 2020a). Some of the enzymes are found as therapeutic targets in different diseases in clinical trial studies, (Casas *et al.,* 2020b). Antioxidant enzymes control oxidative stress and redox equilibrium, which is essential for the survival and growth of cancer cells, (American Journal of Physiology-Cell Physiology, 2024) (Bedard & Krause, 2007). The change in the enzymes and associated protein's behaviours can lead to OS, thereby health problem. In a similar, changing the intracellular oxidative environment significantly affect pathogenic bacteria like *Escherichia coli* (*E. coli*) and potentially fatal infectious pathogens like *Staphylococcus aureus* (SA) and *Mycobacterium tuberculosis* (Mtb). Hence, novel therapies may be created to alter antioxidant levels and/or increase intracellular ROS, (Dharmaraja, 2017). ROS possess the ability to modify the biological environment, resulting in diverse outcomes and potential applications in precision medicine and therapeutic, (Cortese-Krott *et al.*, 2017). This precision medicine approach underscores the need for targeted therapies that specifically modulate disease-relevant ROS pathways, paving the way for more effective and personalized cancer and other disease treatments. Precision medicine comprised of various techniques as illustrated in **Figure 2**

Figure 2. Precision medicine approaches

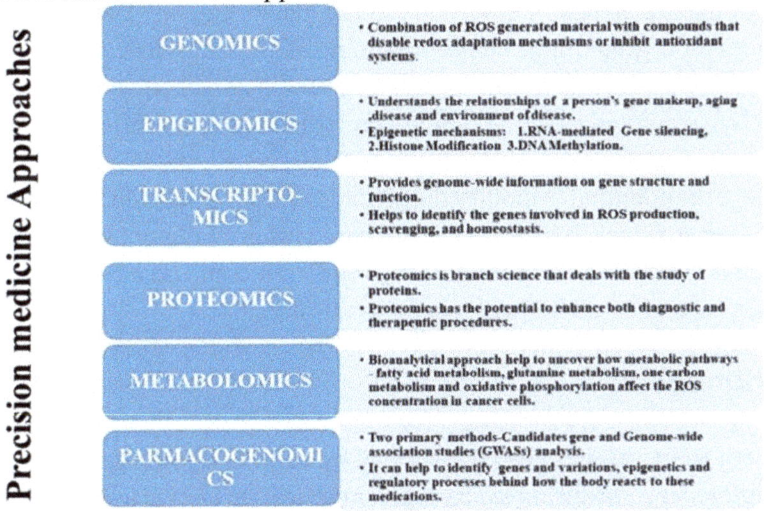

Genomics:

The term genomic medicine is a concept that utilizes the application of genomic information for disease treatment or to provide medicinal therapy based on genetic information that has been gathered since the HGP (human genome project), (Roth, 2019). The genome is nothing but the complete set of information in an organism's DNA, (Bedard & Krause, 2007).

The ROS mediated genomic medicine could be an effective approach for targeting specific gene in various disease. The therapeutic strategy involves the combination of ROS generated material with compounds that disable redox adaptation mechanisms or inhibit antioxidant systems, it is possible to induce significant ROS accumulation in cancer cells, leading to oxidative damage, cell death, and potentially overcoming drug resistance, (American Journal of Physiology-Cell Physiology, 2024).

Epigenomic:

A branch of science that deals with study of genetics that is unrelated to DNA sequence. The DNA sequence of an individual is generally constant but the epigenetic state changes over the course of a lifetime and across different tissues, (Hamilton, 2011). Epigenetic mechanisms allow cells to recall and reprogramme gene activity in response to changes in their internal and external environments. These modifications to the regular reprogramming and programming of gene activity can be remembered by epigenetic processes. There are mainly three main epigenetic mechanisms which include RNA-mediated gene silencing, protein histone modification, and DNA methylation, (Mortaz *et al.,* 2011).

ROS have the potential to modify the genome through epigenetic modifications in addition to genetic ones, which are crucial in the progression of human carcinogenesis, (Ziech *et al.,* 2011). Inhibiting ROS production could be a potential treatment for diseases caused by altered epigenetic modifications.

Transcriptomics:

The set of RNA that has been transcribed by a cell or tissue's genome in a particular physiological state during a developmental stage is known as the transcriptome and studies of transcriptome known as transcriptomic. It emphasis on RNA-level gene expression analysis. Transcriptomic analysis aids in the understanding of how biological processes are regulated, which in turn helps with clinical therapy and disease diagnosis, (Dong & Chen, 2013). Transcriptome (transcriptomic) plays important role in understanding and discovering disease pathways and the development of precise and effective drugs, (Khodadadian *et al.,* 2020). Transcriptomic analysis can help to identify the genes involved in ROS production, scavenging, and homeostasis. For example, in a study performed to reveal reactive oxygen species scavenging mechanisms associated with ferrous iron toxicity in aromatic KetakiJoha rice. Around 69 genes were identified which were associated with the ROS scavenging mechanism, the data from the analysis have information about the mechanism by which the ROS production is taking place on the precision medicine can be developed to target the specific genes responsible for the production of ROS.

Proteomics:

Proteomics is branch of science that deals with the study of proteins and their interactions, functionality, composition, and structural characteristics. In comparison to genomics, proteomics offers a more profound comprehension of organismal structure and function. Despite this, proteomics is more difficult than genomics because of the dynamic changes in protein expression that occur in response to changes in environmental variables, (Al-Amrani *et al.,* 2021). Proteomics holds a significant role in precision medicine as it offers insights into the patient's condition through the analysis of proteins, metabolic factors, and post-translational modifications. The personalized examination of proteomics has the potential to enhance both diagnostic and therapeutic procedures for individual patients by detecting alterations in proteins linked to specific illnesses and choosing appropriate treatment strategies. Furthermore, Proteomics enables prompt detection of infectious diseases by juxtaposing the proteomes of pathogens with the patient's sample, thereby facilitating the prompt identification of infections and mechanisms of antibiotic resistance, (Duarte & Spencer, 2016). In drug discovery, proteomics is very much used for designing novel therapeutics against drug resistance targets also. Therefore, proteomics is important tool for designing of personalized medicine. If alteration in the ROS proteomes, then personalized medicine can be designed against of it.

Metabolomics:

Metabolomics is a bio-analytical approach used to identify the metabolites present in organisms during various biological processes, (Nalbantoglu & Amri, 2019). It serves as a phenotyping tool that aids in the determination of metabolites in biological samples such as tissues, cells, fluids, or organisms, (Klassen *et al.,* 2017).

Metabolomics plays a crucial role in precision medicine by offering metabolic bio-signatures for disease prognosis and surveillance at a molecular level. Metabolomics serves as a functional representation of genomic variations, augmenting the clinical understanding of genetic data for more accurate healthcare strategies. By utilizing pharmaco-metabolomics, it facilitates the discovery of biomarkers for phenotyping drug responses, thus guiding healthcare providers in the choice of suitable medications based on metabolic profiles. In essence, metabolomics is indispensable in precision medicine as it provides valuable insights into disease mechanisms, enables personalized treatment and supports, (Bekri, 2016). It facilitates customized therapeutic approaches through the adaptation of interventions to the specific requirements of individual patients. Metabolome analysis can help to uncover how metabolic pathways like fatty acid metabolism glutamine metabolism, one carbon metabolism and oxidative phosphorylation can affect the ROS concentration in cancer cells. Metabolomics plays an important role in the discovery of novel drugs or combination strategies to target metabolic pathways involved in the ROS regulation which can provide potential advantage for precision medicine in cancer, (Sainero-Alcolado *et al.,* 2022).

Pharmacogenomics:

The goal of genomics is to help implement precision medicine by providing an improved comprehension of the molecular processes regulating blood pressure. This understanding may direct the development of novel drugs, personalize treatment plans, and eventually result in the development of a new taxonomy of hypertension, (Padmanabhan, Aman, & Dominiczak, 2018). Candidates' gene and Genome-wide association studies (GWASs) analysis were the two primary methods employed in the pharmacogenomics research. Finding the most relevant single-nucleotide polymorphisms (SNPs) linked to hypertension and understanding the connections between specific genes and treatment outcomes across various ethnic groups are particularly challenging due to the multifaceted nature of the condition, (Johnson *et al.,* 2019). Multifactorial and multigenic complex characteristics appear to be associated with drug responsiveness, (Zanger, 2010). Thus, more research on the pharmacogenomics of anti-hypertensive medications is necessary to identify more genes and variations as well as the epigenetic and regulatory processes behind how the body reacts to these medications, (Luizon, Pereira, & Sandrim, 2018).

This chapter explores the role of ROS in precision medicine, importance of their impact on pathogenic pathways, biomarkers, therapeutic strategies, drug responsiveness, and lifestyle modifications. By integrating insights on ROS with a thorough understanding of genetic, environmental, and lifestyle factors, precision medicine offers a more refined and effective methodology to healthcare. This personalized approach improves diagnostic accuracy, intervention efficacy, and the comprehensive management of conditions influenced by oxidative stress. Targeted ROS regulation is expected to play an increasingly crucial role in improving patient outcomes and advancing personalized healthcare solutions as precision medicine continues to evolve.

5. PERSONALIZED APPROACHES FOR NEURODEGENERATIVE DISEASES MANAGEMENT:

In many neurodegenerative diseases, such as Alzheimer's disease (AD), Parkinson's disease (PD), amyotrophic lateral sclerosis (ALS), and Huntington's disease (HD), OS is a crucial factor that underlies the death of neuronal cells (Teleanu *et al.,* 2022) (**Figure 3**). Furthermore, OS is implicated in brain and spinal cord damage resulting from stroke and traumatic brain injury (TBI) (89). The pathological accumulation of ROS contributes significantly to mitochondrial dysfunction, (Mandal *et al.,* 2022), which in turn leads to impaired energy production, (Yang & Lian, 2020), disruptions in metal homeostasis, (Sun *et al.,* 2020), and the accumulation of toxic protein aggregates, (Lévy *et al.,* 2019). These molecular disruptions ultimately precipitate the activation of diverse cell death pathways, including apoptosis, necrosis, ferroptosis, and autophagy, (Redza-Dutordoir & Averill-Bates, 2016).

Figure 3. Illustration of involvement of oxidative stress in neurodegeneration

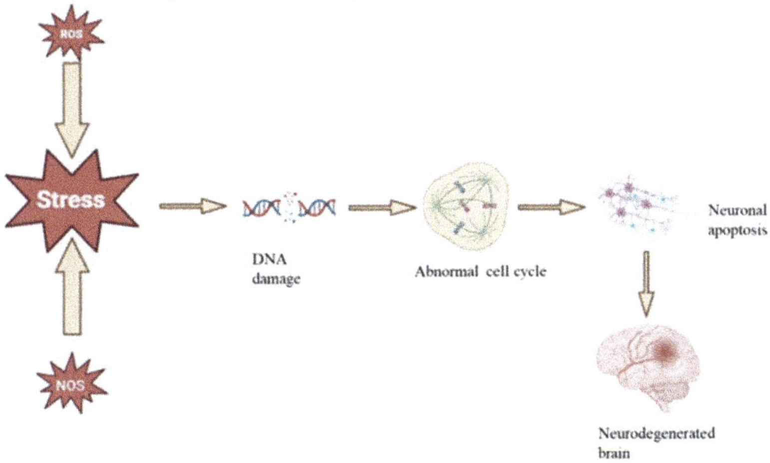

As such, oxidative stress is linked to the development of several metabolic diseases, cancer, and neurological illness as mentioned in detail in **Table 1**.

Table 3. Different types of oxidative stress species and their implications in different neurological diseases

Oxidative stress species	Neurodegenerative disease	Common pathological mechanism	References
Reactive Oxygen Species (ROS)	Alzheimer's Disease (AD)	ROS contribute to the oxidation of lipids, proteins, and DNA, leading to tau hyper phosphorylation and β-amyloid plaque formation.	Perry *et al.*, 1998
	Parkinson's Disease (PD)	ROS induce dopaminergic neuronal death by damaging mitochondrial DNA and inhibiting complex I of the electron transport chain.	Guo *et al.*, 2018
	Amyotrophic Lateral Sclerosis (ALS)	ROS cause oxidative damage to motor neurons, leading to protein aggregation, mitochondrial dysfunction, and cell death	Manfredi and Xu, 2005
	Huntington's Disease (HD)	ROS contribute to neuronal apoptosis through oxidative damage to cellular components and mitochondrial dysfunction.	Nissanka and Moraes, 2018
	Stroke	ROS mediate neuronal injury and cell death by exacerbating inflammation and disrupting cellular redox balance.	Zhou *et al.*, 2022
	Traumatic Brain Injury (TBI)	ROS induce secondary brain damage by amplifying oxidative stress and disrupting cellular repair mechanisms.	Nissanka and Moraes, 2018

In neurodegenerative diseases caused by increase in OS, precision medicinal approaches are beneficial. The various approaches reported in the literatures are discussed below (Summarized in **Table 4**).

A. Genetic and Genomic Techniques

The genomic sequencing and mutations in genetic structure are crucial factor in the development of oxidative species. Targeting these genetic sequencing is found to be very effective in the treatment of neuronal damage associated with the OS.

Whole-Genome Sequencing (WGS) technique is identifying genetic variants, a comprehensive analysis of the entire genome to detect variants linked to OS and neurodegenerative diseases which helps in understanding genetic predispositions and tailoring intervention, (Firdaus & Li, 2024). CRISPR/Cas9 Gene is a key amino acid sequence that involved in the production of SOD1 which is responsible for the endogenous protection from increased ROS. CRISPR/Cas9 Gene Editing related to editing of genes which are associated with OS that are potentially reduces OS by correcting genetic mutations and found to be effective in the management of neuronal damage in Parkinson's disease, (Rahman *et al.*, 2022). Mitochondrial dysfunction in the cell due to OS is leading to apoptosis. Targeting Single-Cell RNA Sequencing is effective in such conditions which are based on analysing gene expression changes. It provides insights into gene expression at the single-cell level focusing on OS responses and identifies cellular mechanisms and potential therapeutic targets, (Cuevas-Diaz *et al.*, 2022). In Huntington's disease the mutations into the exome region is mostly affected. Exome sequencing technique based on detecting coding region mutations that focuses on sequencing the coding regions of the genome. This technique is commonly used to identify mutations affecting OS which is useful for pin-pointing mutations involved in diseases like Huntington's, hence used in management of HD, (Ross, Dion, & Rouleau, 2020). Apart from these techniques, nowadays Gene Expression Profiling which is based on OS-related gene changes are also being used to mitigate the neuronal damage due to OS. This technique analyses gene expression levels related to OS in neurodegenerative conditions and guides targeted therapeutic strategies based on gene expression profiles, (Benetti, Gustincich, & Legname, 2012).

B. Biomarker Discovery and Utilization

Biomarkers are biological molecules that may be detected in bodily fluids such as blood or tissues and serve as indicators of certain conditions or diseases, as well as normal or abnormal processes. Detection of OS biomarkers is commonly utilized in the precisional therapy against the diseases caused by OS. Novel OS biomarkers technique, which is involved in identifying the new biomarkers, such biomarkers, such as advanced glycation end-products (AGEs). This technique also helps in monitoring disease progression and response to treatments, (Meertens *et al.*, 2016). During the early progression of OS some biomarkers like RNA signatures are circulate into the blood stream. Furthermore, Imaging Biomarkers to detect brain changes to elucidate levels of identify OS can contribute to the diagnostic accuracy and treatment planning, (Jain, 2017).

Liquid biopsy techniques are non-invasive method to detect biomarkers related to OS from blood samples which facilitates early diagnosis and ongoing monitoring of neurodegenerative diseases associated with OS, (Malhotra *et al.*, 2023). In the progression of OS some proteins in the cerebrospinal fluid (CSF) are also being unregulated and mutated. Proteomic analysis which analyzes proteins in CSF to understand OS impacts in the brain also provides insights into disease mechanisms, (Bastos *et al.*, 2017).

C. Personalized Drug Development and Pharmacogenomics

Every cell has its own genomic sequence and its own amino acids coding. To maintain the OS in the body, the body produces some endogenous proteins to tackle them called antioxidants. Development of Targeted Antioxidants technique is based on the designing of antioxidants specific to oxidative pathways that provides more effective treatments for diseases like Alzheimer's, (Reddy, 2006). Furthermore, pharmacogenomics of antioxidants technique that is involved in studying genetic variations in antioxidant metabolism investigating how genetic variations affect the metabolism and effectiveness of antioxidant supplements also tailor's antioxidant therapy to individual genetic profiles. Apart from these techniques, Personalized Drug Combinations therapy is commonly used nowadays which is based on the Combining drugs based on genetic and OS profiles. By targeting multiple disease mechanisms, this personalized approach enhances therapeutic efficacy, optimizes treatment outcomes, and improves the quality of life for individuals with neurodegenerative diseases, (Griñán-Ferré *et al.*, 2024).

Table 4. Precision medicine strategies for neurodegenerative diseases

Category	*Approach*	*Example*	*Description*	*Application*	*References*
Genetic and Genomic Techniques	Whole-Genome Sequencing (WGS)	Identifying genetic variants	Comprehensive analysis of the entire genome to detect variants linked to neurodegenerative diseases.	Helps in understanding genetic predispositions and tailoring interventions.	(Firdaus & Li, 2024)
	CRISPR/Cas9 Gene Editing	Correcting SOD1 mutations	Editing genes related to oxidative stress, such as SOD1 in ALS.	Potentially reduces oxidative damage by correcting genetic mutations.	(Rahman *et al.*, 2022)
	Single-Cell RNA Sequencing	Analyzing gene expression changes	Provides insights into gene expression at the single-cell level, focusing on oxidative stress responses.	Identifies cellular mechanisms and potential therapeutic targets.	(Cuevas-Diaz *et al.*, 2022)
	Exome Sequencing	Detecting coding region mutations	Focuses on sequencing the coding regions of the genome to identify mutations affecting oxidative stress.	Useful for pinpointing mutations involved in diseases like Huntington's.	(Ross, Dion, & Rouleau, 2020)
	Gene Expression Profiling	Measuring oxidative stress-related gene changes	Analyzes gene expression levels related to oxidative stress in neurodegenerative conditions.	Guides targeted therapeutic strategies based on gene expression profiles.	(Benetti, Gustincich, & Legname, 2012)

continued on following page

Table 4. Continued

Category	*Approach*	*Example*	*Description*	*Application*	*References*
Biomarker Discovery and Utilization	Novel Oxidative Stress Biomarkers	Advanced glycation end-products (AGEs)	Identifying new biomarkers associated with oxidative stress, such as AGEs.	Helps in monitoring disease progression and response to treatments.	(Meertens *et al.*, 2016)
	Liquid Biopsy Techniques	Detecting RNA signatures in blood	Non-invasive method to detect biomarkers related to oxidative stress from blood samples.	Facilitates early diagnosis and ongoing monitoring of neurodegenerative diseases.	(Malhotra *et al.*, 2023)
	Proteomic Analysis	Identifying oxidative stress-related proteins	Analyzing proteins in cerebrospinal fluid to understand oxidative stress impacts.	Provides insights into disease mechanisms and guides personalized therapies.	(Bastos *et al.*, 2017)
	Metabolomic Profiling	Analyzing metabolic changes	Examining metabolic alterations associated with oxidative stress.	Assists in personalizing dietary and drug interventions.	(Bogdanov et al., 2008) (González-Domínguez, García-Barrera, & Gómez-Ariza, 2015)
	Imaging Biomarkers	Using advanced imaging to detect brain changes	Employing imaging techniques to identify oxidative stress-related brain changes.	Enhances diagnostic accuracy and treatment planning.	(Jain, 2017)

continued on following page

Table 4. Continued

Category	*Approach*	*Example*	*Description*	*Application*	*References*
Personalized Drug Development and Pharmacogenomics	Drug Response Prediction Models	Machine learning algorithms for drug efficacy	Using algorithms to predict individual responses to drugs based on genetic and metabolic profiles.	Optimizes drug selection and dosage for neurodegenerative diseases.	(Rivetti di Val Cervo *et al.*, 2021)
	Development of Targeted Antioxidants	Designing antioxidants specific to oxidative pathways	Creating antioxidants that target specific pathways or cellular compartments affected by oxidative stress.	Provides more effective treatments for diseases like Alzheimer's.	(Reddy, 2006)
	Pharmacogenomics of Antioxidants	Studying genetic variations in antioxidant metabolism	Investigating how genetic variations affect the metabolism and effectiveness of antioxidant supplements.	Tailors antioxidant therapy to individual genetic profiles.	(Griñán-Ferré *et al.*, 2024)
	Personalized Drug Combinations	Combining drugs based on genetic and oxidative stress profiles	Developing combination therapies tailored to individual profiles.	Enhances therapeutic efficacy by targeting multiple disease mechanisms.	
	Metabolic Pathway Targeting	Developing drugs that target specific oxidative pathways	Creating drugs that address metabolic pathways affected by oxidative stress.	Aims to correct imbalances and reduce oxidative damage.	(Millichap *et al.*, 2021)

6. FUTURE PERSPECTIVES:

It's critical to recognize the difference between precision medicine research and its deployment in practice. The goal of research is to find quantitative indicators that can precisely estimate a patient's response to a given course of care. These markers, which are frequently examined using advanced algorithms, might vary from straightforward genetic polymorphisms to difficult combinations of gene expressions or hereditary variations. These markers aid in individual outcome prediction once they have been validated in patient cohorts. Large-scale data sets like the UK Biobank, the Precision Medicine Initiative/All of Us cohort, and the Estonian Genome Project are enabling the discipline to advance quickly.

Furthermore, the effectiveness of prediction models that use digital phenotyping and data from medical records is increasing. This is demonstrated by the VA's Care Assessment Needs ratings, which forecast the risk of hospitalization. Future developments in medical imaging data analysis, such the ability to identify malignant tumors from skin photos, will likely bring with them the addition of socioeconomic, behavioural, and environmental aspects to these models.

In the future, ROS may play a bigger part in precision medicine, which may further change how diseases are treated. We expect significant advancements in the use of ROS as biomarkers for more accurate diagnosis and customized treatments as our understanding of ROS grows. ROS profiles will be used to optimize patient care through advancements in drug administration and individualized treatment strategies. These innovations will also integrate with multi-omics data to refine prognostic models. Personalized therapy will become more proactive and effective as a result of this advancement in our capacity to forecast disease trajectories and treatment responses. But this progress will also necessitate paying close attention to moral and legal issues. Future tailored medicine will be greatly aided by ROS research, which will advance more specialized and efficient medical treatments.

REFERENCES

Afonso, V., Champy, R., Mitrovic, D., Collin, P., & Lomri, A. (2007, July). Reactive oxygen species and superoxide dismutases: Role in joint diseases. *Joint, Bone, Spine*, *74*(4), 324–329. DOI: 10.1016/j.jbspin.2007.02.002 PMID: 17590367

Agnez-Lima, L. F., Melo, J. T., Silva, A. E., Oliveira, A. H. S., Timoteo, A. R. S., Lima-Bessa, K. M., Martinez, G. R., Medeiros, M. H. G., Di Mascio, P., Galhardo, R. S., & Menck, C. F. M. (2012). DNA damage by singlet oxygen and cellular protective mechanisms. *Mutation Research/Reviews in Mutation Research*, *751*(1), 15–28. DOI: 10.1016/j.mrrev.2011.12.005 PMID: 22266568

Ahsan, H., Ali, A., & Ali, R. (2003, February 28). Oxygen free radicals and systemic autoimmunity. *Clinical and Experimental Immunology*, *131*(3), 398–404. DOI: 10.1046/j.1365-2249.2003.02104.x PMID: 12605691

Ahsan, H., Ali, A., & Ali, R. (2003). Oxygen free radicals and systemic autoimmunity. *Clinical and Experimental Immunology*, *131*(3), 398–404. DOI: 10.1046/j.1365-2249.2003.02104.x PMID: 12605691

Al-Amrani, S., Al-Jabri, Z., Al-Zaabi, A., Alshekaili, J., & Al-Khabori, M. (2021, September 27). Proteomics: Concepts and applications in human medicine. *World Journal of Biological Chemistry*, *12*(5), 57–69. DOI: 10.4331/wjbc.v12.i5.57 PMID: 34630910

Alfadda, A. A., & Sallam, R. M. (2012). Reactive Oxygen Species in Health and Disease. *Journal of Biomedicine & Biotechnology*, *2012*, 1–14. DOI: 10.1155/2012/936486 PMID: 22927725

Anastasiou, D., Poulogiannis, G., Asara, J. M., Boxer, M. B., Jiang, J., Shen, M., Bellinger, G., Sasaki, A. T., Locasale, J. W., Auld, D. S., Thomas, C. J., Vander Heiden, M. G., & Cantley, L. C. (2011). Inhibition of pyruvate kinase M2 by reactive oxygen species contributes to cellular antioxidant responses. *Science*, *334*(6060), 1278–1283. DOI: 10.1126/science.1211485 PMID: 22052977

Arsenijevic, D., Onuma, H., Pecqueur, C., Raimbault, S., Manning, B. S., Miroux, B., Couplan, E., Alves-Guerra, M.-C., Goubern, M., Surwit, R., Bouillaud, F., Richard, D., Collins, S., & Ricquier, D. (2000). Disruption of the uncoupling protein-2 gene in mice reveals a role in immunity and reactive oxygen species production. *Nature Genetics*, *26*(4), 435–439. DOI: 10.1038/82565 PMID: 11101840

Bae, Y. S., Oh, H., Rhee, S. G., & Yoo, Y. D. (2011, December). Regulation of Reactive Oxygen Species Generation in Cell Signaling. *Molecules and Cells*, *32*(6), 491–509. DOI: 10.1007/s10059-011-0276-3 PMID: 22207195

Bastos, P., Ferreira, R., Manadas, B., Moreira, P. I., & Vitorino, R. (2017, April 3). Insights into the human brain proteome: Disclosing the biological meaning of protein networks in cerebrospinal fluid. *Critical Reviews in Clinical Laboratory Sciences*, *54*(3), 185–204. DOI: 10.1080/10408363.2017.1299682 PMID: 28393582

Bedard, K., & Krause, K. H. (2007, January). The NOX Family of ROS-Generating NADPH Oxidases: Physiology and Pathophysiology. *Physiological Reviews*, *87*(1), 245–313. DOI: 10.1152/physrev.00044.2005 PMID: 17237347

Bedard, K., & Krause, K. H. (2007, January). The NOX Family of ROS-Generating NADPH Oxidases: Physiology and Pathophysiology. *Physiological Reviews*, *87*(1), 245–313. DOI: 10.1152/physrev.00044.2005 PMID: 17237347

Bekri, S. (2016, November). The role of metabolomics in precision medicine. *Expert Review of Precision Medicine and Drug Development*, *1*(6), 517–532. DOI: 10.1080/23808993.2016.1273067

Benetti, F., Gustincich, S., & Legname, G. (2012, March 1). Gene expression profiling and therapeutic interventions in neurodegenerative diseases: A comprehensive study on potentiality and limits. *Expert Opinion on Drug Discovery*, *7*(3), 245–259. DOI: 10.1517/17460441.2012.659661 PMID: 22468955

Bogdanov, M., Matson, W. R., Wang, L., Matson, T., Saunders-Pullman, R., Bressman, S. S., & Beal, M. F. (2008, February 1). Metabolomic profiling to develop blood biomarkers for Parkinson's disease. *Brain*, *131*(2), 389–396. DOI: 10.1093/brain/awm304 PMID: 18222993

Casas, A. I., Nogales, C., Mucke, H. A. M., Petraina, A., Cuadrado, A., Rojo, A. I., ... On the Clinical Pharmacology of Reactive Oxygen Species. Touyz RM, editor. Pharmacol Rev. 2020 Oct;72(4):801–28. DOI: 10.1124/pr.120.019422

Casas, A. I., Nogales, C., Mucke, H. A. M., Petraina, A., Cuadrado, A., Rojo, A. I., ... On the Clinical Pharmacology of Reactive Oxygen Species. Touyz RM, editor. Pharmacol Rev. 2020 Oct;72(4):801–28. DOI: 10.1124/pr.120.019422

Chang, M. S., Schneider, C., Roberts, R. L., Shappell, S. B., Haselton, F. R., Boeglin, W. E., & Brash, A. R. (2005). Detection and subcellular localization of two 15S-lipoxygenases in human cornea. *Investigative Ophthalmology & Visual Science*, *46*(3), 849–856. DOI: 10.1167/iovs.04-1166 PMID: 15728540

Chen, S. D., Yang, D. I., Lin, T. K., Shaw, F. Z., Liou, C. W., & Chuang, Y. C. (2011, October 21). Roles of Oxidative Stress, Apoptosis, PGC-1α and Mitochondrial Biogenesis in Cerebral Ischemia. *International Journal of Molecular Sciences*, *12*(10), 7199–7215. DOI: 10.3390/ijms12107199 PMID: 22072942

Chen, Y. R., & Zweier, J. L. (2014, January 31). Cardiac Mitochondria and Reactive Oxygen Species Generation. *Circulation Research, 114*(3), 524–537. DOI: 10.1161/CIRCRESAHA.114.300559 PMID: 24481843

Cortese-Krott, M. M., Koning, A., Kuhnle, G. G. C., Nagy, P., Bianco, C. L., Pasch, A., Wink, D. A., Fukuto, J. M., Jackson, A. A., van Goor, H., Olson, K. R., & Feelisch, M. (2017, October). The Reactive Species Interactome: Evolutionary Emergence, Biological Significance, and Opportunities for Redox Metabolomics and Personalized Medicine. *Antioxidants & Redox Signaling, 27*(10), 684–712. DOI: 10.1089/ars.2017.7083 PMID: 28398072

Council, N. R. (2011). *Studies D on E and L, Sciences B on L, Disease C on AF for D a NT of. Toward Precision Medicine: Building a Knowledge Network for Biomedical Research and a New Taxonomy of Disease*. National Academies Press.

Cuevas-Diaz Duran, R., González-Orozco, J. C., Velasco, I., & Wu, J. Q. (2022, October 24). Single-cell and single-nuclei RNA sequencing as powerful tools to decipher cellular heterogeneity and dysregulation in neurodegenerative diseases. [Internet]. *Frontiers in Cell and Developmental Biology, 10*, 884748. Retrieved August 15, 2024, from https://www.frontiersin.org/journals/cell-and-developmental-biology/articles/10.3389/fcell.2022.884748/full. DOI: 10.3389/fcell.2022.884748 PMID: 36353512

DeNicola, G. M., Karreth, F. A., Humpton, T. J., Gopinathan, A., Wei, C., Frese, K., Mangal, D., Yu, K. H., Yeo, C. J., Calhoun, E. S., Scrimieri, F., Winter, J. M., Hruban, R. H., Iacobuzio-Donahue, C., Kern, S. E., Blair, I. A., & Tuveson, D. A. (2011). Oncogene-induced Nrf2 transcription promotes ROS detoxification and tumorigenesis. *Nature, 475*(7354), 106–109. DOI: 10.1038/nature10189 PMID: 21734707

Dharmaraja, A. T. (2017, April 27). Role of Reactive Oxygen Species (ROS) in Therapeutics and Drug Resistance in Cancer and Bacteria. *Journal of Medicinal Chemistry, 60*(8), 3221–3240. DOI: 10.1021/acs.jmedchem.6b01243 PMID: 28135088

Dong, Z., & Chen, Y. (2013, October). Transcriptomics: Advances and approaches. *Science China. Life Sciences, 56*(10), 960–967. DOI: 10.1007/s11427-013-4557-2 PMID: 24091688

Dröge, W. (2002, January 1). Free Radicals in the Physiological Control of Cell Function. *Physiological Reviews, 82*(1), 47–95. DOI: 10.1152/physrev.00018.2001 PMID: 11773609

Duarte, T., & Spencer, C. (2016, October 1). Personalized Proteomics: The Future of Precision Medicine. *Proteomes*, *4*(4), 29. DOI: 10.3390/proteomes4040029 PMID: 27882306

Firdaus, Z., & Li, X. (2024, January). Unraveling the Genetic Landscape of Neurological Disorders: Insights into Pathogenesis, Techniques for Variant Identification, and Therapeutic Approaches. *International Journal of Molecular Sciences*, *25*(4), 2320. DOI: 10.3390/ijms25042320 PMID: 38396996

Foyer, C. H., & Noctor, G. (2013, June). Redox Signaling in Plants. *Antioxidants & Redox Signaling*, *18*(16), 2087–2090. DOI: 10.1089/ars.2013.5278 PMID: 23442120

Fransen, M., Nordgren, M., Wang, B., & Apanasets, O. (2012, September). Role of peroxisomes in ROS/RNS-metabolism: Implications for human disease. Biochimica et Biophysica Acta (BBA) -. *Biochimica et Biophysica Acta. Molecular Basis of Disease*, *1822*(9), 1363–1373. DOI: 10.1016/j.bbadis.2011.12.001 PMID: 22178243

Ghezzi, P., Jaquet, V., Marcucci, F., & Schmidt, H. H. H. W. (2017, June). The oxidative stress theory of disease: Levels of evidence and epistemological aspects. *British Journal of Pharmacology*, *174*(12), 1784–1796. DOI: 10.1111/bph.13544 PMID: 27425643

Ginsburg, G. (2014, April). Medical genomics: Gather and use genetic data in health care. *Nature*, *508*(7497), 451–453. DOI: 10.1038/508451a PMID: 24765668

Ginsburg, G. S., & Phillips, K. A. (2018, May). Precision Medicine: From Science To Value. *Health Affairs*, *37*(5), 694–701. DOI: 10.1377/hlthaff.2017.1624 PMID: 29733705

Giorgio, M., Migliaccio, E., Orsini, F., Paolucci, D., Moroni, M., Contursi, C., Pelliccia, G., Luzi, L., Minucci, S., Marcaccio, M., Pinton, P., Rizzuto, R., Bernardi, P., Paolucci, F., & Pelicci, P. G. (2005). Electron transfer between cytochrome c and p66Shc generates reactive oxygen species that trigger mitochondrial apoptosis. *Cell*, *122*(2), 221–233. DOI: 10.1016/j.cell.2005.05.011 PMID: 16051147

González-Domínguez, R., García-Barrera, T., & Gómez-Ariza, J. L. (2015, March 25). Metabolite profiling for the identification of altered metabolic pathways in Alzheimer's disease. *Journal of Pharmaceutical and Biomedical Analysis*, *107*, 75–81. DOI: 10.1016/j.jpba.2014.10.010 PMID: 25575172

Griñán-Ferré, C., Bellver-Sanchis, A., Guerrero, A., & Pallàs, M. (2024, July 1). Advancing personalized medicine in neurodegenerative diseases: The role of epigenetics and pharmacoepigenomics in pharmacotherapy. *Pharmacological Research*, *205*, 107247. DOI: 10.1016/j.phrs.2024.107247 PMID: 38834164

Guglielmotto, M., Tamagno, E., & Danni, O. (2009). Oxidative Stress and Hypoxia Contribute to Alzheimer's Disease Pathogenesis: Two Sides of the Same Coin. *TheScientificWorldJournal, 9*, 781–791. DOI: 10.1100/tsw.2009.93 PMID: 19705038

Hamilton, J. P. (2011). Epigenetics: Principles and Practice. *Digestive Diseases (Basel, Switzerland), 29*(2), 130–135. DOI: 10.1159/000323874 PMID: 21734376

Holmström, K. M., & Finkel, T. (2014, June). Cellular mechanisms and physiological consequences of redox-dependent signalling. *Nature Reviews. Molecular Cell Biology, 15*(6), 411–421. DOI: 10.1038/nrm3801 PMID: 24854789

Jain, K. K. (2017). Biomarkers of Disorders of the Nervous System. In K. K. Jain (Ed.), The Handbook of Biomarkers [Internet]. (pp. 463–610). Springer., Retrieved August 15, 2024, from, DOI: 10.1007/978-1-4939-7431-3_14

Jelinek, M., Jurajda, M., & Duris, K. (2021, November 25). Oxidative Stress in the Brain: Basic Concepts and Treatment Strategies in Stroke. *Antioxidants, 10*(12), 1886. DOI: 10.3390/antiox10121886 PMID: 34942989

Johnson, R., Dludla, P., Mabhida, S., Benjeddou, M., Louw, J., & February, F. (2019, May). Pharmacogenomics of amlodipine and hydrochlorothiazide therapy and the quest for improved control of hypertension: A mini review. *Heart Failure Reviews, 24*(3), 343–357. DOI: 10.1007/s10741-018-09765-y PMID: 30645721

Khodadadian, A., Darzi, S., Haghi-Daredeh, S., Sadat Eshaghi, F., Babakhanzadeh, E., Mirabutalebi, S. H., & Nazari, M. (2020, September). Genomics and Transcriptomics: The Powerful Technologies in Precision Medicine. *International Journal of General Medicine, 13*, 627–640. DOI: 10.2147/IJGM.S249970 PMID: 32982380

King, M. M., Lai, E. K., & McCay, P. B. (1975). Singlet oxygen production associated with enzyme-catalyzed lipid peroxidation in liver microsomes. *The Journal of Biological Chemistry, 250*(16), 6496–6502. DOI: 10.1016/S0021-9258(19)41092-2 PMID: 169247

Klassen, A., Faccio, A. T., Canuto, G. A. B., da Cruz, P. L. R., Ribeiro, H. C., Tavares, M. F. M., & Sussulini, A. (2017). Metabolomics: definitions and significance in systems biology. Metabolomics: from fundamentals to clinical applications, 3-17.

König, I. R., Fuchs, O., Hansen, G., Von Mutius, E., & Kopp, M. V. (2017, October). What is precision medicine? *The European Respiratory Journal, 50*(4), 1700391. DOI: 10.1183/13993003.00391-2017 PMID: 29051268

Leavy, O. (2014). Regulating ROS.(INFLAMMATION). [Internet]. *Nature Reviews. Immunology*, *14*(6), 357. Retrieved July 20, 2024, from https://go.gale.com/ps/i.do?id=GALE%7CA371843538&sid=googleScholar&v=2.1&it=r&linkaccess=abs&issn=14741733&p=AONE&sw=w. DOI: 10.1038/nri3685 PMID: 24798369

Lévy, E., El Banna, N., Baïlle, D., Heneman-Masurel, A., Truchet, S., Rezaei, H., Huang, M.-E., Béringue, V., Martin, D., & Vernis, L. (2019, August 9). Causative Links between Protein Aggregation and Oxidative Stress: A Review. *International Journal of Molecular Sciences*, *20*(16), 3896. DOI: 10.3390/ijms20163896 PMID: 31405050

Li, J. M., & Shah, A. M. (2004, November). Endothelial cell superoxide generation: Regulation and relevance for cardiovascular pathophysiology. *American Journal of Physiology. Regulatory, Integrative and Comparative Physiology*, *287*(5), R1014–R1030. DOI: 10.1152/ajpregu.00124.2004 PMID: 15475499

Li, J. M., & Shah, A. M. (2004). Endothelial cell superoxide generation: Regulation and relevance for cardiovascular pathophysiology. *American Journal of Physiology. Regulatory, Integrative and Comparative Physiology*, *287*(5), R1014–R1030. DOI: 10.1152/ajpregu.00124.2004 PMID: 15475499

Li, X., Fang, P., Mai, J., Choi, E. T., Wang, H., & Yang, X. (2013, December). Yang X feng. Targeting mitochondrial reactive oxygen species as novel therapy for inflammatory diseases and cancers. *Journal of Hematology & Oncology*, *6*(1), 19. DOI: 10.1186/1756-8722-6-19 PMID: 23442817

Li, X., Fang, P., Yang, W. Y., Chan, K., Lavallee, M., Xu, K., Gao, T., Wang, H., & Yang, X. (2017, March). Mitochondrial ROS, uncoupled from ATP synthesis, determine endothelial activation for both physiological recruitment of patrolling cells and pathological recruitment of inflammatory cells. *Canadian Journal of Physiology and Pharmacology*, *95*(3), 247–252. DOI: 10.1139/cjpp-2016-0515 PMID: 27925481

Liu, Z., Ren, Z., Zhang, J., Chuang, C. C., Kandaswamy, E., Zhou, T., & Zuo, L. (2018, May 17). Role of ROS and Nutritional Antioxidants in Human Diseases. *Frontiers in Physiology*, *9*, 477. DOI: 10.3389/fphys.2018.00477 PMID: 29867535

Luizon, M. R., Pereira, D. A., & Sandrim, V. C. (2018, February 28). Pharmacogenomics of Hypertension and Preeclampsia: Focus on Gene–Gene Interactions. *Frontiers in Pharmacology*, *9*, 168. DOI: 10.3389/fphar.2018.00168 PMID: 29541029

Luo, S., Lei, H., Qin, H., & Xia, Y. (2014). Molecular mechanisms of endothelial NO synthase uncoupling. *Current Pharmaceutical Design*, *20*(22), 3548–3553. DOI: 10.2174/13816128113196660746 PMID: 24180388

Malhotra, S., Miras, M. C. M., Pappolla, A., Montalban, X., & Comabella, M. (2023, January). Liquid Biopsy in Neurological Diseases. *Cells*, *12*(14), 1911. DOI: 10.3390/cells12141911 PMID: 37508574

Mandal, M., Sarkar, M., Khan, A., Biswas, M., Masi, A., Rakwal, R., Agrawal, G. K., Srivastava, A., & Sarkar, A. (2022, July 1). Reactive Oxygen Species (ROS) and Reactive Nitrogen Species (RNS) in plants– maintenance of structural individuality and functional blend. *Advances in Redox Research : an Official Journal of the Society for Redox Biology and Medicine and the Society for Free Radical Research-Europe*, *5*, 100039. DOI: 10.1016/j.arres.2022.100039

Medinas, D. B., Cerchiaro, G., Trindade, D. F., & Augusto, O. (2007, January). The carbonate radical and related oxidants derived from bicarbonate buffer. *IUBMB Life*, *59*(4–5), 255–262. DOI: 10.1080/15216540701230511 PMID: 17505962

Meertens, J. H., Nienhuis, H. L., Lefrandt, J. D., Schalkwijk, C. G., Nyyssönen, K., Ligtenberg, J. J. M., Smit, A. J., Zijlstra, J. G., & Mulder, D. J. (2016, August 16). The Course of Skin and Serum Biomarkers of Advanced Glycation Endproducts and Its Association with Oxidative Stress, Inflammation, Disease Severity, and Mortality during ICU Admission in Critically Ill Patients: Results from a Prospective Pilot Study. *PLoS One*, *11*(8), e0160893. DOI: 10.1371/journal.pone.0160893 PMID: 27529340

Mignolet-Spruyt, L., Xu, E., Idänheimo, N., Hoeberichts, F. A., Mühlenbock, P., Brosché, M., Van Breusegem, F., & Kangasjärvi, J. (2016, June). Spreading the news: Subcellular and organellar reactive oxygen species production and signalling. *Journal of Experimental Botany*, *67*(13), 3831–3844. DOI: 10.1093/jxb/erw080 PMID: 26976816

Millichap, L. E., Damiani, E., Tiano, L., & Hargreaves, I. P. (2021, January). Targetable Pathways for Alleviating Mitochondrial Dysfunction in Neurodegeneration of Metabolic and Non-Metabolic Diseases. *International Journal of Molecular Sciences*, *22*(21), 11444. DOI: 10.3390/ijms222111444 PMID: 34768878

Mittler, R. (2017, January). ROS Are Good. *Trends in Plant Science*, *22*(1), 11–19. DOI: 10.1016/j.tplants.2016.08.002 PMID: 27666517

Mittler, R., Vanderauwera, S., Gollery, M., & Van Breusegem, F. (2004, October). Reactive oxygen gene network of plants. *Trends in Plant Science*, *9*(10), 490–498. DOI: 10.1016/j.tplants.2004.08.009 PMID: 15465684

Moloney, J. N., & Cotter, T. G. (2018, August). ROS signalling in the biology of cancer. *Seminars in Cell & Developmental Biology*, *80*, 50–64. DOI: 10.1016/j.semcdb.2017.05.023 PMID: 28587975

Mortaz, E., Masjedi, M. R., Barnes, P. J., & Adcock, I. M. (2011). Epigenetics and chromatin remodeling play a role in lung disease. *Tanaffos*, *10*(4), 7–16. PMID: 25191382

Nalbantoglu, S., & Amri, H. (2019). *Molecular Medicine*. BoD – Books on Demand. DOI: 10.5772/intechopen.78531

Nishida, M., Maruyama, Y., Tanaka, R., Kontani, K., Nagao, T., & Kurose, H. (2000). Gαi and Gαo are target proteins of reactive oxygen species. *Nature*, *408*(6811), 492–495. DOI: 10.1038/35044120 PMID: 11100733

Noubade, R., Wong, K., Ota, N., Rutz, S., Eidenschenk, C., Valdez, P. A., Ding, J., Peng, I., Sebrell, A., Caplazi, P., DeVoss, J., Soriano, R. H., Sai, T., Lu, R., Modrusan, Z., Hackney, J., & Ouyang, W. (2014). NRROS negatively regulates reactive oxygen species during host defence and autoimmunity. *Nature*, *509*(7499), 235–239. DOI: 10.1038/nature13152 PMID: 24739962

Okada, S. S., de Oliveira, E. M., de Araújo, T. H., Rodrigues, M. R., Albuquerque, R. C., Mortara, R. A., Taniwaki, N. N., Nakaya, H. I., Campa, A., & Moreno, A. C. R. (2016). Myeloperoxidase in human peripheral blood lymphocytes: Production and subcellular localization. *Cellular Immunology*, *300*, 18–25. DOI: 10.1016/j.cellimm.2015.11.003 PMID: 26632272

Onyango, A. N. (2016). Endogenous Generation of Singlet Oxygen and Ozone in Human and Animal Tissues: Mechanisms, Biological Significance, and Influence of Dietary Components. *Oxidative Medicine and Cellular Longevity*, *2016*(1), 1–22. DOI: 10.1155/2016/2398573 PMID: 27042259

Onyango, A. N. (2017). The Contribution of Singlet Oxygen to Insulin Resistance. *Oxidative Medicine and Cellular Longevity*, *2017*(1), 1–14. DOI: 10.1155/2017/8765972 PMID: 29081894

Oyewole, A. O., & Birch-Machin, M. A. (2015, December). Mitochondria-targeted antioxidants. *The FASEB Journal*, *29*(12), 4766–4771. DOI: 10.1096/fj.15-275404 PMID: 26253366

Padmanabhan, S., Aman, A., & Dominiczak, A. F. (2018, June). Recent Findings in the Genetics of Blood Pressure: How to Apply in Practice or Is a Moonshot Required? *Current Hypertension Reports*, *20*(6), 54. DOI: 10.1007/s11906-018-0863-1 PMID: 29881931

Pham, C. G., Bubici, C., Zazzeroni, F., Papa, S., Jones, J., Alvarez, K., Jayawardena, S., De Smaele, E., Cong, R., Beaumont, C., Torti, F. M., Torti, S. V., & Franzoso, G. (2004). Ferritin heavy chain upregulation by NF-κB inhibits TNFα-induced apoptosis by suppressing reactive oxygen species. *Cell*, *119*(4), 529–542. DOI: 10.1016/j.cell.2004.10.017 PMID: 15537542

Popa-Wagner, A., Mitran, S., Sivanesan, S., Chang, E., & Buga, A. M. (2013). ROS and Brain Diseases: The Good, the Bad, and the Ugly. *Oxidative Medicine and Cellular Longevity*, *2013*, 1–14. DOI: 10.1155/2013/963520 PMID: 24381719

Prosser, B. L., Ward, C. W., & Lederer, W. J. (2011, September 9). X-ROS Signaling: Rapid Mechano-Chemo Transduction in Heart. *Science*, *333*(6048), 1440–1445. DOI: 10.1126/science.1202768 PMID: 21903813

Quinlan, C. L., Perevoshchikova, I. V., Hey-Mogensen, M., Orr, A. L., & Brand, M. D. (2013). Sites of reactive oxygen species generation by mitochondria oxidizing different substrates. *Redox Biology*, *1*(1), 304–312. DOI: 10.1016/j.redox.2013.04.005 PMID: 24024165

Radical-free biology of oxidative stress | American Journal of Physiology-Cell Physiology [Internet]. [cited 2024 Jul 18]. Available from: https://journals.physiology.org/doi/full/10.1152/ajpcell.00283.2008

Radisky, D. C., Levy, D. D., Littlepage, L. E., Liu, H., Nelson, C. M., Fata, J. E., Leake, D., Godden, E. L., Albertson, D. G., Angela Nieto, M., Werb, Z., & Bissell, M. J. (2005). Rac1b and reactive oxygen species mediate MMP-3-induced EMT and genomic instability. *Nature*, *436*(7047), 123–127. DOI: 10.1038/nature03688 PMID: 16001073

Rahman, M., Bilal, M., Shah, J. A., Kaushik, A., Teissedre, P.-L., & Kujawska, M. (2022, June). Rahman M ur, Bilal M, Shah JA, Kaushik A, Teissedre PL, Kujawska M. CRISPR-Cas9-Based Technology and Its Relevance to Gene Editing in Parkinson's Disease. *Pharmaceutics*, *14*(6), 1252. DOI: 10.3390/pharmaceutics14061252 PMID: 35745824

Reddy, P. H. (2006). Mitochondrial Oxidative Damage in Aging and Alzheimer's Disease: Implications for Mitochondrially Targeted Antioxidant Therapeutics. *BioMed Research International*, *2006*(1), 031372. DOI: 10.1155/JBB/2006/31372 PMID: 17047303

Redza-Dutordoir, M., & Averill-Bates, D. A. (2016, December). Activation of apoptosis signalling pathways by reactive oxygen species. *Biochimica et Biophysica Acta*, *1863*(12), 2977–2992. DOI: 10.1016/j.bbamcr.2016.09.012 PMID: 27646922

Rinnerthaler, M., Bischof, J., Streubel, M. K., Trost, A., & Richter, K. (2015). Oxidative stress in aging human skin. *Biomolecules*, *5*(2), 545–589. DOI: 10.3390/biom5020545 PMID: 25906193

Rivetti di Val Cervo, P., Besusso, D., Conforti, P., & Cattaneo, E. (2021, June). hiPSCs for predictive modelling of neurodegenerative diseases: Dreaming the possible. *Nature Reviews. Neurology*, *17*(6), 381–392. DOI: 10.1038/s41582-021-00465-0 PMID: 33658662

Roca, F. J., & Ramakrishnan, L. (2013). TNF dually mediates resistance and susceptibility to mycobacteria via mitochondrial reactive oxygen species. *Cell*, *153*(3), 521–534. DOI: 10.1016/j.cell.2013.03.022 PMID: 23582643

Ross, J. P., Dion, P. A., & Rouleau, G. A. (2020). Exome sequencing in genetic disease: Recent advances and considerations. [-Faculty.]. *F1000 Research*, *9*, F1000.

Roth, S. C. (2019, July). What is genomic medicine? *Journal of the Medical Library Association: JMLA*, *107*(3), 442–448. DOI: 10.5195/jmla.2019.604 PMID: 31258451

Sainero-Alcolado, L., Liaño-Pons, J., Ruiz-Pérez, M. V., & Arsenian-Henriksson, M. (2022, July). Targeting mitochondrial metabolism for precision medicine in cancer. *Cell Death and Differentiation*, *29*(7), 1304–1317. DOI: 10.1038/s41418-022-01022-y PMID: 35831624

Sakuma, S., Abe, M., Kohda, T., & Fujimoto, Y. (2015). Hydrogen peroxide generated by xanthine/xanthine oxidase system represses the proliferation of colorectal cancer cell line Caco-2. *Journal of Clinical Biochemistry and Nutrition*, *56*(1), 15–19. DOI: 10.3164/jcbn.14-34 PMID: 25678748

Sandalio, L. M., & Romero-Puertas, M. C. (2015). Peroxisomes sense and respond to environmental cues by regulating ROS and RNS signalling networks. *Annals of Botany*, *116*(4), 475–485. DOI: 10.1093/aob/mcv074 PMID: 26070643

Scortegagna, M., Ding, K., Oktay, Y., Gaur, A., Thurmond, F., Yan, L. J., Marck, B. T., Matsumoto, A. M., Shelton, J. M., Richardson, J. A., Bennett, M. J., & Garcia, J. A. (2003). Multiple organ pathology, metabolic abnormalities and impaired homeostasis of reactive oxygen species in Epas1-/- mice. *Nature Genetics*, *35*(4), 331–340. DOI: 10.1038/ng1266 PMID: 14608355

Sena, L. A., & Chandel, N. S. (2012, October). Physiological Roles of Mitochondrial Reactive Oxygen Species. *Molecular Cell*, *48*(2), 158–167. DOI: 10.1016/j.molcel.2012.09.025 PMID: 23102266

Shiu, Y. T., & Jaimes, E. A. (2018). Transcription factor ETS-1 and reactive oxygen species: Role in vascular and renal injury. *Antioxidants, 7*(7), 84. DOI: 10.3390/antiox7070084 PMID: 29970819

St-Pierre, J., Drori, S., Uldry, M., Silvaggi, J. M., Rhee, J., Jäger, S., Handschin, C., Zheng, K., Lin, J., Yang, W., Simon, D. K., Bachoo, R., & Spiegelman, B. M. (2006). Suppression of reactive oxygen species and neurodegeneration by the PGC-1 transcriptional coactivators. *Cell, 127*(2), 397–408. DOI: 10.1016/j.cell.2006.09.024 PMID: 17055439

Sun, Y., Lu, Y., Saredy, J., Wang, X., Drummer, I. V. C.IV, Shao, Y., Saaoud, F., Xu, K., Liu, M., Yang, W. Y., Jiang, X., Wang, H., & Yang, X. (2020, August 27). ROS systems are a new integrated network for sensing homeostasis and alarming stresses in organelle metabolic processes. *Redox Biology, 37*, 101696. DOI: 10.1016/j.redox.2020.101696 PMID: 32950427

Sun, Y., Lu, Y., Saredy, J., Wang, X., Drummer Iv, C., & Shao, Y.. (2020, October). ROS systems are a new integrated network for sensing homeostasis and alarming stresses in organelle metabolic processes. *Redox Biology, 37*, 101696. DOI: 10.1016/j.redox.2020.101696 PMID: 32950427

Teleanu, D. M., Niculescu, A. G., Lungu, I. I., Radu, C. I., Vladâcenco, O., Roza, E., Costăchescu, B., Grumezescu, A. M., & Teleanu, R. I. (2022, January). An Overview of Oxidative Stress, Neuroinflammation, and Neurodegenerative Diseases. *International Journal of Molecular Sciences, 23*(11), 5938. DOI: 10.3390/ijms23115938 PMID: 35682615

Tretter, V., Hochreiter, B., Zach, M. L., Krenn, K., & Klein, K. U. (2021). Understanding cellular redox homeostasis: A challenge for precision medicine. *International Journal of Molecular Sciences, 23*(1), 106. DOI: 10.3390/ijms23010106 PMID: 35008532

Tsukagoshi, H., Busch, W., & Benfey, P. N. (2010). Transcriptional regulation of ROS controls transition from proliferation to differentiation in the root. *Cell, 143*(4), 606–616. DOI: 10.1016/j.cell.2010.10.020 PMID: 21074051

Umeno, A., Biju, V., & Yoshida, Y. (2017, April 3). In vivo ROS production and use of oxidative stress-derived biomarkers to detect the onset of diseases such as Alzheimer's disease, Parkinson's disease, and diabetes. *Free Radical Research, 51*(4), 413–427. DOI: 10.1080/10715762.2017.1315114 PMID: 28372523

Valko, M., Leibfritz, D., Moncol, J., Cronin, M. T. D., Mazur, M., & Telser, J. (2007, January). Free radicals and antioxidants in normal physiological functions and human disease. *The International Journal of Biochemistry & Cell Biology*, *39*(1), 44–84. DOI: 10.1016/j.biocel.2006.07.001 PMID: 16978905

Yang, S., & Lian, G. (2020). ROS and diseases: Role in metabolism and energy supply. *Molecular and Cellular Biochemistry*, *467*(1), 1–12. DOI: 10.1007/s11010-019-03667-9 PMID: 31813106

Yu, B. P. (1994, January 1). Cellular defenses against damage from reactive oxygen species. *Physiological Reviews*, *74*(1), 139–162. DOI: 10.1152/physrev.1994.74.1.139 PMID: 8295932

Zanger, U. M. Pharmacogenetics – challenges and opportunities ahead. Front Pharmacol [Internet]. 2010 [cited 2024 Jul 19]; Available from: http://journal.frontiersin.org/article/10.3389/fphar.2010.00112/abstract

Zhang, H., Gomez, A. M., Wang, X., Yan, Y., Zheng, M., & Cheng, H. (2013, May 1). ROS regulation of microdomain Ca2+ signalling at the dyads. *Cardiovascular Research*, *98*(2), 248–258. DOI: 10.1093/cvr/cvt050 PMID: 23455546

Ziech, D., Franco, R., Pappa, A., & Panayiotidis, M. I. (2011, June). Reactive Oxygen Species (ROS)—Induced genetic and epigenetic alterations in human carcinogenesis. *Mutation Research. Fundamental and Molecular Mechanisms of Mutagenesis*, *711*(1–2), 167–173. DOI: 10.1016/j.mrfmmm.2011.02.015 PMID: 21419141

Zorov, D. B., Juhaszova, M., & Sollott, S. J. (2014, July). Mitochondrial Reactive Oxygen Species (ROS) and ROS-Induced ROS Release. *Physiological Reviews*, *94*(3), 909–950. DOI: 10.1152/physrev.00026.2013 PMID: 24987008

Zorov, D. B., Juhaszova, M., & Sollott, S. J. (2014, July). Mitochondrial Reactive Oxygen Species (ROS) and ROS-Induced ROS Release. *Physiological Reviews*, *94*(3), 909–950. DOI: 10.1152/physrev.00026.2013 PMID: 24987008

Chapter 15
Future Directions in Reactive Oxygen Species Research:
Translational Opportunities and Challenges

Suyash Saxena
https://orcid.org/0000-0003-3763-6830
Sharda University, India

Rahul Saxena
https://orcid.org/0000-0002-4517-8529
Sharda University, India

Ajit Pal Singh
https://orcid.org/0000-0002-3006-8474
Sharda University, India

Pranav Kumar Prabhakar
https://orcid.org/0000-0001-8130-1822
Lovely Professional University, India & Nagaland University, India

ABSTRACT

Reactive oxygen species (ROS) are crucial for immune defense and cellular signaling, but their overproduction can damage cells and contribute to diseases. Understanding this duality is essential for developing ROS-based therapies. Future treatments might target specific ROS types or their production mechanisms, offering promise for conditions like neurodegeneration, cardiovascular diseases, and certain cancers. Controlled ROS generation could enable targeted cancer cell killing or specific

DOI: 10.4018/979-8-3693-7919-6.ch015

signaling pathway activation. ROS-based biomarkers could aid in early disease diagnosis and treatment monitoring. However, challenges include the complexity of ROS biology, lack of sensitive detection tools, and difficulties in translating research to clinical settings. Advanced techniques for precise ROS measurement and visualization are needed, along with ethical considerations to ensure safety and efficacy. Collaborative efforts are vital to unlock the potential of ROS research for novel therapeutic approaches.

INTRODUCTION

Reactive oxygen species (ROS) are molecules with oxygen atoms that are reactive chemically. They play a dual role in cellular health and disease. On one hand, ROS are essential for normal cellular functions, including cell signaling and immune responses. On the other hand, excessive ROS generation can cause oxidative stress, which can harm lipids, proteins, and DNA in cells. It is assumed that ROS-mediated cellular signaling pathways and ROS homeostasis are old, highly conserved systems that are shared by the majority of organisms. On the other hand, new research emphasizes how intricate and adaptable ROS signaling is, especially in creatures that have evolved in harsh settings, (Hong *et al.,* 2024).

The possibility for creating new therapeutic approaches makes ROS biology research important for the future. Clinical consequences of targeting ROS can be substantial, particularly in the therapy of cancer. Elevated reactive oxygen species (ROS) disrupt cellular equilibrium and are associated with the initiation and progression of several malignancies. Dysregulation of ROS-related signaling pathways can lead to the development of cancer's hallmarks, including unregulated cell proliferation, defective DNA repair systems, and cell death avoidance[2]. It is now essential to research treatment strategies aimed at controlling ROS levels. These tactics include redox-modulating medications, antioxidant-based treatments, and procedures that restore normalcy to cells damaged by oxidative stress. Researchers aim to improve the effectiveness of cancer treatments by reestablishing the redox balance and reducing oxidative damage, (Iqbal *et al.,* 2024).

However, there are obstacles associated with ROS biology translational potential. Because of ROS's dual roles, toxicity, and bioavailability, antioxidant-based therapy has limits. Detailed investigation and clinical validation are required to provide customized treatments that accurately target the impact of oxidative stress on the start and spread of cancer. Furthermore, there is additional work to be done in order to understand the complex mechanisms underlying pathways associated to ROS. Despite these obstacles, there is considerable potential for preventing different forms of cancer by comprehending ROS's function in carcinogenesis and creating efficient strategies, (Iqbal *et al.,* 2024). In the future, scientists will keep delving into the biology of reactive oxygen species (ROS) in an effort to find novel ways to use them therapeutically.

TRANSLATIONAL OPPORTUNITIES IN ROS RESEARCH

Exciting prospects for expanding our knowledge and enhancing health outcomes exist in the realm of oxidative stress and redox biology translational research, (Tucker, Kang, & Williams, n.d.). There is a great deal of promise for new treatments when ROS research is used in the clinic, with an emphasis on ROS-based diagnostics, ROS-mediated therapies, and ROS-modulating medications, (Daiber, Lisa, & Ferdinandy, 2017).

A. ROS-Modulating Drugs:

a) Limitations of Traditional Antioxidant Approaches:

Traditional antioxidant therapies, such as vitamin C and E supplements, aim to neutralize ROS to prevent oxidative damage. However, these approaches have shown limited clinical success due to several reasons:

- **Non-Specific Action:** Traditional antioxidants do not distinguish between harmful and beneficial ROS. ROS play essential roles in cell signaling and homeostasis, and their complete neutralization can disrupt these processes, (Fudge *et al.,* 2016).
- **Bioavailability Issues:** Many antioxidants have poor bioavailability, meaning they do not reach sufficient concentrations in target tissues to exert therapeutic effects, (Fudge *et al.,* 2016).
- **Lack of targeting:** Antioxidants often fail to reach the specific cellular compartments where ROS are produced, leading to suboptimal therapeutic outcomes, (Fudge *et al.,* 2016).
- **Redox Signaling Disruption:** High doses of antioxidants may disrupt normal redox signaling pathways, (Roy *et al.,* 2023).

b) Development of Drugs Targeting Specific ROS Species or Production Pathways

To overcome the limitations of traditional antioxidants, researchers are developing drugs that specifically target ROS production pathways or particular ROS species:

- **NADPH Oxidase Inhibitors:** One of the main sources of ROS in cells is NADPH oxidase. Without interfering with other physiological functions, inhibitors of this enzyme can lower the generation of ROS, (Magnani & Mattevi, 2019). Preclinical research has demonstrated the potential of NADPH oxidase inhibitors such as apocynin and diphenyleneiodonium (DPI).
- **Mitochondria-Targeted Antioxidants:** Mitochondria are significant sources of ROS. Targeting antioxidants to mitochondria can more effectively reduce oxidative stress, (Oyewole & Birch-Machin, 2015). MitoQ and SkQ1 are examples of mitochondria-targeted antioxidants that have shown potential in various disease models, (Jiang *et al.,* 2020).
- **Selective ROS Scavengers:** Developing molecules that selectively scavenge specific ROS, such as superoxide or hydrogen peroxide, can help maintain the balance between beneficial and harmful ROS, (He *et al.,* 2017).

c) Potential Applications in Diseases:

- **Neurodegeneration:** Parkinson's and Alzheimer's disease etiology are linked to reactive oxygen species (ROS), (Rekatsina *et al.,* 2020). By focusing on the processes that produce ROS, one may lessen the oxidative damage that occurs to neurons and possibly decrease the course of the illness. For example, NADPH oxidase inhibitors have demonstrated potential in lowering neuroinflammation and neuronal death, (Subramanian, Kalyanaraman, & Migrino, 2010).
- **Cardiovascular Diseases:** Excessive ROS contribute to cardiovascular diseases by promoting inflammation, endothelial dysfunction, and atherosclerosis. ROS-modulating drugs can help restore vascular health and prevent disease progression. Mitochondria-targeted antioxidants, such as MitoQ, have shown potential in reducing oxidative damage in cardiovascular tissues, (Gorrini, Harris, & Mak, 2013).
- **Cancer:** Cancer cells often exhibit elevated ROS levels, which can be exploited for therapeutic purposes. Drugs that selectively increase ROS in cancer cells can induce apoptosis while sparing normal cells, (Wang *et al.,* 2021). For example, the use of pro-oxidant therapies, such as arsenic trioxide in acute promyelocytic leukemia, leverages ROS to kill cancer cells, (Szymanski, 2017).

B. ROS-Mediated Therapies

a) Controlled ROS Generation for Targeted Cancer Cell Destruction:

Controlled ROS generation is a promising strategy for targeted cancer therapy. Photodynamic therapy (PDT) is one such approach where light-activated compounds generate ROS specifically in cancer cells, leading to their destruction, (Filomeni & Ciriolo, 2006). This method minimizes damage to surrounding healthy tissues and enhances the precision of cancer treatment.

Mechanism: A photosensitizer that accumulates selectively in cancer cells is administered as part of PDT. The photosensitizer produces reactive oxygen species (ROS) in response to certain light wavelengths, which results in localized oxidative damage and cell death, (Filomeni & Ciriolo, 2006).

Advantages: PDT offers several advantages, including minimal invasiveness, targeted action, and the ability to treat tumors that are resistant to conventional therapies.

b) Inducing Specific Signaling Pathways for Therapeutic Purposes:

ROS participate in redox-sensitive signaling pathways (e.g., NF-κB, MAPK) involved in cell survival, proliferation, and apoptosis, (Li *et al.*, 2010). Modulating ROS levels can influence gene expression and cell behavior, (Saccaro *et al.*, 2021). By precisely controlling ROS levels, it is possible to activate specific pathways for therapeutic benefits:

Ischemic Preconditioning: Low levels of ROS can activate protective signaling pathways that enhance cell survival and tissue repair. This concept, known as ischemic preconditioning, has potential therapeutic applications in conditions like myocardial infarction and stroke, (Dunnill *et al.*, 2017).

Wound Healing: Controlled ROS generation can promote wound healing by stimulating cell proliferation and migration. ROS-modulating therapies are being explored to enhance the healing of chronic wounds and ulcers, (Lorente *et al.*, 2020).

C. Biomarkers Based on ROS

a) Value of ROS or Downstream Effects as Biomarkers:

ROS levels correlate with disease states. ROS and their downstream effects can serve as valuable biomarkers for disease diagnosis, monitoring, and treatment response evaluation:

- **Oxidative Stress Markers:** Malondialdehyde (MDA) and 8-oxo-deoxyguanosine (8-oxo-dG) are two examples of biomarkers that signify oxidative DNA and lipid damage, respectively. Increased concentrations of these indicators are linked to a number of illnesses, such as cancer, heart conditions, and neurological conditions, (Carmo de Carvalho *et al.,* 2022).
- **Antioxidant Enzyme Levels:** Superoxide dismutase (SOD) and glutathione peroxidase (GPx) are examples of antioxidant enzymes whose activity might indicate the oxidative state of tissues and aid in tracking the efficacy of antioxidant treatments, (Kontoghiorghes & Kontoghiorghe, 2019).

b) Personalized Medicine Based on Individual ROS Profiles:

Variability in ROS production exists among individuals due to genetics, lifestyle, and environmental factors. Understanding individual ROS profiles can pave the way for personalized medicine. By analyzing a patient's specific ROS-related biomarkers, tailored therapeutic strategies can be developed to optimize treatment efficacy and minimize side effects:

- **Personalized Antioxidant Therapy:** Patients with high oxidative stress levels may benefit from targeted antioxidant therapies, (Sies & Jones, 2020). Personalized approaches can help determine the appropriate type and dosage of antioxidants based on individual ROS profiles.
- **Monitoring Treatment Response:** Monitoring ROS biomarkers on a regular basis can assist in assessing the efficacy of treatments and modifying treatment regimens accordingly. Patients will receive the safest and most effective treatments thanks to this strategy.

CHALLENGES IN TRANSLATING ROS RESEARCH

There are numerous important obstacles to overcome in order to apply research on reactive oxygen species (ROS) to clinical practice. The intricacy of ROS biology is one of the main problems, since ROS can have both positive and negative effects based on their concentration and the cellular environment. This dual nature makes it difficult to develop therapies that can precisely target ROS pathways without causing unintended side effects. Additionally, reproducibility of preclinical findings is a major hurdle; many promising results in laboratory settings fail to translate into effective treatments in humans due to differences in biological systems. Ethical and regulatory considerations also pose challenges, as the long-term effects of manipulating ROS levels are not fully understood. Furthermore, funding and collaboration between basic scientists and clinicians are often insufficient, hindering the progress of translational research. Overcoming these obstacles requires a multidisciplinary approach, robust clinical trials, and enhanced communication between researchers and regulatory bodies.

A. Complexity of ROS biology

The complexity of reactive oxygen species (ROS) biology is a multifaceted challenge in scientific research. One of the primary difficulties lies in the transient and highly reactive nature of ROS. These molecules are produced and consumed rapidly within cells, making it hard to capture their precise roles and effects, (Di Meo & Venditti, 2020). For instance, superoxide anion, hydrogen peroxide, and hydroxyl radicals each have unique biological functions, but they often coexist and interact in ways that complicate the isolation of their individual contributions. This interplay can obscure the specific pathways and mechanisms through which each ROS exerts its influence.

i. Challenges in Isolating and Studying Specific Contributions of Diverse ROS Species

a) Diversity and Reactivity of ROS

ROS encompass a variety of molecules, including superoxide anion (O_2^-), hydrogen peroxide (H_2O_2), hydroxyl radical (OH^-), and singlet oxygen ($1O_2$). Each of these species has distinct chemical properties and biological effects. The high reactivity of ROS, particularly radicals like hydroxyl radical, makes them difficult to study as they react almost instantaneously with nearby molecules, (Di Meo *et al.,* 2016).

b) Overlapping and Redundant Pathways

ROS are involved in numerous overlapping and redundant signaling pathways. For instance, both superoxide and hydrogen peroxide can activate similar signaling cascades, making it challenging to delineate their specific contributions, (Iqbal *et al.,* 2024). Additionally, ROS can be produced by multiple cellular sources, such as mitochondria, NADPH oxidases, and peroxisomes, further complicating the identification of their precise roles, (Olson, 2020).

c) Interactions with Other Reactive Species

Reactive oxygen species (ROS) interact with other reactive species, including reactive nitrogen species (RNS) and reactive sulfur species (RSS), rather than acting alone, (Feelisch *et al.,* 2022). These interactions can lead to the formation of new reactive intermediates with distinct biological activities. Understanding the specific contributions of ROS requires disentangling these complex networks of interactions, (Coutu & Schroeder, 2013).

ii. Need for More Sophisticated Tools for ROS Detection and Quantification In Vivo

Moreover, the current tools for ROS detection and quantification are often inadequate for in vivo studies. Traditional methods, such as fluorescence-based probes, can lack the sensitivity and specificity required to accurately measure ROS levels in real-time within living organisms. These tools may also suffer from issues like photobleaching and interference from other cellular components, leading to unreliable data, (Uusitalo & Hempel, 2012). To address these limitations, there is a pressing need for more sophisticated technologies that can provide precise, real-time measurements of ROS in their native biological contexts.

One notable development is the use of nanoparticles and carbon nanotubes for intracellular and in vivo ROS sensing. These materials offer high sensitivity and specificity, allowing for more accurate detection of ROS levels within living cells, (Herman *et al.,* 2018). For example, nanoparticles can be engineered to respond to specific ROS, providing real-time data on their presence and activity, (Herman *et al.,* 2018).

Additionally, optical spectral detection techniques have seen considerable progress. Advances in this area include the use of self-organizing nanostructures and plasmonic nanoparticles, which enhance the sensitivity and resolution of ROS detection, (Wang, 2016). These technologies allow for the precise measurement of various ROS, such as singlet oxygen, superoxide, and hydrogen peroxide, in complex biological environments.

Another significant advancement is the development of luminescent probes. These probes emit light upon reacting with ROS, enabling researchers to visualize and quantify ROS in real-time, (Zorov, Juhaszova, & Sollott, 2006).

B. Spatial and temporal dynamics of ROS production

Translating ROS research into clinical applications is a multifaceted challenge, particularly when considering the spatial and temporal dynamics of ROS production. The importance of understanding ROS generation and effects within different cellular compartments and over time cannot be overstated. Each cellular compartment, such as mitochondria, peroxisomes, and the cytoplasm, has unique sources and regulatory mechanisms for ROS. For instance, mitochondrial ROS are crucial for signaling pathways that regulate apoptosis and energy metabolism, while cytoplasmic ROS play significant roles in modulating inflammatory responses. This compartmentalization means that ROS can have beneficial or detrimental effects depending on their location and concentration. The temporal dynamics of ROS are equally complex. ROS levels can change rapidly in response to various stimuli, such as oxidative stress, metabolic shifts, or environmental factors.

This temporal and spatial variability adds another layer of complexity to studying ROS. For example, mitochondrial ROS production can fluctuate rapidly in response to changes in cellular metabolism, and localized ROS bursts can occur in specific cellular regions (Yu *et al.,* 2022). Capturing these rapid and localized changes requires advanced techniques that can provide high temporal and spatial resolution.

Advances in imaging techniques, biosensors, and molecular probes are essential to unravel the complex dynamics of ROS and their roles in health and disease.

In addition to improving our knowledge of ROS biology, the development of these cutting-edge instruments and techniques will open the door for the development of fresh treatment approaches that address disorders linked to oxidative stress. This requires a concerted effort from researchers across disciplines, combining expertise in chemistry, biology, and engineering to innovate and refine the tools needed for ROS research.

- **Advancements in the technology to study spatiotemporal dynamics**

Recent breakthroughs in ROS detection technology have significantly advanced our ability to study these reactive molecules in biological systems. Recent innovations include the creation of hydroxyl radical-responsive nanoprobes, which have been used to study the association between ROS and cancer development, (Teixeira, Galland, & Chevrollier, 2024). These luminescent probes are particularly valuable for their ability to provide spatial and temporal information about ROS dynamics in living tissues.

These breakthroughs are paving the way for a deeper understanding of ROS biology and their roles in health and disease. They may also be used to create novel treatment approaches for diseases linked to oxidative stress.

Researchers are developing more sophisticated tools. Super-resolution microscopy allows for the visualization of ROS at a much finer scale than conventional microscopy, enabling the study of ROS dynamics within specific cellular compartments, (Verma *et al.,* 2023). Fluorescence resonance energy transfer (FRET)-based sensors are another promising technology, providing real-time data on ROS interactions and concentrations with high specificity, (Wu *et al.,* 2021). These sensors can be designed to respond to specific ROS, such as hydrogen peroxide or superoxide, allowing for precise measurements in living cells.

Additionally, biosensors and molecular probes are being engineered to improve the detection and quantification of ROS. These tools can provide insights into the spatial and temporal patterns of ROS production, helping to elucidate their roles in various physiological and pathological processes. For example, recent studies have used luminescent probes to track hydroxyl radicals in cancer cells, revealing how ROS contribute to tumor progression and resistance to therapy, (Kostyuk *et al.*, 2020).

Genetically encoded ROS sensors, such as roGFP and HyPer, have emerged as powerful tools for studying ROS in vivo[38]. These sensors can be targeted to specific cellular compartments, providing insights into the spatial dynamics of ROS production. Despite their advantages, genetically encoded sensors can be challenging to use due to potential issues with expression levels, targeting accuracy, and interference with cellular functions, (Seyhan, 2019).

Despite these advancements, significant hurdles remain. The development of these advanced imaging techniques requires interdisciplinary collaboration, combining expertise in chemistry, biology, and engineering. Furthermore, translating these findings into clinical practice involves rigorous validation through preclinical and clinical trials to ensure safety and efficacy.

Overall, understanding the spatial and temporal dynamics of ROS production is essential for developing targeted therapies that can modulate ROS levels in specific cellular contexts. This information will open the door to new treatments for conditions like cancer, neurological problems, and cardiovascular ailments that are brought on by oxidative stress.

C. Translation from In Vitro to In Vivo Models

Translating findings from the ROS researches in in vitro studies to in vivo models presents significant challenges. Let's explore the limitations of cell culture studies and the importance of developing sophisticated animal models to better mimic human diseases.

a) Limitations of Cell Culture Studies

Cell culture studies are invaluable for understanding the basic mechanisms of ROS and their effects on cellular functions. These studies offer a controlled environment where variables can be precisely manipulated, allowing for detailed mechanistic insights. However, several limitations hinder their ability to fully capture the complex interactions within a living organism.

- **Lack of Systemic Interactions:** In vitro models frequently fall short in simulating the complex interactions among many cell types, tissues, and organs.

For example, it is challenging to replicate in a petri dish the immune system's function in controlling ROS levels and reactions, (Long *et al.,* 2022). This lack of systemic interaction can lead to an oversimplified understanding of ROS dynamics.

- **Simplified Microenvironment:** The cellular microenvironment in vitro is vastly different from that in vivo. Factors such as extracellular matrix components, mechanical forces, and nutrient gradients are not accurately represented in cell culture systems, (Bodenstein *et al.*, 2024). These elements are crucial for cellular behavior and ROS signaling pathways.
- **Absence of Metabolic Complexity:** Metabolic interactions and the influence of systemic factors like hormones and cytokines are absent in in vitro models. These factors significantly impact ROS production and scavenging mechanisms, (Mello, Goodman, & Faden, 2012). Consequently, findings from cell culture studies may not translate well to the metabolic complexity of a whole organism.
- **Static Conditions:** In vitro studies often involve static conditions, whereas in vivo environments are dynamic. The constant changes in blood flow, oxygen levels, and other physiological parameters in a living organism can alter ROS behavior, (Mello, Goodman, & Faden, 2012). This discrepancy can lead to differences in ROS effects observed in vitro versus in vivo.

b) Importance of Developing Sophisticated Animal Models

To bridge the gap between in vitro findings and clinical applications, it is essential to develop sophisticated animal models that closely mimic human diseases. These models can provide a more accurate representation of the complex biological systems involved in ROS-related pathologies.

- **Mimicking Human Pathophysiology:** Animal models, particularly genetically engineered ones, can be designed to replicate specific human diseases. For example, mouse models with targeted gene deletions or mutations can mimic human conditions like neurodegenerative diseases or cancer, (Bodenstein *et al.,* 2024). These models help in understanding how ROS contribute to disease progression in a context that is more relevant to human physiology.
- **Systemic Interactions:** Animal models allow for the study of systemic interactions that are impossible to replicate in vitro. The interplay between different organs, the immune system, and metabolic pathways can be observed in a living organism. This all-encompassing method offers a deeper comprehension of ROS dynamics and their function in illness.

- **Translational Relevance:** Findings from animal models are more likely to be translatable to human clinical settings. The effectiveness and safety of prospective treatment strategies that target ROS can be evaluated using these models. By closely mimicking human diseases, animal models increase the likelihood of successful translation from bench to bedside.
- **Technological Advancements:** Advances in imaging and molecular biology techniques have enhanced the ability to study ROS in vivo. Techniques such as live imaging, optogenetics, and CRISPR/Cas9 gene editing allow for precise manipulation and observation of ROS-related processes in animal models. Deeper understanding of the intricate functions ROS play in health and illness is made possible by these technologies.

Although in vitro research has greatly improved our knowledge of ROS, their limitations necessitate the development of sophisticated animal models to capture the complexity of living organisms. By addressing the challenges of translating in vitro findings to in vivo models, researchers can better understand the role of ROS in human diseases and develop more effective therapeutic strategies.

ETHICAL CONSIDERATIONS IN FUTURE DIRECTIONS OF REACTIVE OXYGEN SPECIES RESEARCH

Reactive Oxygen Species (ROS) have garnered significant attention in biomedical research due to their dual role in cellular signaling and oxidative stress. As research progresses towards developing ROS-based therapies, it is crucial to address the ethical considerations associated with their safety, efficacy, and potential side effects. There is a dire need for careful evaluation of ROS-based therapies and the importance of rigorous clinical trials.

A. Need for Careful Evaluation of Safety and Efficacy of ROS-Based Therapies

ROS-based therapeutics are promising for the treatment of a number of illnesses, such as cardiovascular problems, neurological diseases, and cancer. However, the inherent reactivity of ROS necessitates a thorough evaluation of their safety and efficacy.

a) **Balancing Therapeutic Benefits and Risks:** ROS can induce cell death in cancer cells but may also cause damage to healthy tissues. The therapeutic window—the range of doses that are effective without being toxic—must be carefully determined. Ethical research mandates that the potential benefits of ROS-based therapies outweigh the risks of adverse effects, (Jhonson & Smith, 2023).

b) **Long-Term Safety:** It's unclear what the long-term implications of ROS-based treatments will be. Chronic exposure to ROS could lead to cumulative oxidative damage, potentially causing secondary health issues. Ethical considerations require that long-term safety studies be conducted to ensure that therapies do not pose unforeseen risks to patients, (Rivera *et al.,* 2022).

c) **Individual Variability:** Patients may respond differently to ROS-based therapies due to genetic, environmental, and lifestyle factors. Personalized approaches that consider individual variability are essential to maximize efficacy and minimize harm. Ethical research practices must include diverse populations in clinical studies to ensure that findings are generalizable and inclusive, (Goldstein *et al.,* 2018).

B. Importance of Rigorous Clinical Trials and Understanding Potential Side Effects

The transition from preclinical studies to clinical applications involves rigorous testing to ensure that ROS-based therapies are both safe and effective for human use.

a) **Phased Clinical Trials:** Ethical research requires a phased approach to clinical trials, with smaller-scale Phase I trials to determine safety and dosage and larger-scale Phase II and III studies to determine efficacy and track adverse effects. Each phase must be meticulously designed to gather comprehensive data on the therapy's impact.

b) **Informed Consent:** Clinical trial participants need to be thoroughly informed about the possible dangers and advantages of ROS-based therapy. A fundamental component of ethical research is informed consent, which guarantees that subjects freely consent to participate in the study after being fully informed of its nature.

c) **Monitoring and Reporting Side Effects:** Continuous monitoring for adverse effects is crucial during clinical trials. Ethical considerations demand that all side effects, both expected and unexpected, be promptly reported and addressed. This transparency is vital for protecting participants and refining therapeutic approaches· (Goldstein *et al.,* 2018).

d) **Post-Market Surveillance:** Even after therapy is approved, continuous monitoring is required to identify any rare or long-term negative effects that might not have been noticeable during clinical trials. Ethical responsibility extends beyond approval, requiring researchers and healthcare providers to remain vigilant in monitoring patient outcomes· (Goldstein *et al.*, 2018).

The development of ROS-based therapies offers exciting possibilities for treating various diseases. However, ethical considerations are paramount in ensuring that these therapies are safe, effective, and beneficial for patients. Careful evaluation of safety and efficacy, rigorous clinical trials, and a commitment to understanding potential side effects are essential components of ethical ROS research. By adhering to these principles, researchers can advance the field while safeguarding the well-being of patients.

CONCLUSION

The future directions in Reactive Oxygen Species (ROS) research present both exciting translational opportunities and significant challenges. Because ROS are involved in both health and illness in cells, they are an important target for therapeutic therapies. The potential for developing novel therapies, particularly in cancer treatment, underscores the importance of continued research in this field.

Translational opportunities in ROS research include the development of ROS-modulating drugs, ROS-mediated therapies, and the use of ROS-based biomarkers. Traditional antioxidant approaches have shown limited success due to issues such as non-specific action, poor bioavailability, and disruption of redox signaling. However, advancements in targeting specific ROS species or production pathways offer promising alternatives. These targeted interventions can potentially address the limitations of traditional antioxidants and provide more effective treatments for diseases like neurodegeneration, cardiovascular diseases, and cancer.

OS-mediated therapies, such as controlled ROS generation for targeted cancer cell destruction and the induction of specific signaling pathways for therapeutic purposes, represent innovative approaches that leverage the unique properties of ROS. These therapies can enhance the precision and efficacy of treatments, particularly in oncology.

The use of ROS and their downstream effects as biomarkers holds significant promise for disease diagnosis, monitoring, and treatment response evaluation. Personalized medicine, based on individual ROS profiles, can optimize therapeutic strategies and improve patient outcomes.

Despite these opportunities, the complexity of ROS biology presents substantial challenges. Isolating and studying specific ROS species is complicated by their diversity, reactivity, and interactions with other reactive species. The dynamic nature of ROS production and scavenging requires advanced tools for accurate detection and quantification. Recent advances in fluorescent probes, genetically encoded sensors, nanotechnology, and imaging techniques offer promising solutions, but further development and refinement are needed.

In conclusion, the translational potential of ROS research is vast, with the promise of novel therapies and improved health outcomes. However, addressing the inherent challenges in ROS biology is crucial for realizing these opportunities. Continued research, technological advancements, and rigorous clinical validation will be essential in harnessing the therapeutic potential of ROS and advancing the field of redox biology.

REFERENCES

Bodenstein, D. F., Siebiger, G., Zhao, Y., Clasky, A. J., Mukkala, A. N., Beroncal, E. L., Banh, L., Aslostovar, L., Brijbassi, S., Hogan, S. E., McCully, J. D., Mehrabian, M., Petersen, T. H., Robinson, L. A., Walker, M., Zachos, C., Viswanathan, S., Gu, F. X., Rotstein, O. D., & Andreazza, A. C. (2024, May 31). Bridging the gap between in vitro and in vivo models: A way forward to clinical translation of mitochondrial transplantation in acute disease states. *Stem Cell Research & Therapy*, *15*(1), 157. DOI: 10.1186/s13287-024-03771-8 PMID: 38816774

Carmo de Carvalho e Martins. M. D., Martins, da Silva Santos Oliveira, A. S., da Silva, L. A. A., Primo, M. G. S., & de Carvalho Lira, V. B. (2022). Biological indicators of oxidative stress [malondialdehyde, catalase, glutathione peroxidase, and superoxide dismutase] and their application in nutrition. In Biomarkers in Nutrition (pp. 1-25). Cham: Springer International Publishing.

Coutu, D. L., & Schroeder, T. (2013, September 1). Probing cellular processes by long-term live imaging–historic problems and current solutions. *Journal of Cell Science*, *126*(17), 3805–3815. DOI: 10.1242/jcs.118349 PMID: 23943879

Daiber, A., Di Lisa, F., & Ferdinandy, P. (2017, June). Pharmacology of oxidative stress: Translational opportunities. *British Journal of Pharmacology*, *174*(12), 1511–1513. DOI: 10.1111/bph.13795 PMID: 28555779

Di Meo, S., Reed, T. T., Venditti, P., & Victor, V. M. (2016). Role of ROS and RNS sources in physiological and pathological conditions. *Oxidative Medicine and Cellular Longevity*, *2016*(1), 1245049. DOI: 10.1155/2016/1245049 PMID: 27478531

Di Meo, S., & Venditti, P. (2020). Evolution of the knowledge of free radicals and other oxidants. *Oxidative Medicine and Cellular Longevity*, *2020*(1), 9829176. DOI: 10.1155/2020/9829176 PMID: 32411336

Dunnill, C., Patton, T., Brennan, J., Barrett, J., Dryden, M., Cooke, J., Leaper, D., & Georgopoulos, N. T. (2017, February). Reactive oxygen species (ROS) and wound healing: The functional role of ROS and emerging ROS-modulating technologies for augmentation of the healing process. *International Wound Journal*, *14*(1), 89–96. DOI: 10.1111/iwj.12557 PMID: 26688157

Feelisch, M., Cortese-Krott, M. M., Santolini, J., Wootton, S. A., & Jackson, A. A. (2022). Systems redox biology in health and disease. *EXCLI Journal*, *21*, 623. PMID: 35721574

Filomeni, G., & Ciriolo, M. R. (2006, November 1). Redox control of apoptosis: An update. *Antioxidants & Redox Signaling*, *8*(11-12), 2187–2192. DOI: 10.1089/ars.2006.8.2187 PMID: 17034362

Fudge, N., Sadler, E., Fisher, H. R., Maher, J., Wolfe, C. D., & McKevitt, C. (2016, August 4). Optimising translational research opportunities: A systematic review and narrative synthesis of basic and clinician Scientists' perspectives of factors which enable or hinder translational research. *PLoS One*, *11*(8), e0160475. DOI: 10.1371/journal.pone.0160475 PMID: 27490373

Goldstein, C. E., Weijer, C., Brehaut, J. C., Fergusson, D. A., Grimshaw, J. M., Horn, A. R., & Taljaard, M. (2018, December). Ethical issues in pragmatic randomized controlled trials: A review of the recent literature identifies gaps in ethical argumentation. *BMC Medical Ethics*, *19*(1), 1–0. DOI: 10.1186/s12910-018-0253-x PMID: 29482537

Gorrini, C., Harris, I. S., & Mak, T. W. (2013, December). Modulation of oxidative stress as an anticancer strategy. *Nature Reviews. Drug Discovery*, *12*(12), 931–947. DOI: 10.1038/nrd4002 PMID: 24287781

He, L., He, T., Farrar, S., Ji, L., Liu, T., & Ma, X. (2017). Antioxidants maintain cellular redox homeostasis by elimination of reactive oxygen species. *Cellular Physiology and Biochemistry*, *44*(2), 532–553.

Herman, J., Zhang, Y., Castranova, V., & Neal, S. L. (2018, September). Emerging technologies for optical spectral detection of reactive oxygen species. *Analytical and Bioanalytical Chemistry*, *410*(24), 6079–6095. DOI: 10.1007/s00216-018-1233-1 PMID: 30054693

Hong, Y., Boiti, A., Vallone, D., & Foulkes, N. S. (2024, March 1). Reactive Oxygen Species Signaling and Oxidative Stress: Transcriptional Regulation and Evolution. *Antioxidants*, *13*(3), 312. DOI: 10.3390/antiox13030312 PMID: 38539845

Iqbal, M. J., Kabeer, A., Abbas, Z., Siddiqui, H. A., Calina, D., Sharifi-Rad, J., & Cho, W. C. (2024, January 2). Interplay of oxidative stress, cellular communication and signaling pathways in cancer. *Cell Communication and Signaling : CCS*, *22*(1), 7. DOI: 10.1186/s12964-023-01398-5 PMID: 38167159

Jhonson, D., & Smith, J. Upholding Ethics in Medical Science: A Critical Examination of Progress and Patient Safety.

Jiang, Q., Yin, J., Chen, J., Ma, X., Wu, M., Liu, G., Yao, K., Tan, B., & Yin, Y. (2020). Mitochondria-targeted antioxidants: A step towards disease treatment. *Oxidative Medicine and Cellular Longevity*, *2020*(1), 8837893. DOI: 10.1155/2020/8837893 PMID: 33354280

Kontoghiorghes, G. J., & Kontoghiorghe, C. N. (2019, July 3). Prospects for the introduction of targeted antioxidant drugs for the prevention and treatment of diseases related to free radical pathology. *Expert Opinion on Investigational Drugs*, *28*(7), 593–603. DOI: 10.1080/13543784.2019.1631284 PMID: 31185180

Kostyuk, A. I., Panova, A. S., Kokova, A. D., Kotova, D. A., Maltsev, D. I., Podgorny, O. V., Belousov, V. V., & Bilan, D. S. (2020, October 31). In vivo imaging with genetically encoded redox biosensors. *International Journal of Molecular Sciences*, *21*(21), 8164. DOI: 10.3390/ijms21218164 PMID: 33142884

Li, L., Cheung, S. H., Evans, E. L., & Shaw, P. E. (2010, October 15). Modulation of gene expression and tumor cell growth by redox modification of STAT3. *Cancer Research*, *70*(20), 8222–8232. DOI: 10.1158/0008-5472.CAN-10-0894 PMID: 20807804

Long, Y., Xie, B., Shen, H. C., & Wen, D. (2022, November 30). Translation potential and challenges of in vitro and murine models in cancer clinic. *Cells*, *11*(23), 3868. DOI: 10.3390/cells11233868 PMID: 36497126

Lorente, L., Martín, M. M., González-Rivero, A. F., Pérez-Cejas, A., Abreu-González, P., Sabatel, R., Ramos, L., Argueso, M., Cáceres, J. J., Solé-Violán, J., Jiménez, A., & García-Marín, V. (2020, August). High serum DNA and RNA oxidative damage in non-surviving patients with spontaneous intracerebral hemorrhage. *Neurocritical Care*, *33*(1), 90–96. DOI: 10.1007/s12028-019-00864-8 PMID: 31598840

Magnani, F., & Mattevi, A. (2019, December 1). Structure and mechanisms of ROS generation by NADPH oxidases. *Current Opinion in Structural Biology*, *59*, 91–97. DOI: 10.1016/j.sbi.2019.03.001 PMID: 31051297

Mello, M. M., Goodman, S. N., & Faden, R. R. (2012, September 6). Ethical considerations in studying drug safety—The Institute of Medicine report. *The New England Journal of Medicine*, *367*(10), 959–964. DOI: 10.1056/NEJMhle1207160 PMID: 22913661

Oladele, J. O., Oladiji, A. T., Oladele, O. T., & Oyeleke, O. M. (2021). Reactive oxygen species in neurodegenerative diseases: Implications in pathogenesis and treatment strategies. Reactive Oxygen Species, 149-166.

Olson, K. R. (2020, December 1). Are reactive sulfur species the new reactive oxygen species? *Antioxidants & Redox Signaling*, *33*(16), 1125–1142. DOI: 10.1089/ars.2020.8132 PMID: 32586118

Oyewole, A. O., & Birch-Machin, M. A. (2015, December). Mitochondria-targeted antioxidants. *The FASEB Journal*, *29*(12), 4766–4771. DOI: 10.1096/fj.15-275404 PMID: 26253366

Rekatsina, M., Paladini, A., Piroli, A., Zis, P., Pergolizzi, J. V., & Varrassi, G. (2020, January). Pathophysiology and therapeutic perspectives of oxidative stress and neurodegenerative diseases: A narrative review. *Advances in Therapy*, *37*(1), 113–139. DOI: 10.1007/s12325-019-01148-5 PMID: 31782132

Rivera, S. C., Aiyegbusi, O. L., Ives, J., Draper, H., Mercieca-Bebber, R., Ells, C., Hunn, A., Scott, J. A., Fernandez, C. V., Dickens, A. P., & Anderson, N. (2022, May 17). Ethical considerations for the inclusion of patient-reported outcomes in clinical research: The PRO ethics guidelines. *Journal of the American Medical Association*, *327*(19), 1910–1919. DOI: 10.1001/jama.2022.6421 PMID: 35579638

Roy, Z., Bansal, R., Siddiqui, L., & Chaudhary, N. (2023, August 1). Understanding the role of free radicals and antioxidant enzymes in human diseases. *Current Pharmaceutical Biotechnology*, *24*(10), 1265–1276. DOI: 10.2174/1389201024666221121160822 PMID: 36411571

Saccaro, L. F., Aimo, A., Emdin, M., & Pico, F. (2021, October 26). Remote ischemic conditioning in ischemic stroke and myocardial infarction: Similarities and differences. *Frontiers in Neurology*, *12*, 716316. DOI: 10.3389/fneur.2021.716316 PMID: 34764925

Sadoine, M., Ishikawa, Y., Kleist, T. J., Wudick, M. M., Nakamura, M., Grossmann, G., Frommer, W. B., & Ho, C. H. (2021, October 1). Designs, applications, and limitations of genetically encoded fluorescent sensors to explore plant biology. *Plant Physiology*, *187*(2), 485–503. DOI: 10.1093/plphys/kiab353 PMID: 35237822

Seyhan, A. A. (2019, December). Lost in translation: The valley of death across preclinical and clinical divide–identification of problems and overcoming obstacles. *Translational Medicine Communications*, *4*(1), 1–9. DOI: 10.1186/s41231-019-0050-7

Sies, H., & Jones, D. P. (2020, July 15). Reactive oxygen species (ROS) as pleiotropic physiological signalling agents. *Nature Reviews. Molecular Cell Biology*, *21*(7), 363–383. DOI: 10.1038/s41580-020-0230-3 PMID: 32231263

Subramanian, S., Kalyanaraman, B., & Migrino, R. Q. (2010, January 1). Mitochondrially targeted antioxidants for the treatment of cardiovascular diseases. *Recent Advances in Cardiovascular Drug Discovery*, *5*(1), 54–65. DOI: 10.2174/157489010790192601 PMID: 19807685

Szymanski, W. (2017, December 1). Beyond photodynamic therapy: Light-activated cancer chemotherapy. *Current Medicinal Chemistry*, *24*(42), 4905–4950. PMID: 27601187

Teixeira, P., Galland, R., & Chevrollier, A. (2024, June). Super-resolution microscopies, technological breakthrough to decipher mitochondrial structure and dynamic. [). Academic Press.]. *Seminars in Cell & Developmental Biology*, *159*, 38–51.

Tucker, C. M., Kang, S., & Williams, J. L. (2019, December). Translational research to reduce health disparities and promote health equity. *Translational Issues in Psychological Science*, *5*(4), 297–301. DOI: 10.1037/tps0000215

Uusitalo, L. M., & Hempel, N. (2012, August 24). Recent advances in intracellular and in vivo ROS sensing: Focus on nanoparticle and nanotube applications. *International Journal of Molecular Sciences*, *13*(9), 10660–10679. DOI: 10.3390/ijms130910660 PMID: 23109815

Verma, A. K., Noumani, A., Yadav, A. K., & Solanki, P. R. (2023, April 8). FRET based biosensor: Principle applications recent advances and challenges. *Diagnostics (Basel)*, *13*(8), 1375. DOI: 10.3390/diagnostics13081375 PMID: 37189476

Wang, H. S. (2016, December 1). Development of fluorescent and luminescent probes for reactive oxygen species. *Trends in Analytical Chemistry*, *85*, 181–202. DOI: 10.1016/j.trac.2016.09.006

Wang, Y., Qi, H., Liu, Y., Duan, C., Liu, X., Xia, T., Chen, D., Piao, H. L., & Liu, H. X. (2021). The double-edged roles of ROS in cancer prevention and therapy. *Theranostics*, *11*(10), 4839–4857. DOI: 10.7150/thno.56747 PMID: 33754031

Wu, L., Ishigaki, Y., Zeng, W., Harimoto, T., Yin, B., Chen, Y., Liao, S., Liu, Y., Sun, Y., Zhang, X., Liu, Y., Liang, Y., Sun, P., Suzuki, T., Song, G., Fan, Q., & Ye, D. (2021, October 22). Generation of hydroxyl radical-activatable ratiometric near-infrared bimodal probes for early monitoring of tumor response to therapy. *Nature Communications*, *12*(1), 6145. DOI: 10.1038/s41467-021-26380-y PMID: 34686685

Yu, X., Ouyang, W., Qiu, H., Zhang, Z., Wang, Z., & Xing, B. (2022, November 21). Detection of reactive oxygen and nitrogen species by upconversion nanoparticle-based near-infrared nanoprobes: Recent progress and perspectives. *Chemistry (Weinheim an der Bergstrasse, Germany), 28*(65), e202201966. DOI: 10.1002/chem.202201966 PMID: 35989566

Zorov, D. B., Juhaszova, M., & Sollott, S. J. (2006, May 1). Mitochondrial ROS-induced ROS release: An update and review. Biochimica et Biophysica Acta (BBA)-. *Biochimica et Biophysica Acta. Bioenergetics, 1757*(5-6), 509–517. DOI: 10.1016/j.bbabio.2006.04.029

About the Contributors

Pranav Kumar Prabhakar is currently working as Professor and Head at Research and Development Cell, Lovely Professional University, Punjab, India. He is one among the World's Top 2% Scientists (list published by Stanford University, USA, 2021 and 2022). He has completed his Ph.D. in Biotechnology from IIT Madras. His main area of research interest focuses on elucidating molecular mechanism and strategies for oral insulin delivery and mimics the signaling pathways in metabolic disorders (diabetes) with natural products. He is Member of Royal Society of Chemistry, and Asia-Pacific Chemical, Biological & Environmental Engineering Society. He is also serving as Editorial Board Member and Reviewer for many reputed national and international journals. He received honors and is Recipient of travel grant toward attending ATTD 2009 in Greece from Indian Institute of Technology Madras & Council for Scientific and Industrial Research (CSIR) and approved by Department of Science and Technology (DST). He has published 75+ research articles in journals, authored 3 books, and 14 chapters. He also delivered 9 oral and poster presentations in scientific meeting.

***

Banashree Bondhopadhyay The corresponding author has 13 years of experience working in the field of breast cancer biology and etiology. The PI has an in vitro breast cancer cell project with basic cell culture equipment like an inverted phase contrast microscope, UV-Vis spectrophotometer, and centrifuge. Additionally, the PI gets practical training in the following areas: * Seek noble genes beyond BRCA1/BRCA2 using genetic landscape studies. • Use meta-analysis methods to explore tumour suppressors in TNBCs. TNBCs are rare breast cancer subtypes with few treatments. Indian breast cancer findings were validated with METABRICS and TCGA data. We reported an inverse connection between PML and RB1 gene expression in TNBCs in Meta Gene [IF 3.913]. * Interested in studying the impact of commercialized OCPs (Levonorgestral) on breast cancer cell lines. * Researched Ets-1 transcription factor as a possible biomarker for breast cancer and published findings in Journal of Cellular Biochemistry [IF 4.48]. * We published a review in Cancers [IF 6.639] on tiny extracellular vesicles and their function in breast cancer. The review discussed how exosomes contribute to breast carcinogenesis and how current immunotherapeutics can target them for promising interventions. •Another American Journal of Blood Research review [IF 5.942] highlights a promising and futuristic immunotherapeutic method in combination with conventional treatment. We investigated the potential cell membrane-bound action of TLR-3 in breast cancer to support its pro-tumor effect. Poly (I:C) triggered TLR-3 in MCF-7 and MDA-MB-231 breast cancer cells. The delivery of Poly (I:C) in the cytoplasm via lipid transfecting medium lowered cell survival. Our study provided the first functional proof that TLR-3 increases cell survival and metastasis via cell surface. * Researched that knocking down uPA and MMP9 simultaneously can slow breast cancer progression by enhancing cell-cell adhesion and influencing EMT genes. •Additional studies included small compounds and anti-cancer drugs in vitro cancer cell lines.

Meenakshi Dhanawat, PhD, M.S. (Pharm.)-NIPER, Mohali, PhD- IIT, BHU, Varanasi Dr Meenakshi have a background in Pharmaceutical Sciences with a specialization in Medicinal Chemistry. She has done Masters from NIPER, Mohali and the project was based on Glycosylamino acids. Her PhD is from IIT-BHU, Varanasi and she worked on CNS drug delivery targeting 5HT1A/2A receptor. Apart from that she is having a vast experience in extraction of herbal drugs. Area of interest are Synthetic Chemistry, Heterocyclic Chemistry, Nanotechnology, Drug Delivery, Prodrug and Co-Drug Drug Design Complexation, Polymer Chemistry, and their application in drug delivery. She is currently working as Associate Professor in Amity Institute of Pharmacy, Amity University Haryana. She has 17 years of teaching and research experience. She has attended several national and international conferences in india and abroad the recipient of the travelship award in DDD section of AAPS-2010, National Biotechnology Conference, California, USA, She has been able to grab the international travel grant from DBT and DST for attending the International Conference at California, USA and Munich, Germany Respectively. She has also grabbed many other fellowships awards, which includes Rajiv Gandhi National Fellowship. She has Published 67 Research/Reviews along with Patent-05, Books 6, Book Chapter 09, Book Editor 02, Guest Editor Thematic Issue 01, Copyright 01.

Simon Akpomemera Erho is a lecturer and Examination Record Officer, Department of Human Physiology, Faculty of Basic Medical Sciences, Bayelsa

Medical University, Yenagoa, Bayelsa State, Nigeria. He holds a bachelor of medicine bachelor of surgery (MBBS) degree. He is licensed by the Medical and Dental Council of Nigeria (MDCN) as a medical practitioner. He has authored some peer-reviewed scientific articles in reputable journals. His research interests is in blood, cardiovascular, and reproduction Physiology .He is currently taking a post graduate program in human Physiology. He is a member of the Nigerian Medical Association, and Physiological Society Of Nigeria.

Pawan Kumar Gupta is currently working as an Associate Professor at Department of Pharmaceutical Chemistry, SVKMs Institute of Pharmacy, Dhule, Maharashtra. He did Bchelor degree in pharmacy, from BN College of Pharmacy, Udaipur and masters and PhD degree in Pharmacoinformatics from NIPER, Mohali, India. He did post-doc from Latvian Institute of Organic Synthesis, Latvia, Europe. He has more than 10 years research (post-doc) and teaching experiences in India and abroad. His current research interests are identification of novel small molecules/ leads/hits against biological targets of the life-threatening diseases (Cancer and virus) using computer aided drug design and discovery methods. He has completed a DST-SERB project under TARE scheme with mentorship of the Prof. Anjana Munshi, Dean Research, Central University of Punjab, Bathinda, India. He has published more than 40 papers in international reputed journals, some of are like European Journal of Medicinal Chemistry (Elsevier), Archives of Toxicology (Springer), Bioorganic Medicinal Chemistry (Elsevier), Chemical Biology & Drug Design (Wiley), Molecular Diversity (Springer), Cancer Biomarkers (IOP), Food and Function (RSC), Journal of Ethnopharmacology (Elsevier), Plants (MDPI), Molecules (MDPI), Phytomedicine (Elsevier), Current drug target (Bentham), and many more. He has presented more than 10 papers in national and international conferences (including, Greece, Switzerland and Latvia). He also awarded the international travel grants by different funding agencies (DST, DBT and CICS) to present his research outcomes in different international conference abroad. He has also attended 17 different workshops/symposium/ seminars/conferences held in India and abroad. He has delivered various lectures/talks in different institutions in India on the topic of his expertise. He has guided the 8 masters and 20 bachelor students in their projects. Currently, 2 PhD students are working under his guidance. He is having life-time membership of the different professional bodies (like APTI, ISCA, BIDDS). His research expertise has guided the different discipline researchers in their research projects like biotech, chemistry, botany, management, etc.

Odangowei Inetiminebi Ogidi is a Lecturer and Acting Head, Department of Biochemistry, Bayelsa Medical University, Yenagoa, Bayelsa State, Nigeria. He holds a Bachelor of Science degree in Biochemistry, Master of Science and

Doctor of Philosophy degrees in Biotechnology. He has also served as a Lecturer, Head of Departments of Science Laboratory Technology and Biochemistry, and Deputy Director, Centre for Continuing Education in Federal Polytechnic Ekowe, Bayelsa State. He has authored over 80 peer-reviewed academic articles in reputable local and international journals, published over 40 book chapters and presented over 35 conference papers. He is a member of various Journal Editorial Boards. He is indexed in World Scientist and University Ranking (AD Scientific Index ID 4306173) amongst other international bodies. He have supervised over 30 ND and HND students and currently supervising over 10 undergraduate students. His research interest has been Environmental Biotechnology, Phytomedicine, Phytochemistry, Human and Environmental health risk assessments and ecotoxicology amongst others. He is a member of West Africa Society of Toxicology, Nigerian Society of Biochemistry and Molecular Biology, Foundation for African Development through International Biotechnology and Nigerian Environmental Society.

Rashmi Pathak I am currently working as an Assistant Professor in the Department of Pharmacy at Invertis University Bareilly. I have completed my B. Pharm from Invertis University, Bareilly, and M.Pharm from the IFTM University, Moradabad. I was the silver medalist in B.Pharm and gold medalist in M.Pharm. I have 2 years of academic experience. I have published 4 research articles, 15 review articles, 7 book chapters and 1 Patent. I have received the Young Scientist Award and Best Research Scholar Award.

Reyed M. Reyed: A Brief Academic and Professional Journey Dr. REYED M REYED, He works at the Genetic Engineering and Biotechnology Research Institute (GEBRI) at the City of Scientific Research and Technological Applications, Alexandria, Egypt (egyhumanmicrobiome@inbox.ru &r.mohamed@srtacity.sci .eg). Dr. Reyed M. Reyed is an Associate Professor of Microbial Biotechnology with extensive expertise in microbial genomics, computational enzymes, and environmental sustainability. He holds multiple academic credentials, including Dia.M.C, RN, PAND, PhD, MSc, and BSc in Microbial Biotechnology. His research spans a wide range of fields, from the nutritional human gut microbiome and food microbiology to water treatment and food safety. From 2008 to 2019, Dr. Reyed served as an expert in water treatment technology, applying his knowledge to critical challenges such as biofilm formation and marine biofouling in reverse osmosis systems. From 2020 to 2023, he led the Egyptian-Indian joint project on water treatment in collaboration with Indian leader Asifa Qureshi at the Environmental Biotechnology and Genomics Division (EBGD), CSIR-National Environmental Engineering Research Institute (CSIR-NEERI) in Nagpur, India, contributing to practical environmental solutions. In the field of food microbiology, Dr. Reyed has made notable contributions to food safety and has founded a winter school on the subject. He is currently the program coordinator at SRTA-CITY SCI EG. His interdisciplinary research includes advances in computational enzymes and bioinformatics, which complement his expertise in microbial biotechnology, as well as the integration of synthetic biology and systems biology in exploring gut health and environmental sustainability. Dr. Reyed has participated in the Diabetes Master Class (Dia.M.C) organized by Novartis Egypt and the Arabic Society for Diabetes and Metabolism in 2022, under the supervision of the Egyptian Medical Syndicates in Cairo, Egypt. His leadership extends beyond research; he serves as the Academic Editor of the International Journal of Environment and Climate Change and as the Editor-in-Chief of the Journal of Climate Change and Renewable Energy. Dr. Reyed has participated in major international conferences, including events in the UAE, such as the International Conference on Bacteria, Viruses, and Infectious Diseases (Dubai, 2018), and the Food Safety Exhibition in collaboration with Emirates Universities and ADNOC (Abu Dhabi, 2023). He will also serve on the Research Evaluation Committee at the 8th International Conference on Genetic Disorders in Dubai in May 2024. Dr. Reyed holds several certifications in Clinical Nutrition, including Code B N 0458 from Baheya Hospital for Breast Cancer Treatment in Giza, Egypt, in collaboration with Medix Foundation for Training and Medical Education. He is a Registered Clinical Nutritionist (Registration No. 44320/6068) in the Egyptian Syndicate of Scientific Professions and a Registered Physical and Health Nutritionist with the Egyptian Professional Syndicate Committee for Specialists in Cosmetic Professions and Health Services (900444). Additionally, he is a member of the General Syndicate of Egyptian Ambulance Workers (No. 2283) and the Egyptian Red Crescent (No. 1352). His commitment to environmental sustainability, public health, and microbial biotechnology is further reflected through his volunteer work in the UAE, where he holds multiple licenses, including General Volunteer No. 523273 (Abu Dhabi) and Sharjah Volunteer Center No. 61318. Reyed M. Reyed has participated in several workshops related to drug design. These include the 4th Computer-Assisted Drug Design Workshop (CADD-W) held at Kafr El-Sheikh University on November 6-7, 2021, and the InSilico Drug Design Course at Ain Shams Center of Genetic Engineering, Faculty of Agriculture, from September 28-30, 2021. Furthermore, he is scheduled to participate in the Drug Discovery, Delivery, and Therapeutics Workshop at New York University Abu Dhabi campus from May 6-8, 2024, in collaboration with the International Center for Genetic Engineering and Biotechnology (ICGEB), Italy. Reyed M. Reyed participated in several funded training programs and courses, notably in 2007 at Sandia National Laboratories, USA, focusing on biosafety, and at the International Centre for Genetic Engineering and Biotechnology (ICGEB) in Italy. He also contributed to a workshop on CRISPR/Cas9 Genome Editing. He is an active member of several professional societies, including the Arab Society for Diabetes and Metabolism, the Egyptian Society of Functional Medicine, and ESPGHAN (European Society for Pediatric Gastroenterology, Hepatology, and Nutrition). Reyed holds several certificates and has completed numerous courses. He earned a Certificate in Entrepreneurship through the Knowledge Project in cooperation with the Regional Office for Arab States of the United Nations Development Corporation and the Mohammed bin Rashid Al Maktoum Foundation for Knowledge, from Copenhagen Business School in March 2024. He completed a course on Understanding Obesity at the University of Edinburgh in May 2023, and a course on Nutrition and Lifestyle in Pregnancy at Ludwig-Maximilians-Universität München (LMU) in August 2022. Additionally, he completed two bioinformatics-related courses: Bacterial Genomes: From DNA to Protein Function Using Bioinformatics (Wellcome Connecting Science, January 2021) and Train the Trainer: Design Genomics and Bioinformatics Training (Wellcome Connecting Science, October 2021). Most recently, he took a course on Nutrition for Health and Sustainability with EIT Food in September 2023. Reyed also holds SCOPE Certification (the international standard of obesity management expertise) with 16 SCOPE points, having passed 20 modules. He was awarded the SCOPE Core Learning Path Scholarship for Healthcare Professionals in 2021. As an international arbitrator and reviewer, Reyed contributes his expertise to several prominent scientific journals, including the European Journal of Nutrition, B. SNF (Swedish Nutrition Foundation), Cureus Journal of Medical Science, Microbiology Research Journal International, Journal of Advances in Microbiology, Plant Cell Biotechnology and Molecular Biology, and Food and Health. Reyed has also been involved in various bioinformatics courses and workshops. These include the NCBI Virtual Workshop: An Introduction to NCBI Cloud Computing for Biologists in November 2021, and the Data Science Summer School of 2022 held at Hertie School Data Science Lab in Berlin, Germany. He attended the Virtual VeuPathDB Bioinformatics Workshop in December 2021, and the Ensembl Workshop organized by Erasmus Postgraduate School of Molecular Medicine, Rotterdam, in April 2021. He further participated in the 2nd Ensembl Workshop (Virtual Ensembl REST API Course) from June 15-17, 2021, held by the European Bioinformatics Institute (EMBL-EBI) in the UK. In December 2020, Reyed contributed to the 3rd GED-PerMed Project: Big Data Meets Precision Medicine virtual workshop, organized by the DAAD German Egyptian Progress Partnership Program and Nile University, Egypt.

Himanshu Sharma is working as an Assistant Professor, in the Department of Pharmacology, Teerthanker Mahaveer College of Pharmacy, Teerthanker Mahaveer University, Moradabad, (U.P.), India. He has done his B. Pharm. from Teerthanker Mahaveer University, Moradabad (Uttar Pradesh), India, and completed his M.Pharm. (Pharmacology) from IFTM University, Moradabad (Uttar Pradesh), India. He has published 37 research/review papers in National and International journals and has 6 patents. He is the author of one book and 68 book chapters with different publishers of international repute.

Index

A

B

C

D

E

O

P

R

S

T

V

Z

www.ingramcontent.com/pod-product-compliance
Ingram Content Group UK Ltd.
Pitfield, Milton Keynes, MK11 3LW, UK
UKHW051036091026
15624UKWH00011B/451

9 798369 379196